Lecture Notes in Computer Science 15829

Founding Editors

Gerhard Goos
Juris Hartmanis

Editorial Board Members

Elisa Bertino, *Purdue University, West Lafayette, IN, USA*
Wen Gao, *Peking University, Beijing, China*
Bernhard Steffen, *TU Dortmund University, Dortmund, Germany*
Moti Yung, *Columbia University, New York, NY, USA*

The series Lecture Notes in Computer Science (LNCS), including its subseries Lecture Notes in Artificial Intelligence (LNAI) and Lecture Notes in Bioinformatics (LNBI), has established itself as a medium for the publication of new developments in computer science and information technology research, teaching, and education.

LNCS enjoys close cooperation with the computer science R & D community, the series counts many renowned academics among its volume editors and paper authors, and collaborates with prestigious societies. Its mission is to serve this international community by providing an invaluable service, mainly focused on the publication of conference and workshop proceedings and postproceedings. LNCS commenced publication in 1973.

Ipek Oguz · Shaoting Zhang ·
Dimitris N. Metaxas
Editors

Information Processing in Medical Imaging

29th International Conference, IPMI 2025
Kos, Greece, May 25–30, 2025
Proceedings, Part I

Springer

Editors
Ipek Oguz ⓘ
Vanderbilt University
Nashville, TN, USA

Shaoting Zhang ⓘ
Shanghai AI Laboratory
Shanghai, China

Dimitris N. Metaxas ⓘ
Rutgers University
Piscataway, NJ, USA

ISSN 0302-9743　　　　　　　　ISSN 1611-3349 (electronic)
Lecture Notes in Computer Science
ISBN 978-3-031-96627-9　　　ISBN 978-3-031-96628-6 (eBook)
https://doi.org/10.1007/978-3-031-96628-6

© The Editor(s) (if applicable) and The Author(s), under exclusive license to Springer Nature Switzerland AG 2026

This work is subject to copyright. All rights are solely and exclusively licensed by the Publisher, whether the whole or part of the material is concerned, specifically the rights of translation, reprinting, reuse of illustrations, recitation, broadcasting, reproduction on microfilms or in any other physical way, and transmission or information storage and retrieval, electronic adaptation, computer software, or by similar or dissimilar methodology now known or hereafter developed.
The use of general descriptive names, registered names, trademarks, service marks, etc. in this publication does not imply, even in the absence of a specific statement, that such names are exempt from the relevant protective laws and regulations and therefore free for general use.
The publisher, the authors and the editors are safe to assume that the advice and information in this book are believed to be true and accurate at the date of publication. Neither the publisher nor the authors or the editors give a warranty, expressed or implied, with respect to the material contained herein or for any errors or omissions that may have been made. The publisher remains neutral with regard to jurisdictional claims in published maps and institutional affiliations.

This Springer imprint is published by the registered company Springer Nature Switzerland AG
The registered company address is: Gewerbestrasse 11, 6330 Cham, Switzerland

If disposing of this product, please recycle the paper.

Preface

It is our pleasure to present the proceedings of the 29th International Conference on Information Processing in Medical Imaging (IPMI 2025), held on the island of Kos, Greece, May 25–30, 2025. The conference was organized by co-chairs Ipek Oguz from Vanderbilt University and Dimitris N. Metaxas from Rutgers University.

Since 1969, IPMI has become widely recognized as a premier international forum for the presentation of cutting-edge methodological advancements in the medical image computing field. The conference is known for its unique format that fosters in-depth paper discussion, interdisciplinary collaboration, and rigorous scientific exchange in a retreat-style setting.

For the IPMI 2025 conference, we received 196 submissions, of which 143 papers went through the double-blind peer review. The majority of the papers was evaluated by three reviewers and few received two reviews. Following the initial review process, each manuscript was further evaluated by two members of the Paper Selection Committee and subsequently discussed by the full committee at the paper selection meeting. Papers with conflicting scores from reviewers (both accept and reject recommendations) were discussed in detail at this meeting. Finally, 51 submissions were accepted for publication and presentation at the conference. Selected papers were invited for a Special Issue of the Machine Learning for Biomedical Imaging (MELBA) journal following the conference. We are deeply grateful to the Paper Selection Committee for their efforts and commitment. We also thank the 88 reviewers listed below, who each provided thoughtful feedback on an average of 5 papers.

The scientific program was punctuated by two excellent scientific keynotes delivered by Leon Axel from NYU Grossman School of Medicine and Shaoting Zhang from Shanghai Jiao Tong University. We are also grateful to Jim Duncan, Julia Schnabel, Herve Lombaert, and Miaomiao Zhang, who participated in the Future Directions for IPMI: A Soul Searching Panel event at the conference, which was a thought-provoking session for our community.

The François Erbsmann Prize is awarded during each IPMI to a young scientist of age 35 or below, the first author of a paper, giving their first oral presentation at IPMI. The prizes winners were determined by the IPMI Board. NVIDIA sponsored the awards for the François Erbsmann Prize, the runner-up award, as well as the best poster award. NVIDIA also led a virtual workshop at IPMI 2025 focusing on the latest upcoming technology trends with an AWS cloud platform hands-on session titled: "Holoscan × MONAI: Next-Gen Multi-Modal AI & Co-pilots". We are grateful for their support of IPMI 2025.

We were honored to organize IPMI 2025 and present the excellent set of scientific contributions documented in these proceedings. IPMI continues to be the premier venue for methodological advancements in medical image computing, and we are excited to pass the baton to the IPMI 2027 team to organize the next exciting meeting in the series.

May 2025

Ipek Oguz
Shaoting Zhang
Dimitris N. Metaxas

Organization

General Chairs

Ipek Oguz	Vanderbilt University, USA
Dimitris N. Metaxas	Rutgers University, USA

Program Chairs

Ipek Oguz	Vanderbilt University, USA
Shaoting Zhang	Shanghai Jiao Tong University, China

Paper Selection Committee

Ismail Ben Ayed	École de Technologie Superieure, Canada
Shireen Elhabian	University of Utah, USA
Sarah Frisken	Brigham and Women's Hospital, USA
Sharon Huang	Pennsylvania State University, USA
Marc Niethammer	University of California San Diego, USA
Baba Vemuri	University of Florida, USA
Carl-Fredrik Westin	Harvard University, USA

Information Processing in Medical Imaging Board

Gary Christensen	University of Iowa, USA
Albert Chung	University of Exeter, UK
Marleen de Bruijne	Erasmus MC, The Netherlands
James S. Duncan	Yale University, USA
Alejandro Frangi	University of Leeds, UK
Polina Golland	Massachusetts Institute of Technology, USA
Richard Leahy	University of Southern California, USA
Alison Noble	University of Oxford, UK
Ipek Oguz	Vanderbilt University, USA
Sebastien Ourselin	King's College London, UK
Stephen M. Pizer	University of North Carolina at Chapel Hill, USA
Jerry Prince	Johns Hopkins University, USA

Stefan Sommer   University of Copenhagen, Denmark
Martin Styner   University of North Carolina at Chapel Hill, USA
Gabor Szekely   ETH Zurich, Switzerland
Chris Taylor   University of Manchester, UK
Andrew Todd-Pokropek   University College London, UK
William M. Wells III   Harvard Medical School, USA

Reviewers

Aasa Feragen
Albert C. S. Chung
Alexis Arnaudon
Alison Noble
Alison Pouch
Alvin Chen
Andrew P. King
Ario Sadafi
Aristeidis Sotiras
Arrate Munoz Barrutia
Baiying Lei
Bennett A. Landman
Bjoern Menze
Carole H. Sudre
Chris Taylor
Christian Desrosiers
Chunfeng Lian
Chuyang Ye
Daniel C. Moyer
Daniel Rueckert
Dou Qi
Dustin Scheinost
Ender Konukoglu
Gary E. Christensen
Gianfranco Cortes
Guotai Wang
Hao Li
Haomiao Ni
Hoel Kervadec
Hong Xu
Hongmin Cai
Ilker Hacihaliloglu
Ilwoo Lyu
Islem Rekik

Jack Noble
Jadie Adams
James S. Duncan
Jan Kybic
Jerry L. Prince
Jiazhou Chen
John Ashburner
Jon Sporring
Jonghye Woo
Jose Dolz
Julia A. Schnabel
Julio Silva-Rodriguez
Junzhou Huang
Kaleem Siddiqi
Karl Rohr
Karthik Gopinath
Krithika Iyer
Linwei Wang
Marco Lorenzi
Maria A. Zuluaga
Martin Styner
Matthew Toews
MattiasPaul Heinrich
Meng Ye
Miaomiao Zhang
Minjeong Kim
Nicha Dvornek
Ninon Burgos
P. Thomas Fletcher
Peirong Liu
Peng Jin
Pengcheng Shi
Pew-Thian Yap
Qingyu Zhao

Reuben Dorent
Reyer Zwiggelaar
Sara Garbarino
Steve Pizer
Suyash P. Awate
Tim Cootes
Tobias Heimann
Tuo Zhang
Ulas Bagci
Vivek Gopalakrishnan

Wenzheng Tao
Xiang Li
Xiaodong Wu
Yi Hong
Yixuan Yuan
Yonggang Shi
Yuan Xue
Yuan Zhou
Yuankai Huo
Yunhe Gao

Contents – Part I

Classification/Detection

SpectMamba: Integrating Frequency and State Space Models for Enhanced Medical Image Detection 3
Yao Wang, Dong Yang, Zhi Qiao, Wenjian Huang, Liuzhi Yang, and Zhen Qian

Hierarchical Neural Cellular Automata for Lightweight Microscopy Image Classification 19
Chen Yang, Michael Deutges, Nassir Navab, Ario Sadafi, and Carsten Marr

PathTTT: Test-Time Training with Meta-auxiliary Learning for Pathology Image Classification 33
Haoyu He, Mahdi S. Hosseini, and Yang Wang

Registration

Bi-invariant Geodesic Regression with Data from the Osteoarthritis Initiative 49
Johannes Schade, Christoph von Tycowicz, and Martin Hanik

GSSD: A Self-distillation Paradigm with Gradient Surgery for End-to-End Deformable Image Registration 64
Yuxi Zheng, Yansong Bai, and Yuchuan Qiao

Medical Image Registration Meets Vision Foundation Model: Prototype Learning and Contour Awareness 79
Hao Xu, Tengfei Xue, Jianan Fan, Dongnan Liu, Yuqian Chen, Fan Zhang, Carl-Fredrik Westin, Ron Kikinis, Lauren J. O'Donnell, and Weidong Cai

Vascular-Topology-Aware Deep Structure Matching for 2D DSA and 3D CTA Rigid Registration 94
Xiaosong Xiong, Caiwen Jiang, Peng Wu, Xiao Zhang, Yanli Song, Xinyi Zhang, Ze Tao, Dijia Wu, and Dinggang Shen

Unsupervised Deformable Image Registration with Structural
Nonparametric Smoothing 108
    Hang Zhang, Renjiu Hu, Xiang Chen, Min Liu, Yaonan Wang,
    Rongguang Wang, Jinwei Zhang, Gaolei Li, Xinxing Cheng,
    and Jinming Duan

Reconstruction

Unsupervised Accelerated MRI Reconstruction via Ground-Truth-Free
Flow Matching 127
    Xinzhe Luo, Yingzhen Li, and Chen Qin

Optimization of Acquisition Schemes Towards a Better Estimation
of Microstructure Parameters in Multidimensional Diffusion MRI 142
    Constance Bocquillon, Isabelle Corouge, and Emmanuel Caruyer

Bilinear Projector: Mitigating Discretization Artifacts in Model Based
Iterative Reconstruction for X-Ray CT 154
    Ke Chen and Alireza Entezari

Subspace Implicit Neural Representations for Real-Time Cardiac Cine
MR Imaging 168
    Wenqi Huang, Veronika Spieker, Siying Xu, Gastao Cruz,
    Claudia Prieto, Julia A. Schnabel, Kerstin Hammernik,
    Thomas Kuestner, and Daniel Rueckert

Image Synthesis

3D Shape-to-Image Brownian Bridge Diffusion for Brain MRI Synthesis
from Cortical Surfaces 187
    Fabian Bongratz, Yitong Li, Sama Elbaroudy, and Christian Wachinger

Cascaded Diffusion Model and Segment Anything Model for Medical
Image Synthesis via Uncertainty-Guided Prompt Generation 203
    Haowen Pang, Xiaoming Hong, Peng Zhang, and Chuyang Ye

DIReCT: Domain-Informed Rectified Flow for Controllable Brain MRI
to PET Translation 218
    Tuo Liu, Haifeng Wang, Heng Chang, Fan Wang, Chunfeng Lian,
    and Jianhua Ma

IGG: Image Generation Informed by Geodesic Dynamics in Deformation
Spaces 232
    Nian Wu, Nivetha Jayakumar, Jiarui Xing, and Miaomiao Zhang

Image Enhancement

Cycle-Consistent Zero-Shot Through-Plane Super-Resolution
for Anisotropic Head MRI 249
   *Samuel W. Remedios, Shuwen Wei, Aaron Carass, Blake E. Dewey,
and Jerry L. Prince*

Bayesian Learning with Stochastic Perturbations and Langevin
Expectation Maximization for Unsupervised DNN Image Quality
Enhancement 265
   Vatsala Sharma and Suyash P. Awate

Segmentation

MC-NuSeg: Multi-Contour Aware Nuclei Instance Segmentation
with Segment Anything Model 283
   Hyun Namgung, Siwoo Nam, Soopil Kim, and Sang Hyun Park

Pitfalls of Topology-Aware Image Segmentation 297
   *Alexander H. Berger, Laurin Lux, Alexander Weers, Martin J. Menten,
Daniel Rueckert, and Johannes C. Paetzold*

GeoT: Geometry-Guided Instance-Dependent Transition Matrix
for Semi-supervised Tooth Point Cloud Segmentation 313
   Weihao Yu, Xiaoqing Guo, Chenxin Li, Yifan Liu, and Yixuan Yuan

RemInD: Remembering Anatomical Variations for Interpretable Domain
Adaptive Medical Image Segmentation 327
   *Xin Wang, Yin Guo, Kaiyu Zhang, Niranjan Balu,
Mahmud Mossa-Basha, Linda Shapiro, and Chun Yuan*

Dynamic Allocation Hypernetwork with Adaptive Model Recalibration
for Federated Continual Learning 342
   *Xiaoming Qi, Jingyang Zhang, Huazhu Fu, Guanyu Yang, Shuo Li,
and Yueming Jin*

Skelite: Compact Neural Networks for Efficient Iterative Skeletonization 357
   *Luis D. Reyes Vargas, Martin J. Menten, Johannes C. Paetzold,
Nassir Navab, and Mohammad Farid Azampour*

VerSe: Integrating Multiple Queries as Prompts for Versatile Cardiac MRI Segmentation 373
  Bangwei Guo, Meng Ye, Yunhe Gao, Bingyu Xin, Leon Axel, and Dimitris Metaxas

Author Index 389

Contents – Part II

Computer-Aided Diagnosis/Surgery

Concepts from Neurons: Building Interpretable Medical Image Diagnostic
Models by Dissecting Opaque Neural Networks 3
   Shizhan Gong, Huayu Wang, Xiaofan Zhang, and Qi Dou

BioSonix: Can Physics-Based Sonification Perceptualize Tissue
Deformations From Tool Interactions? 19
   *Veronica Ruozzi, Sasan Matinfar, Laura Schütz, Benedikt Wiestler,
Alberto Redaelli, Emiliano Votta, and Nassir Navab*

Brain

Explainable Deep Model for Understanding Neuropathological Events
Through Neural Symbolic Regression 37
   Tingting Dan and Guorong Wu

A Multi-layer Neural Transport Model for Characterizing Pathology
Propagation in Neurodegenerative Diseases 51
   Haifeng Huang, Yi Wang, Tingting Dan, Yang Yang, and Guorong Wu

Enhancing Alzheimer's Diagnosis: Leveraging Anatomical Landmarks
in Graph Convolutional Neural Networks on Tetrahedral Meshes 65
   *Yanxi Chen, Mohammad Farazi, Zhangsihao Yang, Yonghui Fan,
Nicholas Ashton, Eric M. Reiman, Yi Su, and Yalin Wang*

Hierarchical Variable Importance with Statistical Control for Medical
Data-Based Prediction 79
   *Joseph Paillard, Antoine Collas, Denis A. Engemann,
and Bertrand Thirion*

Disentangle Disease-Relevant Patterns from Irrelevant Patterns in fMRI
Analysis Using Equivariant and Contrastive Learning 94
   Xin Shen, Shengjie Zhang, Wenbin Liu, and Yuan Zhou

Diffusion Models

Continuous Diffusion Model for Self-supervised Denoising
and Super-Resolution on Fluorescence Microscopy Images 111
    Colin S. C. Tsang and Albert C. S. Chung

Self-supervised Denoising of Diffusion MRI Data with Efficient
Collaborative Diffusion Model 125
    Xiaoyu Bai, Haotian Jiang, and Geng Chen

MAD-AD: Masked Diffusion for Unsupervised Brain Anomaly Detection 139
    Farzad Beizaee, Gregory Lodygensky, Christian Desrosiers,
    and Jose Dolz

Self-supervised Learning

Taming Masked Image Modeling for Chest X-ray Diagnosis
by Incorporating Clinical Visual Priors 157
    Zihao Zhao, Mei Wang, Zhiming Cui, Sheng Wang, Qian Zhou, Li Fan,
    Qian Wang, and Dinggang Shen

Diffusion MAE: Paving the Way for Representation Learning of Diffusion
MRI 172
    Haotian Jiang and Geng Chen

Resolving Quantitative MRI Model Degeneracy in Self-supervised
Machine Learning 186
    Giulio V. Minore, Louis Dwyer-Hemmings, Timothy J. P. Bray,
    and Hui Zhang

Vision-Language Models

Knowledge-Enhanced Hyperbolic Language-Image Pretraining
for Zero-Shot Learning 203
    Linbin Han, Zhi Qiao, Xiantong Zhen, Jiahong Gao, and Zhen Qian

Structure Observation Driven Image-Text Contrastive Learning
for Computed Tomography Report Generation 218
    Hong Liu, Dong Wei, Qiong Peng, Yawen Huang, Xian Wu,
    Yefeng Zheng, and Liansheng Wang

Hierarchical CLIPs for Fine-Grained Anatomical Lesion Localization
from Whole-Body PET/CT Images 234
    Mingyang Yu, Yaozong Gao, Yiran Shu, Yanbo Chen, Jingyu Liu,
    Caiwen Jiang, Kaicong Sun, Weifang Zhang, Yiqiang Zhan,
    Xiang Sean Zhou, Shaonan Zhong, Xinlu Wang, Meixin Zhao,
    and Dinggang Shen

Multi-view and Multi-scale Alignment for Contrastive Language-Image
Pre-training in Mammography 247
    Yuexi Du, John A. Onofrey, and Nicha C. Dvornek

Interpretable Few-Shot Retinal Disease Diagnosis with Concept-Guided
Prompting of Vision-Language Models 263
    Deval Mehta, Yiwen Jiang, Catherine Jan, Mingguang He,
    Kshitij Jadhav, and Zongyuan Ge

Full Conformal Adaptation of Medical Vision-Language Models 278
    Julio Silva-Rodríguez, Leo Fillioux, Paul-Henry Cournède,
    Maria Vakalopoulou, Stergios Christodoulidis, Ismail Ben Ayed,
    and Jose Dolz

A Reality Check of Vision-Language Pre-training in Radiology: Have We
Progressed Using Text? 294
    Julio Silva-Rodríguez, Jose Dolz, and Ismail Ben Ayed

Shape Analysis

ToothForge: Automatic Dental Shape Generation Using Synchronized
Spectral Embeddings 313
    Tibor Kubík, François Guibault, Michal Španěl, and Hervé Lombaert

LEDA: Log-Euclidean Diffeomorphism Autoencoder for Efficient
Statistical Analysis of Diffeomorphisms 327
    Krithika Iyer, Shireen Elhabian, and Sarang Joshi

CoRLD: Contrastive Representation Learning of Deformable Shapes
in Images 342
    Tonmoy Hossain and Miaomiao Zhang

Time-Series Image Analysis

4DRGS: 4D Radiative Gaussian Splatting for Efficient 3D Vessel
Reconstruction from Sparse-View Dynamic DSA Images 361
    Zhentao Liu, Ruyi Zha, Huangxuan Zhao, Hongdong Li, and Zhiming Cui

Brightness-Invariant Tracking Estimation in Tagged MRI 375
    *Zhangxing Bian, Shuwen Wei, Xiao Liang, Yuan-Chiao Lu,
    Samuel W. Remedios, Fangxu Xing, Jonghye Woo, Dzung L. Pham,
    Aaron Carass, Philip V. Bayly, Jiachen Zhuo, Ahmed Alshareef,
    and Jerry L. Prince*

SafeTriage: Facial Video De-identification for Privacy-Preserving Stroke
Triage 390
    *Tongan Cai, Haomiao Ni, Wenchao Ma, Yuan Xue, Qian Ma,
    Rachel Leicht, Kelvin Wong, John Volpi, Stephen T. C. Wong,
    James Z. Wang, and Sharon X. Huang*

Author Index 405

Classification/Detection

SpectMamba: Integrating Frequency and State Space Models for Enhanced Medical Image Detection

Yao Wang[1], Dong Yang[2], Zhi Qiao[1], Wenjian Huang[3], Liuzhi Yang[2], and Zhen Qian[1(✉)]

[1] United-Imaging Research Institute of Intelligent Imaging, Beijing, China
{yao.wang,zhen.qian}@cri-united-imaging.com
[2] Nankai University, Tianjin, China
[3] Southern University of Science and Technology, Shenzhen, China

Abstract. Abnormality detection in medical imaging is a critical task requiring both high efficiency and accuracy to support effective diagnosis. While convolutional neural networks (CNNs) and Transformer-based models are widely used, both face intrinsic challenges: CNNs have limited receptive fields, restricting their ability to capture broad contextual information, and Transformers encounter prohibitive computational costs when processing high-resolution medical images. Mamba, a recent innovation in natural language processing, has gained attention for its ability to process long sequences with linear complexity, offering a promising alternative. Building on this foundation, we present SpectMamba, the first Mamba-based architecture designed for medical image detection. A key component of SpectMamba is the Hybrid Spatial-Frequency Attention (HSFA) block, which separately learns high- and low-frequency features. This approach effectively mitigates the loss of high-frequency information caused by frequency bias and correlates frequency-domain features with spatial features, thereby enhancing the model's ability to capture global context. To further improve long-range dependencies, we propose the Visual State-Space Module (VSSM) and introduce a novel Hilbert Curve Scanning technique to strengthen spatial correlations and local dependencies, further optimizing the Mamba framework. Comprehensive experiments show that SpectMamba achieves state-of-the-art performance while being both effective and efficient across various medical image detection tasks.

Keywords: State space model · Hilbert curve scan · Frequency domain

1 Introduction

Image detection has become a cornerstone of medical image analysis, with significant research dedicated to enhancing its accuracy and applicability in clinical settings [24]. Various methods have been proposed, such as YOLO [9], Cascade R-CNN [4], DETR [49], and DINO [48], each aimed at improving detection performance. These methods can generally be classified into two categories: convolutional neural network (CNN)-based approaches and Transformer-based approaches. CNN-based methods are limited by their reliance on local receptive fields, which hinder their ability to capture global features effectively. In contrast, Transformer-based methods excel at modeling long-range dependencies through the self-attention mechanism. Yet, this advantage comes at a significant computational cost, with complexity scaling quadratically with spatial resolution.

Among the recent innovations, the Mamba framework has garnered attention as a promising alternative. Mamba models [25,51] integrate state-space models [14] to reduce the quadratic complexity typically associated with traditional Transformer architectures, achieving linear complexity instead. This is accomplished by using compressed hidden states to effectively capture long-range dependencies. Mamba has already demonstrated superior performance in natural image detection and segmentation [10,52], with results indicating its potential for medical image segmentation as well [36]. However, despite its potential, the Mamba model faces challenges in 2D medical image detection tasks.

Two primary issues arise when applying Mamba to medical image detection. First, state-space models exhibit a frequency bias [46], a concept first described in over-parameterized multilayer perceptrons (MLPs) [34]. Frequency bias occurs when low-frequency components are learned faster than high-frequency components. It hinders the ability to capture high-frequency details, which are crucial for radiologists who rely on fine anatomical features, particularly edges and textures, when detecting and diagnosing pathological regions [3,39,43]. Second, the Mamba model, designed for 1D sequences, struggles with local dependencies in 2D images. Previous attempt to adapt Mamba to 2D tasks, such as bidirectional and cascade scans [25,51], involve flattening 2D patches into 1D sequences. This transformation compromises the model's ability to interpret spatial relationships, limiting its effectiveness in abnormality detection.

In this paper, we present SpectMamba, a Mamba-based architecture specifically designed to integrate frequency and state-space models for abnormality detection in medical images. SpectMamba integrates Hybrid Spatial-Frequency Attention (HSFA) blocks and Visual State-Space Module (VSSM). The HSFA block uses parallel convolutions to capture spatial information while employing a Low-High Frequency Domain Information Separator to enhance high-frequency components. By processing the feature map in two branches: one for spatial information and the other for high- and low-frequency separation. SpectMamba reduces the impact of frequency bias, enabling it to learn high-frequency information, which distinguishes it from prior models such as FreqMamba [5,53].

To address the issue of local dependencies in 2D image processing [19], we introduce the Hilbert scan curve, which preserves locality and spatial neighborhood relationships. Unlike bidirectional and cascade scans, which flatten features into 1D and disrupt spatial relationships, the Hilbert curve [6] maintains local clustering and spatial proximity dependencies. Additionally, the integration of feature pyramids [41] enables multiscale representations for lesion localization, ensuring accurate detection across varying target sizes.

We validate the effectiveness of SpectMamba on three benchmark datasets, demonstrating its superior performance compared to current state-of-the-art models. SpectMamba achieves a 1.07% relative improvement in the mean Average Precision (mAP) for pneumonia detection on X-ray images [12], a 3.7% increase for brain tumor detection on MRI scans [1,2,29], and a 0.48% improvement for bone fracture detection on X-ray images [31]. Furthermore, SpectMamba significantly enhances computational efficiency, operating at twice the speed of VMamba while delivering higher accuracy. Notably, it also outperforms the ViT-L-based model with a 2.52% improvement in mAP, despite using only 32% of its parameter count.

In summary, our contributions are as follows.

– We propose the SpectMamba model, which integrates HSFA blocks to extract spatial and high-frequency details, reducing frequency bias, and incorporates a VSSM for capture long-range memory with linear complexity.
– For the challenge of local dependency of the 1D Mamba scan, we introduce a novel Hilbert curve scanning technique to further enhance the spatial perception.
– Extensive experiments demonstrate that SpectMamba outperforms state-of-the-art methods in medical abnormality detection tasks.

2 Related Work

State-Space Models. State-Space Models (SSMs) [28,42] have gained significant attention for their efficiency in modeling long-range dependencies while maintaining linear scalability with respect to sequence length. Inspired by continuous state-space models in control systems, LSSL [17], initialized with HiPPO [15], demonstrated the potential of SSMs to address long-range dependency issues. However, the high computational and memory demands of LSSL make it impractical for real-world applications. To overcome these limitations, S4 [16] introduced diagonal parameter normalization, improving efficiency, while further advancements integrated gated units to optimize performance [14]. The Mamba model, a data-dependent SSM with a selection mechanism and hardware-optimized design, is claimed to outperform traditional Transformers in natural language tasks while maintaining linear scalability with input length.

Building upon these advancements, selective state-space blocks have been integrated into vision backbones to enhance visual representation learning. Notable models like Vmamba [25] and Vision Mamba [51] incorporate SSM modules into image processing pipelines. Both models demonstrate the effectiveness of Mamba-based approaches in handling diverse visual tasks. Vision Mamba employs bidirectional scanning, while Vmamba utilizes cascade scanning with four-directional coverage, both improving task performance in natural image.

Frequency Domain Analysis. The Fourier Transform, a fundamental technique for converting signals from the time or spatial domain to the frequency domain, is widely used in computer vision for analyzing statistical properties and frequency-domain information. GFNet [35] leverages frequency domain analysis to learn long-range spatial dependencies with log-linear complexity, capturing fine-grained image attributes. Similarly, Octave Convolution [7] and HiLo [33] separate high- and low-frequency components through convolution or attention mechanisms, enabling the capture of both fine-grained and global features. FreqMamba [53] explores correlations across different frequency bands for image deraining. In medical imaging, HF-ResDif [39] employs high-frequency information to guide models in reconstructing image details.

These techniques underscore the importance of distinguishing between high- and low-frequency components to improve model performance by detecting multi-scale and fine-grained features, which is particularly beneficial for medical image analysis.

Space-Filling Curves. Space-filling curves, such as the Hilbert curve [18], Z-order curve [32], and sweep curve, are fractal structures that traverse every point in a multi-dimensional space without repetition [30]. Widely used in point cloud processing [38], they reduce dimensionality while preserving spatial topology and locality. Among them, the Hilbert curve is particularly notable for its strong locality-preserving properties, making it a promising tool for spatial data processing. In contrast, many Mamba-based models in computer vision rely on scanning techniques like bidirectional scanning (Bidi-Scan) [51] and cascade scanning (Cascade-Scan) [25]. While these methods achieve reasonable performance [37], these methods struggle to align with the human eye's foveal vision, often losing critical local information and compromising fine-grained spatial details. The Hilbert curve, by preserving locality, maintains proximity between image features, enabling visual models to better simulate the human eye's focus on local regions and handle detailed information more effectively.

This paper explores integrating spatial-frequency domain and state-space models with the Hilbert curve to create a robust baseline for medical image detection. By combining these techniques, we aim to enhance local feature preservation and long-range dependency modeling, thereby improving the detection of medical abnormalities.

3 Methods

3.1 Preliminaries

State Space Model. In the vanilla Mamba framework, the token mixer is implemented as a selective SSM. The SSM [8,14,16] is a continuous-time latent state model that maps a 1D input signal $p(t) \in \mathbb{R}^L$ to an output signal $q(t) \in \mathbb{R}^L$ via a hidden state $h(t) \in \mathbb{R}^N$.

This model defines four input-dependent parameters: $(\Delta, \mathbf{A}, \mathbf{B}, \mathbf{C})$, including a timescale parameter Δ to transform the continuous parameters A, B to discrete

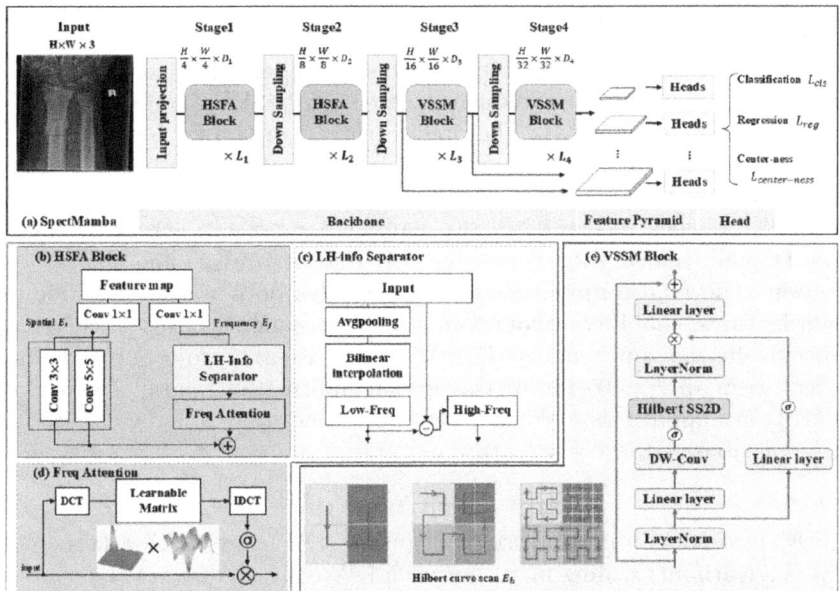

Fig. 1. Top: (a) The overall architecture of SpectMamba. Bottom: (b) is the structure of the Hybrid Spatial-Frequency Attention Block. (c) and (d) depict the Low-High Frequency Domain Information Separator (LH-Info Separator) and Freq attention Module, respectively. (e) presents the Visual State-Space Blocks, which incorporate the Hilbert SS2D module.

parameters $\overline{\mathbf{A}}, \overline{\mathbf{B}}$. The commonly used method for transformation is zero-order hold (ZOH), which is defined as follows:

$$\overline{\mathbf{A}} = \exp(\Delta \mathbf{A}), \quad \overline{\mathbf{B}} = (\Delta \mathbf{A})^{-1} \left(\exp(\Delta \mathbf{A}) - \mathbf{I} \right) \cdot \Delta \mathbf{B} \tag{1}$$

Then the SSM's sequence-to-sequence transformation is defined as:

$$h'(t) = \overline{\mathbf{A}} h(t) + \overline{\mathbf{B}} p(t), \quad q(t) = \mathbf{C} h(t), \tag{2}$$

where $\mathbf{A} \in \mathbb{R}^{N \times N}$ is the evolution parameter, and $\mathbf{B} \in \mathbb{R}^{N \times 1}$, $\mathbf{C} \in \mathbb{R}^{1 \times N}$ are the projection parameters. Finally, the model computes the output q through a global convolution. The convolutional kernel $\overline{\mathbf{K}} \in \mathbb{R}^L$ is formulated as:

$$\begin{aligned} \overline{K} &= (\mathbf{C}\overline{\mathbf{B}}, \mathbf{C}\overline{\mathbf{A}}\overline{\mathbf{B}}, \ldots, \mathbf{C}\overline{\mathbf{A}}^{L-1}\overline{\mathbf{B}}), \\ q &= p * \overline{\mathbf{K}}, \end{aligned} \tag{3}$$

where L is the length of the input sequence p, and $*$ represents the convolution operation. This approach introduces a more efficient convolution mode that bypasses state computation, enabling robust sequence-to-sequence transformations.

3.2 Hybrid Spatial-Frequency Attention Block

In Fig. 1, we present the architecture of SpectMamba, which comprises two key modules: the Hybrid Spatial-Frequency Attention (HSFA) block and the Visual State-Space Module (VSSM). The architecture is organized into four network stages, each preceded by down-sampling layers that facilitate the creation of hierarchical feature representations.

The HSFA block incorporates two parallel components: the Low-High Frequency Domain Information Separator (LH-Info Separator) and spatial feature extraction. The final output seamlessly integrates both spatial and frequency-domain features, enabling a more comprehensive understanding of the image.

Specifically, an input image $I \in \mathbb{R}^{H \times W \times C}$ is first projected and mapped to a feature map F_s. To retain the spatial information, parallel convolutions (Fig. 1(b)) are applied to extract feature information at different scales, which is then concatenated into the feature map Q_s:

$$Q_s = \oplus_{k \in K} \{\gamma_i E_s(F_s, k)\}, \tag{4}$$

where E_s represents a depth-wise convolution (DW conv) with a kernel size k, and γ_i is a learnable scaling factor for each DW conv. $\oplus$ denotes the concatenation operator, and $K = \{3, 5\}$ is the set of kernel sizes used in this context.

Inspired by the high- and low-frequency separation mechanism in OctConv [7], we employ average pooling to extract the low-frequency component, which subsamples each region through averaging rather than selectively shifting the window. This approach effectively addresses the misalignment issues commonly encountered during multi-scale information aggregation, in contrast to downsampling via convolution with stride. The downsampling rate and step size are both set to 2, while the upsampling operation is performed using bilnear interpolation.

Building on this foundation, the feature vector is decomposed into low- and high-frequency components, F_f^l and F_f^h, using the FH-info module, which is designed to preserve high-frequency details.

To further enhance feature representation, we leverage the Discrete Cosine Transform (DCT), renowned for its high energy concentration and efficient conductibility, to perform domain conversion from the spatial domain to the frequency domain. Adaptive learnable matrix is then applied to the DCT-based frequency domain representation, enabling a more robust and expressive feature learning process.

$$Q_f^h, Q_f^l = \text{IDCT}(E_f(\text{DCT}(F_f^h, F_f^l))), \tag{5}$$

where E_f is a learnable matrix as shown in Fig 1 (d). Finally, the HSFA block combines the spatial domain and high-low frequency domain representations as:

$$Q_m = \Gamma\Big(\text{Conv}\big(F_s \oplus Q_s \oplus (\sigma Q_f^h \oplus \sigma Q_f^l)\big)\Big), \tag{6}$$

Q_m is the output of the HSFA Block and Γ represents the GELU activation function [20]. $\oplus$ denotes the concatenation of the feature maps and σ is the Sigmoid activation function.

3.3 Visual State-Space Module

After obtaining the mixed spatial and frequency domain features, the Visual State-Space Module (VSSM) performs further information fusion and feature extraction. Leveraging its long-range dependency capabilities, the VSSM captures global information effectively. However, the SSM, originally designed as a continuous-time latent state model for 1D scans, faces challenges when applied to 2D images. Previous methods, such as bidirectional and cascade scanning [25,51], fail to preserve local dependencies between neighboring pixels effectively.

To overcome these limitations, we propose the use of the Hilbert curve as a scanning method. This method improves the capture of spatial locality by enhancing the connections between neighboring pixels [18]. As illustrated in Fig. 1(e), Hilbert space-filling curves traverse all elements in a space without repetition or gaps, thereby preserving spatial dependencies. The mapping of the Hilbert curve of order n can be represented as a mapping function from a 1D interval to a 2D space $e: [0,1] \to [0,1] \times [0,1]$.

$$E_h(\beta) = \lim_{n \to \infty} e_n(\beta), \quad \text{for } \beta \in [0,1], \tag{7}$$

Subsequently, all pixels are sorted into a 1D sequence based on their traversal position h:

$$Q_c = E_c(E_h(Q_m)), \tag{8}$$

where E_c indicates the VSSM block as illustrated in Fig 1 (e), and Q_c represents the feature vector after feature extraction by the VSSM block. Hilbert-SS2D in the VSSM block involves three steps: first, the 2D feature map is processed by traversing it along scanning paths (Hilbert-Scan); second, each sequence is independently processed by distinct S6 blocks [14,25]; and finally, the results are merged to generate the final 2D feature map.

3.4 Loss Function

Building on feature pyramid networks (FPN) [23], we detect objects of varying sizes across multiple levels of feature maps. Specifically, we utilize five levels of feature maps and share the detection heads across these levels. This strategy enhances parameter efficiency while simultaneously improving detection performance.

The loss function used during training is based on the FCOS [41], defined as follows:

$$\begin{aligned} L(\{p_{x,y}\}, \{t_{x,y}\}, \{o_{x,y}\}) &= \frac{1}{N_{pos}} \sum_{x,y} L_{cls}(p_{x,y}, c^*_{x,y}) \\ &+ \frac{\lambda_1}{N_{pos}} \sum_{x,y} \mathbf{1}_{\{c^*_{x,y} > 0\}} L_{reg}(t_{x,y}, t^*_{x,y}) \\ &+ \frac{\lambda_2}{N_{pos}} \sum_{x,y} \mathbf{1}_{\{c^*_{x,y} > 0\}} L_{center-ness}(o_{x,y}, o^*_{x,y}), \end{aligned} \tag{9}$$

where L_{cls} is the focal loss for classification, L_{reg} is the intersection over union (IoU) loss for regression, as in UnitBox [47] and $L_{center-ness}$ is the binary cross-entropy loss that reflects the distance from the target center. $p_{x,y}, t_{x,y}, o_{x,y}$ represent the predicted categories, positions, and center-ness at feature map location (x, y), respectively, and $*$ denotes the corresponding ground truth. $\mathbf{1}_{\{d^*_{x,y}>0\}}$ is equal to 1 when the specified condition is met and 0 otherwise. N_{pos} denotes the number of positive samples, and λ_1, λ_2, set to 1 [40].

4 Experiments

4.1 Experimental Settings

Dataset. To evaluate SpectMamba's performance in medical image detection, we conducted experiments on three widely-used public datasets from different medical domains: the RSNA Pneumonia (PenD) dataset [12], the Brain Tumor (Brats) dataset [1,2,29], and the Bone Fracture GRAZPEDWRI-DX dataset (Graz) [31].

PenD is a public dataset for chest pneumonia detection, consisting of 6,012 clinical chest X-ray images. Brats is a MRI dataset containing 3D images of gliomas in the brain, comprising 259 cases. Following standard protocols [45], we sliced the 3D images along the Z axis, resulting in an average of 50 slices per case. Images in both datasets are standardized to 256×256. Graz is a dataset of 20,327 pediatric wrist trauma X-ray images, with image size standardized to 512×512.

We randomly split datasets into training, test and validation sets with a 70%, 20%, and 10% split, respectively. To ensure fairness, the same data augmentation techniques were applied uniformly during training. All models were optimized using the adam algorithm with a weight decay of $1e^{-5}$ and trained for 200 epochs.

Baseline Methods. We evaluate SpectMamba's performance in medical image detection using three datasets. Our analysis involves two comparisons: first, we assess the impact of different backbone architectures, and second, we compare SpectMamba with state-of-the-art methods.

For the first comparison, models are categorized by their network structures: CNN, Transformer, and Mamba-based frameworks. CNN-based models include FCOS with ResNet [41] and YOLOv3 with Darknet [50]. Transformer-based models include PCViT [21], which employs Vision Transformer (ViT) in both base and large configurations. Mamba-based models, VMamba [25] and Vision Mamba [51], leverage the Mamba framework for visual state-space modeling. All models are processed through FPN and trained with a consistent loss function for fair comparison. In the second comparison, we compare SpectMamba with the latest methods, including Dense Distinct Queries (DDQ) [49] and DETR with Improved deNoising anchOrboxes (DINO) [48], which a complex architecture.

Evaluation Metrics. The evaluation metrics include average precision (AP) and average recall (AR), computed across three IoU thresholds: 0.5, 0.6, and 0.7. To derive the mean average precision (mAP) and mean average recall (mAR), the AP and AR values are averaged across these thresholds (Table 1).

Table 1. Results Across Three Datasets. The best results are highlighted in **bold**, and the second-best results are underlined.

Method	AP_{50}	AP_{60}	AP_{70}	mAP	AR_{50}	AR_{60}	AR_{70}	mAR
PenD detection results								
CNN-based model								
ResNet[41]	40.06	23.71	8.79	24.19	68.67	47.35	24.53	**46.85**
Darknet[50]	39.19	24.07	**9.38**	24.21	49.30	38.70	**25.10**	37.70
transformer-based model								
Vit-B[22]	38.96	20.62	7.60	22.39	65.79	42.84	22.89	43.84
Vit-L[22]	39.17	22.75	6.93	22.95	66.58	44.47	22.04	44.36
Mamba-based model								
Vim-base[51]	34.49	19.46	6.93	20.29	62.66	41.92	20.99	41.86
VMamba-base[25]	36.49	21.68	8.43	22.20	65.08	44.93	24.20	44.74
SpectMamba	**43.11**	**24.71**	7.95	**25.26**	**70.11**	**47.61**	22.76	46.83
Brats detection results								
CNN-based model								
ResNet[41]	85.29	69.52	43.01	65.94	92.93	79.15	55.91	76.00
Darknet[50]	59.70	49.46	27.96	45.70	77.23	69.24	51.96	66.14
transformer-based model								
Vit-B[22]	87.39	72.71	49.71	69.94	93.59	81.07	61.48	78.71
Vit-L[22]	88.35	74.56	52.17	71.69	93.62	82.80	63.36	79.93
Mamba-based model								
Vim-base[51]	84.18	66.90	43.10	64.72	92.32	78.49	57.83	76.22
VMamba-base[25]	87.96	74.43	52.52	71.64	94.32	82.64	63.36	80.11
SpectMamba	**89.83**	**78.64**	**57.54**	**75.34**	**95.35**	**85.64**	**68.24**	**83.08**
Graz detection results								
CNN-based model								
ResNet[41]	87.77	78.58	56.15	74.17	96.77	89.16	70.89	85.61
Darknet[50]	63.53	54.82	40.41	52.92	70.48	61.53	49.37	60.46
transformer-based model								
Vit-B[22]	86.45	77.39	53.23	72.36	95.42	87.53	68.57	83.84
Vit-L[22]	89.11	79.86	57.14	75.37	95.92	88.55	71.36	85.27
Mamba-based model								
Vim-base[51]	86.17	76.66	54.09	72.31	96.27	88.36	69.90	84.84
VMamba-base[25]	**90.55**	82.36	59.33	77.41	97.24	**91.42**	74.01	87.56
SpectMamba	90.27	82.37	**61.03**	**77.89**	**97.54**	90.98	**74.89**	**87.80**

4.2 Comparison with Different Backbones

PenD Results. While ResNet excels at capturing local features, SpectMamba outperforms it by 3.05% in AP@50 and 1.07% in mAP. Its advantage stems from the long-distance dependencies of VSSM, enabling it to better capture the globally diffuse nature of pneumonia. **Brats Results.** SpectMamba surpasses VMamba by 1.87% in AP@50 and 3.7% in mAP, owing to its spatial and hilbert scan module that enhance abnormality detection by capturing local details. **Graz Results.** SpectMamba outperforms VMamba by 0.48% in mAP, leveraging its capacity to capture spatial and high-frequency dependencies. This

Table 2. Computational efficiency on Graz dataset. Throughput values are measured with an L40 GPU, following the protocol proposed in [26].

Method	mAP	mAR	Params(M)	FLOPs.(G)	Train Throughput
ResNet[41]	74.17	85.61	32	51.3	914
Darknet[50]	52.92	60.46	62	49.9	2642
Vit-B[22]	72.36	83.84	94	77.3	691
Vit-L[22]	75.37	85.27	312	226.1	542
Vim-base[51]	72.31	84.84	45	78.2	230
VMamba-base[25]	77.41	87.56	110	38.0	136
SpectMamba	**77.89**	**87.80**	100	77.5	275

strengthens local feature extraction, providing an advantage in detecting complex fracture patterns Quantitative and qualitative results are shown in Table 1 and Fig. 3, respectively.

Computational Efficiency. As shown in Table 2, SpectMamba achieves a higher mAP (77.89) compared to VMamba (77.41), while running twice as fast. Additionally, SpectMamba outperforms ViT-L (mAP 77.89 vs. 75.37) with only 32% of the parameters.

4.3 Comparison with SOTA Methods

Recent DETR-based methods, such as DDQ [49] and DINO [48], have achieved SOTA performance in object detection for natural images. However, medical imaging presents unique challenges that differ significantly from natural image analysis. To rigorously assess SpectMamba's capabilities, we conducted comparative experiments under controlled conditions: no pre-trained models, consistent

Table 3. The results of SpectMamba, DDQ, and DETR across the three datasets.

Method	AP_{50}	AP_{60}	AP_{70}	mAP	AR_{50}	AR_{60}	AR_{70}	mAR
PenD detection results								
DDQ [49]	13.83	6.33	2.24	7.47	35.12	23.94	14.13	24.4
DINO [48]	14.26	6.38	1.99	7.55	30.92	20.34	11.05	20.77
SpectMamba	**43.11**	**24.71**	**7.95**	**25.26**	**70.11**	**47.61**	**22.76**	**46.83**
Brats detection results								
DDQ [49]	48.02	29.87	12.75	30.22	66.24	51.27	32.41	49.97
DINO [48]	58.31	43.23	19.7	40.41	68.51	58.14	38.25	54.97
SpectMamba	**89.83**	**78.64**	**57.54**	**75.34**	**95.35**	**85.64**	**68.24**	**83.08**
Graz detection results								
DDQ [49]	52.84	43.33	27.22	41.13	71.5	64.35	50.35	62.06
DINO [48]	61.63	45.02	21	42.55	74.01	62.37	41.33	59.24
SpectMamba	**90.27**	**82.37**	**61.03**	**77.89**	**97.54**	**90.98**	**74.89**	**87.80**

Table 4. The results of the ablation experiment on the brats dataset. The best results are highlighted in **bold** and the second-best results are underlined.

Bidi-scan	Hilbert curve	LH-info	spatial	AP_{50}	mAP	AR_{50}	mAR
	✓			87.96	71.64	<u>94.32</u>	80.11
	✓	✓		87.73	71.38	<u>94.32</u>	81.44
	✓		✓	<u>89.64</u>	73.28	94.14	81.24
✓		✓	✓	89.35	<u>74.81</u>	94.23	<u>82.58</u>
	✓	✓	✓	**89.83**	**75.33**	**95.35**	**83.08**

data augmentation strategies, 200 training epochs, and an identical learning rate. As illustrated in Table 3, SpectMamba consistently outperforms these SOTA models. This difference may stem from challenges unique to medical images, which differ from natural images. Medical images have high resolution and often contain small yet critical regions, such as lesions and masses. They also typically feature fewer objects per image and exhibit dataset imbalances, with a focus on a limited range of abnormalities. This results in a high class imbalance, with positive cases (e.g., unhealthy subjects) being less frequent than negative ones [13], and fewer objects of interest overall [11].

Consistent with our findings, prior work [44] has showed that standard practices from natural image processing—such as complex encoder architectures, multi-scale feature fusion, query initialization, and iterative bounding box refinement—often fail to enhance performance in medical imaging. In some cases, these techniques may even hinder detection accuracy.

These observations underscore the need to reconsider traditional approaches for transformer-based models and to explore more specialized, efficient frameworks tailored to the demands of medical imaging, such as SpectMamba.

4.4 Ablation Study

SpectMamba consists of three main components: Hilbert curve scanning, the LH-info block, and the spatial layer. The ablation results in Table 4 for these modules and different scanning modes. The baseline includes results from two reverse Hilbert curve scans. The network's performance improves when either the HSFA or spatial layer is integrated separately. Prior work [25] has shown that bidirectional scanning is more computationally efficient and yields better classification accuracy than cascade scanning. The results indicate that replacing Hilbert curve scanning with bidirectional scanning preserves the effectiveness and robustness of both the LH-info and spatial components.

4.5 Hilbert Curve Scanning Variants

We explored different 2D scanning paths, with Table 5 and Fig 2 presenting the results and effective receptive fields (ERFs) for each Hilbert scanning mode. The top section of Fig 2 illustrates schematic diagrams and scanning paths for the

Table 5. The results of the different Hilbert curve filling on the brats dataset. The best results are highlighted in **bold** and the second-best results are underlined.

Methods	AP_{50}	AP_{60}	AP_{70}	mAP	AR_{50}	AR_{60}	AR_{70}	mAR
Hilbert-UniDir	88.42	75.86	53.58	72.62	94.09	83.68	65.44	81.07
Hilbert-FourDir1	88.92	<u>78.36</u>	55.54	74.27	94.28	84.68	66.59	81.85
Hilbert-FourDir2	89.19	77.82	<u>56.36</u>	<u>74.46</u>	<u>94.66</u>	84.25	<u>67.55</u>	<u>82.16</u>
Hilbert-FourDir3	<u>89.33</u>	78.02	55.39	74.25	94.24	<u>85.18</u>	66.01	81.81
Hilbert-BiDir	**89.83**	**78.64**	**57.54**	**75.34**	**95.35**	**85.64**	**68.24**	**83.08**

unidirectional Hilbert method, while the bottom section visualizes the ERFs [27] for the three methods prior to training. The left image shows the unidirectional Hilbert scan curve, while the center and right images depict curves with varying start and end points.

"Hilbert-UniDir" represents a one-way Hilbert scan, where the scan only proceeds in a forward direction without reversing, as shown in Fig. 2(a). "Hilbert-FourDir1" involves four scanning directions: one forward and its reverse (Fig. 2(b)), and another by flipping the matrix along the Y-axis and its reverse (Fig. 2(c)). "Hilbert-FourDir2" also involves four scanning directions: one forward and its reverse (Fig. 2(b)), and the other by transposing the matrix and its reverse. "Hilbert-FourDir3" includes scanning in four directions: forward, X-axis flip, Y-axis flip, and matrix transpose.This scanning method makes the scanning path of the network symmetrical. Finally, "Hilbert-BiDir" represents scanning in both forward and reverse directions.

The results show that the "Hilbert-UniDir" method already performed well, while "Hilbert-BiDir" achieved the best overall performance, outperforming all other scanning methods. Interestingly, "Hilbert-FourDir" underperformed compared to "Hilbert-BiDir," likely due to excessive overlapping paths, causing the network to overemphasize certain regions. These findings suggest that Hilbert-BiDir can replace four-directional scanning, reducing memory usage by 50% and improving computational efficiency (Fig. 3).

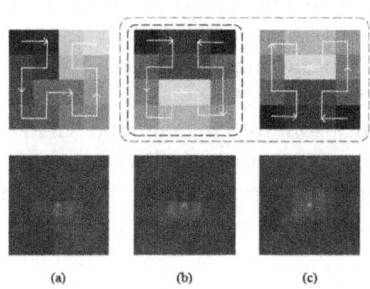

Fig. 2. The figure illustrates the effective receptive fields of three Hilbert scanning methods.

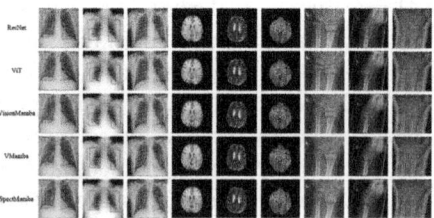

Fig. 3. Visualized examples of Spect-Mamba and comparative methods, where red boxes indicate ground truths and blue boxes represent predictions. (Color figure online)

5 Conclusion

SpectMamba addresses the need for efficient and accurate pathology detection in medical imaging by overcoming the limitations of traditional CNNs and Transformer-based models. By leveraging Mamba's linear complexity for handling long sequences and the Hilbert curve scanning technique, the Visual State-Space Module captures global dependencies, while the Hybrid Spatial-Frequency Attention Block processes both spatial and high-frequency information. Experiments demonstrate SpectMamba's superior performance in tasks such as brain tumor detection in MRI scans and pneumonia and bone fracture detection in X-rays, highlighting its effectiveness in advancing medical diagnostics.

References

1. Bakas, S., et al.: Advancing the cancer genome atlas glioma MRI collections with expert segmentation labels and radiomic features. Sci. Data **4**(1), 1–13 (2017)
2. Bakas, S., et al.: Identifying the best machine learning algorithms for brain tumor segmentation, progression assessment, and overall survival prediction in the brats challenge. arXiv preprint arXiv:1811.02629 (2018)
3. Brigham, E.O.: The Fast Fourier Transform and Its Applications. Prentice-Hall Inc., Hoboken (1988)
4. Cai, Z., Vasconcelos, N.: Cascade R-CNN: delving into high quality object detection. In: Proceedings of the IEEE Conference on Computer Vision and Pattern Recognition, pp. 6154–6162 (2018)
5. Cao, Y., et al.: Remote sensing image segmentation using vision mamba and multiscale multi-frequency feature fusion. arXiv preprint arXiv:2410.05624 (2024)
6. Chen, J., Yu, L., Wang, W.: Hilbert space filling curve based scan-order for point cloud attribute compression. IEEE Trans. Image Process. **31**, 4609–4621 (2022)
7. Chen, Y., et al.: Drop an octave: reducing spatial redundancy in convolutional neural networks with octave convolution. In: Proceedings of the IEEE/CVF International Conference on Computer Vision, pp. 3435–3444 (2019)
8. Dao, T., Gu, A.: Transformers are SSMs: Generalized models and efficient algorithms through structured state space duality. arXiv preprint arXiv:2405.21060 (2024)
9. Diwan, T., Anirudh, G., Tembhurne, J.V.: Object detection using YOLO: challenges, architectural successors, datasets and applications. Multimedia Tools Appl. **82**(6), 9243–9275 (2023)
10. Dong, W., et al.: Fusion-mamba for cross-modality object detection. arXiv preprint arXiv:2404.09146 (2024)
11. Dosovitskiy, A., et al.: An image is worth 16x16 words: transformers for image recognition at scale. arXiv preprint arXiv:2010.11929 (2020)
12. Gabruseva, T., Poplavskiy, D., Kalinin, A.: Deep learning for automatic pneumonia detection. In: Proceedings of the IEEE/CVF Conference on Computer Vision and Pattern Recognition Workshops, pp. 350–351 (2020)
13. Galdran, A., Carneiro, G., González Ballester, M.A.: Balanced-mixup for highly imbalanced medical image classification. In: de Bruijne, M., et al. (eds.) MICCAI 2021. LNCS, vol. 12905, pp. 323–333. Springer, Cham (2021). https://doi.org/10.1007/978-3-030-87240-3_31

14. Gu, A., Dao, T.: Mamba: linear-time sequence modeling with selective state spaces. arXiv preprint arXiv:2312.00752 (2023)
15. Gu, A., Dao, T., Ermon, S., Rudra, A., Ré, C.: Hippo: recurrent memory with optimal polynomial projections. In: Advances in Neural Information Processing Systems, vol. 33, pp. 1474–1487 (2020)
16. Gu, A., Goel, K., Ré, C.: Efficiently modeling long sequences with structured state spaces. arXiv preprint arXiv:2111.00396 (2021)
17. Gu, A., et al.: Combining recurrent, convolutional, and continuous-time models with linear state space layers. In: Advances in Neural Information Processing Systems, vol. 34, pp. 572–585 (2021)
18. He, C., Li, R., Li, S., Zhang, L.: Voxel set transformer: a set-to-set approach to 3D object detection from point clouds. In: Proceedings of the IEEE/CVF Conference on Computer Vision and Pattern Recognition, pp. 8417–8427 (2022)
19. Hu, V.T., et al.: Zigma: a dit-style zigzag mamba diffusion model. In: European Conference on Computer Vision, pp. 148–166. Springer, Cham (2025)
20. Lee, M.: Gelu activation function in deep learning: a comprehensive mathematical analysis and performance. arXiv preprint arXiv:2305.12073 (2023)
21. Li, J., Tian, P., Song, R., Xu, H., Li, Y., Du, Q.: PCViT: a pyramid convolutional vision transformer detector for object detection in remote-sensing imagery. IEEE Trans. Geosci. Remote Sens. **62**, 1–15 (2024). https://doi.org/10.1109/TGRS.2024.3360456
22. Li, Y., Mao, H., Girshick, R., He, K.: Exploring plain vision transformer backbones for object detection. In: European Conference on Computer Vision, pp. 280–296. Springer, Cham (2022)
23. Lin, T.Y., Dollár, P., Girshick, R., He, K., Hariharan, B., Belongie, S.: Feature pyramid networks for object detection. In: Proceedings of the IEEE Conference on Computer Vision and Pattern Recognition, pp. 2117–2125 (2017)
24. Litjens, G., et al.: A survey on deep learning in medical image analysis. Med. Image Anal. **42**, 60–88 (2017)
25. Liu, Y., et al.: Vmamba: visual state space model (2024). https://arxiv.org/abs/2401.10166
26. Liu, Z., et al.: Swin transformer: hierarchical vision transformer using shifted windows. In: Proceedings of the IEEE/CVF International Conference on Computer Vision, pp. 10012–10022 (2021)
27. Luo, W., Li, Y., Urtasun, R., Zemel, R.: Understanding the effective receptive field in deep convolutional neural networks. In: Advances in Neural Information Processing Systems, vol. 29 (2016)
28. Mehta, H., Gupta, A., Cutkosky, A., Neyshabur, B.: Long range language modeling via gated state spaces. arXiv preprint arXiv:2206.13947 (2022)
29. Menze, B.H., et al.: The multimodal brain tumor image segmentation benchmark (BRATS). IEEE Trans. Med. Imaging **34**(10), 1993–2024 (2014)
30. Mokbel, M.F., Aref, W.G., Kamel, I.: Analysis of multi-dimensional space-filling curves. GeoInformatica **7**, 179–209 (2003)
31. Nagy, E., Janisch, M., Hržić, F., Sorantin, E., Tschauner, S.: A pediatric wrist trauma X-ray dataset (grazpedwri-dx) for machine learning. Sci. Data **9**(1), 222 (2022)
32. Orenstein, J.A.: Spatial query processing in an object-oriented database system. In: Proceedings of the 1986 ACM SIGMOD International Conference on Management of Data, pp. 326–336 (1986)

33. Pan, Z., Cai, J., Zhuang, B.: Fast vision transformers with hilo attention. In: Advances in Neural Information Processing Systems, vol. 35, pp. 14541–14554 (2022)
34. Rahaman, N., et al.: On the spectral bias of neural networks. In: International Conference on Machine Learning, pp. 5301–5310. PMLR (2019)
35. Rao, Y., Zhao, W., Zhu, Z., Lu, J., Zhou, J.: Global filter networks for image classification. In: Advances in Neural Information Processing Systems, vol. 34, pp. 980–993 (2021)
36. Ruan, J., Xiang, S.: VM-Unet: vision mamba Unet for medical image segmentation. arXiv preprint arXiv:2402.02491 (2024)
37. Shi, D.: TransNext: robust foveal visual perception for vision transformers. In: Proceedings of the IEEE/CVF Conference on Computer Vision and Pattern Recognition, pp. 17773–17783 (2024)
38. Sun, P., et al.: RSN: range sparse net for efficient, accurate lidar 3D object detection. In: Proceedings of the IEEE/CVF Conference on Computer Vision and Pattern Recognition, pp. 5725–5734 (2021)
39. Tang, Z., Jiang, C., Cui, Z., Shen, D.: HF-resdiff: high-frequency-guided residual diffusion for multi-dose PET reconstruction. In: International Conference on Medical Image Computing and Computer-Assisted Intervention, pp. 372–381. Springer, Cham (2024)
40. Tian, Z., Chu, X., Wang, X., Wei, X., Shen, C.: Fully convolutional one-stage 3D object detection on lidar range images. In: Advances in Neural Information Processing Systems, vol. 35, pp. 34899–34911 (2022)
41. Tian, Z., Shen, C., Chen, H., He, T.: FCOS: a simple and strong anchor-free object detector. IEEE Trans. Pattern Anal. Mach. Intell. **44**(4), 1922–1933 (2020)
42. Wang, J., et al.: Selective structured state-spaces for long-form video understanding. In: Proceedings of the IEEE/CVF Conference on Computer Vision and Pattern Recognition, pp. 6387–6397 (2023)
43. Wei, K., et al.: CT synthesis from MR images using frequency attention conditional generative adversarial network. Comput. Biol. Med. **170**, 107983 (2024)
44. Xu, Y., Shen, Y., Fernandez-Granda, C., Heacock, L., Geras, K.J.: Understanding differences in applying DETR to natural and medical images. arXiv preprint arXiv:2405.17677 (2024)
45. Xu, Z., Zhang, X., Zhang, H., Liu, Y., Zhan, Y., Lukasiewicz, T.: EFPN: effective medical image detection using feature pyramid fusion enhancement. Comput. Biol. Med. **163**, 107149 (2023)
46. Yu, A., Lyu, D., Lim, S.H., Mahoney, M.W., Erichson, N.B.: Tuning frequency bias of state space models. arXiv preprint arXiv:2410.02035 (2024)
47. Yu, J., Jiang, Y., Wang, Z., Cao, Z., Huang, T.: Unitbox: an advanced object detection network. In: Proceedings of the 24th ACM International Conference on Multimedia, pp. 516–520 (2016)
48. Zhang, H., et al.: Dino: DETR with improved denoising anchor boxes for end-to-end object detection. arXiv preprint arXiv:2203.03605 (2022)
49. Zhang, S., et al: Dense distinct query for end-to-end object detection. In: Proceedings of the IEEE/CVF Conference on Computer Vision and Pattern Recognition, pp. 7329–7338 (2023)
50. Zhao, L., Li, S.: Object detection algorithm based on improved YOLOv3. Electronics **9**(3), 537 (2020)
51. Zhu, L., Liao, B., Zhang, Q., Wang, X., Liu, W., Wang, X.: Vision mamba: efficient visual representation learning with bidirectional state space model. arXiv preprint arXiv:2401.09417 (2024)

52. Zhu, Q., et al.: Samba: semantic segmentation of remotely sensed images with state space model. Heliyon **10**(19) (2024)
53. Zou, Z., Yu, H., Huang, J., Zhao, F.: Freqmamba: viewing mamba from a frequency perspective for image deraining. In: Proceedings of the 32nd ACM International Conference on Multimedia, pp. 1905–1914 (2024)

Hierarchical Neural Cellular Automata for Lightweight Microscopy Image Classification

Chen Yang[1,3], Michael Deutges[1], Nassir Navab[2,3], Ario Sadafi[1,2,3], and Carsten Marr[1(✉)]

[1] Institute of AI for Health, Computatioal Health Center, Helmholtz Munich, Munich, Germany
carsten.marr@helmholtz-munich.de
[2] Computer Aided Medical Procedures, Technical University of Munich, Munich, Germany
[3] TUM School of Computation, Information and Technology, Technical University Munich, Munich, Germany

Abstract. Classification of cells in microscopy images is an essential step in the diagnostic workflows for various medical conditions. These diagnostic processes benefit from emerging deep learning solutions, which make them more accessible, reliable, and scalable. However, their extensive deployment is hindered by limited generalizability and high computational demands of such architectures. We address this issue by introducing a lightweight, general-purpose hierarchical classification model based on Neural Cellular Automata (NCA). Our approach utilizes NCA to extract features at multiple resolutions, combining the advantages of NCA-based methods with those of convolutional architectures. We evaluate our model on six microscopy datasets from different modalities and demonstrate that it consistently outperforms existing NCA-based approaches. With significantly fewer parameters than conventional deep learning methods, our model is suitable for deployment in resource-constrained areas, such as remote clinics with limited computational infrastructure or mobile devices with lower computational capacities. Our results highlight the potential of NCA-based models as an effective, lightweight alternative for image classification, addressing critical barriers to the equitable distribution of automated diagnostic tools.

Keywords: Neural Cellular Automata · Image Classification · Microscopy Image Analysis · Robustness · General Purpose Solutions · Lightweight

1 Introduction

Microscopy is an essential and widely used imaging modality in medicine due to its affordability, accessibility, and ability to provide detailed visual information

---
C. Yang and M. Deutges—Equal contribution.
C. Marr—Shared last authorship.

at the cellular level. The classification of cells in microscopy images is a critical step in diagnostic workflows for a wide range of medical conditions, including infectious diseases, certain cancer types, and hematological disorders. Accurate and efficient classification enables early diagnosis and improves treatment outcomes. However, this process typically relies on expert pathologists to manually analyze microscopy images, which can be time-consuming, labor-intensive, and prone to variability between observers.

In recent years, automated solutions have shown significant potential in supporting workflows that rely on microscopy image analysis for a variety of medical conditions. For instance, hematological diseases such as leukemia can be diagnosed and monitored through the analysis of blood cell morphology, leveraging advanced image classification techniques [6,10,17,18]. Similarly, malaria detection benefits from solutions that identify parasitized red blood cells, a crucial step in providing timely treatment in endemic regions [16]. Automated screening of pap smears has also been developed to detect abnormal cells, enabling early identification of cervical cancer and significantly improving prevention efforts [8]. Furthermore, urine sediment analysis plays a vital role in diagnosing metabolic and infectious conditions, such as kidney disorders, diabetes, and urinary tract infections, by classifying sediment particles in microscopic images [23]. These advancements demonstrate the broad applicability of automated image analysis across diverse diagnostic workflows.

These advancements in deep learning have shown potential to transform medical diagnostics by automating tasks, including the classification of cells in microscopy images. Models can achieve expert-level accuracy, offering faster and more consistent results compared to manual analysis. Importantly, deep learning-based tools hold promise for improving equity in healthcare by providing access to reliable diagnostics in remote regions.

Despite their potential, conventional deep learning methods face significant challenges that hinder their widespread deployment in real-world medical diagnostics. These models are often large and computationally demanding, making them impractical for widely accessible but resource-constrained devices such as smartphones. Additionally, their performance is heavily dependent on careful dataset-specific fine-tuning of parameters. As a result, these models tend to generalize poorly to new datasets or unseen variations, limiting their applicability across diverse clinical settings. While there exist methods for domain adaptation [19], they rely on large pretraining datasets and often compromise performance. This reliance on extensive computational resources and dataset-specific optimization creates barriers to deploying automated solutions, particularly in remote or underserved areas where flexibility and efficiency are essential.

Neural Cellular Automata (NCA) have emerged as a promising alternative to conventional deep learning methods, offering a lightweight and robust approach to image processing tasks [4]. NCA are based on the iterative application of local update rules, which allows them to operate with significantly fewer parameters, making them highly efficient in terms of memory requirements. Additionally, their compact design inherently reduces the risk of overfitting, improving generalization. While NCA offer notable advantages in terms of efficiency and robustness, existing NCA models face limitations in capturing global context. Their reliance on local interactions, where updates are based solely on the immediate pixel neighborhood, restricts their ability to aggregate information across the entire image. As a result, NCA often require many iterative updates for information to propagate, which can be ineffective for tasks that require a broader understanding of spatial relationships. This inherent limitation hinders their performance on complex classification problems, where global patterns or context play a critical role, preventing them from achieving state-of-the-art accuracy.

Introducing hierarchical structures has been shown to improve CA-based models in various domains. Qin et al. [14] applied a hierarchical approach for visual saliency using standard CA, while more recent works have extended NCA to enable self-organization [12] and morphogenesis [3]. However, these methods focus on structured growth or feature refinement rather than feature extraction for classification.

To address the limitations of existing NCA classification models, we propose a novel architecture that integrates both local and global feature extraction through a multi-stage hierarchical design. By combining iterative local updates with hierarchical processing, our method effectively captures both detailed and high-level patterns, overcoming the limitations of standard NCA models. This design allows our model to outperform existing NCA methods while retaining the advantages of NCA, making it particularly suitable for deployment in resource-constrained settings.

Our code is available at https://github.com/marrlab/hNCA.

2 Methods

Our model combines the capabilities of Convolutional Neural Networks (CNNs) with the strengths of recently emerging NCA-based models. CNNs follow hierarchical architectures that process images through a series of layers, which gradually transforms low-level features such as edges and textures into high-level semantic representations. This hierarchical design has been proven effective at capturing diverse patterns in images, making it a foundational concept in computer vision. Inspired by this principle, we introduce a hierarchical architecture based on NCA.

NCA are models that iteratively update pixels in feature maps based on local interactions. This dynamic approach allows information to propagate across the image over multiple steps. Unlike CNNs, which rely on numerous convolution kernels for processing, NCA achieve effective information aggregation by iterative processes, allowing them to remain lightweight.

2.1 NCA Backbone

NCA are defined by iterative updates of the cells' states based on their local neighborhoods (Fig. 1). A cell refers to an individual pixel within the image, but

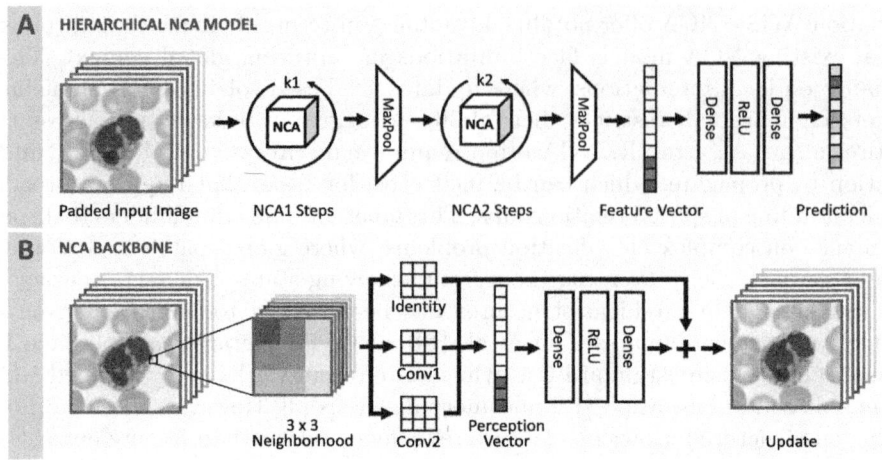

Fig. 1. Model architecture of our proposed hierarchical Neural Cellular Automata (hNCA). **A:** The hNCA consists of two NCA stages operating at different scales, connected via a max pooling operation. This allows the model to capture both local and global patterns and reduces training resources. After the feature extraction by two NCA backbones iterated with k1 and k2 steps, respectively, the feature vector is propagated through a fully connected network for the class predictions. **B:** The architecture of the NCA backbone consists of iterative updates of each cell based on their local neighborhoods. Information from the cell's neighborhood is perceived via two convolutions and the resulting perception vector is propagated through a fully connected network and added to the cell's state.

extended across all its channels. For example, in an RGB image with three color channels, each cell is represented as a three-dimensional vector corresponding to the values in the red, green, and blue channels.

The updates of a cell c are defined by a transition function that maps the state of a cell's 3×3 neighborhood N_c to a new state that is added to the cell's current state

$$c^{t+1} = c^t + \delta f_u(f_p(N_c)). \qquad (1)$$

Here, δ is a stochastic binary variable, randomly set to 0 or 1 for each cell at each step. This ensures that only approximately 50% of the cells are updated in each iteration, acting as regularization that improves robustness.

The perception function f_p aggregates the information from the cell's neighborhood by applying two convolutions defined by kernels κ_1 and κ_2. This information is concatenated with the cell's state to form the perception vector $f_p(N_c)$. The update function processes this information by propagating it through a two layer fully connected network with Rectified Linear Unit (ReLU) activation, parameterized by weights $\tilde{W}_1, \tilde{W}_2$ and biases $\tilde{b}_1, \tilde{b}_2$

$$f_p : \mathbf{R}^{3 \times 3 \times n} \to \mathbf{R}^{3n}, \qquad N_c \mapsto (c, N_c * \kappa_1, N_c * \kappa_2)^T \qquad (2)$$
$$f_u : \mathbf{R}^{3n} \to \mathbf{R}^n, \qquad f_p(N_c) \mapsto \tilde{W}_2 \max(\tilde{W}_1 f_p(N_c) + \tilde{b}_1, 0) + \tilde{b}_2. \qquad (3)$$

We denote the collection of cell states at a given time step t as $S_t \in \mathbf{R}^{H \times W \times n}$ for an image with n channels. We denote the single step for the collection of cells as $\text{NCA}_\phi : \mathbf{R}^{H \times W \times n} \to \mathbf{R}^{H \times W \times n}$, where $\phi = (\kappa_1, \kappa_2, \tilde{W}_1, \tilde{W}_2, \tilde{b}_1, \tilde{b}_2)$ is the collection of trainable parameters. A single step for one cell is illustrated in Fig. 1B.

2.2 Hierarchical NCA (hNCA)

We are here proposing an architecture based on a cascade of NCA units operating at different scales. As Fig. 1 illustrates, the first NCA stage processes the input image at its original resolution, iteratively extracting fine-grained, local features. These features are then downsampled and processed by a following NCA stage, which operates on a coarser resolution. This hierarchical design enables the model to capture both local and global patterns.

A general NCA unit at stage i in our model consists of three components: i) Padding, ii) NCA steps, and iii) Downsampling.

i) Padding: Before processing an image with NCA, zero padding is added to increase the number of channels to c_i. This padded array is referred to as the seed, $S_i^0 \in \mathbf{R}^{H_i \times W_i \times c_i}$.

ii) NCA steps: Next, k_i NCA steps are applied to extract low-level features into the additional channels

$$S_i^{t+1} = \text{NCA}_{\phi_i}(S_i^t), \qquad (4)$$

where S_i^t represents the state of the cells in stage i at iteration t.

iii) Downsampling: The features extracted by the NCA after k_i steps are downsampled by a factor F_i according to

$$\begin{aligned}&\text{pool}_i(S_i) : \mathbf{R}^{H_i \times W_i \times c_i} \to \mathbf{R}^{\frac{H_i}{F_i} \times \frac{W_i}{F_i} \times c_i} \\ &\text{pool}_i(S_i)[j, k, c] = \max_{x, y \in R_{j,k}} S_i[x, y, c],\end{aligned} \qquad (5)$$

where $R_{j,k}$ is the $F_i \times F_i$ region corresponding to the downsampled pixel (j, k).

This process of NCA steps followed by downsampling can be repeated for m stages. In the final NCA stage, a global maximum pooling operation is performed on each channel to obtain the final feature vector v.

$$\begin{aligned}&\text{pool}_m(S_m) : \mathbf{R}^{H_m \times W_m \times c_m} \to \mathbf{R}^{c_m} \\ &\text{pool}_m(S_m)[c] = \max_{x, y \leq H_m, W_m} S[x, y, c].\end{aligned} \qquad (6)$$

This feature vector v is passed through a two-layer fully connected network, referred to as the classifier g, to generate the model's logits

$$g : \mathbf{R}^{c_m} \to \mathbf{R}^C, \quad g(v) = W_2 \max(W_1 v + b_1, 0) + b_2, \qquad (7)$$

where W_1, W_2, b_1, b_2 are the weights and biases. Predicted probabilities for the C different classes in the respective dataset can be obtained with a Softmax function on the model's logits.

We denote the full model including several NCA stages and the classifier as hNCA$_\theta$: $\mathbf{R}^{H \times W \times 3} \to \mathbf{R}^C$, where $\theta = (\phi_1, ..., \phi_m, W_1, W_2, b_1, b_2)$ are all trainable parameters of the model.

We train the model end-to-end with focal loss [9] by backpropagating the gradient through the classifier and all steps of the NCA stages. The parameters of our model hNCA are obtained via gradient descent for minimizing the loss over the training dataset $\mathcal{D} = \{(x_i, y_i)\}_{i=1}^{N}$, where x_i is an input sample, y_i is its corresponding ground-truth label, and N is the total number of training samples:

$$\theta^* = \arg\min_\theta \frac{1}{N} \sum_{i=1}^{N} \mathcal{L}_{\text{focal}}(\text{hNCA}_\theta(x_i), y_i; \gamma) \tag{8}$$

$$\mathcal{L}_{\text{focal}}(p_i, y_i; \gamma) = -\sum_{c=1}^{C} y_{i,c} (1 - p_{i,c})^\gamma \log(p_{i,c}) \tag{9}$$

The introduction of γ in focal loss downweights the contribution of easy training samples and shifts the focus towards harder ones, making it particularly effective for imbalanced datasets.

Figure 1A shows a model consisting of two NCA stages. Figure 1B shows the architecture of a NCA backbone, which is identical for both scales for simplicity.

3 Experiments

In this section, we design an experiment involving six microscopy datasets from different clinics and laboratories. The proposed hNCA model is trained on each dataset individually using a general-purpose configuration for both architecture design and training process. We evaluate and compare our method against established baselines for each dataset and conduct an ablation study to demonstrate the effectiveness of various components in our design.

3.1 Datasets

We evaluate our method on six diverse microscopy datasets to assess its effectiveness in microscopic cell classification. Except for INT-20, all datasets are publicly available. The details of these datasets are as follows:

Matek-19 [10] contains 18,365 expert-annotated images of white blood cells, collected from 200 individuals at the Munich University Hospital. Half of the subjects are diagnosed with acute myeloid leukemia. The dataset is publicly available and features 15 classes. Each image has a resolution of 400×400 pixels, corresponding to 29×29 micrometers.

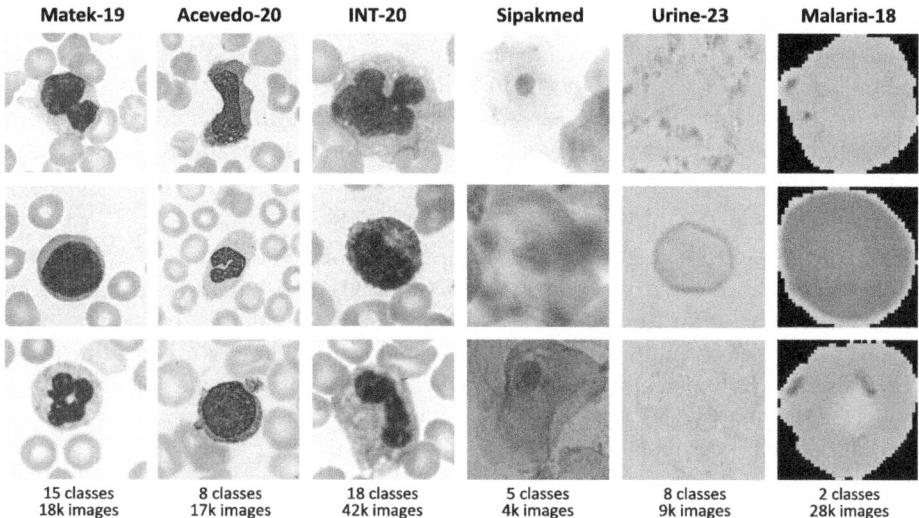

Fig. 2. Sample images from the six microscopy datasets used in this study. The datasets vary significantly in terms of domain, class distributions, and dataset sizes.

Acevedo-20 [2] includes 17,092 single-cell images collected from healthy donors at the Hospital Clinic of Barcelona. The images are grouped into 8 classes and have a resolution of 360×363 pixels, corresponding to 36×36.3 micrometers.

INT-20 comprises approximately 42,000 single-cell images spanning 18 different classes. The images have a resolution of 288×288 pixels, equivalent to 25×25 micrometers, providing detailed visual information for classification.

Sipakmed [13] is a collection of 4049 annotated cell images cropped from Pap smear slides. These images, captured using an optical microscope with a CCD camera, are categorized into five classes based on cytomorphological features.

Urine-23 [21] consists of 8,509 annotated particle images from urine sediment samples, collected from 409 patients at the Biochemistry Clinics of Elazig Fethi Sekin Central Hospital. The images were captured using an Optika B293PLi microscope and are categorized into 8 classes.

Malaria-18 [15] consists of 27,560 Giemsa-stained red blood cell images collected from 193 patients at Chittagong Medical College Hospital, Bangladesh. The dataset is evenly split into two classes: parasitized and uninfected cells (Fig. 2).

3.2 Implementation Details

In this section we introduce the settings for implementing and training hNCA. This general-purpose configuration shows good results across all six datasets.

Model Parameters: The proposed standard model used for the experiments consists of two NCA stages with a downscaling factor of 4. We chose $c_1 = 128$ channels with $k_1 = 32$ steps for the first NCA stage, and $c_2 = 128$ channels with $k_2 = 16$ steps for the second stage. Using 128 channels in both stages ensures a robust information retrieval. The second stage requires fewer iterations compared to the first because it operates in smaller spatial size. The fully connected hidden layers in both NCA and the classifier have a size of 128.

Training: Adam optimizer with a learning rate of 0.0002 and a β_1 of 0.9 and β_2 of 0.999 together with a weight decay of 0.0001 is used for training with an exponential learning rate decay with weight 0.9999. Input images are resized to 64×64 pixels and normalized with dataset-specific mean and standard deviation. For training, the images are augmented by random rotation and flipping. Each model is trained for 50 epochs with a batch size of 16. All experiments are conducted with stratified five folds cross validation.

3.3 Results

Using the discussed general-purpose configuration of our hNCA, the model is trained on six datasets under the specified training settings. The model's performance is evaluated against a single stage NCA method (WBC-NCA [4]), as well as a lightweight CNN (MobileNetV3 [7]), and a lightweight ViT (Mobile-ViT [11]). For WBC-NCA, we adhered to the training setup described in [4]. For MobileNetV3 and MobileViT, we maintained the same optimizer, batch size, augmentations, and loss function as the hNCA training setup for consistency, only adjusting the learning rate to 0.003.

Table 1. Classification accuracy of hNCA, WBC-NCA, MobileNetV3, MobileViT$_{xxs}$, and dataset-specific baselines is reported across six datasets, with their model sizes in parentheses. Our hNCA consistently outperforms the lightweight baselines across all datasets. The means and standard deviations are reported across five different folds. The highest accuracies among lightweight models are highlighted in **bold**.

Dataset	Method				
	hNCA	WBC-NCA	MobileNet	MobileViT	Published
	(150k)	(86k)	(580k)	(955k)	(21M-139M)
Matek-19	**95.3 ± 0.1**	92.6 ± 0.4	90.1±0.3	92.2±0.2	96.1 [10]
Acevedo-20	**91.0 ± 0.4**	89.8 ± 0.7	79.6±0.6	85.6±0.3	96.2 [1]
INT-20	**92.0 ± 0.1**	88.0 ± 0.3	86.3±0.4	89.7±0.3	88.7 [19]
Sipakmed	**95.2 ± 1.0**	92.5 ± 0.7	82.9±2.0	87.8±1.7	99.1 [8]
Urine-23	**92.7 ± 0.4**	87.9 ± 1.3	79.6±1.5	85.4±1.5	96.0 [23]
Malaria-18	**96.7 ± 0.2**	96.5± 0.3	93.4±1.7	95.2±0.4	98.9 [16]

For additional reference, we report the performance of state-of-the-art solutions published for each dataset. For the white blood cell datasets,

Matek et al. [10] introduced a model based on ResNeXt [22] architecture when publishing Matek-19 dataset. This model achieves an accuracy of 96.1% with 25M parameters. Building on this work, Salehi et al. [19] trained the same ResNeXt architecture on the INT-20 dataset, reporting an accuracy of 88.7%. Acevedo et al. [1] achieved an accuracy of 96.2% using a VGG-16 architecture [20] on the Acevedo-20 dataset. For the Sipakmed [13] dataset, Karri et al. [8] attained a remarkable accuracy of 99.12% using a 19-layer CNN with a model size of 24M parameters. On the Urine-23 dataset, Muhammed et al. [23] evaluated various well-known CNN backbones and proposed a hybrid model combining textural features with a ResNet50 [5] backbone. This model achieved an accuracy of 96%. Although the model's complexity was not explicitly documented, it is expected to exceed the 25M parameters of a standard ResNet50. For the Malaria-18 dataset, Rajaraman et al. [16] utilized a VGG-16 [20] backbone, achieving an accuracy of 98.9%, and the size should exceed the 139M parameters of a standard VGG-16.

Table 1 compares accuracy of the proposed hNCA model with the mentioned baselines and published state-of-the-art solutions. It is evident that NCA-based models offer a clear advantage in complexity over traditional models, while outperforming existing lightweight solutions. Figure 3 shows the confusion matrices of our model for each dataset.

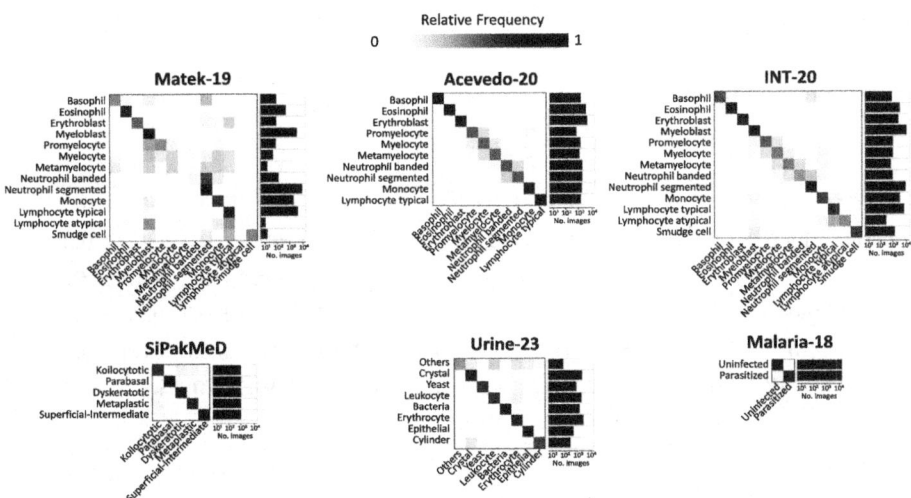

Fig. 3. Our proposed hNCA performs well on all six datasets: Matek-19, Acevedo-20, INT-20, Sipakmed, Urine-23, and Malaria-18, achieving average F1-scores of 94.9, 91.0, 91.8, 95.2, 92.6, and 96.7, respectively. Confusion matrices are provided for further details.

3.4 Ablation Study

To evaluate the effectiveness of different components in the proposed hNCA model, we conducted an ablation study focusing on three critical design choices. **i) Pooling method:** Comparing max pooling (max) and average pooling (avg) between the spatial dimensions in hNCA. **ii) Architecture:** hNCA versus a single-stage NCA. **iii) Loss function:** Cross entropy (CE) loss versus focal (Focal) loss. These combinations resulted in six distinct models, each trained individually on all datasets for comparative analysis. The accuracies obtained from these 36 experiments are presented in Table 2.

Focal loss consistently leads to better results across most datasets, and the hNCA architecture consistently outperforms the single-stage NCA. Max pooling generally leads to better performance compared to average pooling. In conclusion, the best results for most datasets are achieved with the hNCA model configured with max pooling between the two spatial stages and focal loss, effectively addressing imbalanced class distributions in certain datasets.

Table 2. Classification accuracy of basic single-stage NCA (single) and hNCA using max pooling (max) and average pooling (avg) between spatial stages, each with cross entropy (CE) loss and focal (Focal) loss. Results are reported across six datasets. The highest accuracy for each dataset is highlighted in **bold**, the second highest is underlined. The mean and standard deviation of accuracy are calculated across five different folds.

Dataset	Method					
	single CE	single Focal	max CE	max Focal	avg CE	avg Focal
Matek-19	92.6 ± 0.4	93.8 ± 0.2	<u>94.3 ± 0.2</u>	**95.3 ± 0.1**	92.3 ± 0.3	**95.3 ± 0.3**
Acevedo-20	89.8 ± 0.7	90.2 ± 0.4	**91.4 ± 0.2**	<u>91.0 ± 0.4</u>	90.6 ± 0.3	90.3 ± 0.3
INT-20	88.0 ± 0.3	91.0 ± 0.2	89.6 ± 0.4	<u>92.0 ± 0.1</u>	89.2 ± 0.4	**92.4 ± 0.2**
Sipakmed	92.5 ± 0.7	92.8 ± 1.1	<u>94.9 ± 0.4</u>	**95.2 ± 1.0**	93.6 ± 0.6	93.6 ± 1.3
Urine-23	87.9 ± 1.3	88.7 ± 1.6	<u>92.5 ± 0.3</u>	**92.7 ± 0.4**	91.9 ± 0.6	91.5 ± 0.5
Malaria-18	96.5 ± 0.3	96.5 ± 0.1	<u>96.6 ± 0.2</u>	**96.7 ± 0.2**	**96.7 ± 0.2**	**96.7 ± 0.2**

3.5 Discussion

With only a single setting of model implementation and training parameters hNCA achieves strong performance across datasets with diverse characteristics, including variations in image resolution, class distributions, and classification complexity (Table 1). Notably, this strong performance holds consistently across all six datasets, which highlights the flexibility and robustness of our method.

Easy Deployment: The compact size and simple architecture of NCA-based models make parameter tuning significantly easier compared to other deep learning models. Their smaller parameter count reduces the risk of overfitting, allowing them to generalize effectively even on datasets with limited sample sizes. Unlike larger models, NCA models do not rely on extensive hyperparameter searches, pretraining on large datasets, or complex regularization techniques.

Robustness: In contrast, traditional baseline models are often tailored specifically to a single dataset and finely optimized for a particular task. This specialization can explain the performance gap observed in certain cases between hNCA and state-of-the-art. For instance, the ResNeXt model, used as a baseline performs significantly worse on INT-20, [19] compared to its performance on Matek-19, presented in [10] (Table 1). This suggests how highly tuned, dataset-specific parameters can deliver strong results on individual datasets but often struggle to generalize across new datasets and domains. As a result, their applicability to real-world tasks remains limited.

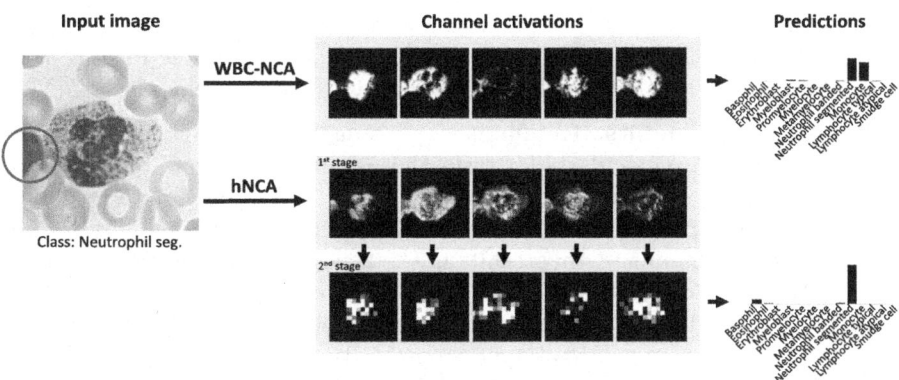

Fig. 4. Example of how hNCA achieves global context more effectively compared to WBC-NCA. The input cell image contains a partially visible second white blood cell (red circle). Channel activations from WBC-NCA show highlighted areas in the regions of both cells. This indicates its limited ability to differentiate between the two cells due to its reliance on local updates. In contrast, the hNCA progressively builds global context through two stages of processing, as seen in the first and second NCA activations. (Color figure online)

Hierarchical Feature Aggregation: The results in Table 1 show that hNCA consistently outperforms WBC-NCA across all datasets. This improvement highlights the advantage of a hierarchical architecture, which effectively addresses the limitations of WBC-NCA. The key limitation of WBC-NCA lies in its reliance on local iterative updates, where each cell's state is updated based only on its immediate neighborhood. While this approach captures fine-grained local details, it struggles to propagate information across the image, making it less suited for tasks requiring an understanding of complex spatial relationships.

hNCA addresses this limitation by incorporating global context through its multi-stage hierarchical design. In the initial stages, it processes the input at high resolution, extracting fine local features. These features are then pooled to coarser representations and passed to later stages. This hierarchical approach allows the model to progressively integrate both local and global patterns, effectively overcoming the constraints of purely local processing.

Figure 4 illustrates this process. WBC-NCA struggles to differentiate a white blood cell in the center of the image from a partially visible secondary cell, which requires a degree of global understanding that a single NCA cannot easily achieve. In comparison, the first-stage activations of hNCA capture fine-grained details, much like WBC-NCA. However, in the second stage, these features are aggregated and refined into a higher-level representation. This progression from local to global understanding allows hNCA to resolve ambiguities and make more robust predictions, successfully identifying the relevant cell in the image.

The consistent accuracy observed across different experiments for the Malaria dataset (Table 2) shows that the global context is not relevant since classification in this dataset is driven primarily by small, localized regions of relevance within the images. In this dataset, the presence or absence of parasites is determined by small, localized features that are typically confined to specific regions of the image. As a result, for this task a single-stage NCA is just as effective as hNCA.

Training Improvements: Training a single stage NCA model can be computationally intensive, particularly for tasks that require many update steps. Each NCA step updates every pixel in the image based on its immediate neighbors, increasing the receptive field by one pixel. For problems that require a global understanding of the image, this results in the need of many iterations, as information propagates only through these successive local updates. A cell at step k has been influenced by $(2 \cdot k - 1)^2$ surrounding cells. For images of dimension $H \times W$ with k NCA steps, this implies complexity growing with $O(H \cdot W \cdot k^2)$. Our method addresses this bottleneck of the NCA model by reducing both the image size during processing and the number of necessary steps.

Memory Consumption: We performed inference on one 64×64 RGB image using an NVIDIA A100-SXM4-40GB both with hNCA and ResNeXt50_32x4d [22], the architecture of the Matek-19 baseline. While ResNeXt50_32x4d allocates a maximum of 98.19 MB of VRAM (Video Random Access Memory), hNCA requires only 30.72 MB, achieving comparable accuracy.

4 Conclusion

In this work, we introduced hNCA, a lightweight and efficient classification model that combines the strengths of NCA with a hierarchical architecture to address the limitations of existing NCA-based approaches. By incorporating both local and global feature extraction through a multi-stage design, hNCA effectively captures fine-grained details and broader spatial context.

Our method represents a step toward rethinking the design of neural networks. By demonstrating that NCA can replace key components of traditional architectures while maintaining strong performance, this work highlights the potential for broader adoption of NCA principles. These principles could inspire alternative neural network designs, replacing convolutional layers with NCA operations to create lightweight, robust models that offer reduced computational complexity and enhanced adaptability.

Our evaluation on six clinical microscopy datasets demonstrates that hNCA achieves a high accuracy on diverse tasks, consistently outperforming existing NCA models, while using significantly fewer parameters than conventional methods. Importantly, these results were obtained through a single configuration across all datasets, highlighting the model's robustness and adaptability without the need for extensive dataset-specific tuning.

This combination of efficiency, robustness, and ease of deployment makes hNCA particularly well-suited for real-world applications in resource-constrained settings, such as remote clinics or mobile devices. By minimizing computational demands and simplifying deployment, hNCA advances equitable and scalable access to automated diagnostic tools.

References

1. Acevedo, A., Alférez, S., Merino, A., Puigví, L., Rodellar, J.: Recognition of peripheral blood cell images using convolutional neural networks. Comput. Methods Programs Biomed. **180**, 105020 (2019)
2. Acevedo, A., Merino, A., Alférez, S., Molina, Á., Boldú, L., Rodellar, J.: A dataset of microscopic peripheral blood cell images for development of automatic recognition systems. Data Brief **30**, 105474 (2020)
3. Bielawski, K., Gaylinn, N., Lunn, C., Motia, K., Bongard, J.: Evolving hierarchical neural cellular automata. In: Proceedings of the Genetic and Evolutionary Computation Conference, pp. 78–86 (2024)
4. Deutges, M., Sadafi, A., Navab, N., Marr, C.: Neural cellular automata for lightweight, robust and explainable classification of white blood cell images. In: International Conference on Medical Image Computing and Computer-Assisted Intervention, pp. 693–702. Springer, Cham (2024)
5. He, K., Zhang, X., Ren, S., Sun, J.: Deep residual learning for image recognition. arXiv preprint arXiv:1512.03385 (2015)
6. Hehr, M., et al.: Explainable AI identifies diagnostic cells of genetic AML subtypes. PLOS Digit. Health **2**(3), e0000187 (2023)
7. Howard, A., et al.: Searching for mobilenetv3. In: Proceedings of the IEEE/CVF International Conference on Computer Vision, pp. 1314–1324 (2019)
8. Karri, M., Annavarapu, C., Mallik, S., Zhao, Z., Acharya, U.R.: Multi-class nucleus detection and classification using deep convolutional neural network with enhanced high dimensional dissimilarity translation model on cervical cells. Biocybern. Biomed. Eng. **42**(3), 797–814 (2022). https://doi.org/10.1016/j.bbe.2022.06.003
9. Lin, T.Y., Goyal, P., Girshick, R., He, K., Dollár, P.: Focal loss for dense object detection. IEEE Trans. Pattern Anal. Mach. Intell. **42**(2), 318–327 (2020). https://doi.org/10.1109/TPAMI.2018.2858826
10. Matek, C., Schwarz, S., Spiekermann, K., Marr, C.: Human-level recognition of blast cells in acute myeloid leukaemia with convolutional neural networks. Nat. Mach. Intell. **1**(11), 538–544 (2019)

11. Mehta, S., Rastegari, M.: MobileViT: light-weight, general-purpose, and mobile-friendly vision transformer. arXiv preprint arXiv:2110.02178 (2021)
12. Pande, R., Grattarola, D.: Hierarchical neural cellular automata. In: Artificial Life Conference Proceedings, vol. ALIFE 2023: Ghost in the Machine: Proceedings of the 2023 Artificial Life Conference, p. 20 (2023).https://doi.org/10.1162/isal_a_00601
13. Plissiti, M.E., Dimitrakopoulos, P., Sfikas, G., Nikou, C., Krikoni, O., Charchanti, A.: Sipakmed: a new dataset for feature and image based classification of normal and pathological cervical cells in pap smear images. In: 2018 25th IEEE International Conference on Image Processing (ICIP), pp. 3144–3148. IEEE (2018)
14. Qin, Y., Feng, M., Lu, H., Cottrell, G.W.: Hierarchical cellular automata for visual saliency. Int. J. Comput. Vision **126**(7), 751–770 (Jul 20https://doi.org/10.1007/s11263-017-1062-2
15. Rajaraman, S., et al.: Pre-trained convolutional neural networks as feature extractors toward improved malaria parasite detection in thin blood smear images. PeerJ **6**, e4568 (2018)
16. Rajaraman, S., et al.: Understanding the learned behavior of customized convolutional neural networks toward malaria parasite detection in thin blood smear images. J. Med. Imaging **5**(3), 034501–034501 (2018)
17. Sadafi, A., Bordukova, M., Makhro, A., Navab, N., Bogdanova, A., Marr, C.: Redtell: an AI tool for interpretable analysis of red blood cell morphology. Front. Physiol. **14**, 1058720 (2023)
18. Sadafi, A., et al.: Multiclass deep active learning for detecting red blood cell subtypes in brightfield microscopy. In: Shen, D., et al. (eds.) MICCAI 2019. LNCS, vol. 11764, pp. 685–693. Springer, Cham (2019). https://doi.org/10.1007/978-3-030-32239-7_76
19. Salehi, R., et al.: Unsupervised cross-domain feature extraction for single blood cell image classification. In: Medical Image Computing and Computer Assisted Intervention–MICCAI 2022: 25th International Conference, Singapore, 18–22 September 2022, Proceedings, Part III, pp. 739–748. Springer, Cham (2022)
20. Simonyan, K., Zisserman, A.: Very deep convolutional networks for large-scale image recognition (2015). https://arxiv.org/abs/1409.1556
21. Tuncer, T., Erkuş, M., Çınar, A., Ayyıldız, H., Tuncer, S.A.: Urine dataset having eight particles classes. arXiv preprint arXiv:2302.09312 (2023)
22. Xie, S., Girshick, R., Dollár, P., Tu, Z., He, K.: Aggregated residual transformations for deep neural networks. In: Proceedings of the IEEE Conference on Computer Vision and Pattern Recognition, pp. 1492–1500 (2017)
23. Yildirim, M., Bingol, H., Cengil, E., Aslan, S., Baykara, M.: Automatic classification of particles in the urine sediment test with the developed artificial intelligence-based hybrid model. Diagnostics **13**(7) (2023). https://doi.org/10.3390/diagnostics13071299

PathTTT: Test-Time Training with Meta-auxiliary Learning for Pathology Image Classification

Haoyu He[✉], Mahdi S. Hosseini, and Yang Wang

Concordia University, Montreal, QC, Canada
haoyu.he@mail.concordia.ca

Abstract. Detecting cancer through pathology imaging is crucial for early diagnosis and effective treatment. However, the performance of deep learning (DL) models is often undermined by domain shifts between training and test data—variations caused by differences in imaging acquisition device, staining protocols, and patient populations in real-world scenarios. To address this challenge, we propose PathTTT, a novel framework that combines Test-Time Training (TTT) with Model-Agnostic Meta-Learning (MAML) to enhance model robustness under domain shift conditions. Our experimental results demonstrate significant improvements over SOTA methods on benchmark datasets, highlighting PathTTT's potential for robust cancer detection in real-world dataset applications.

Keywords: Computational pathology · Test-time-training · Meta learning

1 Introduction

Cancer remains one of the leading cause of death globally, with breast and colorectal cancers being the most common among all [1]. Both cancers have witnessed more than 310K and 152K new cases in the USA in 2024, resulting in more than 42K and 53K deaths, respectively [2]. Whole slide imaging (WSI) from tissue pathology plays a primary role in the early detection and diagnosis of cancer, which continues to pose significant challenges for clinicians and researchers have been collaborating to develop Computer Aided Diagnosis (CAD) systems to automate the diagnostic evaluations. Despite the growing demand for Computational Pathology (CPath), analyzing WSIs to accurately assess the disease's progression remains a complex and time-consuming task. The development of CAD systems in CPath is prone to data distribution shift from one cohort to another due to different protocols used for tissue preparation from different pathology labs, and (b) different WSI scanners used for digitization [1,3].

For the past few years, many CPath algorithms in deep learning [4–7] have been introduced for representational learning of pathology images and the performance accuracy of these algorithms remains consistently high as long as these algorithms are evaluated on similar data distribution. However, their performance deteriorates under domain shifts, where training and testing data differ. For example, the detection and classification [8] of biomedical signals and images showcase remarkable performance when trained and tested on the same datasets. However, when the data encounters a distribution shift, i.e. the test data differs from training data distribution, a significant challenge is imposed to maintain a similar level of accuracy [9,10].

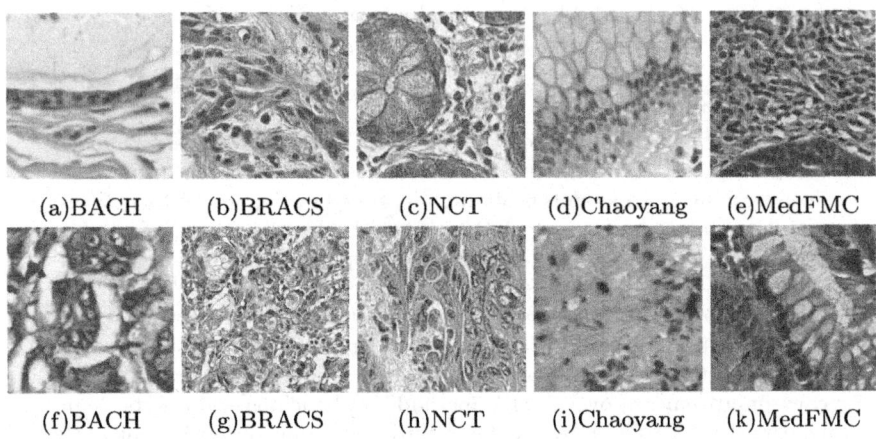

Fig. 1. Sample images from two different classes of the BACH, BRACS, NCT-CRC-HE-100K, Chaoyang, and MedFMC datasets. The first row represents the benign class, while the second row represents the malignant class. The breast cancer datasets, BACH and BRACS, display distinct visual differences between these classes. Similarly, the colorectal cancer datasets, NCT-CRC-HE-100K, Chaoyang, and MedFMC, also exhibit clear visual differences between the benign and malignant cases, though some overlap is present.

Figure 1 demonstrates the visual differences and similarities between breast cancer datasets (BACH [11] and BRACS [12]) and colorectal cancer datasets (NCT-CRC-HE-100K (NCT) [13], Chaoyang [14], and MedFMC [15]), highlighting domain shift challenges. The t-Distributed Stochastic Neighbor Embedding (t-SNE) diagrams [16] in Fig. 2 visualize distribution shifts in high-dimensional feature spaces. The t-SNE visualization of the BACH and BRACS breast cancer datasets reveals a noticeable distribution shift between the two datasets. While there is some overlap, the points representing BACH (in blue) and BRACS (in orange) largely form separate clusters. This separation underscores the domain shift issue, where the same class labels (benign and malignant) are present in both datasets, but their visual representations differ significantly, as shown in Fig. 1(a), (b), (f) and (g). Similarly, the t-SNE plot for three colorectal cancer datasets NCT (in blue), Chaoyang (in orange), and MedFMC (in green)—shows significant overlap, suggesting that these datasets share more similar visual features, although distinct clusters are still present. These results align with the

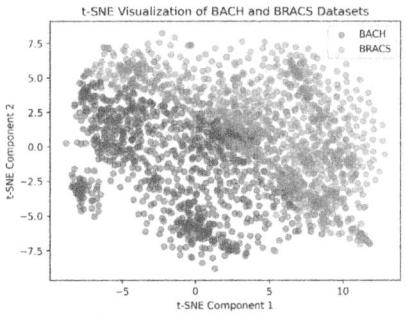
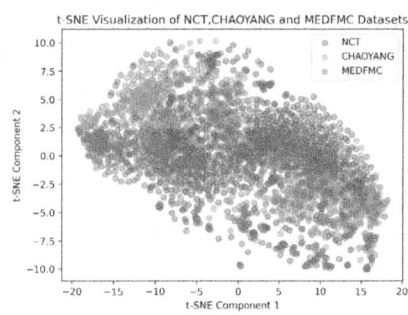

(a) t-SNE visualization of BACH and BRACS

(b) t-SNE visualization of NCT, Chaoyang and MedFMC

Fig. 2. Two t-SNE diagrams visualize the distribution shift in high-dimensional features for BACH, BRACS, NCT, Chaoyang, and MedFMC datasets. The t-SNE visualization shows a distribution shift between the BACH (blue) and BRACS (orange) breast cancer datasets, with some overlap but mostly distinct clusters. In contrast, the colorectal cancer datasets NCT (blue), Chaoyang (orange), and MedFMC (green)—show more overlap, indicating shared visual features, though distinct clusters are still evident. These results are consistent with the visualization shown in Fig. 1 (Color figure online).

visual findings presented in Fig. 1. The scales of X and Y in Fig. 2 do not carry specific units, they are arbitrary and depend on the t-SNE optimization process.

To address the challenge of distribution shifts, research has focused on techniques like domain adaptation, domain generalization, and meta-learning to enhance model robustness across diverse datasets. Our work builds on the TTT paradigm, which uses self-supervised auxiliary tasks to update model parameters dynamically during testing, and MAML, which optimizes models for quick adaptation to unseen tasks. We propose PathTTT, a novel framework that integrates the strengths of TTT and MAML to improve the robustness and generalization of deep learning models in cancer detection under domain shift conditions. PathTTT treats each training instance as an individual task within the MAML framework, employing a self-supervised Masked Autoencoder (MAE) loss to extract richer latent features. The model uses bi-level optimization, where the auxiliary task enhances primary task performance, allowing dynamic adaptation to new data distributions at test time. This approach addresses variability in imaging protocols and patient populations, bridging performance gaps across datasets and ensuring reliable cancer diagnosis in diverse clinical settings.

2 Related Work

Cancer Detection. Deep learning has significantly advanced cancer detection, especially with convolutional neural networks like ResNet [5,17], DenseNet [18], and EfficientNetV2 [19]. These models achieve high accuracy but often struggle with domain shifts when applied to unseen data from different sources. Transfer learning and domain adaptation techniques have been proposed to address this, but they require extensive tuning and struggle with significant distribution shifts.

Masked Autoencoders. MAE [20] are a self supervised learning method that reconstructs masked portions of input images to learn informative representations. Unlike traditional autoencoders, MAEs mask a significant portion of the image, forcing the model to learn meaningful global structures rather than local details.

Test-Time Training. TTT [21] addresses distribution shifts by updating model parameters during inference using a self-supervised auxiliary task. This allows models to adapt to test-time data characteristics without requiring labeled test samples, enhancing robustness in real-world clinical settings. Integrating TTT with MAE reconstruction can further improve model generalization across diverse cancer datasets.

Meta Learning. Meta-learning, or "learning to learn," is a framework that focuses on designing models capable of learning new tasks quickly with limited data. MAML [22] designed to find an optimal initialization of model parameters that can be fine-tuned to new tasks with just a few gradient updates. MAML has been widely used in various fields, including computer vision, natural language processing, and reinforcement learning.

Application of Classification in Patch-Level Analysis. Deep learning has been widely employed for pathology classification in patch-level analysis. Among recent advancements, FCCS-Net (FCCS) [23] introduces a multi-level fully convolutional network with dual attention mechanisms, specifically designed for breast cancer classification. CRCCN-Net (CRCCN) [10], a lightweight convolutional neural network, addresses automated colorectal tissue classification in histopathological imaging by balancing computational efficiency with diagnostic accuracy. EfficientNetV2 (EV2) [19] represents a family of convolutional neural networks optimized for faster training and improved parameter efficiency.

Test-Time Adaptation Methods. Recent advancements in test-time adaptation (TTA) have yielded frameworks tailored to dynamic clinical environments. SAR [24] ensures robust model updates through entropy minimization and sharpness-aware optimization, mitigating distributional uncertainties during inference. EATA [25] integrates parameter-efficient updates with anti-forgetting mechanisms, preserving model reliability across diverse domain shifts while minimizing computational overhead.

3 Methodology

Our method builds on the standard TTT framework [21,26], as originally proposed. TTT involves adapting the model to each test sample during the test phase before making a prediction. We use the MAE [20] reconstruction task for the adaptation process. Unlike conventional training methods, TTT does not rely on the availability of the entire test dataset; instead, the model adapts to each test sample independently as it is encountered. After the network parameters are adjusted for a specific sample, the updated weights are immediately used to predict its class label. A key characteristic of the traditional TTT approach is that it does not modify the initial weights of the encoder, decoder, or classifier. However, optimizing these initial weights could significantly enhance the model's adaptability and performance. To address this, we integrate meta-learning into our training process. This integration equips the model with the ability to adapt more quickly and effectively to unseen samples, resulting in more accurate and robust predictions during the test phase.

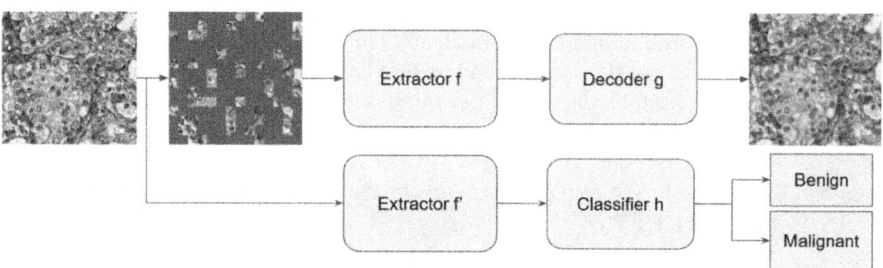

Fig. 3. Illustration of the model architecture. The input images are first masked and processed through the shared feature extractor f and decoder g for the reconstruction task. This task updates the extractor, resulting in an enhanced version f', which is then used by the classification head h for cancer detection.

3.1 Architecture

The overall architecture is shown in Fig. 3 consisting of three components: a feature extractor f, a classification head h for the primary supervised classification, and a decoder head g for an auxiliary self-supervised task. Here, The feature extractor f acts as the encoder for the MAE, and the decoder head g corresponds to the decoder component of the MAE. Both the encoder and decoder are implemented using Vision Transformers (ViTs). The classification head h is implemented as a linear layer that maps the high-dimensional feature space generated by the encoder to the specific number of target classes. During TTT, the shared features extractor f is updated to enhance the quality of the extracted features, thereby improving the accuracy of the classification task. This adaptability allows the model to fine-tune its parameters for each test sample, optimizing performance dynamically.

3.2 Training

Following standard practice, we begin with the MAE model [20], specifically using a ViT-Base encoder that has been pre-trained for reconstruction on the

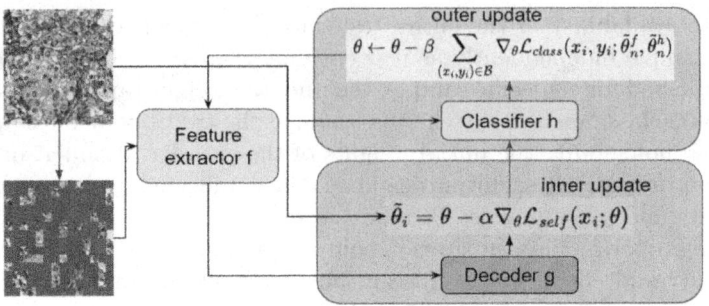

Fig. 4. Illustration of the Meta-Auxiliary training. For a training example x_i with ground truth y_i, we update the model parameters θ using a self-supervised MAE loss in the inner loop to obtain an adapted model $\tilde{\theta}$. The adapted model is then used for the classification task. We use the supervised loss for the primary task as the meta-objective in the outer loop to update the model parameters θ.

ImageNet-1k dataset. We explore four methods: supervised learning, joint training, TTT, and PathTTT.

Supervised Learning (SL). We use the same architecture as in Sect. 3.1, excluding the decoder. The model is trained by minimizing the cross-entropy loss function for classification, formulated as $\mathcal{L}_{\text{class}}(x_i, y_i; \theta) = l(\theta^h \circ \theta^f(x_i), y_i)$, where θ^f represents the parameters of the feature extractor, θ^h denotes the parameters of the classification head, x_i denotes the input images and the function l computes the cross-entropy loss between the predicted class probabilities and the true labels y_i.

Joint Training (JT). We compare PathTTT with joint training, using the same architecture. The self-supervised reconstruction loss is calculated by computing the pixel-wise mean squared error (MSE) between the decoded patches and their corresponding ground truth patches. The model is trained by minimizing the sum of the classification loss and self-supervised loss. The self-supervised reconstruction loss can be expressed as: $\mathcal{L}_{self}(x_i; \theta) = l(\theta^g \circ \theta^f(M(x_i)), x_i))$ Here θ^f represents the parameters of the feature extractor (encoder), and θ^g denotes the parameters of the decoder. $M(x_i)$ is a masked version of the input image x_i, where portions of the image are masked out to enable self-supervised learning.

During joint training, the model jointly optimizes both the main task (classification) and the auxiliary task (self-supervised reconstruction) via gradient descent: $\theta \leftarrow \theta - \alpha \nabla_\theta (L_{self}(x_i; \theta_0) + L_{class}(x_i, y_i; \theta_0))$

Test-Time Training. At test time, we start with the primary task head h_1 produced by joint training, as well as the encoder f_1 and decoder g_1. For each test input x, we perform test-time training (TTT) by optimizing the following loss function, focusing on the self-supervised reconstruction task. This updated model is then used to make predictions: $\tilde{\theta}_i = \theta - \alpha \nabla_\theta \mathcal{L}_{self}(x_i; \theta)$. Here, $\tilde{\theta}_i$ represents the adapted model parameters after TTT, where α is the learning

Algorithm 1. Test-Time Training

Require: Test dataset $\mathcal{D}_{\text{test}}$, adaptation criterion l_{self}, learning rate η.
1: Initialize the model with fine-tune weights:
2: $\theta = \{\theta^f, \theta^g, \theta^h\}$
3: **for** each test sample $x_i \in \mathcal{D}_{\text{test}}$ **do**
4:     Evaluate self-supervised reconstruction loss $\mathcal{L}_{self}$
5:     Update model parameters with gradient descent: $\tilde{\theta}_i = \theta - \alpha \nabla_\theta \mathcal{L}_{self}(x_i; \theta)$
6:     Make prediction $\hat{y}_i = f(x_i; \tilde{\theta}_t^f; \tilde{\theta}_t^h)$
7: **end for**

rate, and L_{self} is the self-supervised reconstruction loss. The Test-Time Training algorithm is outlined in Algorithm 1.

Meta Auxiliary Training. PathTTT enhances the model's adaptability during test-time by framing the self-supervised task as a form of "learning to learn," or meta-learning. The objective of meta-training is to discover meta-parameters that enable the model to quickly adapt to new, unseen samples during testing, even in the face of unknown distribution shifts. This approach aims to ensure that the model can achieve more accurate classifications under varying conditions. The meta auxiliary training process consists of a bi-level optimization strategy:

1. Inner Loop: During the inner loop, the model is fine-tuned on individual training instances by minimizing the self-supervised loss l_{self} This process refines the model's feature extraction capabilities, making it more adept at handling the self-supervised task.

2. Outer Loop: Concurrently, the outer loop optimizes the meta-parameters across the entire training set. This alignment between the self-supervised task and the main task is crucial for ensuring that the model's meta-knowledge is robust and can generalize effectively to new data and tasks. The overall process as Fig. 4 shown. This bi-level optimization ensures the model will have the ability to rapidly adapt during test-time, allowing it to deal with distribution shifts and varying data complexities more effectively. The overall procedure for meta-training is summarized in Algorithm 2. This approach combines the benefits of meta-learning with test-time training, making it highly versatile in dealing with distribution shifts and varying data complexities.

Meta Testing. During the test phase, the PathTTT procedure mirrors the Test-Time Training process, where the model is continually adapted to each test sample using the learned meta-parameters.

Algorithm 2. Meta Auxiliary Training

Require: Train dataset $\mathcal{D}_{\text{train}}$, learning rates β, α.
1: Initialize the model with fine-tune weights: $\theta = \{\theta^f, \theta^g, \theta^h\}$
2: **while** not converged **do**
3: Sample a batch $\mathcal{B}$ of data from $\mathcal{D}_{train}$
4: **for** each training instance $x_i, y_i \in \mathcal{B}$ **do**
5:   Evaluate self-supervised reconstruction loss $\mathcal{L}_{self}(x_i;\theta)$
6:   Compute adapted parameters with gradient descent: $\tilde{\theta}_n = \theta - \alpha \nabla_\theta \mathcal{L}_{self}(x_i;\theta)$
7: **end for**
8: Compute task-specific classification loss and update the global model:
9: $\theta \leftarrow \theta - \beta \sum_{(x_i,y_i) \in \mathcal{B}} \nabla_\theta \mathcal{L}_{class}(x_i, y_i; \tilde{\theta}_n^f, \tilde{\theta}_n^h)$
10: **end while**
11: **return** Optimized model parameters θ

4 Experiment

To demonstrate the performance of the PathTTT with MAE in addressing domain shift, we performed binary classification experiments (benign vs. malignant) on two prevalent cancer types: Breast Cancer and Colon Cancer. The datasets utilized in this study, described in detail below and summarized in Table 1, were selected to assess cross-domain generalization.

A critical consideration in processing pathology whole slide images (WSIs) is ensuring a consistent field of view (FoV) across varying microns per pixel (Mpp) resolutions.

$$\text{FOV} = \text{Mpp} \times \text{Patch Size}$$

WSIs are often scanned at different Mpps, and extracting fixed pixel-sized patches without accounting for resolution differences would yield inconsistent physical coverage. To maintain fairness in model training and evaluation, we standardized patches to represent the same real-world area (FOV = $112\mu m$) across all datasets, as shown in Table 1.

For each dataset, we applied a standard split, allocating 70% of the data for training, 10% for validation, and 20% for testing. All experiments were conducted on NVIDIA A100 GPUs.

4.1 Breast Cancer Datasets

The BACH dataset [11] comprises H&E-stained breast histology images with Normal and Invasive Carcinoma classes used as benign and malignant categories (2048 × 1536 pixels at 0.42 m/pixel). Images were cropped to 267 × 267 pixels, with noise and background removed. Similarly, the BRACS dataset [12] includes 4539 annotated H&E-stained ROIs (up to 4000 × 4000 pixels at 0.25 m/pixel), focusing on Normal and Invasive Carcinoma classes, cropped to 448 × 448 pixels after artifact removal.

4.2 Colorectal Cancer Datasets

The NCT-CRC-HE-100K dataset [13] contains 100,000 H&E-stained patches (224 × 224 pixels at 0.5 m/pixel), with Normal Colon Mucosa and Colorectal Adenocarcinoma as benign/malignant classes. The Chaoyang dataset [14] provides ×20-magnified colon patches resized to 224 × 224 pixels, using normal and adenoma samples. The MedFMC dataset [15] from Ruijin Hospital includes 10,009 patches (1024 × 1024 pixels at 0.475 m/pixel), cropped to 236 × 236 pixels and labeled as tumor/non-tumor after preprocessing.

Table 1. Overview of datasets used in the experiments. Mpp (μm/pixel) stands for "Microns per Pixel". It refers to the spatial resolution of the images in the dataset, indicating how much real-world distance (in microns) corresponds to a single pixel in the image.

Dataset	Patch Size	Mpp(μm/pixel)	FOV(μm)	#Patch
BACH	267 × 267	0.42	112	26000
BRACS	448 × 448	0.25	112	30000
NCT	224 × 224	0.5	112	17400
Chaoyang	224 × 224	0.5	112	24000
MedFMC	236 × 236	0.475	112	200000

4.3 Implementation Detail

Both supervised learning and joint training follow [20], using the AdamW optimizer (base lr = 1e−5, momentum = 0.9/0.95, weight decay = 0.05) with a cosine decay schedule. Training runs for 100 epochs (batch size = 32) with early stopping after 15 epochs of no validation loss improvement. Test-time training uses SGD (lr = 1e−4, momentum = 0.9), while meta-auxiliary training applies AdamW (lr = 1e−4). Following [27], we apply data augmentations: random vertical flipping (p = 0.5), color dropping (p = 0.2), and weak color jittering (p = 0.8) with brightness, contrast, saturation, and hue adjustments (0.2, 0.2, 0.2, 0.1).

4.4 Main Results

Our main results for Breast Cancer and Colon Cancer datasets are presented in Tables 2 and 3. These experiments were conducted using the default training setup discussed in the previous section. In these experiments, models were trained on one dataset and tested on a different dataset to evaluate their robustness to domain shifts. PathTTT, applied on top of our baseline, demonstrates significant improvements in performance, effectively addressing the challenges posed by domain shifts.

The results presented in Table 2 and Table 3 provide a detailed comparison of classification performance under these conditions. Each row in the tables corresponds to a dataset transition, demonstrating how well the models handle different data distributions. These transitions simulate real-world scenarios where

Table 2. Classification accuracy (%) for breast cancer classification. "BACH → BRACS" means training on the BACH dataset and testing on the BRACS dataset. The highest accuracy is in bold.

	SOTA		TTA			Ours			
Top1-ACC	FCCS	CRCCN	EV2	SAR	EATA	SL	JT	TTT	PathTTT
BACH → BRACS	56.93	57.9	65.93	68.37	56.8	66.8	66.42	67.00	**69.73**
BRACS → BACH	67.46	60.56	69.77	73.82	74.75	75.19	75.46	75.57	**76.07**
Avg	62.20	59.23	67.85	71.10	65.78	71.00	70.94	71.29	**72.9**

Table 3. Classification accuracy (%) for colorectal cancer classification. "NCT → Chaoyang" means training on the NCT dataset and testing on the Chaoyang dataset. The highest accuracy is in bold.

	SOTA		TTA			Ours			
Top1-ACC	FCCS	CRCCN	EV2	SAR	EATA	SL	JT	TTT	PathTTT
NCT → Chaoyang	51.08	52.9	53.33	51.21	50.17	52.79	51.83	55.06	**56.33**
NCT → MedFMC	72.84	66.47	71.98	76.02	69.57	72.9	75.91	76.07	**77.17**
Chaoyang → NCT	66.72	76.06	71.67	81.09	75.57	82.41	88.45	88.47	**88.59**
Chaoyang → MedFMC	69.68	68.06	59.89	67.48	55.39	65.29	67.30	68.40	**69.74**
MedFMC → NCT	68.65	79.25	55.57	76.24	77.18	75.09	73.22	80.89	**83.10**
MedFMC → Chaoyang	85.21	86.15	80.62	84.44	84.73	84.35	85.69	86.02	**86.17**
Avg	69.03	71.48	65.51	72.75	68.77	72.14	73.73	75.81	**76.85**

the testing data comes from a distribution distinct from the training data, providing insights into the robustness of each method. These results were obtained using various SOTA methods, TTA strategies, baseline methods and our proposed PathTTT method. Across all transitions, PathTTT consistently outperforms other methods under domain shift scenarios. This suggests that PathTTT is particularly well-suited for tasks where the testing data distribution differs significantly from the training data distribution, highlighting its potential for improving generalization in real-world applications.

Further evidence supporting PathTTT's effectiveness is provided by the t-SNE visualizations in Fig. 2, there is noticeable overlap between clusters from different datasets, distinct dataset-specific structures remain evident. These differences reduce the generalization capability of models trained on one dataset when applied to another. PathTTT, however, demonstrates its ability to bridge these gaps effectively, as evidenced by its superior performance in transitions like BACH → BRACS (69.73%) and BRACS → BACH (76.07%). This ability to adapt to unseen test distributions, even under significant domain shifts, underscores the robustness and adaptability of PathTTT.

A deeper examination of the datasets also reveals varying levels of cross-dataset compatibility. For example, the relatively high cross-dataset performance

Table 4. Classification accuracy (%) for non-cross domain. The highest accuracy is in bold.

ACC	SOTA		TTA			Ours			
	FCCS	CRCCN	EV2	SAR	EATA	SL	JT	TTT	PTTT
BACH	99.33	96.46	88.60	99.59	97.98	**99.75**	99.54	99.59	99.61
BRACS	**89.00**	81.52	74.80	86.38	83.63	81.00	88.66	88.51	88.4
NCT	99.22	98.30	90.52	99.36	**99.54**	99.51	99.37	99.19	99.25
Chaoyang	98.04	91.52	94.77	98.22	99.18	**98.66**	98.10	98.19	98.17
MedFMC	98.92	95.90	95.77	99.52	97.57	99.46	**99.59**	98.82	98.85
Avg	96.90	92.74	88.89	96.61	95.58	95.68	**97.05**	96.86	**97.05**

between NCT and MedFMC can be attributed to the significant overlap in their feature distributions. In contrast, the limited overlap between Chaoyang and NCT highlights the difficulty of adapting to Chaoyang, where domain shifts are more pronounced.

Table 1 and Table 3 show larger datasets like MedFMC improve generalization by capturing diverse features, which capture more diverse feature representations, tend to improve generalization. However, even with such datasets, distinct clusters observed in the t-SNE plots and the corresponding performance drops emphasize the persistent challenge of adapting to domain shifts. PathTTT effectively addresses this by dynamically updating the model at test time, enabling robust performance across datasets with varying feature distributions.

When compared to existing SOTA methods, PathTTT provides a significant performance boost. PathTTT improves generalization across different datasets. For example, it outperforms EV2 with an accuracy of 69.73% in the BACH → BRACS setup and 76.07% in the BRACS → BACH setup. This superior performance highlights PathTTT's ability to adapt during testing, addressing domain shift issues more effectively than the SOTA methods.

We also compare our method to other test time adaptation (TTA) methods such as EATA and SAR, as shown in Table 2 and 3. PathTTT consistently outperforms all other methods across various datasets, proving its effectiveness in handling domain shifts. TTT often ranks second, highlighting its adaptability. In contrast, SAR and EATA sometimes show lower performance, further emphasizing the advantages of PathTTT in handling complex domain shifts.

4.5 Ablation Study

In this section, we conduct ablation studies on PathTTT to analyze various components of the proposed method.

Contribution of Each Component. We isolate the contributions of MAE, TTT and MAML to determine their individual impact.

As Tables 2 and 3 show, JT (equipped with MAE), TTT and PathTTT (Combines JT, TTT and MAML) demonstrate progressive improvements, validating the distinct roles of each component. For breast cancer classification, JT achieves 66.42% and 75.46%, surpassing SL and highlighting MAE's reconstruction-driven alignment for domain invariance. PathTTT, combining TTT with MAML, further improves performance to 69.73% and 76.07%, demonstrating MAML's rapid adaptation to domain shifts. For colorectal cancer, JT shows strong generalization (e.g., 75.91% for NCT → MedFMC, 88.45% for Chaoyang → NCT), while PathTTT outperforms it with 77.17% and 88.59%, confirming MAML's adaptability. In the challenging MedFMC → NCT setting, PathTTT achieves 83.10% (+2.21% over TTT), emphasizing MAML's role in few-shot generalization.

Furthermore, the results in Table 4 demonstrate that PathTTT maintains competitive performance in non-cross-domain classification tasks. Across five datasets, PathTTT achieves accuracy comparable to or slightly lower than the best-performing methods. Notably, PathTTT attains the highest average accuracy (97.05%). This indicates that PathTTT does not negatively impact performance in scenarios without domain shift, ensuring its effectiveness across diverse settings.

Ablation studies further confirm the significance of our method's components. JT, integrating MAE, enhances adaptation through self-supervised learning, improving feature representations under distribution shifts. TTT further enhances adaptation by updating the encoder during inference, outperforming static baselines. Finally, PathTTT, which integrates MAML, optimizes the model to be more adaptable, ensuring rapid fine-tuning in novel domains. The combined effect of these components in PathTTT leads to superior performance compared to their individual contributions. While PathTTT requires approximately four additional hours of meta-training on the NCT dataset compared to JT, this computational cost is justified by its substantial gains in adaptability and performance.

5 Conclusion

We present a novel approach to addressing distribution shifts during test-time by adapting to individual samples through a meta-learning framework combined with self-supervised learning. Our method, PathTTT, leverages the principles of meta-learning and self-supervision to enable robust adaptation to new data distributions. Unlike previous methods that relied solely on joint training, PathTTT explicitly optimizes meta-parameters for rapid adaptation, as demonstrated through extensive experiments. Experiments were conducted on Breast Cancer and Colon Cancer datasets, simulating real-world scenarios where training and testing data come from distinct distributions. The results demonstrate that PathTTT consistently outperforms SOTA methods and TTA strategies. This highlights the effectiveness of PathTTT in improving model performance across diverse datasets and domain shifts, demonstrating its potential to enhance adaptability and generalization in complex real-world scenarios.

References

1. Hosseini, M.S., et al.: Computational pathology: a survey review and the way forward. J. Pathol. Inform. 100357 (2024)
2. Epidemiology Surveillance and End Results (SEER) Program. Cancer stat facts (2024). https://seer.cancer.gov/statfacts/. Accessed 14 Aug 2024
3. Jahanifar,M., et al.: Domain generalization in computational pathology: survey and guidelines. arXiv preprint arXiv:2310.19656 (2023)
4. Bychkov, D., et al.: Deep learning based tissue analysis predicts outcome in colorectal cancer. Sci. Rep. **8**(1), 3395 (2018)
5. Kather, J.N., et al.: Predicting survival from colorectal cancer histology slides using deep learning: a retrospective multicenter study. PLOS Med. **16**(1), e1002730 (2019)
6. LeCun, Y., Bengio, Y., Hinton, G.: Deep learning. Nature **521**(7553), 436–444 (2015)
7. Liang, M., Ren, Z., Yang, J., Feng, W., Li, B.: Identification of colon cancer using multi-scale feature fusion convolutional neural network based on shearlet transform. IEEE Access **8**, 208969–208977 (2020)
8. Ananda Kumar, K.S., Prasad, A.Y., Metan, J.: A hybrid deep cnn-cov-19-res-net transfer learning architype for an enhanced brain tumor detection and classification scheme in medical image processing. Biomed. Sig. Process. Control **76**, 103631 (2022)
9. Baroni, G.L., Rasotto, L., Roitero, K., Siraj, A.H., Della Mea, V.: Vision transformers for breast cancer histology image classification. In: International Conference on Image Analysis and Processing, pp. 15–26. Springer, Cham (2023)
10. Kumar, A., Vishwakarma, A., Bajaj, V.: CRCCN-net: automated framework for classification of colorectal tissue using histopathological images. Biomed. Sig. Process. Control **79**, 104172 (2023)
11. Aresta, G., et al.: Bach: grand challenge on breast cancer histology images. Med. Image Anal. **56**, 122–139 (2019)
12. Brancati, N., et al.: Bracs: a dataset for breast carcinoma subtyping in H&E histology images. Database **2022**, baac093 (2022)
13. Kather, J.N., Halama, N., Marx, A.: 100,000 histological images of human colorectal cancer and healthy tissue (2018)
14. Zhu, C., Chen, W., Peng, T., Wang, Y., Jin, M.: Hard sample aware noise robust learning for histopathology image classification. IEEE Trans. Med. Imaging **41**(4), 881–894 (2022)
15. Wang, D., et al.: A real-world dataset and benchmark for foundation model adaptation in medical image classification. Sci. Data **10**(1), 9 (2023)
16. van der Maaten, L., Hinton, G.: Visualizing data using t-SNE. J. Mach. Learn. Res. **9**(86), 2579–2605 (2008)
17. Maleki, A., Raahemi, M., Nasiri, H.: Breast cancer diagnosis from histopathology images using deep neural network and XGBoost. Biomed. Sig. Process. Control **86**, 105152 (2023)
18. Gomathi, P., Muniraj, C., Periasamy, P.S.: Digital infrared thermal imaging system based breast cancer diagnosis using 4D U-net segmentation. Biomed. Sig. Process. Control **85**, 104792 (2023)
19. Prezja, F., et al.: Improving performance in colorectal cancer histology decomposition using deep and ensemble machine learning. arXiv preprint arXiv:2310.16954 (2023)

20. He, K., Chen, X., Xie, S., Li, Y., Dollár, P., Girshick, R.: Masked autoencoders are scalable vision learners. In: Proceedings of the IEEE/CVF Conference on Computer Vision and Pattern Recognition, pp. 16000–16009 (2022)
21. Sun, Y., Wang, X., Liu, Z., Miller, J., Efros, A., Hardt, M.: Test-time training with self-supervision for generalization under distribution shifts. In: International Conference on Machine Learning, pp. 9229–9248. PMLR (2020)
22. Finn, C., Abbeel, P., Levine, S.: Model-agnostic meta-learning for fast adaptation of deep networks. In: International Conference on Machine Learning, pp. 1126–1135. PMLR (2017)
23. Maurya, R., Pandey, N.N.: Dutta, M.K., Karnati, M.: FCCS-net: breast cancer classification using multi-level fully convolutional-channel and spatial attention-based transfer learning approach. Biomed. Sig. Process. Control **94**, 106258 (2024)
24. Niu, S., et al.: Towards stable test-time adaptation in dynamic wild world. arXiv preprint arXiv:2302.12400 (2023)
25. Niu, S., et al.:. Efficient test-time model adaptation without forgetting. In: International Conference on Machine Learning, pp. 16888–16905. PMLR (2022)
26. Yossi Gandelsman, Yu., Sun, X.C., Efros, A.: Test-time training with masked autoencoders. In: Advances in Neural Information Processing Systems, vol. 35, pp. 29374–29385 (2022)
27. Kang, M., Song, H., Park, S., Yoo, D., Pereira, S.: Benchmarking self-supervised learning on diverse pathology datasets. In: Proceedings of the IEEE/CVF Conference on Computer Vision and Pattern Recognition, pp. 3344–3354 (2023)

Registration

Bi-invariant Geodesic Regression with Data from the Osteoarthritis Initiative

Johannes Schade[1,3], Christoph von Tycowicz[3], and Martin Hanik[2,3(✉)]

[1] Freie Universität Berlin, Berlin, Germany
[2] Technical University Berlin, Berlin, Germany
[3] Zuse Institute Berlin, Berlin, Germany
{johannes.schade,vontycowicz,hanik}@zib.de

Abstract. Many phenomena are naturally characterized by measuring continuous transformations such as shape changes in medicine or articulated systems in robotics. Modeling the variability in such datasets requires performing statistics on Lie groups, that is, manifolds carrying an additional group structure. As the Lie group captures the symmetries in the data, it is essential from a theoretical and practical perspective to ask for statistical methods that respect these symmetries; this way they are insensitive to confounding effects, e.g., due to the choice of reference coordinate systems. In this work, we investigate geodesic regression—a generalization of linear regression originally derived for Riemannian manifolds. While Lie groups can be endowed with Riemannian metrics, these are generally incompatible with the group structure. We develop a nonmetric estimator using an affine connection setting. It captures geodesic relationships respecting the symmetries given by left and right translations. For its computation, we propose an efficient fixed point algorithm requiring simple differential expressions that can be calculated through automatic differentiation. We perform experiments on a synthetic example and evaluate our method on an open-access, clinical dataset studying knee joint configurations under the progression of osteoarthritis.

Keywords: Geometric statistics · Lie group · Affine connection space

1 Introduction

We begin with the mathematical background before introducing the subject of this work: regression for geometric data using geodesic models. A discussion of related work and our contributions is deferred to the end of this introduction.

Basics of Lie Groups. We briefly review Lie group theory; refer to [26] for more detail. In the following, "smooth" means "infinitely differentiable".

A Lie group G is a smooth manifold that has a compatible group structure; that is, it comes with a smooth, associative (often not commutative) group operation $G \times G \ni (g, h) \mapsto gh \in G$ with identity element e and a smooth inversion map $G \ni g \mapsto g^{-1}$.

The group operation defines two automorphisms for each g in a general Lie group G: the left and right translation $L_h : g \mapsto hg$ and $R_h : g \mapsto gh$. Their derivatives $d_g L_h$ and $d_g R_h$ map tangent vectors $v \in T_g G$ bijectively to the tangent spaces $T_{hg} G$ and $T_{gh} G$ at hg and gh, respectively.

Each $v \in T_e G$ defines a left-invariant vector field X by $X_g = d_e L_g(v)$ for all $g \in G$. Right invariant vector fields are constructed analogously. The integral curve $\alpha_v : \mathbb{R} \to G$ of a (left or right) invariant vector field X with $v = X_e$ determines a 1-parameter subgroup of G through e. The *group exponential* exp is then defined by $\exp(v) := \alpha_v(1)$; being a diffeomorphism around e, it has a local inverse: the *group logarithm* log.

Given two vector fields X, Y on G, a so-called connection ∇ enables differentiation of Y along X, yielding a vector field $\nabla_X Y$. With $\gamma' := \frac{d\gamma}{dt}$, a *geodesic* $\gamma : [0,1] \to G$ is defined by $\nabla_{\gamma'} \gamma' = 0$, creating a curve without acceleration. Every $f \in G$ has a so-called normal convex neighborhood U, in which any two points $f, h \in U$ can be connected by a unique geodesic $[0,1] \ni t \mapsto \gamma(t; f, h)$ that lies within U. With $v := \gamma'(0; g, h)$, the *connection exponential* $\text{Exp}_g : T_g G \to G$ at g is defined by $\text{Exp}_g(v) := \gamma(1; g, h)$; it is also a local diffeomorphism with the *connection logarithm* $\text{Log}_g(h) = \gamma'(0; g, h)$ as its inverse. These maps are called *Riemannian exponential* and *logarithm*, respectively, if ∇ is the Levi-Civita connection of a Riemannian metric.

Translated group and Riemannian maps coincide if a bi-invariant Riemannian metric is used. Unfortunately, many Lie groups do not allow for such a metric [23]. If, on the other hand, ∇ is the canonical Cartan-Schouten (CCS) connection [8,26], then geodesics and left (or right) translated 1-parameter subgroups *always* coincide. More precisely, there is a maximal normal convex neighborhood U for every $g \in G$ such that [25, Cor. 5.1]

$$\text{Exp}_g(v) = g \exp\left(d_g L_{g^{-1}}(v)\right) = \exp\left(d_g R_{g^{-1}}(v)\right) g, \quad v \in T_g G,$$
$$\text{Log}_g(f) = d_e L_g\left(\log(g^{-1} f)\right) = d_e R_g\left(\log(f g^{-1})\right), \quad f \in U. \tag{1}$$

Geodesics in U are then of the form

$$t \mapsto \gamma(t; g, f) = \text{Exp}_g(t \, \text{Log}_g(f)) = g \exp(t \log(g^{-1} f)) = \exp(t \log(f g^{-1})) g. \tag{2}$$

This fact makes the CCS connection the canonical connection on G.

From the above, we have two equivalent ways to parametrize geodesics in U: We can use boundary values (g, f) or initial values (g, v), which are connected by $(g, f) \leftrightarrow (g, \text{Log}_g(f))$. We will use the former convention in this work, as its inherent symmetry simplifies, e.g., the implementation of derivatives of geodesics with respect to their parameters. Particularly, estimators of regression problems will be defined by the start and endpoint of the optimal geodesic.

Riemannian Geodesic Regression. Geodesic regression was proposed for Riemannian manifolds by Fletcher in [13]. We now summarize the approach for Lie groups, adapting our parametrization of geodesics with initial values for consistency of exposition.

For Riemannian geodesic regression, the Lie group G must have a Riemannian metric—a smoothly varying inner product on the tangent spaces. The metric determines the corresponding distance function dist and Levi-Civita connection of G; the latter defines the geodesics.

For a G-valued random variable Y and a non-random, real-valued variable t, the generalization of the linear regression model to the manifold setting is the geodesic model

$$Y = \mathrm{Exp}_{\gamma(t;g_0,g_1)}(\epsilon), \qquad (3)$$

where ϵ is a random variable taking values in the tangent space at $\gamma(t;g_0,g_1)$. The points $g_0, g_1 \in G$ are the model's parameters that must be estimated.

Let $U \subset G$ be a normal convex neighborhood. Without loss of generality, we assume in this work that the variable t takes values in the interval $[0,1]$. Given data $(f_1,t_1),\ldots,(f_N,t_N) \in U \times [0,1]$, the *least-squares estimator* ϑ^{LS} : $(U \times [0,1])^N \to U^2$ is the minimizer of the *sum-of-squared error*

$$\vartheta^{LS}\left((f_i,t_i)_{i=1}^N\right) = \underset{(g_0,g_1)\in U^2}{\arg\min} \; E(g_0,g_1) := \underset{(g_0,g_1)\in U^2}{\arg\min} \; \frac{1}{2}\sum_{i=1}^N \mathrm{dist}^2\left(\gamma(t_i;g_0,g_1), f_i\right). \qquad (4)$$

It can be computed using gradient descent. Denoting the adjoint operator with respect to the Riemannian metric by "$*$", the gradients of E with respect to g_0 and g_1 are given by [17, Section 3]

$$\mathrm{grad}_{g_0} E(\,\cdot\,,g_1) = -\sum_{i=1}^N \left(\mathrm{d}_{g_0}\gamma(t_i;\,\cdot\,,g_1)\right)^* \mathrm{Log}_{\gamma(t_i)}(f_i) \quad \text{and} \qquad (5)$$

$$\mathrm{grad}_{g_1} E(g_0,\,\cdot\,) = -\sum_{i=1}^N \left(\mathrm{d}_{g_1}\gamma(t_i;g_0,\,\cdot\,)\right)^* \mathrm{Log}_{\gamma(t_i)}(f_i), \qquad (6)$$

respectively.[1] Thus, the desired optima in (4) are roots to both (5) and (6).

The Problem with Riemannian Geodesic Regression in Lie Groups. A natural property we can expect from an estimator $\vartheta : (U \times [0,1])^N \to U^2$ for (3) is that it is equivariant under joint translations of the data; that is, for every $h \in G$,

$$\vartheta\left((\mathrm{T}_h(f_i),t_i)_{i=1}^N\right) = \mathrm{T}_h\left(\vartheta\left((f_i,t_i)_{i=1}^N\right)\right), \qquad \mathrm{T}_h \in \{\mathrm{L}_h, \mathrm{R}_h\}.$$

Indeed, translations are the symmetries of G, and, intuitively, the solution to a symmetry-transformed version of the regression problem should be the symmetry-transformed original solution.

The following example shows that equivariance under translations trivially holds for ordinary linear regression.

---

[1] The adjoint operators in (5) and (6) are given in terms of so-called Jacobi fields; see [15, Section 5.1] for information on when and how they can be computed explicitly.

Example 1. Let $M = \mathbb{R}^n$ with the Euclidean metric. It is also a Lie group with vector addition as the group operation. We have $\mathrm{L}_z(x) = \mathrm{R}_z(x) = x + z$, so the Euclidean metric is bi-invariant. Consequently, the CCS connection coincides with the Levi-Civita connection, which is just the ordinary directional derivative. With the identification $T_x \mathbb{R}^n \cong \mathbb{R}^n$, we have $\mathrm{Exp}_x(v) = x + v$, implying that geodesics are straight lines. With this, we see that (4)) is the standard least-squares estimator of linear regression, which is equivariant under translations.

The equivariance of the least-squares estimator for linear regression has fundamental consequences in applications. There is, e.g., the helpful effect that an experimenter does not need to consider the placement of the origin of the coordinate system; if it were not equivariant, then the quality of the result would depend on it.

Because of the raised points, it is a major drawback that, whenever G does not possess a bi-invariant metric, the Riemannian least-squares estimator (4) is not equivariant under all joint translations of the data. Indeed, for such G either left or right translations do not preserve distances, leading to not-only-translated least-squares estimators. (See Sect. 4.1 for an example.) Two important examples of Lie groups without a bi-invariant metric are the general linear group $\mathrm{GL}(n)$ and the special Euclidean group $\mathrm{SE}(n)$ including the group of rigid-body motions. There are many applications where data from these groups is considered [4,7,9,29,31]. In the next section, we thus propose an alternative estimator that does not suffer from this issue.

Related Work and Contribution. Regression for geometric data has gained increasing attention in recent years, driven by significant empirical improvements [1,10,13,14,16,19] that have been obtained using intrinsic approaches for such data. Most approaches in this line of work employ the Riemannian framework to generalize regression methods from multivariate statistics. Beyond geodesic regression introduced by Fletcher [13], there are Riemannian doppelgangers for non-parametric regression [11,22,28] as well as higher-order models including polynomial [20] and spline [1,17] approaches. However, all methods suffer from the discussed problem when data from Lie groups without a bi-invariant metric is considered.

An incompatibility of the group and Riemannian structure was also recognized in other works, where the authors investigate more general notions of metrics. One option is to consider bi-invariant pseudo-Riemannian metrics [30]; these are again only available for special cases [23]. Another option poses Cartan-Schouten metrics [12], that is, (pseudo-)Riemannian metrics whose Levi-Civita connection agrees with CCS connection, albeit at the possible expense of invariance. For both options, the potential for generalizing statistical procedures under the lack of invariance or positive definiteness remains to be explored.

Beyond the Riemannian statistical framework, a promising line of work follows a more general approach based on an affine connection structure [25]. With such a non-metric structure, statistical estimators cannot be defined by optimizing residuals or variance as in Riemannian manifolds. However, one may obtain generalizations in terms of weaker notions. In particular, Pennec and Arsigny [24] characterized the mean as exponential barycenters (the critical points of the variance in the Riemannian case) allowing in turn for bi-invariant notions of higher order moments and Mahalanobis distance. Building on these, Hanik et al. [18] derived dissimilarity measures for sample distributions allowing for bi-invariant two-sample tests.

In this work, we generalize linear regression by modeling the relationships between a Lie group-valued response and a scalar explanatory variable via the geodesics of the CCS connection. Employing the affine connection structure of the group, we thus obtain a geodesic estimator that is completely consistent with left and right translation, while reproducing the well-known least-squares approach for the Euclidean case. To the best of our knowledge, this is the first bi-invariant estimator for geodesic regression. We further propose an efficient numerical scheme for estimation that exploits a reformulation using Jacobi fields computable in a single pass of automatic differentiation. The source code of our prototype implementation is published as part of the open-source library Morphomatics [2].

2 Bi-invariant Geodesic Regression

From now on, we consider G endowed with the CCS connection. The geodesics in G are thus given by (2). In the following, we will need the fact that the start point map $g \mapsto \gamma(t; g, f)$ is differentiable in U with an invertible derivative; that is, the inverse of the partial derivative $\mathrm{d}_g \gamma(t; \cdot, f) : T_g G \to T_{\gamma(t;g,f)} G$ exists for all $t \in [0,1]$ and $f \in U$. Because of the symmetry $\gamma(t; g, f) = \gamma(1 - t; f, g)$, this is also true for the derivative with respect to the endpoint.

Our proposition for an equivariant estimator starts with (5) and (6) but replaces the metric-dependent adjoint with the weighted connection-dependent inverted differential. Both operators are intimately related via Jacobi fields [6, Sections 3-4], but the inverted differential does not depend on a given metric.

Definition 1. *Let $U \subseteq G$ be a normal convex neighborhood. Furthermore, let $(f_1, t_1), \ldots, (f_N, t_N) \in U \times [0,1]$. The bi-invariant estimator for (3) is the geodesic between the points $\vartheta^{\mathrm{BI}}((f_i, t_i)_{i=1}^N) := (\hat{g}_0, \hat{g}_1) \in U^2$ that satisfy*

$$\sum_{i=1}^N (1-t_i)^2 \big(\mathrm{d}_{\hat{g}_0} \gamma(t_i; \cdot, \hat{g}_1)\big)^{-1} \Big(\mathrm{Log}_{\gamma(t_i)}(f_i)\Big) = 0 \quad \text{and} \quad (7)$$

$$\sum_{i=1}^N t_i^2 \big(\mathrm{d}_{\hat{g}_1} \gamma(t_i; \hat{g}_0, \cdot)\big)^{-1} \Big(\mathrm{Log}_{\gamma(t_i)}(f_i)\Big) = 0. \quad (8)$$

A generalization should always coincide with the notion it is motivated by. The following remark shows this is the case for the new estimator: It reduces to the standard least-squares estimator in the Euclidean case.

Remark 1. Denoting the n-by-n identity matrix by I, we find for $t \in [0,1]$ and $x, y \in \mathbb{R}^n$

$$d_{x_0}\gamma(t; \,\cdot\, , x_1) = d_{x_0}(z \mapsto z + t(x_1 - z)) = (1-t)I \quad \text{and}$$
$$d_{x_1}\gamma(t; x_0, \,\cdot\,) = d_{x_1}(z \mapsto x_0 + t(z - x_0)) = tI.$$

Hence, since the adjoint is the matrix transpose, we find

$$\left(d_{x_0}\gamma(t; \,\cdot\, , x_1)\right)^* = (1-t)I = (1-t)^2 \left(d_{x_0}\gamma(t; \,\cdot\, , x_1)\right)^{-1} \quad \text{and}$$
$$\left(d_{x_1}\gamma(t; x_0, \,\cdot\,)\right)^* = tI = t^2 \left(d_{x_1}\gamma(t; x_0, \,\cdot\,)\right)^{-1}.$$

It follows that $\vartheta^{\mathrm{BI}} = \vartheta^{\mathrm{LS}}$ in Euclidean space $G = \mathbb{R}^n$ since we look for the (unique) zero of the same system of equations.

We will now show that the new estimator does not suffer from the same problem as the least-squares estimator. To this end, we need the following Lemma:

Lemma 1. *Geodesics $\gamma(\,\cdot\,; g_0, g_1)$ are equivariant under joint translations of their endpoints g_0 and g_1; that is, for all $t \in [0,1]$ and $h \in G$,*

$$\gamma(t; T_h(g_0, g_1)) = T_h\left(\gamma(t; g_0, g_1)\right), \qquad T_h \in \{L_h, R_h\}.$$

Proof. Because of the characterization of geodesics with left translations in (2), we find

$$\gamma(t; hg_0, hg_1) = hg_0 \exp\left(t \log\left(g_0^{-1} g_1\right)\right) = h\gamma(t; g_0, g_1).$$

The proof works analogously for right translations using the characterization of geodesics with right translations in (2). □

We also need the following facts: Clearly, $(L_g)^{-1} = L_{g^{-1}}$ and $(R_g)^{-1} = R_{g^{-1}}$; hence, for all $g, h \in G$,

$$(d_g L_h)^{-1} = d_{hg} L_{h^{-1}} \quad \text{and} \quad (d_g R_h)^{-1} = d_{gh} R_{h^{-1}}. \tag{9}$$

Furthermore, translations also induce *joint* translations on the product group[2] G^m by $L_h((g_i)_{i=1}^m) := ((L_h(g_i))_{i=1}^m)$ and $R_h((g_i)_{i=1}^m) := ((R_h(g_i))_{i=1}^m)$, which are also automorphisms. Note that we use the same symbols L and R for translations and (real) joint translations. We are now ready for the following theorem:

Theorem 1. *Let $(f_1, t_1), \ldots, (f_N, t_N) \in U \times \mathbb{R}$. Then, for all $h \in G$,*

$$\vartheta^{\mathrm{BI}}\left((T_h(f_i), t_i)_{i=1}^N\right) = T_h\left(\vartheta^{\mathrm{BI}}\left((f_i, t_i)_{i=1}^N\right)\right), \qquad T_h \in \{L_h, R_h\}.$$

---

[2] A product of Lie groups is also a Lie group with operations working entry-wise.

Proof. In the following, we consider the geodesic boundary map $\gamma_t : U^2 \to U$, $(g_0, g_1) \mapsto \gamma(t; g_0, g_1)$, to avoid always writing two cases. (Note that t appears as a subscript in this proof as it can be interpreted as arbitrary but fixed.) Remember that $T_{(g_0,g_1)}G^2 \cong T_{g_0}G \times T_{g_1}G$. Let $s : [0,1] \times T_{(g_0,g_1)}G^2 \to T_{(g_0,g_1)}G^2$, $(t,(v,w)) \mapsto s_t(v,w) := ((1-t)^2 v, t^2 w)$, be multiplication of tangent vectors of G^2 with the scalars from Definition 1.

We set $b_i = \gamma_{t_i}(g_0, g_1)$ for $i = 1, \ldots, n$. Our goal is to show that the vector

$$v\left(g_0, g_1, (f_i)_{i=1}^N\right) := \sum_{i=1}^N s_{t_i} \left(\mathrm{d}_{(g_0,g_1)} \gamma_{t_i}\right)^{-1} \left(\mathrm{Log}_{b_i}(f_i)\right) \in T_{(g_0,g_1)}G^2 \quad (10)$$

fulfills $v \circ L_h = dL_h \circ v$ as a function on U^{N+2}. The derivative $\mathrm{d}_{(g_0,g_1)}L_h$ being an invertible, linear operator (with inverse $\mathrm{d}_{(hg_0,hg_1)}L_{h^{-1}}$) at each $(g_0, g_1) \in G^2$ then implies that $v(g_0, g_1, (f_i)_{i=1}^N) = 0$ if and only if $v(hg_0, hg_1, (hf_i)_{i=1}^N) = 0$.

From Lemma 1, we get for all t,

$$\gamma_t \circ L_h = L_h \circ \gamma_t;$$

so differentiating at (g_0, g_1) and applying the chain rule[3] yields

$$\mathrm{d}_{(hg_0,hg_1)}\gamma_t \circ \mathrm{d}_{(g_0,g_1)}L_h = \mathrm{d}_{\gamma_t(g_0,g_1)}L_h \circ \mathrm{d}_{(g_0,g_1)}\gamma_t.$$

Together with (9), this implies

$$\left(\mathrm{d}_{(hg_0,hg_1)}\gamma_t\right)^{-1} = \mathrm{d}_{(g_0,g_1)}L_h \circ \left(\mathrm{d}_{(g_0,g_1)}\gamma_t\right)^{-1} \circ \mathrm{d}_{h\gamma_t(g_0,g_1)}L_{h^{-1}}. \quad (11)$$

Combining (10), (1), and (11) gives

$$(v \circ L_h)\left(g_0, g_1, (f_i)_{i=1}^N\right)$$
$$= \sum_{i=1}^N s_{t_i} \left(\mathrm{d}_{(hg_0,hg_1)}\gamma_{t_i}\right)^{-1} \left(\mathrm{Log}_{hb_i}(hf_i)\right)$$
$$= \sum_{i=1}^N s_{t_i} \left(\mathrm{d}_{(g_0,g_1)}L_h \circ \left(\mathrm{d}_{(g_0,g_1)}\gamma_{t_i}\right)^{-1} \circ \mathrm{d}_{hb_i}L_{h^{-1}} \circ \mathrm{d}_e L_{hb_i}\right)\left(\log\left(b_i^{-1}f_i\right)\right).$$

We now apply the chain rule again and pull the first differential out of the sum. The latter can be done since there is no cross-talk between the two copies of G under joint translations. We thereby obtain

$$(v \circ L_h)\left(g_0, g_1, (f_i)_{i=1}^N\right)$$
$$= \mathrm{d}_{(g_0,g_1)}L_h \left(\sum_{i=1}^N \left(s_{t_i} \left(\mathrm{d}_{(g_0,g_1)}\gamma_{t_i}\right)^{-1} \circ \mathrm{d}_e\left(L_{h^{-1}} \circ L_{hb_i}\right)\right)\left(\log\left(b_i^{-1}f_i\right)\right)\right)$$
$$= \mathrm{d}_{(g_0,g_1)}L_h \left(\sum_{i=1}^N \left(s_{t_i} \left(\mathrm{d}_{(g_0,g_1)}\gamma_{t_i}\right)^{-1} \circ \mathrm{d}_e L_{b_i}\right)\left(\log\left(b_i^{-1}f_i\right)\right)\right)$$
$$= \mathrm{d}_{(g_0,g_1)}L_h \left(\sum_{i=1}^N \left(s_{t_i} \left(\mathrm{d}_{(g_0,g_1)}\gamma_{t_i}\right)^{-1}\right)\left(\mathrm{Log}_{b_i}(f_i)\right)\right)$$
$$= \left(\mathrm{d}_{(g_0,g_1)}L_h \circ v\right)\left(g_0, g_1, (f_i)_{i=1}^N\right),$$

---

[3] Let M_1, M_2, M_3 be smooth manifolds and $F : M_1 \to M_2$, $H : M_2 \to M_3$ smooth. The chain rule yields the differential $\mathrm{d}_p(H \circ F) = \mathrm{d}_{F(p)}H \circ \mathrm{d}_p F$ where $(H \circ F)(p)$ is defined.

which is what we wanted to show.

The proof for right translation works analogously by replacing left with right translations everywhere. □

The theorem shows that the novel estimator is indeed equivariant under left and right translations. This and the fact that it coincides with linear regression in Euclidean space make it a natural choice for regression in Lie groups.

3 Computation

Note that the components of the vector $v(g_0, g_1, (f_i)_{i=1}^N)$ in the proof act like a "net-force", which the data points apply (through the geodesic) to g_0 and g_1. Therefore, we propose the following algorithm to approximate solutions of (7) and (8): Starting with initial guesses $\hat{g}_0^{(0)}$ and $\hat{g}_1^{(0)}$, we use the iteration

$$\hat{g}_j^{(k+1)} := \mathrm{Exp}_{\hat{g}_j^{(k)}}(\lambda v_j) \quad \text{with} \quad (v_0, v_1) := v\left(\hat{g}_0^{(k)}, \hat{g}_1^{(k)}, (f_i)_{i=1}^N\right).$$

with a tuned stepsize λ; convergence can be determined, e.g., by observing the length of the update vector v according to some auxiliary norm. Similar algorithms have been used with great success. Indeed, the algorithm for computing the equivariant mean of Pennec and Arsigny [25, Algorithm 1] can be seen as a special case, in which only constant geodesics are considered.

In each step, we must compute the vectors v_0 and v_1 at the current iterate. Luckily, we need not explicitly invert the differential operators in (7) and (8) for this. Consider a general geodesic γ in U to see this. When viewed as a function in t and with $v \in T_{g_0}G$,

$$J : t \mapsto \mathrm{d}_{g_0}\gamma(t;\,\cdot\,,g_1)(v)$$

is a so-called Jacobi (vector) field along the geodesic $\gamma(\,\cdot\,;g_0,g_1)$ that fulfills the Jacobi equation—a linear second-order differential equation [6]. Importantly, $J(0) = v$ and $J(1) = 0$. A Jacobi field also encodes the derivative with respect to the endpoint because of the symmetry of geodesics under the exchange of the start and endpoint [6, Section 3.1].

Let now $w := \mathrm{d}_{g_0}\gamma(t_0;\,\cdot\,,g_1)(v)$ for some $t_0 \in (0,1)$. Then, v can be determined from w by finding the Jacobi field J and evaluating at $t = 0$. To this end, set $f := \gamma(t_0; g_0, g_1)$. The linear reparametrization $s(t) := (1-t)/(1-t_0)$ yields the geodesic $\eta : [0, 1/(1-t_0)] \to U, s \mapsto \gamma(s; g_1, f)$. Note that $\eta(1/(1-t_0)) = g_0$. The Jacobi field

$$\widetilde{J} : t \mapsto \mathrm{d}_f\eta(t; g_1,\,\cdot\,)(w)$$

fulfills $\widetilde{J}(0) = 0$ and $\widetilde{J}(1) = w$. Therefore, we have two Jacobi fields J and $\widetilde{J}$ along the same—albeit differently parametrized—geodesic that coincide at $f = \gamma(t_0) = \eta(1)$ and $g_1 = \gamma(1) = \eta(0)$.

Solutions to the Jacobi equation are well-known to be equivariant under reflections and linear reparametrizations. Therefore, and since solutions to second-order equations are uniquely determined by the values at two different points, we have $\tilde{J}((1-t)/(1-t_0)) = J(t)$ for all $t \in [0,1]$. In particular, it follows that

$$(d_{g_0}\gamma(t_0;\cdot,g_1))^{-1}(w) = d_f\eta\left(\frac{1}{1-t_0};g_1,\cdot\right)(w) = d_f\gamma\left(\frac{1}{1-t_0};g_1,\cdot\right)(w).$$

The right-hand side can be efficiently computed using, e.g., automatic differentiation on (2) and projecting the result to $T_{g_0}G$.

For differentials with respect to endpoints, everything works analogously when replacing $(1-t)$ by t.

4 Experiments

In the following, we evaluate the proposed regression scheme experimentally both for a synthetic example and in an application to knee joint configurations during the progression of osteoarthritis.

Throughout this section, the dependent variables will be elements in the Lie group SE(3) of rigid-body motions. We denote the group of 3-by-3 rotation matrices and their identity matrix by SO(3) and I, respectively. Remember that SE(3) = SO(3) ⋉ $\mathbb{R}^3$ is a semidirect product with the group operation $(R,x)(Q,y) = (RQ, x+Ry)$. Its tangent space $\mathfrak{se}(3)$ at the identity element $(I,0)$ is the product of the set of skew-symmetric 3-by-3 matrices and $\mathbb{R}^3$. The inverse of an element (R,x) is given by $(R^\intercal, -R^\intercal x)$. As the name suggests, SE(3) consists of the motions a rigid body can perform in 3-space. Naturally, it appears in many real-world applications. It is well known that SE(3) does *not* possess a bi-invariant metric [32]. Screw motions are 1-parameter subgroups and, thus, the geodesics of the CCS connection.

4.1 Simulation Study

We start with a collection of synthetic experiments demonstrating the equivariance of our proposed scheme on the one hand and, on the other, juxtaposing it to the variance of estimators based on the Riemannian framework. In particular, we compare our algorithm to the Riemannian geodesic regression by Fletcher [13]. To this end, we equip the group of rigid motions with a Riemannian structure based on the product of the standard bi-invariant metrics of SO(3) and $\mathbb{R}^3$—a frequent choice in the literature [5,32].

We simulated data according to model (3) by determining a geodesic in terms of a random start and endpoint and, subsequently, perturbing it at $N = 10$ equidistant time points in the unit interval. To generate random points and vectors we pseudo-randomly sample centered normal distributions in $\mathfrak{se}(3)$ with isotropic variance of 1 and 0.01, respectively. Based on these samples, tangent vectors are obtained via left-translation, whereas points are computed via the group exponential. To avoid a bias towards a specific geometric structure in our comparison, we use the random samples to derive two datasets from model

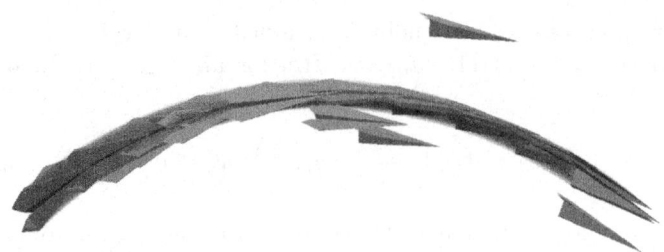

Fig. 1. Equivariance of bi-invariant regression under symmetries visualized by the action of SE(3)-valued data on a reference object (paper plane). Green: $R_h(\mathcal{D}_{CCS})$, that is, sample data $\mathcal{D}_{CCS}$ after right translation with random h; Red: Estimated geodesic by bi-invariant regression on $R_h(\mathcal{D}_{CCS})$; Orange: $R_h(\gamma)$, that is, the geodesic γ from bi-invariant regression on $\mathcal{D}_{CCS}$ right-translated by h. (Color figure online)

(3): The first ($\mathcal{D}_{CCS}$) using the CCS connection for exponentiation and geodesic interpolation, while the second ($\mathcal{D}_{LC}$) is based on the Levi-Civita connection.

In Fig. 1, we show estimates of the geodesic trend underlying $\mathcal{D}_{CCS}$ as determined by our proposed regression scheme (with stepsize $\lambda = 0.1$) after right translation by a random motion h. Alongside the translated dataset $R_h(\mathcal{D}_{CCS})$, this figure shows the translated estimate on the untransformed data in orange together with the estimate for $R_h(\mathcal{D}_{CCS})$. While both estimates show a high degree of agreement, we observe a minor difference that can be attributed to finite precision arithmetic.

Next, we examine the same construction for Riemannian geodesic regression with the dataset $\mathcal{D}_{LC}$ using the same random h for the right translation as in the previous example. In Fig. 2, we again show the estimate for the translated data and the translated estimate for the original data in red and orange, respectively. Both curves are quite different, illustrating the lack of equivariance of the Riemannian estimator. In particular, we can observe a bending of the data and estimator under translation causing a violation of geodesicity in the Riemannian sense. A naïve approach to obtain a Riemannian geodesic via translation would be to apply R_h to the parametrization, that is, the endpoints. The resulting geodesic—shown in yellow—is far from the estimator for $R_h(\mathcal{D}_{LC})$ with a goodness of fit decreasing with the distance to the endpoints.

We further provide a quantitative assessment of the deviation from Riemannian geodesicity under the right translation and hence diminishing adequacy of the Riemannian geodesic regression. To this end, we draw 100 random motions as before but with a variance of 100. For each of the motions, we right-translate the dataset $\mathcal{D}_{LC}$ and perform Riemannian geodesic regression. We then assess the goodness of fit for the estimator in terms of the Riemannian coefficient of determination [13] denoted R^2. For the untranslated dataset, the estimated Riemannian geodesic achieves a fit of $R^2 = 0.93$. A histogram showing the

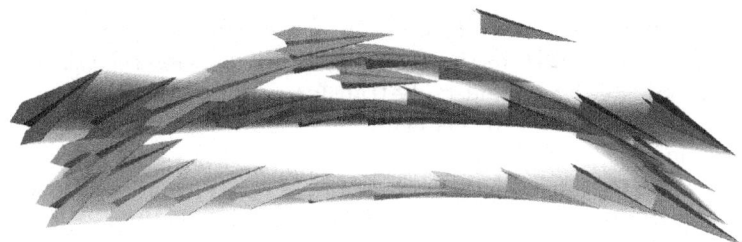

Fig. 2. Variance of Riemannian regression under symmetries visualized by the action of SE(3)-valued data on a reference object (paper plane). Green: $R_h(\mathcal{D}_{LC})$, that is, sample data $\mathcal{D}_{LC}$ after right translation with random h; red: Estimated geodesic by Riemannian regression on $R_h(\mathcal{D}_{LC})$; orange: $R_h(\gamma)$, that is, the geodesic γ from Riemannian regression on $\mathcal{D}_{LC}$ right-translated by h; yellow: Geodesic connecting endpoints of $R_h(\gamma)$. (Color figure online)

distribution of this coefficient under random right translation is given in Fig. 3. Note that, while in most cases the Riemannian model can fit the translated data well (relative to the untranslated case), we can observe instances that are very poorly replicated. This result implies that the Riemannian geodesic model can be arbitrarily unfavorable even if a reference frame exists for which the model assumption of Riemannian geodesicity applies.

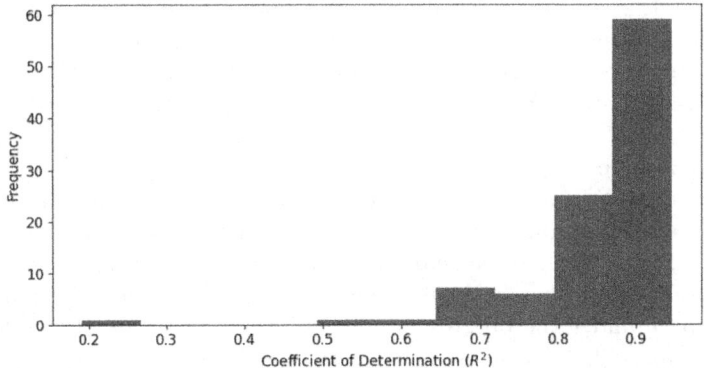

Fig. 3. Histogram of R^2 values for Riemannian estimators under random translation.

4.2 Skeletal Configurations of the Knee Joint

A possible application of bi-invariant geodesic regression is studying the progression of osteoarthritis (OA) in knee joints. OA is a degenerative disease of the joints. It is characterized by a loss of cartilage leading, among others, to a narrowing of the joint space between the femur and tibia. A method of quantifying

OA is the *Kellgren-Lawrence (KL) grade*, which assigns integers between 0 and 4 to indicate the severity [21]. In this experiment, we describe the relative position of the femur and tibia by a rigid-body transformation as in [18, Section 4.1]. We then regress the relative positions against OA grades to (re)discover the joint space narrowing.

Our data set stems from 50 subjects (10 per grade) randomly selected from the Osteoarthritis Initiative (a longitudinal, prospective study of knee OA) for which femur and tibia segmentations of the respective magnetic resonance images and KL grades are publicly available.[4] From these segmentations, triangular meshes were extracted; in a supervised post-process, the quality of segmentations was ensured.

For each femur and tibia mesh, a *frame* $(R, x) \in \mathrm{SE}(3)$ is computed to represent its location and orientation in space: The center of gravity of the vertex set is picked as the origin x; the 3 positively oriented, normalized principal directions obtained by principal component analysis for the vertices are chosen as orientation. With this setup, we can express the relative position of the tibia frame $F_\mathrm{T}^{(i)} = (R_\mathrm{T}^{(i)}, x_\mathrm{T}^{(i)})$ of the i-th subject to its femur frame $F_\mathrm{F}^{(i)} = (R_\mathrm{F}^{(i)}, x_\mathrm{F}^{(i)})$ by the rigid-body motion

$$M^{(i)} := F_\mathrm{F}^{(i)}(F_\mathrm{T}^{(i)})^{-1} = \left(R_\mathrm{F}^{(i)}(R_\mathrm{T}^{(i)})^\mathsf{T}, x_\mathrm{F}^{(i)} - R_\mathrm{F}^{(i)}(R_\mathrm{T}^{(i)})^\mathsf{T} x_\mathrm{T}\right).$$

The construction principle is depicted on the left of Fig. 4.

Assuming equidistant KL grades, we choose $t = 0, 1/4, 1/2, 3/4, 1$ to represent the grades 0, 1, 2, 3, and 4, respectively, and assign the corresponding value t_i to each subject. We now compute $\vartheta^{\mathrm{BI}}((M_i, t_i)_{i=1}^{50})$. To this end, the geodesic start and endpoints are initialized with a sample of a subject with KL grade 0 and 4, respectively; a stepsize $\lambda = 0.01$ is used. Evaluating the obtained geodesic at $t = 0, 1/4, 1/2, 3/4, 1$ gives the "denoised" knee configurations $\hat{M}^{(j)} = (\hat{R}^{(j)}, \hat{x}^{(j)})$ for the 5 KL grades.

Using the Euclidean norm on the translations $\hat{x}^{(j)}$, we can measure the distance between the origins for the estimated geodesic; the results are shown on the right of Fig. 4. We can see that the centers of gravity of the femur and tibia move towards each other with increasing KL grade showing the narrowing of the joint space under the progression of OA. The magnitude of the decrease in width is comparable to what is reported in the literature; see, e.g., [27].[5]

---

[4] See [3]; the data can be found at https://doi.org/10.12752/4.ATEZ.1.0.
[5] The decrease reported in [27] is slightly larger. However, their data comes from radiographic images usually taken from standing subjects with strained joints. Our MRI data comes from lying subjects, whose joints were relaxed.

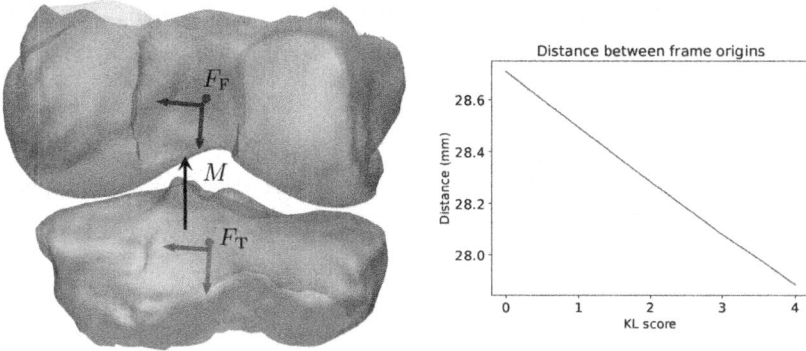

Fig. 4. Left: Knee-constituting parts of the tibia (bottom) and femur (top) with their respective frames (one arrow pointing into the background). The rigid motion M moves the former's frame onto the latter's. Right: Euclidean norm of the translational part of the estimator $\vartheta^{\mathrm{BI}}((M_i, t_i)_{i=1}^{50})$ plotted against the KL grade.

Acknowledgements. We thank Ana Djurdjevac for her support. Martin Hanik is funded by BIFOLD—the Berlin Institute for the Foundations of Learning and Data (ref. 01IS18025A and ref. 01IS18037A). We are grateful for the open-access data set of the Osteoarthritis Initiative, which is a public-private partnership comprised of five contracts (N01-AR-2-2258; N01-AR-2-2259; N01-AR-2-2260; N01-AR-2-2261; N01-AR-2-2262) funded by the National Institutes of Health, a branch of the Department of Health and Human Services, and conducted by the OAI Study Investigators. Private funding partners include Merck Research Laboratories; Novartis Pharmaceuticals Corporation, GlaxoSmithKline; and Pfizer, Inc. Private sector funding for the OAI is managed by the Foundation for the National Institutes of Health. This manuscript was prepared using an OAI public use data set and does not necessarily reflect the opinions or views of the OAI investigators, the NIH, or the private funding partners.

Disclosure of Interests. The authors have no competing interests to declare that are relevant to the content of this article.

References

1. Adouani, I., Samir, C.: Regression and Fitting on Manifold-valued Data. Springer, Berlin (2024). https://doi.org/10.1007/978-3-031-61712-6
2. Ambellan, F., Hanik, M., von Tycowicz, C.: Morphomatics: geometric morphometrics in non-Euclidean shape spaces (2021). https://doi.org/10.12752/8544, https://morphomatics.github.io/
3. Ambellan, F., Tack, A., Ehlke, M., Zachow, S.: Automated segmentation of knee bone and cartilage combining statistical shape knowledge and convolutional neural networks: Data from the Osteoarthritis Initiative. Med. Image Anal. **52**, 109–118 (2019). https://doi.org/10.1016/j.media.2018.11.009
4. Ambellan, F., Zachow, S., von Tycowicz, C.: An as-invariant-as-possible $GL^+(3)$-based statistical shape model. In: Zhu, D., et al. (eds.) MBIA/MFCA -2019. LNCS,

vol. 11846, pp. 219–228. Springer, Cham (2019). https://doi.org/10.1007/978-3-030-33226-6_23
5. Belta, C., Kumar, V.: Euclidean metrics for motion generation on SE(3). Proc. Inst. Mech. Eng., Part C **216**(1), 47–60 (2002). https://doi.org/10.1243/0954406021524909
6. Bergmann, R., Gousenbourger, P.Y.: A variational model for data fitting on manifolds by minimizing the acceleration of a Bézier curve. Front. Appl. Math. Stat. **4**, 1–16 (2018). https://doi.org/10.3389/fams.2018.00059
7. Boisvert, J., Cheriet, F., Pennec, X., Labelle, H., Ayache, N.: Geometric variability of the scoliotic spine using statistics on articulated shape models. IEEE Trans. Med. Imaging **27**(4), 557–568 (2008). https://doi.org/10.1109/TMI.2007.911474
8. Cartan, E., Schouten, J.: On the geometry of the group-manifold of simple and semi-groups. Proc. Akad. Wetensch., Amsterdam **29**, 803–815 (1926)
9. Ćesić, J., Marković, I., Cvišić, I., Petrović, I.: Radar and stereo vision fusion for multitarget tracking on the special Euclidean group. Rob. Auton. Syst. **83**, 338–348 (2016). https://doi.org/10.1016/j.robot.2016.05.001
10. Cornea, E., Zhu, H., Kim, P., Ibrahim, J.G., Initiative, A.D.N.: Regression models on Riemannian symmetric spaces. J. R. Stat. Soc. Series B Stat. Methodol. **79**(2), 463–482 (2017). https://doi.org/10.1111/rssb.12169
11. Davis, B.C., Fletcher, P.T., Bullitt, E., Joshi, S.: Population shape regression from random design data. Int. J. Comput. Vis. **90**, 255–266 (2010). https://doi.org/10.1007/s11263-010-0367-1
12. Diatta, A., Manga, B., Sy, F.: On Dual Quaternions, Dual Split Quaternions and Cartan-Schouten Metrics on Perfect Lie Groups, pp. 317–339. Springer, Berlin (2024). https://doi.org/10.1007/978-3-031-52681-7_15
13. Fletcher, P.T.: Geodesic regression and the theory of least squares on Riemannian manifolds. Int. J. Comput. Vis. **105**, 171–185 (2013). https://doi.org/10.1007/s11263-012-0591-y
14. Giovanis, D.G., Shields, M.D.: Data-driven surrogates for high dimensional models using Gaussian process regression on the Grassmann manifold. Comput. Methods Appl. Mech. Eng. **370**, 113269 (2020). https://doi.org/10.1016/j.cma.2020.113269
15. Hanik, M.: Geometric data analysis: advancements of the statistical methodology and applications. Ph.D. thesis, Freie Universität Berlin (2023). https://doi.org/10.17169/refubium-39809
16. Hanik, M., Ducke, B., Hege, H.C., Fless, F., Tycowicz, C.V.: Intrinsic shape analysis in archaeology: a case study on ancient sundials. ACM J. Comput. Cult. Herit. **16**(4), 1–26 (2023). https://doi.org/10.1145/3606698
17. Hanik, M., Hege, H.-C., Hennemuth, A., von Tycowicz, C.: Nonlinear regression on manifolds for shape analysis using intrinsic Bézier splines. In: Martel, A.L., et al. (eds.) MICCAI 2020. LNCS, vol. 12264, pp. 617–626. Springer, Cham (2020). https://doi.org/10.1007/978-3-030-59719-1_60
18. Hanik, M., Hege, H.C., von Tycowicz, C.: Bi-invariant dissimilarity measures for sample distributions in Lie groups. SIAM J. Math. Data Sci. **4**(4), 1223–1249 (2022). https://doi.org/10.1137/21m1410373
19. Hanik, M., Nava-Yazdani, E., von Tycowicz, C.: De Casteljau's algorithm in geometric data analysis: theory and application. Comput. Aided Geom. Des. **110**, 102288 (2024). https://doi.org/10.1016/j.cagd.2024.102288
20. Hinkle, J., Muralidharan, P., Fletcher, P.T., Joshi, S.: Polynomial regression on riemannian manifolds. In: Fitzgibbon, A., Lazebnik, S., Perona, P., Sato, Y., Schmid, C. (eds.) ECCV 2012. LNCS, vol. 7574, pp. 1–14. Springer, Heidelberg (2012). https://doi.org/10.1007/978-3-642-33712-3_1

21. Kellgren, J.H., Lawrence, J.S.: Radiological assessment of osteo-arthrosis. Ann. Rheum. Dis. **16**(4), 494–502 (1957). https://doi.org/10.1136/ard.16.4.494
22. Mallasto, A., Feragen, A.: Wrapped Gaussian process regression on Riemannian manifolds. In: Proceedings of the IEEE Conference on Computer Vision and Pattern Recognition, pp. 5580–5588 (2018)
23. Miolane, N., Pennec, X.: Computing bi-invariant pseudo-metrics on Lie groups for consistent statistics. Entropy **17**, 1850–1881 (2015). https://doi.org/10.3390/e17041850
24. Pennec, X., Arsigny, V.: Exponential barycenters of the canonical Cartan connection and invariant means on Lie groups. In: Matrix Information Geometry, pp. 123–166. Springer, Berlin (2013). https://doi.org/10.1007/978-3-642-30232-9_7
25. Pennec, X., Lorenzi, M.: 5 - Beyond Riemannian geometry: the affine connection setting for transformation groups. In: Riemannian Geometric Statistics in Medical Image Analysis, pp. 169–229. Academic Press, London (2020). https://doi.org/10.1016/B978-0-12-814725-2.00012-1
26. Postnikov, M.: Geometry VI: Riemannian Geometry. Springer, Berlin (2013). https://doi.org/10.1007/978-3-662-04433-9
27. Ratzlaff, C., Ashbeck, E., Guermazi, A., Roemer, F., Duryea, J., Kwoh, C.: A quantitative metric for knee osteoarthritis: reference values of joint space loss. Osteoarthritis Cartilage **26**(9), 1215–1224 (2018). https://doi.org/10.1016/j.joca.2018.05.014
28. Steinke, F., Hein, M., Schölkopf, B.: Nonparametric regression between general Riemannian manifolds. SIAM J. Imaging Sci. **3**(3), 527–563 (2010). https://doi.org/10.1137/080744189
29. Vemulapalli, R., Arrate, F., Chellappa, R.: Human action recognition by representing 3D skeletons as points in a Lie group. In: Proceedings of the IEEE Conference on Computer Vision and Pattern Recognition, pp. 588–595 (2014). https://doi.org/10.1109/CVPR.2014.82
30. Woods, R.P.: Characterizing volume and surface deformations in an atlas framework: theory, applications, and implementation. Neuroimage **18**(3), 769–788 (2003). https://doi.org/10.1016/s1053-8119(03)00019-3
31. Wu, G., et al.: ISB recommendation on definitions of joint coordinate systems of various joints for the reporting of human joint motion—part ii: shoulder, elbow, wrist and hand. J. Biomech. **38**(5), 981–992 (2005). https://doi.org/10.1016/j.jbiomech.2004.05.042
32. Zefran, M., Kumar, V., Croke, C.: Metrics and connections for rigid-body kinematics. Int. J. Robot. Res. **18**(2), 243–258 (1999). https://doi.org/10.1177/027836499901800208

GSSD: A Self-distillation Paradigm with Gradient Surgery for End-to-End Deformable Image Registration

Yuxi Zheng, Yansong Bai, and Yuchuan Qiao(✉)

Institute of Science and Technology for Brain-Inspired Intelligence, Fudan University, Shanghai, China
yuchuanqiao@fudan.edu.cn

Abstract. Deformable Image Registration (DIR) is crucial for various medical image analysis tasks. However, deep learning methods struggle to balance registration accuracy and computational complexity. Incorporating knowledge distillation into registration emerges as a promising approach; however, these methods often rely on pre-designed or heuristically chosen teacher networks. Their efficiency is not optimal, primarily because they fail to account for gradient conflicts between student and teacher networks. In this paper, we propose a novel self-distillation paradigm with gradient surgery for end-to-end deformable image registration, named **GSSD**. Specifically, to design a universally applicable knowledge distillation paradigm, the teacher network is directly cloned from the student network and is removed after training, reducing hardware requirements upon deployment. To resolve potential gradient conflicts between the student and teacher networks, we introduce a two-stage gradient surgery optimization strategy by projecting the conflicting gradient into the normal plane of the dominant gradient, ensuring the distillation efficacy. Extensive experiments conducted on three publicly available datasets demonstrate consistent improvements over various methods, with no increase in inference time and parameters, especially more than 3% increase in Dice score for liver CT.

Keywords: Deformable Image Registration · End-to-End · Self-Distillation · Gradient Surgery

1 Introduction

Deformable image registration (DIR) is fundamental and crucial in various medical image analysis tasks. Traditional methods [4,6,15,27] address the registration task as an optimization problem, but their time-consuming nature and high computational costs often render them impractical for clinical applications.

Recently, deep learning-based unsupervised methods [5,8,21,29] have accelerated DIR, with the U-shape networks [5,7,9] being the most popular. Nevertheless, their simpler network architectures allow for improvements in registration accuracy. In recent years, cascade networks [33] and pyramid networks [13] have employed progressive structures to gradually generate deformation fields from coarse to fine. These approaches enhance registration accuracy at the cost of a more complex training process and increased model parameters. Enhancing the registration performance of the simple and light U-shape networks which have lower computational costs remains an unresolved challenge.

Some research has introduced knowledge distillation into registration networks, enabling a lighter student network to achieve the comparable performance of a heavier teacher network. The pioneering LDR [28] utilizes an adversarial distillation loss to transfer the knowledge from the teacher network to the student. Other typical approaches [12,34] adopt the concept of self-distillation, where lower-resolution layers/iterations are trained under the guidance of higher-resolution layers/iterations. However, these works still have some limitations. *Firstly*, they require the pre-design or careful selection of a suitable teacher network and are only applicable to network structures with multiple resolution layers/iterations. *Secondly*, they only constrain the similarity of the deformation fields between the teacher and student through a simple Mean Squared Error (MSE) loss, which neglects the potential gradient conflicts between them and leads to sub-optimal registration accuracy. Integrating the knowledge distillation into registration networks confronts the following *major challenges*:

- *How to* design a universally applicable distillation paradigm that enables various student networks to achieve higher registration accuracy? (▷ Sect. 2.1)
- *How to* reduce potential gradient conflicts between the student and teacher networks, ensuring effective distillation? (▷ Sect. 2.2)

In this paper, we address the above challenges and present a novel self-distillation paradigm with gradient surgery for end-to-end deformable image registration, named **GSSD**. We highlight the main contributions as follows:

- **We design a universal self-distillation registration paradigm.** The teacher network is cloned and cascaded after the student to obviate the need for pre-designing or selection, making it compatible with any end-to-end registration backbone architecture. Once the model is trained, the teacher is removed to reduce the deployment requirements.
- **We introduce a two-stage gradient surgery optimization strategy in the distillation process.** To guarantee that the teacher network effectively enhances the performance of the student, we assign specific dominant gradients at different training stages and project the conflicting gradient onto the dominant gradients' normal plane, reinforcing distillation effectiveness.
- **We improve the registration performance without additional parameters and inference time compared to the corresponding baseline student.** Three representative classical baselines CNN-based (VoxelMorph [5]), Transformer-based (TransMorph [7]) and Mamba-based (MambaMorph [9]) are evaluated as student networks. Experiments conducted on two brain MR datasets and one liver CT dataset show consistent improvements in the proposed paradigm over the baselines.

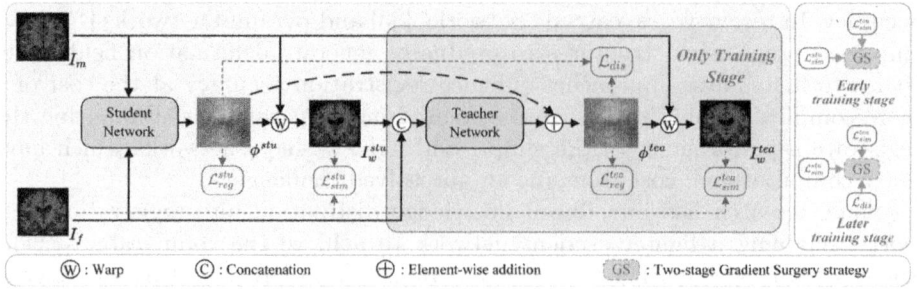

Fig. 1. The overall framework of our proposed **GSSD**. The teacher network is cloned then cascaded after the student network for joint training and distillation. GSSD incorporates a two-stage Gradient Surgery strategy for computation of the gradient of the loss function in the distillation process, which can reduce potential gradient conflicts. The teacher network is removed after training.

2 Methods

Figure 1 illustrates the framework of our proposed **GSSD**. Given a pair of moving and fixed images $\{I_m, I_f\}$, our objective is to estimate the deformation field ϕ^{stu} generated by the student network such that the warped image predicted by the student network $I_w^{stu} = I_m \circ \phi^{stu}$ can be aligned with I_f. To enhance the registration performance of the student network without increasing extra parameters or inference time, we design a simple and general self-distillation paradigm. To ensure the efficacy of self-distillation, we also incorporate the gradient surgery strategy in the distillation process to reduce potential gradient conflicts. Further details are provided in the following sections.

2.1 Self-distillation Registration Paradigm

Instead of pre-designing or selecting a suitable teacher network [12,28,34], we adopt the concept of "*self-distillation*" [31] by directly cloning the student network as the teacher, which is then cascaded after the student network for joint training and is removed during the validation and inference stages.

As depicted in Fig. 1, the teacher network takes the concatenation of I_w^{stu} and I_f as input to enhance alignment in regions where the student network has been underestimated. The predicted result is then element-wise added to ϕ^{stu}, outputting the final ϕ^{tea}. To effectively transfer helpful knowledge from the teacher to the student, we introduce the distillation loss $\mathcal{L}_{dis}$ to ensure similarity between ϕ^{stu} and ϕ^{tea}:

$$\mathcal{L}_{dis} = \sum_{\mathbf{p} \in \Omega} \left\| \phi^{stu}(\mathbf{p}) - \phi^{tea}(\mathbf{p}) \right\|_2^2. \tag{1}$$

In addition, we also introduce a similarity loss $\mathcal{L}_{sim}$ to measure the alignment between the warped and fixed images, as well as a regularization loss $\mathcal{L}_{reg}$ to

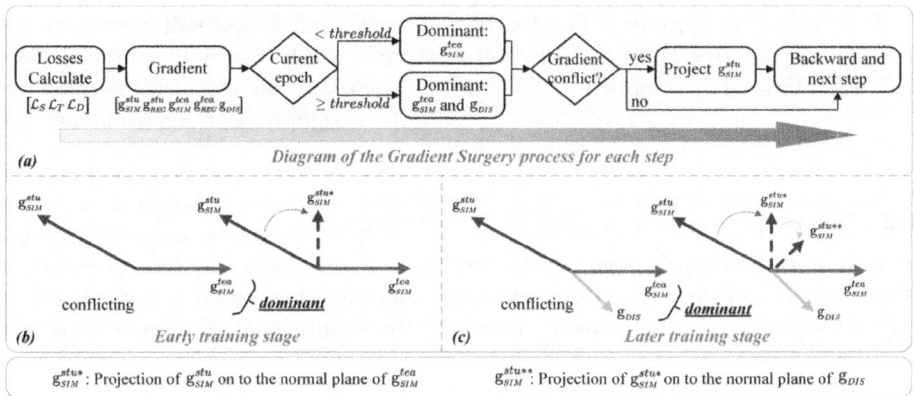

Fig. 2. (a) Diagram of the Gradient Surgery process for each training step. (b) Gradient surgery of **early training stage** where the dominant gradient is the gradient of similarity loss from teacher network $\mathbf{g}_{SIM}^{tea}$. (c) Gradient surgery of **later training stage** where the dominant gradient is both the gradient of teacher network's similarity loss $\mathbf{g}_{SIM}^{tea}$ and the weighted distillation loss $\mathbf{g}_{DIS}$.

ensure the smoothness of the deformation field. The total loss of the student network $\mathcal{L}_S$ and the teacher network $\mathcal{L}_T$ can be formulated as:

$$\mathcal{L}_S = \mathcal{L}_{sim}^{stu}(I_w^{stu}, I_f) + \lambda_r \mathcal{L}_{reg}^{stu}(\phi^{stu}), \tag{2}$$

$$\mathcal{L}_T = \mathcal{L}_{sim}^{tea}(I_w^{tea}, I_f) + \lambda_r \mathcal{L}_{reg}^{tea}(\phi^{tea}), \tag{3}$$

where λ_r is the weight of the regularization term.

In our approach, the Mean Squared Error (MSE) is chosen as the similarity loss, and the regularization loss is defined as the L2-norm of the gradient of the deformation field. The overall loss of GSSD can be written as:

$$\mathcal{L}_{TOTAL} = \mathcal{L}_S + \mathcal{L}_T + \mathcal{L}_D, \tag{4}$$

where $\mathcal{L}_D = \lambda_d \mathcal{L}_{dis}(\phi^{stu}, \phi^{tea})$ is the distillation loss $\mathcal{L}_{dis}$ weighted by the hyperparameter λ_d.

Diffeomorphic Self-distillation. To highlight our superior registration performance while maintaining topological preservation and invertibility of the deformation, we follow [5,19] and propose a diffeomorphic variant of GSSD, named **GSSD-diff**. Similar to Fig. 1, both the student and teacher networks of GSSD-diff produce velocity fields v, which are subsequently using the scaling and squaring rule [3,16] to form the deformation field ϕ. Therefore, the overall loss of GSSD-diff is:

$$\mathcal{L}_{TOTAL} = \mathcal{L}_S + \mathcal{L}_T + \mathcal{L}_D = \mathcal{L}_{sim}^{stu}(I_w^{stu}, I_f) + \lambda_r \mathcal{L}_{reg}^{stu}(v^{stu}) + \\ \mathcal{L}_{sim}^{tea}(I_w^{tea}, I_f) + \lambda_r \mathcal{L}_{reg}^{tea}(v^{tea}) + \lambda_d \mathcal{L}_{dis}(v^{stu}, v^{tea}). \tag{5}$$

It is worth mentioning that for both **GSSD** and **GSSD-diff**, the teacher network *is removed* after the model is trained. Only the lightweight student network, with high inference speed and competitive accuracy, is used during the validation and inference stage. Therefore, the deployment requirements and complexity of the model are reduced greatly.

2.2 Two-Stage Gradient Surgery Strategy

In DIR tasks, existing optimization methods [2,11,22,23,25] mostly focus on efficiency or automatic hyperparameter selection, but almost none consider the potential conflicts between gradients. To ensure the effectiveness of self-distillation, we introduce the gradient surgery (GS) optimization strategy to deconflict gradients inspired by [30] for their PCGrad applied in multi-task learning.

We first define that the gradient of the student network's similarity loss conflicts with the dominant gradient when the angle between them is the obtuse angle (their cosine similarity is negative). In such cases, we project the former onto the normal plane of the dominant gradient. When no conflicts are present, we update the student network using its original gradient directly, representing its gradient has no incompatible component along the direction of the dominant.

To select the dominant gradient, existing studies [17,30] generally apply GS by sequentially considering each loss as the dominant gradient throughout the training process. However, this approach is only suitable for multi-task learning where tasks are of equal importance. It does not apply to our self-distillation registration paradigm, as in our task, there is an inherent priority among different losses (student and teacher). In this paper, we propose a ***two-stage GS strategy*** that assigns specific dominant gradients at different training stages:

Early Training Stage: The dominant gradient is the gradient of the teacher network's similarity loss $\mathbf{g}_{SIM}^{tea}$ (Fig. 2(b)), guiding the student network to follow the teacher's optimization direction. In this stage, the gradient of the weighted distillation loss $\mathbf{g}_{DIS}$ cannot directly serve as the dominant since the teacher is not yet well-trained, leading to large errors in the deformation field it predicts.

Later Training Stage: As the teacher network's predictions become more accurate, we add $\mathbf{g}_{DIS}$ as a dominant gradient alongside $\mathbf{g}_{SIM}^{tea}$ (Fig. 2(c)).

A pictorial description of each training step is shown in Fig. 2(a): **(1)** We first calculate the total loss of the student network $\mathcal{L}_S$ and teacher network $\mathcal{L}_T$, and the weighted distillation loss $\mathcal{L}_D$. Their gradients are computed separately as $\mathbf{g}_{SIM}^{stu}, \mathbf{g}_{REG}^{stu}, \mathbf{g}_{SIM}^{tea}, \mathbf{g}_{REG}^{tea}, \mathbf{g}_{DIS}$. **(2)** If the current epoch number is smaller than the threshold epoch number, the dominant gradient is $\mathbf{g}_{SIM}^{tea}$. Otherwise, both $\mathbf{g}_{SIM}^{tea}$ and $\mathbf{g}_{DIS}$ are the dominant gradient. **(3)** It is determined whether $\mathbf{g}_{SIM}^{stu}$ conflicts with the dominant gradient by computing the cosine similarity between them. **(4)** If the cosine similarity is negative, we replace $\mathbf{g}_{SIM}^{stu}$ by its projection onto the normal plane of the dominant gradient. Otherwise, the original gradient $\mathbf{g}_{SIM}^{stu}$ remains unaltered. **(5)** Gradient backpropagation and move to the next step.

Our two-stage GS strategy mitigates the conflicting gradient problem, guaranteeing that the gradient of the student network interferes minimally with that of the teacher, further ensuring the effectiveness of self-distillation.

3 Experiments

Data and Pre-processing. We evaluate our method on two brain MRI datasets (OASIS[1] [18] and LPBA40[2] [24]) and one liver CT dataset (SLIVER[3] [10]) with standard pre-processing steps, including center cropping, resampling to a uniform size (160×192×224 for OASIS, 128×128×128 for LPBA40 and SLIVER) and intensity normalization to [0,1]. The voxel spacing of all datasets is 1×1×1mm. For OASIS, we randomly select 350 (350×349 pairs), 10 (10×9 pairs), and 11 (11×10 pairs) images for training, validation, and test sets, respectively. The segmentation map of 35 structures is used for evaluation. For LPBA40, we split it into 30 (30×29 pairs) and 10 (10×9 pairs) images for training and testing, respectively. The segmentation map of 56 structures is used for evaluation. For SLIVER, we follow [32] to choose both the MSD[4] [1] (933 scans) and BFH [33] (92 scans) datasets for training (1025×1024 pairs). We then split SLIVER into 5 (5×4 pairs) and 15 (15×14 pairs) scans for validation and testing.

Implementation Details. All methods are implemented in PyTorch [20], using an NVIDIA L40 GPU with 48 GB. We choose MSE loss as our similarity loss for all datasets. We employ Adam [14] optimizer with parameters β_1 as 0.9, β_2 as 0.99. We train our models with a fixed learning rate of 1e−04 and a batch size of 1. Based on the results of the validation set, the parameter λ_d is set as 0.001 for all datasets, while λ_r is set as 0.01 for OASIS and LPBA40, and 0.1 for SLIVER. We set the number of iteration steps as 100 in each epoch within the 1000-epoch training period. The threshold epoch is set as 500.

Baseline Methods. We apply and compare GSSD on three classical baselines: VoxelMorph (VM) [5], TransMorph (TM) [7] and MambaMorph (MM) [9], denoted as GSSD-VM, GSSD-TM and GSSD-MM. The diffeomorphic variant of all the baselines and our corresponding methods are also provided. We also compare two recent image registration distillation methods (LDR [28] and SDHNet [34]) to better demonstrate the advantages of our distillation paradigm. All baseline methods are implemented using their official implementation online with default parameters. Our corresponding methods keep all the hyperparameters as similar as possible with them for a fair comparison.

Evaluation Metrics. We evaluate the registration performance using: **(1)** the Dice score of the subcortical segmentation maps (DSC(%)) [5] and **(2)** the Average Symmetric Surface Distance (ASSD) [26] to evaluate the similarities between images, **(3)** the percentage of voxels with non-positive Jacobian determinant ($|J_\phi \leq 0|(\%)$) [19] to access the smoothness of the deformation field. Higher DSC, lower ASSD and $|J_\phi \leq 0|$ indicate a better registration performance.

---

[1] https://www.oasis-brains.org.
[2] https://resource.loni.usc.edu/resources/atlases-downloads.
[3] https://sliver07.grand-challenge.org.
[4] http://medicaldecathlon.com.

Table 1. Quantitative results over three baselines and our corresponding methods on SLIVER and LPBA40 datasets. Different colors correspond to the optimal performance under various methods: green for VM, blue for TM, and red for MM. Symbol ∗ marks results where GSSD significantly outperforms its baseline ($p < 0.05$, Wilcoxon signed-rank test). The results for the comparing distillation methods are also listed in the last two rows. Best viewed in color.

Methods	SLIVER (128×128×128)			LPBA40 (128×128×128)			Time	Params	FLOPs
	DSC(%)↑	ASSD (mm)↓	$\|J_\phi \leq 0\|$(%)↓	DSC(%)↑	ASSD (mm)↓	$\|J_\phi \leq 0\|$(%)↓	(ms)	(M)	(G)
VM	87.51±3.13	2.76±0.79	1.42±0.41	68.43±1.91	1.31±0.11	0.08±0.05	20	0.33	156
VM$_{diff}$	85.95±3.41	3.09±0.85	< 0.001	68.83±1.84	1.12±0.09	< 0.001	21	0.33	156
GSSD-VM	90.10$^*_{\pm 2.42}$	2.23$^*_{\pm 0.68}$	2.62±0.98	70.55$^*_{\pm 1.17}$	1.19$^*_{\pm 0.08}$	0.09±0.06	20	0.33	156
GSSD-VM$_{diff}$	88.90$^*_{\pm 3.13}$	2.43$^*_{\pm 0.78}$	< 0.001	71.03$^*_{\pm 1.14}$	1.05$^*_{\pm 0.07}$	< 0.001	21	0.33	156
TM	88.42±3.47	2.60±0.96	1.65±0.62	69.80±1.78	1.25±0.10	0.07±0.07	27	11.71	91
TM$_{diff}$	86.01±3.46	3.08±0.87	< 0.001	70.05±1.69	1.11±0.09	< 0.001	28	11.71	91
GSSD-TM	90.80$^*_{\pm 2.67}$	2.08$^*_{\pm 0.76}$	2.69±1.37	71.76$^*_{\pm 1.09}$	1.16$^*_{\pm 0.08}$	0.11±0.09	27	11.71	91
GSSD-TM$_{diff}$	89.57$^*_{\pm 3.15}$	2.28$^*_{\pm 0.82}$	< 0.001	71.93$^*_{\pm 1.04}$	1.04$^*_{\pm 0.07}$	< 0.001	28	11.71	91
MM	88.12±3.21	2.63±0.84	1.54±0.51	68.87±1.93	1.14±0.09	0.06±0.05	35	1.53	85
MM$_{diff}$	85.39±3.48	3.23±0.89	< 0.001	68.91±1.62	1.13±0.09	< 0.001	37	1.53	85
GSSD-MM	90.33$^*_{\pm 2.56}$	2.18$^*_{\pm 0.72}$	2.24±1.47	70.82$^*_{\pm 1.51}$	1.07$^*_{\pm 0.08}$	0.09±0.05	35	1.53	85
GSSD-MM$_{diff}$	88.53$^*_{\pm 3.16}$	2.51$^*_{\pm 0.81}$	< 0.001	70.89$^*_{\pm 1.36}$	1.07$^*_{\pm 0.07}$	< 0.001	37	1.53	85
LDR	87.68±3.47	4.22±1.40	17.94±0.63	68.27±1.72	1.28±0.09	0.91±0.91	4	0.73	11
SDHNet	91.57±2.81	1.96±0.70	0.41±0.27	71.64±0.94	1.12±0.07	0.03±0.01	525	18.29	57

4 Results

Tables 1 and 2 give a comprehensive summary of the results. Our method consistently improves performance over the original model on both the brain and liver datasets. Specifically, GSSD-TM achieves a Dice score of 78.64% on the OASIS dataset and 71.76% on LPBA40, surpassing TM by margins of 1.57% and 1.96%, respectively. Similar improvements are also observed in GSSD-VM and GSSD-MM, with Dice score outperforming the baseline by 1.43% and 1.40% on the OASIS dataset, 2.12% and 1.95% on the LPBA40 dataset. For liver CT, our GSSD can also outperform the corresponding baselines by 2.59%, 2.38%, and 2.21% on the Dice score. We highlight that this consistent improvement did not introduce any additional inference time, parameters or FLOPs, achieving a balance between registration accuracy and computational complexity.

Diffeomorphic. To explore the advantages of GSSD in cases where no folding in the deformation field (i.e. $|J_\phi \leq 0|$ (%) approximately equals to 0), we also present the results of diffeomorphic variants in Table 1 and 2. Under the condition of preserving the topology, GSSD-VM$_{diff}$ achieves a 2.20% higher Dice score and a 0.07mm lower ASSD on the LPBA40 dataset compared to VM$_{diff}$, while GSSD-TM$_{diff}$ and GSSD-MM$_{diff}$ improve Dice score by 1.88% and 1.98%, respectively. Similar improvements are also observed in the OASIS and SLIVER datasets. These results demonstrate the effectiveness of GSSD, confirming that the improvements in registration accuracy are not accompanied by a compromise in the smoothness of the deformation field.

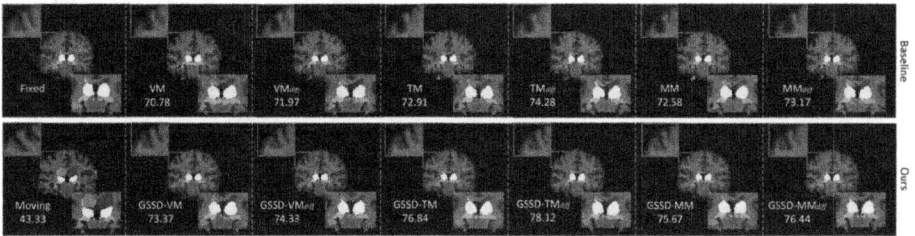

Fig. 3. Visualization of the warped image produced by baselines and our corresponding methods, with dice scores shown in the bottom left corner.

Table 2. Quantitative results on the OASIS dataset. Best viewed in color.

Methods	OASIS (160×192×224)		
	DSC(%)↑	ASSD (mm)↓	$\|J_\phi\| \leq 0\|(\%)\downarrow$
VM	75.89±4.67	0.60±0.24	0.09±0.08
VM$_{diff}$	76.05±4.48	0.59±0.22	< 0.001
GSSD-VM	77.32*±4.20	0.55*±0.21	0.22±0.10
GSSD-VM$_{diff}$	77.52*±3.74	0.54*±0.19	< 0.001
TM	77.07±4.00	0.55±0.19	0.09±0.06
TM$_{diff}$	77.32±3.75	0.54±0.18	< 0.001
GSSD-TM	78.64*±3.38	0.50*±0.16	0.19±0.08
GSSD-TM$_{diff}$	78.77*±3.09	0.49*±0.14	< 0.001
MM	77.07±4.03	0.55±0.19	0.10±0.06
MM$_{diff}$	77.21±3.78	0.54±0.18	< 0.001
GSSD-MM	78.47*±3.52	0.51*±0.17	0.22±0.09
GSSD-MM$_{diff}$	78.57*±3.44	0.49*±0.17	< 0.001
LDR	75.61±4.32	0.58±0.20	3.70±0.63
SDHNet	78.62±2.93	0.50±0.14	0.71±0.13

Visualization. Fig. 3 illustrates the example of image slices with large initial misalignment in the OASIS dataset. It is noteworthy that our method not only outperforms the baselines in the cortical areas (enlarged in the top left corner) but also demonstrates sustained improvement in certain specific anatomical structures (highlighted by the red arrows). These regions are often challenging to register due to low imaging quality and a small number of voxels. Our approach takes a step further beyond the baseline.

Performance on Anatomical Structures. To verify the robust registration performance, we additionally provide boxplots showing the Dice score on specific anatomical structures in the OASIS dataset in Fig. 4. The boxplots show that our method performs better on both large structures such as white matter or cortex and small structures such as the ventricle or vessel.

4.1 Ablation Studies

As GSSD is the first to incorporate gradient surgery into self-distillation deformable registration networks, we conduct comprehensive ablation studies on the LPBA40 dataset to verify the effectiveness of both the GS strategy and the Self-Distillation paradigm.

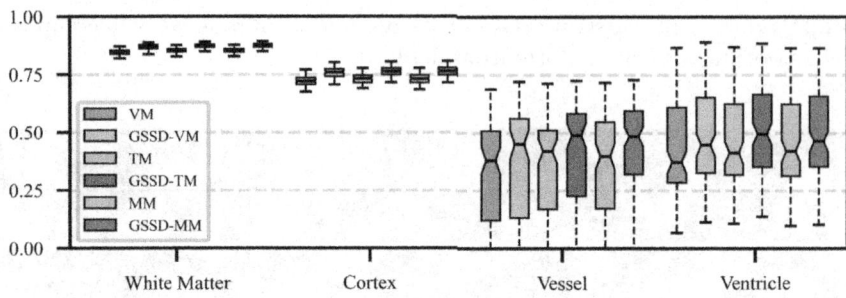

Fig. 4. The comparable Dice boxplot of different methods on the top and bottom two anatomical structures (ranked by the average numbers of the region voxels).

Gradient Surgery in Self-Distillation. We design and test four variants to explore their impact on registration performance. Results are shown in Table 3.

- **Origin**: The initial baseline network configuration, without the teacher network and distillation paradigm.
- **noConflict+noGS**: Incorporating the teacher network and distillation paradigm, but gradient conflicts are eliminated by removing the student network's total loss $\mathcal{L}_S$. GS is also removed as it is no longer necessary.
- **Conflict+noGS**: We introduce gradient conflicts by maintaining $\mathcal{L}_S$, without the application of GS.
- **Conflict+GS**: GS is applied to reduce the conflicts.

For **noConflict+noGS** with distillation paradigm, VM, TM, and MM show a slight decline in performance in Dice scores and ASSD, but a better smoothness compared to **Origin**. This is because both $\mathcal{L}_{sim}^{stu}$ and $\mathcal{L}_{reg}^{stu}$ are removed, with the original weight of $\mathcal{L}_{sim}^{stu}$ being much higher than that of $\mathcal{L}_{reg}^{stu}$, and the network tends to ensure smoothness. Thus, in the absence of the student network's total loss $\mathcal{L}_S$, **noConflict+noGS** avoids gradient conflicts but at the cost of registration accuracy. To enhance registration accuracy, addressing the issue of gradient conflicts is thus inevitable.

When removing GS when the gradient conflicts exist (**Conflict+noGS**), VM, TM, and MM exhibit a marginal increase in Dice scores compared to **Origin**, albeit with a slight decrease compared to **Conflict+GS**. This indicates that self-distillation has the potential to enhance the performance of the student network but always results in sub-optimal accuracy. The possible reason is that the network fails to resolve potential gradient conflicts without gradient surgery.

Our **Conflict+GS** achieves the highest registration accuracy, which strongly validates the efficiency of our GS strategy. We notice that $|J_\phi \leq 0|\,(\%)$ slightly increased after adding GS. This might be due to the dominance of the similarity loss over the regularization loss. As a result, the update of $\mathcal{L}_{reg}$ is affected by $\mathcal{L}_{sim}$, leading to a slight decrease in the smoothness of the deformation field.

To better demonstrate the effectiveness of the proposed two-stage GS strategy, we calculate the percentage of gradient conflicts within each epoch

Table 3. Ablations on removing the gradient conflicts and gradient surgery.

Network	Ablations				Metrics		
	Stu	Tea	Conflict	GS	DSC(%)↑	ASSD (mm)↓	$\|J_\phi \leq 0\|(\%)\downarrow$
GSSD -VM	✓	✗	✗	✗	$68.43_{\pm1.91}$	$1.31_{\pm0.11}$	$0.08_{\pm0.05}$
	✓	✓	✗	✗	$68.15_{\pm1.92}$	$1.38_{\pm0.08}$	$0.01_{\pm0.01}$
	✓	✓	✓	✗	$69.02_{\pm1.41}$	$1.25_{\pm0.09}$	$0.09_{\pm0.04}$
	✓	✓	✓	✓	$70.55_{\pm1.17}$	$1.19_{\pm0.08}$	$0.09_{\pm0.06}$
GSSD -TM	✓	✗	✗	✗	$69.80_{\pm1.78}$	$1.25_{\pm0.10}$	$0.07_{\pm0.07}$
	✓	✓	✗	✗	$68.50_{\pm2.03}$	$1.26_{\pm0.08}$	$0.01_{\pm0.01}$
	✓	✓	✓	✗	$70.15_{\pm1.52}$	$1.21_{\pm0.09}$	$0.08_{\pm0.11}$
	✓	✓	✓	✓	$71.76_{\pm1.09}$	$1.16_{\pm0.08}$	$0.11_{\pm0.09}$
GSSD -MM	✓	✗	✗	✗	$68.87_{\pm1.93}$	$1.14_{\pm0.09}$	$0.06_{\pm0.05}$
	✓	✓	✗	✗	$67.81_{\pm2.03}$	$1.27_{\pm0.08}$	$0.02_{\pm0.03}$
	✓	✓	✓	✗	$68.94_{\pm1.76}$	$1.13_{\pm0.09}$	$0.07_{\pm0.08}$
	✓	✓	✓	✓	$70.82_{\pm1.51}$	$1.07_{\pm0.08}$	$0.09_{\pm0.05}$

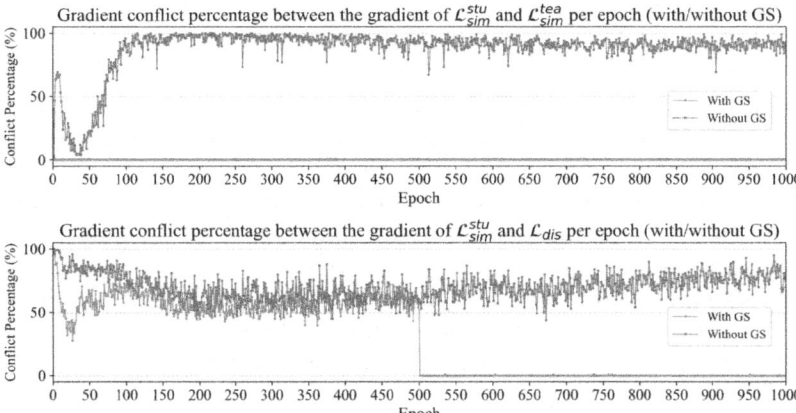

Fig. 5. Gradient conflict percentage per epoch: between the gradient of the student's similarity loss and the teacher's similarity loss (top), and between the gradient of the student's similarity loss and the distillation loss (bottom).

during training (using VM as the baseline on the LPBA40 dataset). The results are shown in Fig. 5. It can be observed that the introduction of GS eliminates gradient conflicts between the student's similarity loss $\mathcal{L}_{sim}^{stu}$ and the teacher's similarity loss $\mathcal{L}_{sim}^{tea}$, resulting in a conflict ratio of 0. In contrast, without GS, nearly 90% of epochs exhibit a high gradient conflict ratio. In addition, during the early training stage (epoch<=500), gradient conflicts between $\mathcal{L}_{sim}^{stu}$ and $\mathcal{L}_{dis}$

Table 4. Ablation on the dominant gradient and the stages of gradient surgery.

Network	Dominant	DSC(%)	ASSD (mm)	$\|J_\phi \leq 0\|$ (%)
GSSD -VM	Baseline (no GS)	$68.43_{\pm 1.91}$	$1.31_{\pm 0.11}$	$0.08_{\pm 0.05}$
	Only $\mathbf{g}_{DIS}$	$67.61_{\pm 2.15}$	$1.34_{\pm 0.11}$	$0.29_{\pm 0.21}$
	Only $\mathbf{g}_{SIM}^{tea}$	$70.15_{\pm 1.52}$	$1.26_{\pm 0.09}$	$0.11_{\pm 0.06}$
	$\mathbf{g}_{SIM}^{tea}$ then $\mathbf{g}_{DIS}$	$69.97_{\pm 1.53}$	$1.27_{\pm 0.10}$	$0.11_{\pm 0.08}$
	$\mathbf{g}_{SIM}^{tea}$ then ($\mathbf{g}_{SIM}^{tea}$, $\mathbf{g}_{DIS}$)	$70.55_{\pm 1.17}$	$1.19_{\pm 0.08}$	$0.09_{\pm 0.06}$
GSSD -TM	Baseline (no GS)	$69.80_{\pm 1.78}$	$1.25_{\pm 0.10}$	$0.07_{\pm 0.07}$
	Only $\mathbf{g}_{DIS}$	$69.49_{\pm 1.88}$	$1.27_{\pm 0.11}$	$0.30_{\pm 0.23}$
	Only $\mathbf{g}_{SIM}^{tea}$	$71.47_{\pm 1.26}$	$1.22_{\pm 0.08}$	$0.12_{\pm 0.10}$
	$\mathbf{g}_{SIM}^{tea}$ then $\mathbf{g}_{DIS}$	$71.35_{\pm 1.33}$	$1.24_{\pm 0.08}$	$0.13_{\pm 0.10}$
	$\mathbf{g}_{SIM}^{tea}$ then ($\mathbf{g}_{SIM}^{tea}$, $\mathbf{g}_{DIS}$)	$71.76_{\pm 1.09}$	$1.16_{\pm 0.08}$	$0.11_{\pm 0.09}$
GSSD -MM	Baseline (no GS)	$68.87_{\pm 1.93}$	$1.14_{\pm 0.09}$	$0.06_{\pm 0.05}$
	Only $\mathbf{g}_{DIS}$	$68.60_{\pm 1.99}$	$1.19_{\pm 0.10}$	$0.28_{\pm 0.21}$
	Only $\mathbf{g}_{SIM}^{tea}$	$70.43_{\pm 1.45}$	$1.15_{\pm 0.08}$	$0.12_{\pm 0.11}$
	$\mathbf{g}_{SIM}^{tea}$ then $\mathbf{g}_{DIS}$	$70.00_{\pm 1.28}$	$1.16_{\pm 0.08}$	$0.15_{\pm 0.20}$
	$\mathbf{g}_{SIM}^{tea}$ then ($\mathbf{g}_{SIM}^{tea}$, $\mathbf{g}_{DIS}$)	$70.82_{\pm 1.51}$	$1.07_{\pm 0.08}$	$0.09_{\pm 0.05}$

exist regardless of whether GS is introduced (as $\mathcal{L}_{dis}$ is not yet dominant in the GS). However, the conflict ratio is slightly lower with GS than without GS. This is due to the fact that, during the early training stage, GS operates on $\mathcal{L}_{sim}^{stu}$ and $\mathcal{L}_{sim}^{tea}$, forcing the student to align more closely with the teacher. This alignment reduces the difficulty for the teacher to distill knowledge to the student, thereby leading to a slightly lower conflict ratio between $\mathcal{L}_{sim}^{stu}$ and $\mathcal{L}_{dis}$ compared to the case without GS. During the later training stage (epoch>500), as the distillation loss $\mathcal{L}_{dis}$ becomes dominant, the conflict ratio between $\mathcal{L}_{sim}^{stu}$ and $\mathcal{L}_{dis}$ drops to 0 with GS. In contrast, without GS, conflicts persist during this stage.

Dominance and Stages in Gradient Surgery. In Table 4, we explore specific strategies for gradient surgery, focusing on whether to divide different stages and how to select the dominant gradient.

We first assess the single-stage GS strategy (**only $\mathbf{g}_{DIS}$** and **only $\mathbf{g}_{SIM}^{tea}$**), which respectively represent using $\mathbf{g}_{DIS}$ or $\mathbf{g}_{SIM}^{tea}$ as the dominant gradient throughout the training process. For the **only $\mathbf{g}_{DIS}$** strategy, both $\mathbf{g}_{SIM}^{tea}$ and $\mathbf{g}_{SIM}^{stu}$ are projected to $\mathbf{g}_{DIS}$, which forces the network learning to minimize the difference between the teacher and student rather than enhance the student's registration performance. As a result, both the Dice score and ASSD are even worse than the original baseline. Furthermore, the teacher network is not yet well-trained during the early training iterations, leading to large errors in ϕ^{tea}. Directly distilling from such a deformation field can 'mislead' the student,

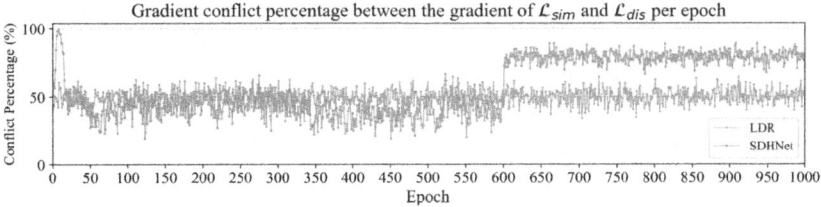

Fig. 6. Gradient conflict percentage in two comparison distillation-based methods.

resulting in poor registration performance. For the **only g_{SIM}^{tea}** strategy, the student is forced to be aligned with the teacher, leading to a superior result over the original baseline. However, since g_{DIS} is not introduced as the dominant gradient, such accuracy is not optimal. Further improvement can still be explored in the distillation efficiency.

For the two-stage GS strategy, we considered when g_{SIM}^{tea} dominates in the early training stage, followed by whether g_{DIS} or a combination of g_{SIM}^{tea} and g_{DIS} dominance in the later training stage. For the g_{SIM}^{tea} **then** g_{DIS} strategy, only g_{DIS} is dominant in the later training stage aiming at reducing the difference between the student and teacher. In such cases, it is unclear whether the teacher network improves the student's performance, or vice versa, since the teacher is no longer dominant in the later stage. As a result, the student's final registration performance of g_{SIM}^{tea} **then** g_{DIS} strategy is inferior to the **only g_{SIM}^{tea}** strategy.

We ultimately choose the g_{SIM}^{tea} **then** (g_{SIM}^{tea}, g_{DIS}) strategy, where the teacher is dominant throughout the training stage. In the later stage, g_{DIS} is introduced as a dominant gradient on the condition that the teacher is well-trained. Our approach ensures that the teacher network enhances the student's performance with optimized distillation efficiency.

Comparison with Other Distillation Methods. We also compare GSSD with two distillation methods (LDR [28] and SDHNet [34]) in Tables 1 and 2. LDR employs adversarial learning for distillation, demonstrating lightweight parameters and minimal computational cost, but limited registration accuracy. This may result from its offline distillation (pre-training a teacher network) and indirect distillation using the discriminator's output features instead of the deformation field. Consequently, this approach (1) constrains the student's accuracy to the specific chosen teacher network (upper bound); (2) lacks distillation efficiency. In contrast, SDHNet shows impressive performance using the final deformation field as the teacher to distill intermediate deformation fields during 6 iterations (reported in [34]). Such distillation strategy leads to (1) complex training and longer inference time (525ms); (2) a relatively large parameters (18.29M). Compared to these methods, our GSSD is simpler and more universally applicable to all end-to-end registration networks, and the final accuracy depends on the student's original performance. Notably, even the U-shape GSSD-TM without any iteration can achieve competitive accuracy compared to SDHNet while maintaining a reduced inference time. Furthermore, Fig. 6 shows the gradient conflict percentages between $\mathcal{L}_{sim}$ and $\mathcal{L}_{dis}$ for LDR and SDHNet during training, where most epochs exhibit high conflict ratios. This indicates that such conflicts are commonly present in distillation-based registration methods, and our GS strategy is a promising way to deconflict gradients and improve distillation efficiency.

5 Conclusions

This paper proposes a novel self-distillation deformable image registration paradigm with gradient surgery, named **GSSD**. GSSD jointly trains the student and teacher network (clone of the student), and removes the teacher after training. To ensure the distillation efficiency, we introduce a two-stage gradient surgery strategy to deconflict gradients during training. Extensive experiments demonstrate stable improvements for various student networks without increasing extra parameters or inference time. To conclude, we have successfully introduced gradient surgery into self-distillation DIR tasks, and our GSSD demonstrates promising results on various datasets. The more streamlined distillation methods and more effective gradient optimization strategies are left to be explored in the future, especially for multimodal and low-quality images.

Acknowledgments. This work is supported by the National Natural Science Foundation of China under Grant 82102002.

References

1. Antonelli, M., et al.: The medical segmentation decathlon. Nat. Commun. **13**(1), 4128 (2022)
2. Arora, P., et al.: An adaptive medical image registration using hybridization of teaching learning-based optimization with affine and speeded up robust features with projective transformation. Clust. Comput. **27**(1), 607–627 (2024)
3. Arsigny, V., Commowick, O., Pennec, X., Ayache, N.: A log-Euclidean framework for statistics on diffeomorphisms. In: Larsen, R., Nielsen, M., Sporring, J. (eds.) MICCAI 2006. LNCS, vol. 4190, pp. 924–931. Springer, Heidelberg (2006). https://doi.org/10.1007/11866565_113
4. Ashburner, J.: A fast diffeomorphic image registration algorithm. Neuroimage **38**(1), 95–113 (2007)
5. Balakrishnan, G., et al.: Voxelmorph: a learning framework for deformable medical image registration. IEEE Trans. Med. Imaging **38**(8), 1788–1800 (2019)
6. Beg, M.F., et al.: Computing large deformation metric mappings via geodesic flows of diffeomorphisms. Int. J. Comput. Vision **61**, 139–157 (2005)
7. Chen, J., et al.: Transmorph: transformer for unsupervised medical image registration. Med. Image Anal. **82**, 102615 (2022)
8. De Vos, B.D., et al.: A deep learning framework for unsupervised affine and deformable image registration. Med. Image Anal. **52**, 128–143 (2019)

9. Guo, T., et al.: Mambamorph: a mamba-based backbone with contrastive feature learning for deformable MR-CT registration. arXiv Preprint arxiv:2401.13934 (2024)
10. Heimann, T., et al.: Comparison and evaluation of methods for liver segmentation from CT datasets. IEEE Trans. Med. Imaging **28**(8), 1251–1265 (2009)
11. Hoopes, A., et al.: Hypermorph: amortized hyperparameter learning for image registration. In: International Conference on Information Processing in Medical Imaging, pp. 3–17 (2021)
12. Hu, B., et al.: Cross-resolution distillation for efficient 3D medical image registration. IEEE Trans. Circuits Syst. Video Technol. **32**(10), 7269–7283 (2022)
13. Kang, M., et al.: Dual-stream pyramid registration network. Med. Image Anal. **78**, 102379 (2022)
14. Kingma, D.P., Ba, J.: Adam: a method for stochastic optimization. arXiv Preprint arxiv:1412.6980 (2014)
15. Klein, S., et al.: Elastix: a toolbox for intensity-based medical image registration. IEEE Trans. Med. Imaging **29**(1), 196–205 (2009)
16. Krebs, J., et al.: Unsupervised probabilistic deformation modeling for robust diffeomorphic registration. In: Deep Learning in Medical Image Analysis and Multimodal Learning for Clinical Decision Support: 4th International Workshop, DLMIA 2018, and 8th International Workshop, ML-CDS 2018, Held in Conjunction with MICCAI 2018, Granada, Spain, 20 September 2018, Proceedings 4, pp. 101–109. Springer, Cham (2018)
17. Mansilla, L., et al.: Domain generalization via gradient surgery. In: Proceedings of the IEEE/CVF International Conference on Computer Vision, pp. 6630–6638 (2021)
18. Marcus, D.S., et al.: Open access series of imaging studies (OASIS): cross-sectional MRI data in young, middle aged, nondemented, and demented older adults. J. Cogn. Neurosci. **19**(9), 1498–1507 (2007)
19. Mok, T.C., Chung, A.C.: Large deformation diffeomorphic image registration with Laplacian pyramid networks. In: International Conference on Medical Image Computing and Computer-Assisted Intervention, pp. 211–221 (2020)
20. Paszke, A., et al.: Automatic differentiation in pytorch (2017)
21. Qiao, Y., Shi, Y.: Unsupervised deep learning for FOD-based susceptibility distortion correction in diffusion MRI. IEEE Trans. Med. Imaging **41**(5), 1165–1175 (2021)
22. Qiao, Y., et al.: Fast automatic step size estimation for gradient descent optimization of image registration. IEEE Trans. Med. Imaging **35**(2), 391–403 (2015)
23. Qiao, Y., et al.: An efficient preconditioner for stochastic gradient descent optimization of image registration. IEEE Trans. Med. Imaging **38**(10), 2314–2325 (2019)
24. Shattuck, D.W., et al.: Construction of a 3D probabilistic atlas of human cortical structures. Neuroimage **39**(3), 1064–1080 (2008)
25. Sun, S., et al.: Medical image registration via neural fields. Med. Image Anal. **97**, 103249 (2024)
26. Taha, A.A., Hanbury, A.: Metrics for evaluating 3D medical image segmentation: analysis, selection, and tool. BMC Med. Imaging **15**(1), 1–28 (2015)
27. Thirion, J.P.: Image matching as a diffusion process: an analogy with Maxwell's demons. Med. Image Anal. **2**(3), 243–260 (1998)
28. Tran, M.Q., et al.: Light-weight deformable registration using adversarial learning with distilling knowledge. IEEE Trans. Med. Imaging **41**(6), 1443–1453 (2022)
29. Wang, H., et al.: Recursive deformable pyramid network for unsupervised medical image registration. IEEE Trans. Med. Imaging (2024)

30. Yu, T., et al.: Gradient surgery for multi-task learning. In: Advances in Neural Information Processing Systems, vol. 33, pp. 5824–5836 (2020)
31. Zhang, L., Song, J., Gao, A., Chen, J., Bao, C., Ma, K.: Be your own teacher: improve the performance of convolutional neural networks via self distillation. In: Proceedings of the IEEE/CVF International Conference on Computer Vision, pp. 3713–3722 (2019)
32. Zhao, S., et al.: Recursive cascaded networks for unsupervised medical image registration. In: Proceedings of the IEEE/CVF International Conference on Computer Vision, pp. 10600–10610 (2019)
33. Zhao, S., et al.: Unsupervised 3D end-to-end medical image registration with volume tweening network. IEEE J. Biomed. Health Inform. **24**(5), 1394–1404 (2019)
34. Zhou, S., et al.: Self-distilled hierarchical network for unsupervised deformable image registration. IEEE Trans. Med. Imaging (2023)

Medical Image Registration Meets Vision Foundation Model: Prototype Learning and Contour Awareness

Hao Xu[1], Tengfei Xue[1], Jianan Fan[1], Dongnan Liu[1], Yuqian Chen[2,4], Fan Zhang[3], Carl-Fredrik Westin[2,4], Ron Kikinis[2,4], Lauren J. O'Donnell[2,4], and Weidong Cai[1(✉)]

[1] The University of Sydney, Sydney, Australia
tom.cai@sydney.edu.au
[2] Harvard Medical School, Boston, USA
[3] University of Electronic Science and Technology of China, Chengdu, China
[4] Brigham and Women's Hospital, Boston, USA

Abstract. Medical image registration is a fundamental task in medical image analysis, aiming to establish spatial correspondences between paired images. However, existing unsupervised deformable registration methods rely solely on intensity-based similarity metrics, lacking explicit anatomical knowledge, which limits their accuracy and robustness. Vision foundation models, such as the Segment Anything Model (SAM), can generate high-quality segmentation masks that provide explicit anatomical structure knowledge, addressing the limitations of traditional methods that depend only on intensity similarity. Based on this, we propose a novel SAM-assisted registration framework incorporating prototype learning and contour awareness. The framework includes: (1) Explicit anatomical information injection, where SAM-generated segmentation masks are used as auxiliary inputs throughout training and testing to ensure the consistency of anatomical information; (2) Prototype learning, which leverages segmentation masks to extract prototype features and aligns prototypes to optimize semantic correspondences between images; and (3) Contour-aware loss, a contour-aware loss is designed that leverages the edges of segmentation masks to improve the model's performance in fine-grained deformation fields. Extensive experiments demonstrate that the proposed framework significantly outperforms existing methods across multiple datasets, particularly in challenging scenarios with complex anatomical structures and ambiguous boundaries. Our code is available at https://github.com/HaoXu0507/IPMI25-SAM-Assisted-Registration.

Keywords: Medical Image Registration · Vision Foundation Model · Prototype Learning · Contour Awareness

1 Introduction

Medical image registration refers to establishing spatial correspondence between fixed and moving images, ensuring that their anatomical structures align with each other [1,4,5,21,30]. Existing unsupervised deformable registration methods [1,4,5,21,24,26,30] primarily rely on intensity-based similarity metrics, which lack explicit anatomical knowledge and often result in suboptimal performance, especially in complex scenarios. These limitations underline the need for incorporating additional structural information to enhance registration accuracy and robustness.

However, due to the difficulty in obtaining segmentation masks of medical images and their inconsistency with clinical scenarios, registration methods with auxiliary segmentation masks have not gained significant attention.

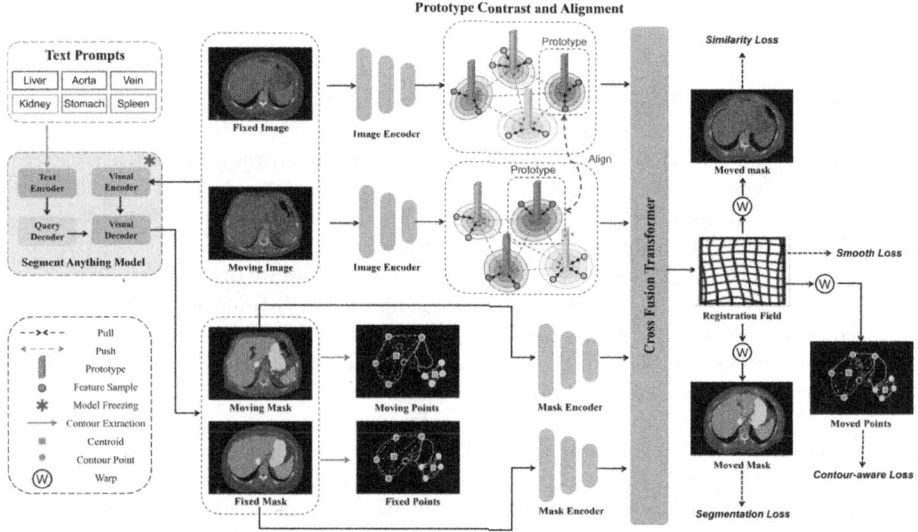

Fig. 1. The overview of the proposed SAM-assisted medical image registration framework includes the use of the Segment Anything Model (SAM) to generate segmentation masks, optimization of the feature space through prototype contrast and alignment, enhancement of contour alignment via contour-aware loss, and a cross fusion transformer for image and mask feature fusion registration.

Segment Anything Model (SAM) and its variants [6,10,12,19,20,27,33] gain attention for their powerful segmentation capabilities in both natural and medical images. Current medical-based SAM [6,10,19,27,33] can segment whole-body anatomical structures (e.g., different brain regions, heart partitions, abdominal organs) according to the prompts. The common prompts include point, box, mask, and text. Among these, text prompts yield segmentation results comparable in accuracy to the other three types and are more stable. Image registration could gain benefits from SAM assistance due to SAM's ability to generate high-quality segmentation masks [13]. These masks provide explicit anatomical structure knowledge, which can complement traditional intensity-based methods. However, current mask-assisted registration methods have two flaws: i) Since the segmentation mask is inaccessible during the test phase, the segmentation mask is only used to calculate the segmentation loss during the training phase and cannot be integrated into the model, resulting in insufficient utilization of valuable semantic information; ii) Current methods only use segmentation loss, making it difficult to accurately align the contours of organs, especially those with complex shapes.

To address these issues, we propose a novel SAM-assisted registration framework that leverages segmentation masks to explicitly guide the registration process through prototype learning and contour awareness, as shown in Fig. 1. The proposed framework incorporates three key innovations: i) Explicit anatomical information injection: the segmentation mask generated by SAM is directly used as auxiliary input during the training and testing phases to provide explicit anatomical guidance for the model and ensure the consistency and alignment of anatomical information during the registration process. ii) Prototype contrast and alignment: the segmentation mask is used to extract the prototype features of each anatomical region, and the features of the same anatomical region are gathered to their corresponding prototypes through contrast learning, thereby improving the consistency of features within the region. At the same time, the prototypes in the image pairs are aligned to ensure that the same anatomical structure across images maintains semantic consistency, and optimizes the registration effect from a global and local level. iii) Contour-awareness: we design a contour-aware loss function based on the mask contour, focusing on the precise alignment of complex anatomical shapes. This loss function significantly improves the registration accuracy of the model in boundary-sensitive areas. Extensive experiments show that our model exceeds state-of-the-art methods on the Abdomen CT dataset and ACDC Cardiac MRI dataset.

The main contributions of this work are as follows: i) We propose a novel SAM-assisted deformable medical image registration framework that utilizes anatomical masks explicitly. ii) We introduce prototype learning to optimize the semantic consistency of anatomical structures at local and global levels, which enhances the feature disentanglement capability of the model for different anatomical structures. iii) We design a contour-aware loss to better improve the registration accuracy of the model in the boundary area. iv) We evaluate the effectiveness of our method on the Abdomen CT dataset and the ACDC Cardiac MRI dataset.

2 Related Work

2.1 Segment Anything Model

SAM and its variants [10,14–16,20,27,31,33] receive widespread attention from the community due to their superior segmentation performance and good generalization. SAM [12] consists of an image encoder, a prompt decoder, and a lightweight mask decoder. SAM segments the target according to the given prompts, and the prompt's quality strongly affects SAM's segmentation performance. A good prompt can achieve good results and vice versa. Common prompt methods include point, box, mask, and text. Compared with the first three prompts, the text prompt has the most stable segmentation performance while maintaining the segmentation performance. MedSAM [19] is the first medical-image-based SAM, which freezes the prompt encoder and fine-tunes the image encoder and mask decoder of the original SAM. Medical SAM Adapter (Med-SA) [19] fine-tunes the SAM for medical image segmentation with a learnable adapter layer. SAM-Med2D [6] freezes the image encoder of the original SAM and introduces a learnable adapter layer to fine-tune the prompt encoder and update the mask decoder during training. SAM-Med3D [27] extends SAM-Med2D to a 3D network, further improving its performance. SAT [33] adopts contrastive learning to align medical text and images and uses text prompts to guide the model to segment the corresponding targets. It also establishes a multi-organ and multi-modality dataset for medical image segmentation and achieves good results. However, the provision of text prompts requires some medical background knowledge, which hinders its scope of use.

2.2 Deformable Medical Image Registration

Unsupervised deformable registration is to find the spatial correspondence between a pair of images based on their similarity measure. VoxelMorph [1] is one of the most widely used benchmark registration methods, which is a CNN-based U-Net architecture. Spatial transform is introduced to register the moving image to the moved image according to the field output by the network. TransMorph [4] adopts the paradigm of voxel morph, and the network consists of a vision transformer encoder and a CNN decoder. However, TransMorph simply uses the vision transformer as an encoder, ignoring that the attention mechanism can be approximated as a similarity measure of image patches, which coincides with the image registration. Therefore, TransMatch [5] adopts the paradigm of voxel morph, proposes to model the spatial correspondence of images using model attention mechanism, and achieves promising results. The above three methods are single-step registration methods, and it is difficult to achieve good results when the image pairs have large gaps. The multi-step image registration methods split single-step registration into an iterative process. CorrMLP [21] is a multi-step registration method, which includes a feature pyramid encoder and a coarse-to-fine MLP registration decoder. By using MLP to model the fine-grained long-distance dependencies of images, it achieves state-of-the-art performance. However, the unsupervised methods model the spatial correspondence of image pairs based only on image-level metrics, and the performance of the methods is limited due to the lack of medical anatomical information.

Deformable registration with segmentation masks introduces anatomical information, not only relying on the similarity metric of a pair of images but also using the overlap rate of the registered moving mask and the fixed mask as the loss function for model training. The current common paradigm comes from the auxiliary registration module of VoxelMorph. However, this paradigm is limited by the inaccessibility of the segmentation mask during inference, which cannot introduce anatomical information during inference, resulting in limited improvement.

2.3 Prototype Learning

Prototype learning [8,32,34] refers to enhancing the model's ability to represent category semantics by building category prototypes in feature space. In semantic segmentation, prototype-based methods usually leverage masks to aggregate features in feature space, pulling features of the same category closer and pushing features of different categories farther apart to learn the distinguishable category representation space. Zhou et al. [34] extract non-learnable pixel prototypes (i.e. cluster centers) from pixels of the same class, and improve the segmentation accuracy by bringing the same class pixels and prototypes closer and pushing different pixels and prototypes farther away. Zhang et at. [32] presents prototypical information bottlenecking and disentangling for multimodal survival analysis using pathology and genomics, which extends previous progress thus far on co-attention-based early-based fusion and learning information bottleneck. These methods show the great potential of prototypical learning for optimizing category feature space.

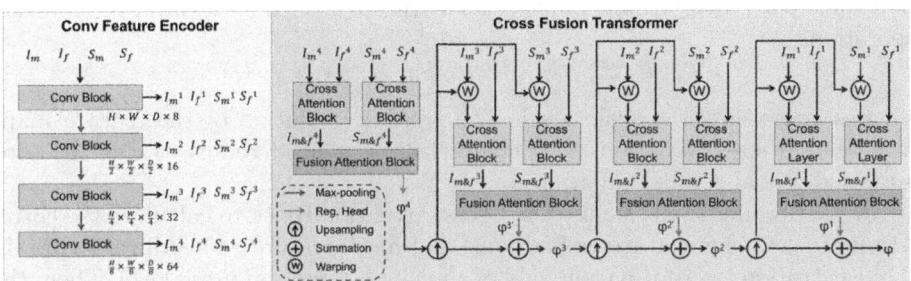

Fig. 2. The overall architecture of our Encoder and Cross Fusion Decoder.

3 Method

The goal of medical image registration is to find a spatial transformation that aligns a moving image I_m with a fixed image I_f in terms of anatomical structures. Formally, given a pair of images I_m and I_f, the registration task aims to learn a deformation field ϕ that transform the moving image I_m to the fixed image I_f:

$$I'_m = I_m \circ \varphi$$

where I'_m is the moved image, φ is the spatial transformation function, and ○ denotes the warp operation. As illustrated in Fig. 1, our SAM-assisted framework aims to improve deformable medical image registration by integrating image-level and mask-level features via pairs of images and their corresponding SAM masks. We are going to sequentially introduce the core components of our framework: SAM Mask Generation, Encoder and Cross Fusion Decoder, Prototype Contrastive Learning and Alignment, Contour-Aware Loss, and the Overall Loss Function.

3.1 SAM Segmentation Mask Generation

We utilize the SAM (Segment Anything Model) with text prompts to generate high-quality segmentation masks, which provide explicit anatomical information for the registration process. Given the input image pair I_m (moving image) and I_f (fixed image), SAM produces the corresponding segmentation masks S_f and S_m according to text prompts $T = \{t1, ..., tn\}$. These masks serve as auxiliary inputs to guide the feature extraction and semantic alignment.

3.2 Encoder and Cross Fusion Decoder

As shown in Fig. 2, our network consists of a CNN feature encoder and a coarse-to-fine cross fusion decoder. The encoder receives the input image and segmentation mask pairs and extracts multi-scale features through multi-layer convolution operations:

$$I_m^l, I_f^l, S_m^l, S_f^l = Encoder(I_m, I_f, S_m, S_f), \qquad (1)$$

where l represents the scale level. The spatial resolution of the feature maps decreases progressively while the channel dimension increases to capture both local and global context information.

In the decoder stage, we use the Cross-Attention Block to calculate the correlation of the image feature embeddings and mask feature embeddings to obtain the registration transformation field of the image level and mask level. Then, the Fusion Attention Block is used to aggregate multi-scale features of the image and mask to gradually optimize the deformation field. The Cross-Attention block [18] consists of four sub-modules: the Window-based Multi-Head Cross-Attention (W-MCA) module, the Fully Connected Network module (MLP), the Shifted Window Multi-Head Cross-Attention (SW-MCA) module, and another MLP module. The cross-attention mechanism [25] is shown as follows:

$$Cross\text{-}Attention(Q, K, V) = Softmax(\frac{QK^T}{\sqrt{d}})V, \qquad (2)$$

where Q is the moving image (mask) feature, and K and V are the fixed image (mask) features.

The Fusion Attention Block differs from the Cross Attention Block because it introduces bi-directional feature interactions. In the bi-directional computation, the image features are first used as the query Q and the mask features as the key K and value V. Then, the roles are reversed: the mask features are used as the query (q), and the image features as the key K and value V. The results of these two interactions are averaged to achieve bi-directional information fusion between the image and the mask. Finally, the crossed image feature $I_{m\&f}$ and the crossed mask feature $S_{m\&f}$ are fed into the Fusion Attention Block to obtain the registration domain φ^4. Then, φ^4 will be upsampled, with one side used to transform $I_m{}^3$ and $S_m{}^3$, and the other side used to superposition $\varphi^{3'}$ as the current scaled transformation field φ^3. Similarly, we can obtain φ^2 and φ^1, where φ^1 is the ultimate registration field φ.

3.3 Prototype Contrastive Learning and Alignment

To improve the semantic consistency of anatomical regions, we introduce Prototype Feature Extraction, Contrast, and Alignment. First, to enhance the feature consistency within the same anatomical region, we extract prototype features for each anatomical region using Masked Average Pooling based on segmentation masks and align all features to the prototype of the same category. Second, to improve the separability of different prototype features while maintaining semantic consistency across images, we push away different prototypes and align the same prototype between different images.

Prototype Feature Extraction. We leverage the segmentation mask S_m to extract prototype features for each anatomical region in the moving image I_m using Masked Average Pooling:

$$P_m^k = \frac{\sum_{i,j} I_{m(i,j)} \cdot S_{m(i,j)}}{\sum_{i,j} S_{m(i,j)}}, \tag{3}$$

where P_m^k is the prototype feature for the k-th region, $I_{m(i,j)}$ represents the feature at position (i,j) in the moving image, and $S_{m(i,j)}$ is the corresponding mask value. Similarly, we get P_f^k for the fixed image I_f and the corresponding mask S_f.

Prototype Contrast and Alignment. A contrastive loss is applied to pull features in I_m corresponding to the same anatomical structure closer to their prototype while pushing features of different structures apart:

$$\mathcal{L}_{\text{contrast}} = -\log \frac{\exp(\text{sim}(I_{m(i,j)}, P_k))}{\sum_{k'} \exp(\text{sim}(I_{m(i,j)}, P_{k'}))}, \tag{4}$$

where $\text{sim}(x,y)$ denotes cosine similarity. Prototypes from the moving image I_m and fixed image I_f are aligned to ensure semantic consistency across the corresponding anatomical regions. The alignment loss based on cosine similarity can be defined as:

$$\mathcal{L}_{\text{align}} = \sum_k \left(1 - \text{sim}(P_f^k, P_m^k)\right), \tag{5}$$

where k iterates over all anatomical regions. We use the sum of $L_{contrast}$ and L_{align} as the prototype loss:

$$L_{prototype} = L_{contrast} + L_{align}. \tag{6}$$

3.4 Contour-Aware Loss

One of the key challenges in medical image registration is achieving precise alignment of anatomical contours, especially in regions with complex structures or ambiguous edges [28]. To address this issue, we introduce the Chamfer loss [7] as the contour-aware loss to optimize contour alignment. Specifically, first, we sample point sets C_f and C_m from the contours of the fixed mask S_f and the moved mask S'_m. Second, we calculate the Chamfer loss between C_f and C_m. The Chamfer Loss is defined as:

$$\mathcal{L}_{contour} = \frac{1}{|C_m|} \sum_{i \in C_m} \min_{j \in C_f} \|i - j\|^2 + \frac{1}{|C_f|} \sum_{j \in C_f} \min_{i \in C_m} \|j - i\|^2, \tag{7}$$

where $\|i - j\|^2$ denotes the squared Euclidean distance between points i and j. By minimizing this loss, the model learns to align the contour of S_m with S_f, thereby improving the precision of contour alignment.

3.5 Overall Loss

In addition to the above prototype loss and contour loss, the overall loss also includes the following common registration losses: similarity loss, smooth loss, and segmentation loss.

Similarity Loss. Similarity loss is to measure the similarity of the fixed image I_f and the deformed moving image $I_{m'} = Warp(I_m, \varphi^1)$. We adopt a widely used image similarity loss called local normalized cross-correlation (LNCC) loss [1,5,25]:

$$L_{sim}(I_f, I_{m'}) = \sum_{P \in \Omega} \frac{(\sum_{P_i}(I_f(P_i) - \hat{I}_f(P))(I_{m'}(P_i) - \hat{I_{m'}}(P)))^2}{(\sum_{P_i}(I_f(P_i) - \hat{I}_f(P))^2)(\sum_{P_i}(I_{m'}(P_i) - \hat{I_{m'}}(P))^2)}. \tag{8}$$

Smooth Loss. We use the spatial gradient of diffusion regularization as the deformation field of smoothing loss:

$$L_{smooth} = \sum_{p \in \Omega} \|\nabla \varphi(p)\|^2, \tag{9}$$

where p denotes the voxel location.

Segmentation Loss. Dice loss is adopted as the segmentation loss:

$$L_{seg} = 1 - \frac{2\,|S_{m'} \cap S_f|}{|S_{m'}| + |S_f|}, \tag{10}$$

where $S_{m'}$ is the deformed moving mask.

Finally, the overall loss is as follows:

$$L = \omega_1 L_{sim} + \omega_2 L_{smooth} + \omega_3 L_{seg} + \omega_4 L_{prototype} + \omega_5 L_{contour}, \tag{11}$$

where ω_1, ω_2, ω_3, ω_4, and ω_5 are hyperparameters.

4 Experiments and Results

4.1 Datasets

We evaluate our method on two widely validated datasets of different parts of different modalities, inter-patient Abdomen CT (3D Abdomen Organs) [29], and ACDC MRI (4D Cardiac) [2]. The inter-patient Abdomen CT dataset is from Learn2Reg [9], which contains 30 CT scans. We follow the previous settings [3,17] and randomly select 20 scans for training and 10 scans for testing, pairing them to generate 190 pairs of training sets and 45 pairs of testing sets. Each scan contains manual annotations of 12 anatomical structures: spleen, right kidney, left kidney, esophagus, liver, stomach, aorta, inferior vena cava, portal and splenic vein, pancreas, left adrenal gland, and right adrenal gland. These anatomical regions are also used as text prompts to guide SAM in generating masks. The voxel resolution of all images is $2\,\mathrm{mm}^3$, and the spatial resolution is $160 \times 160 \times 160$.

The 4D ACDC Cardiac dataset contains end-systolic (ES) and end-diastolic (ED) frames of cardiac MRI images of 100 patients, including 20 healthy patients, 20 patients with previous myocardial infarction, 20 patients with dilated cardiomyopathy, 20 patients with hypertrophic cardiomyopathy, and 20 patients with abnormal right ventricle. Each image contains three anatomical structures: left ventricular (LV), right ventricular (RV), and myocardium (Myo), which are used as text prompts for SAM. We aim to register the ED and ES frames of the same patient. We follow the experimental settings of previous works [11,23] and randomly divide them into 90 pairs for training and 10 pairs for testing. The voxel resolution of all images is $1.5 \times 1.5 \times 3.15\,\mathrm{mm}^3$, and the spatial resolution is $160 \times 160 \times 160$.

4.2 Implementation Details

We utilize the SAT-Nano model as our SAM model [33]. Our experiments are implemented using PyTorch [22] 1.10 on a Nvidia RTX-4090 GPU with 24 GB Memory. Our experiments and all comparison experiments are trained with a batch size of 1 for 500 epochs, using the Adam optimizer with a learning rate of 0.0001. The hyperparameters ω_1, ω_2, ω_3, ω_4, and ω_5 are 1, 4, 1, 1, and 0.1, respectively.

Table 1. Quantitative comparisons on the Abdomen CT dataset. ↑: higher is better, ↓: lower is better. Bold indicates the highest DSC. Spl, Kid R, Kid L, Eso, Liv, Sto, Aor, IVC, Vei, Pan, Adr R, and Adr L represent Spleen, Kidney Right, Kidney Left, Esophagus, Liver, Stomach, Aorta, Inferior Vena Cava, Portal Vein and Splenic Vein, Pancreas, Adrenal Gland Left, and Adrenal Gland Right, respectively. AVG is the average DSC of all 12 organs.

Methods	DSC (%) ↑													SDlogJ ↓
	Spl	Kid R	Kid L	Eso	Liv	Sto	Aor	IVC	Vei	Pan	Adr R	Adr L	AVG	
Initial	30.5	23.0	27.2	10.9	50.2	19.2	21.9	20.1	1.9	7.1	3.7	3.8	17.1	-
VoxelMorph [1]	61.1	55.3	55.6	30.7	70.1	31.2	44.4	44.0	16.7	19.3	18.3	14.4	38.4 ± 16.6	0.143
TransMorph [4]	60.3	54.3	54.0	30.1	70.7	33.5	45.4	46.1	19.3	18.5	18.2	16.3	39.0 ± 16.2	0.254
TransMatch [5]	65.3	58.4	56.3	33.4	72.3	36.4	49.3	54.5	21.3	20.6	19.9	18.7	42.2 ± 14.7	0.101
CorrMLP [21]	68.2	60.1	63.7	38.4	73.4	45.6	51.2	56.9	24.7	29.1	24.6	24.0	46.7 ± 13.2	0.099
SAM Masks [33]	92.8	88.1	89.5	72.2	95.5	89.4	88.4	84.5	68.5	79.7	64.2	62.8	81.2	-
VoxelMorph [1] + SAM	64.2	61.1	59.6	40.7	70.3	44.4	52.1	53.5	19.2	34.1	24.8	25.9	45.8 ± 12.7	0.065
TransMorph [4] + SAM	71.0	68.7	68.8	49.2	73.1	47.4	66.3	61.1	21.6	38.2	33.9	31.7	52.6 ± 12.2	0.088
TransMatch [5] + SAM	74.0	70.4	69.6	**70.2**	74.3	54.3	67.1	67.9	32.8	52.9	**46.3**	47.1	60.5 ± 11.4	0.095
CorrMLP [21] + SAM	77.1	72.8	75.7	55.1	77.7	57.2	**76.1**	73.1	35.1	50.0	44.6	41.8	61.3 ± 11.3	0.103
Ours	**83.9**	**79.3**	**76.0**	55.2	**83.8**	**82.5**	74.5	**73.2**	**36.9**	**56.4**	44.8	**48.5**	**66.3 ± 10.6**	0.091

4.3 Evaluate Metrics

We adopt two widely used evaluation metrics, dice score (DSC) and standard deviation of the Jacobian determinant (SDlogJ). The DSC metric is used to evaluate the overlap of important anatomical regions before and after registration, and SDlogJ is used to evaluate the smoothness of the registration transform field generated by the registration model. The larger the DSC, the higher the overlap and the better the registration result. The smaller the SDlogJ, the smoother the registration transform domain and the better the registration result.

4.4 Comparison with the SOTA Methods

Comparison Methods. We compare our method with state-of-the-art deformable image registration methods, including the unsupervised methods VoxelMoprh [1], TransMoph [4], TransMatch [5], and CorrMLP [21], and their weakly supervised methods using the SAM segmentation masks. For fair comparison, all weakly supervised methods use the same SAM masks.

Abdomen CT Inter-Patients Registration. As shown in Table 1, on the Abdomen CT dataset, our method achieves superior performance of 12 organs compared to all SOTA methods under both unsupervised and weakly supervised setups. Specifically, our method achieves an average DSC of 66.3%, outperforming the SOTA method CorrMLP by a significant margin while maintaining competitive SDlogJ. The reason is that other SOTA methods discard the SAM mask in the inference phase, resulting in poor performance due to the lack of explicit information of the anatomical structure. In contrast, our method leverages SAM masks in both training and inference phases to ensure that the model can fully

Table 2. Quantitative comparisons on the ACDC Cardiac MRI dataset. ↑: higher is better, and ↓: lower is better. Bold indicates the highest DSC. LV, Myo, and RV represent the left ventricular, myocardium, and right ventricular, respectively. AVG is the average DSC of LV, Myo, and RV.

Methods	DSC (%) ↑				SDlogJ ↓
	LV	Myo	RV	AVG	
Initial	58.1	35.8	74.5	56.1	-
VoxelMorph [1]	83.2	60.1	80.5	74.6 ± 7.1	0.041
TransMorph [4]	82.5	58.8	80.4	73.9 ± 7.4	0.031
TransMatch [5]	82.3	58.6	82.1	74.4 ± 6.9	0.037
CorrMLP [21]	82.8	72.9	83.2	79.7 ± 4.6	0.054
SAM Masks [33]	89.2	79.2	76.2	81.6	-
VoxelMorph [1] + SAM	86.2	57.9	82.0	75.4 ± 6.3	0.077
TransMorph [4] + SAM	86.0	60.7	81.3	76.0 ± 5.5	0.026
TransMatch [5] + SAM	87.6	61.8	82.8	77.5 ± 5.1	0.073
CorrMLP [21] + SAM	83.4	74.2	**84.2**	80.6 ± 4.2	0.047
Ours	**91.9**	**77.6**	83.9	**84.6 ± 3.7**	0.049

Table 3. Analysis of voxel-based contrastive loss on Abdomen CT and ACDC Cardiac MRI datasets. ↑: higher is better, ↓: lower is better.

Dataset	$L_{prototype}$	$L_{contour}$	DSC (%) ↑	SDlogJ ↓
Abdomen	✗	✗	62.8	0.089
	✓	✗	63.3	0.086
	✓	✓	66.3	0.091
ACDC	✗	✗	83.0	0.053
	✓	✗	83.5	0.045
	✓	✓	84.6	0.049

utilize anatomical information, thereby improving the accuracy of image registration.

ACDC Cardiac MRI Registration of Diastolic and Systolic Phase. Table 2 compares our methods against several SOTA methods on the ACDC dataset. As seen, *Ours* outperforms the CorrMLP by 4.0%. Moreover, *Ours* achieves the highest DSC in both LV and Myo, verifying the effectiveness of our method.

4.5 Ablation Study

To evaluate the effectiveness of the proposed prototype loss and contour loss, we conduct an ablation study on the Abdomen CT and ACDC MRI dataset, as

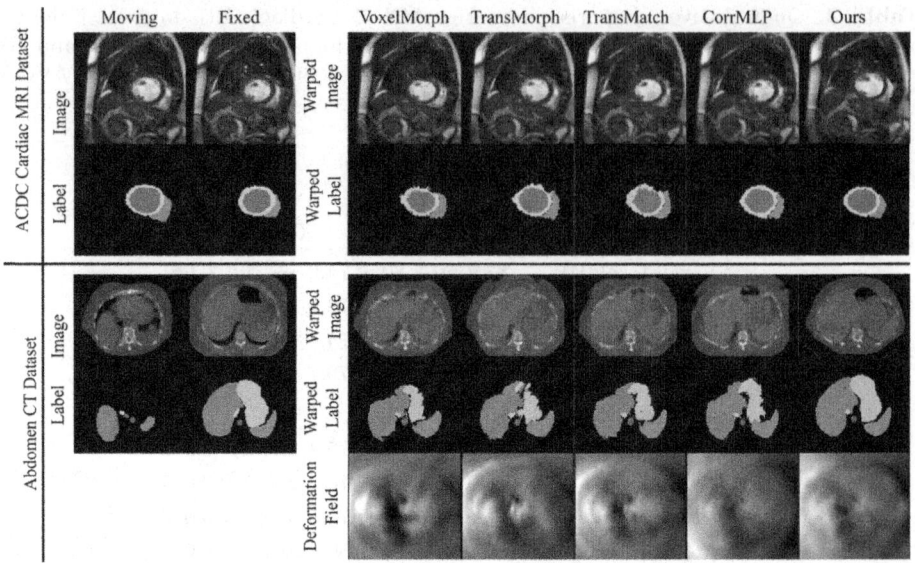

Fig. 3. Visualization of registration results for our proposed method and compared methods on ACDC Cardiac MRI dataset and Abdomen CT dataset. The comparison methods are all weakly supervised registration methods utilizing SAM masks. (Color figure online)

shown in Table 3. Specifically, using only the $L_{prototype}$, the DSC of our method increases by 0.5% from 62.8% to 63.3% on the Abdomen dataset and by 0.5% from 83.0% to 83.5% on the ACDC dataset, proving that the prototype loss can improve the semantic consistency to improve the registration accuracy. Furthermore, with $L_{prototype}$ and $L_{contour}$, the DSC of our method further increases by 3.0% up to 66.3% on the Abdomen dataset and by 1.1% up to 84.6% on the ACDC dataset, evaluating that $L_{contour}$ can enhance the contour alignment ability of the model and thus improve the registration performance. The ablation study shows that $L_{prototype}$ and $L_{contour}$ significantly enhance both registration accuracy and contour alignment while maintaining smooth deformations.

4.6 Visualization

Figure 3 illustrates the qualitative comparison of our method with state-of-the-art (SOTA) registration methods w/ SAM masks on the ACDC MRI and Abdomen CT datasets. On the ACDC dataset, our method achieves the most precise alignment of anatomical regions compared to other approaches. The warped labels generated by our method align closely with the fixed labels, particularly for the myocardium (brown) and right ventricle (yellow). In contrast, the warped labels of other methods are misaligned, especially in the myocardial region, indicating their limitations in capturing complex anatomical details. On the Abdomen CT dataset, our method demonstrates superior registration accuracy for challenging anatomical structures such as the liver (orange) and spleen (dark green). The warped labels produced by our method exhibit that our shapes are more complete, and the borders are more rounded compared with other methods. These results validate the effectiveness of our contour loss in enhancing contour alignment and improving the overall quality of the registration.

5 Conclusion

In this study, we propose a SAM-assisted medical image registration framework that achieves an effective balance between global semantic consistency and local detail alignment through the integration of explicit anatomical knowledge, prototype learning, and contour-aware loss. The framework first leverages segmentation masks generated by SAM to provide explicit anatomical region annotations, ensuring the consistency of anatomical information during the training and testing phases. Next, prototype learning is employed to optimize semantic consistency across paired images. Additionally, contour Loss is introduced to align the contour points of the fixed and moved masks, further improving boundary alignment precision. Extensive experiments demonstrate that our model outperforms SOTA methods on the Abdomen CT and ACDC Cardiac MRI datasets by effectively addressing the challenges of complex anatomical structures through explicit anatomical guidance and contour-aware optimization.

References

1. Balakrishnan, G., Zhao, A., Sabuncu, M.R., Guttag, J., Dalca, A.V.: Voxelmorph: a learning framework for deformable medical image registration. IEEE Trans. Med. Imaging **38**(8), 1788–1800 (2019)
2. Bernard, O., et al.: Deep learning techniques for automatic MRI cardiac multi-structures segmentation and diagnosis: is the problem solved? IEEE Trans. Med. Imaging **37**(11), 2514–2525 (2018)
3. Bigalke, A., Hansen, L., Mok, T.C., Heinrich, M.P.: Unsupervised 3D registration through optimization-guided cyclical self-training. In: International Conference on Medical Image Computing and Computer-Assisted Intervention, pp. 677–687. Springer (2023)
4. Chen, J., Frey, E.C., He, Y., Segars, W.P., Li, Y., Du, Y.: Transmorph: transformer for unsupervised medical image registration. Med. Image Anal. **82**, 102615 (2022)
5. Chen, Z., Zheng, Y., Gee, J.C.: Transmatch: a transformer-based multilevel dual-stream feature matching network for unsupervised deformable image registration. IEEE Trans. Med. Imaging **43**(1), 15–27 (2024)
6. Cheng, J., et al.: Sam-med2d. arXiv preprint arXiv:2308.16184 (2023)
7. Fan, H., Su, H., Guibas, L.J.: A point set generation network for 3D object reconstruction from a single image. In: Proceedings of the IEEE/CVF Conference on Computer Vision and Pattern Recognition, pp. 605–613 (2017)
8. Guo, M.H., Zhang, Y., Mu, T.J., Huang, S.X., Hu, S.M.: Tuning vision-language models with multiple prototypes clustering. IEEE Trans. Pattern Anal. Mach. Intell. (2024)

9. Hering, A., et al.: Learn2reg: comprehensive multi-task medical image registration challenge, dataset and evaluation in the era of deep learning. IEEE Trans. Med. Imaging **42**(3), 697–712 (2022)
10. Huang, Y., et al.: Segment anything model for medical images? Med. Image Anal. **92**, 103061 (2024)
11. Kim, B., Han, I., Ye, J.C.: Diffusemorph: unsupervised deformable image registration using diffusion model. In: European Conference on Computer Vision, pp. 347–364. Springer (2022)
12. Kirillov, A., et al.: Segment anything. arXiv preprint arXiv:2304.02643 (2023)
13. Kögl, F., et al.: General vision encoder features as guidance in medical image registration. In: International Workshop on Biomedical Image Registration, pp. 265–279. Springer (2024)
14. Lei, W., Xu, W., Li, K., Zhang, X., Zhang, S.: MedLSAM: localize and segment anything model for 3D CT images. Med. Image Anal. **99**, 103370 (2025)
15. Li, C., et al.: LLAVA-med: training a large language-and-vision assistant for biomedicine in one day. In: Advances in Neural Information Processing Systems, vol. 36 (2024)
16. Li, S., Cao, J., Ye, P., Ding, Y., Tu, C., Chen, T.: ClipSAM: CLIP and SAM collaboration for zero-shot anomaly segmentation. arXiv preprint arXiv:2401.12665 (2024)
17. Li, Z., et al.: Samconvex: fast discrete optimization for ct registration using self-supervised anatomical embedding and correlation pyramid. In: International Conference on Medical Image Computing and Computer-Assisted Intervention, pp. 559–569. Springer (2023)
18. Liu, Z., et al.: Swin transformer: hierarchical vision transformer using shifted windows. In: Proceedings of the IEEE/CVF International Conference on Computer Vision, pp. 10012–10022 (2021)
19. Ma, J., He, Y., Li, F., Han, L., You, C., Wang, B.: Segment anything in medical images. Nat. Commun. **15**(1), 654 (2024)
20. Mazurowski, M.A., Dong, H., Gu, H., Yang, J., Konz, N., Zhang, Y.: Segment anything model for medical image analysis: an experimental study. Med. Image Anal. **89**, 102918 (2023)
21. Meng, M., Feng, D., Bi, L., Kim, J.: Correlation-aware coarse-to-fine MLPs for deformable medical image registration. In: Proceedings of the IEEE/CVF Conference on Computer Vision and Pattern Recognition, pp. 9645–9654 (2024)
22. Paszke, A., et al.: Pytorch: an imperative style, high-performance deep learning library. In: Advances in Neural Information Processing Systems, vol. 32 (2019)
23. Qin, Y., Li, X.: FSDiffReg: feature-wise and score-wise diffusion-guided unsupervised deformable image registration for cardiac images. In: International Conference on Medical Image Computing and Computer-Assisted Intervention, pp. 655–665. Springer (2023)
24. Sotiras, A., Davatzikos, C., Paragios, N.: Deformable medical image registration: a survey. IEEE Trans. Med. Imaging **32**(7), 1153–1190 (2013)
25. Vaswani, A., et al.: Attention is all you need. In: Advances in Neural Information Processing Systems, vol. 30 (2017)
26. Viergever, M.A., Maintz, J.A., Klein, S., Murphy, K., Staring, M., Pluim, J.P.: A survey of medical image registration-under review. Med. Image Anal. **33**, 140–144 (2016)
27. Wang, H., et al.: Sam-med3d. arXiv preprint arXiv:2310.15161 (2023)

28. Xu, H., et al.: Sat-morph: unsupervised deformable medical image registration using vision foundation models with anatomically aware text prompt. In: MICCAI Workshop on Foundation Models for General Medical AI, pp. 71–80. Springer (2024)
29. Xu, Z., et al.: Evaluation of six registration methods for the human abdomen on clinically acquired CT. IEEE Trans. Biomed. Eng. **63**(8), 1563–1572 (2016)
30. Zhang, F., Wells, W.M., O'Donnell, L.J.: Deep diffusion MRI registration (DDM-REG): a deep learning method for diffusion MRI registration. IEEE Trans. Med. Imaging **41**(6), 1454–1467 (2021)
31. Zhang, S., Metaxas, D.: On the challenges and perspectives of foundation models for medical image analysis. Med. Image Anal. **91**, 102996 (2024)
32. Zhang, Y., Xu, Y., Chen, J., Xie, F., Chen, H.: Prototypical information bottlenecking and disentangling for multimodal cancer survival prediction. In: The Twelfth International Conference on Learning Representations (2024)
33. Zhao, Z., et al.: One model to rule them all: towards universal segmentation for medical images with text prompts. arXiv preprint arXiv:2312.17183 (2023)
34. Zhou, T., Wang, W., Konukoglu, E., Van Gool, L.: Rethinking semantic segmentation: a prototype view. In: Proceedings of the IEEE/CVF Conference on Computer Vision and Pattern Recognition, pp. 2582–2593 (2022)

Vascular-Topology-Aware Deep Structure Matching for 2D DSA and 3D CTA Rigid Registration

Xiaosong Xiong[1], Caiwen Jiang[1], Peng Wu[2], Xiao Zhang[3], Yanli Song[2], Xinyi Zhang[2], Ze Tao[2], Dijia Wu[2(✉)], and Dinggang Shen[1,2,4(✉)]

[1] School of Biomedical Engineering and State Key Laboratory of Advanced Medical Materials and Devices, ShanghaiTech University, Shanghai, China
dinggang.shen@gmail.com
[2] Shanghai United Imaging Intelligence Co., Ltd., Shanghai, China
dijia.wu@uii-ai.com
[3] School of Information Science and Technology, Northwest University, Xi'an, China
[4] Shanghai Clinical Research and Trial Center, Shanghai, China

Abstract. Accurate 2D/3D integration of Digital Subtraction Angiography (DSA) and Computed Tomography Angiography (CTA) images holds significant potential for reducing risk for navigation in percutaneous coronary intervention (PCI). Rigid registration is a crucial step in achieving this integration. However, existing rigid registration methods based on manually designed features and subsequent rule-based graph matching exhibit limited generalization and often fail in complex surgical scenarios, such as vessel overlapping from 3D-to-2D projection and branch missing due to Chronic Total Occlusion (CTO). To address these challenges, we propose a vascular-topology-aware deep structure matching framework for 3D CTA and 2D DSA rigid registration. Our framework includes key-point extraction, where 2D/3D topological priors are used to extract key points and their descriptors, and a matching stage that employs a 3D spatial-aware hybrid attention mechanism to capture vessel structures while mitigating the impact of vessel overlap and branch missing on feature-based matching. We also designed a data simulation strategy to generate a large set of paired data for network training, using various rigid transformations and random branch trimming to simulate complex and variable real-world scenarios, especially the vessel overlap and branch missing. Extensive evaluations conducted on the simulated dataset and 1,016 pairs of real CTA and DSA samples demonstrate the effectiveness and robustness of our method, highlighting its strong performance and potential for real-world surgical applications. The code is available at https://github.com/xxsxxsxxs666/2D-3DCoronary.

Keywords: Percutaneous coronary intervention · 2D/3D registration for DSA and CTA · Deep structure matching · Vascular topology

1 Introduction

Percutaneous Coronary Intervention (PCI) is a common treatment for severe atherosclerotic plaques that obstruct blood flow, often involving stent placement to restore vessel patency [5]. Digital Subtraction Angiography (DSA) is typically used to provide real-time surgical guidance. However, 2D DSA offers limited information regarding the position and phenotype of plaques, such as vulnerable or calcified plaques, which complicates accurate stent placement and related decision-making (e.g., whether to perform additional intravascular imaging for potentially non-obstructive vulnerable plaques). Additionally, DSA lacks the ability to visualize Chronic Total Occlusion (CTO) vessels where the contrast agent cannot flow, posing a risk of vascular perforation. In contrast, Computed Tomography Angiography (CTA), commonly used in preoperative planning, provides 3D information on plaque morphology and occluded vessel trajectories, offering critical guidance for navigating CTO segments [18], as shown in Fig. 1. However, CTA cannot provide real-time imaging during surgery. Therefore, integrating DSA and CTA via 2D/3D registration can enable real-time intraoperative 3D navigation, enhancing procedural precision and reducing risks. Given challenges such as cardiac motion and instrument errors, accurate rigid registration between 2D DSA and 3D CTA is essential for this integration.

For 2D/3D rigid registration, existing methods can be broadly categorized into intensity-based [8,11] and feature-based approaches [1,2,12,22]. Intensity-based approaches align images by computing the similarity of voxel or pixel values between images. However, they often perform poorly when the structures in both modalities are sparse, as is common in vessel imaging [2,21]. Additionally, the non-invasive injection of contrast agents in CTA, which enhances not only the coronary arteries but also non-coronary structures, leads to significant disparities between raw CTA and DSA images. This further limits the applicability of intensity-based techniques in this context. In contrast, feature-based methods, leveraging geometric representations such as vessel centerlines or bifurcations, are more effective for vessel registration than intensity-based approaches [2]. These methods establish correspondences between features in DSA and CTA, often using sampling points along vessel centerlines and optimizing rigid transformations through algorithms like Perspective-n-Point (PnP)

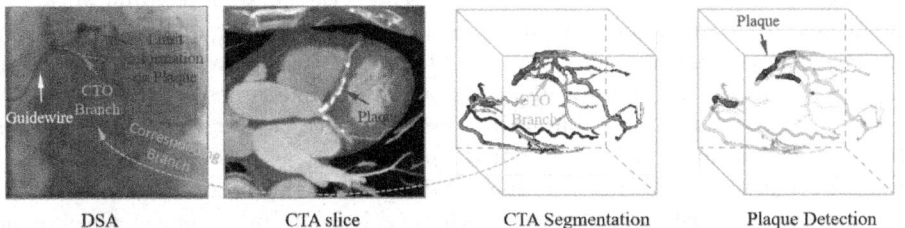

Fig. 1. Differences between DSA and CTA images. From left to right: DSA, CTA, coronary segmentation in CTA, and plaque detection in CTA.

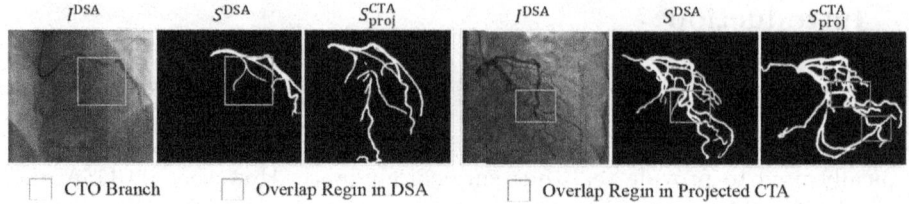

Fig. 2. Branch missing due to CTO, with overlap in both DSA and projected CTA.

[7]. Early techniques, such as Iterative Closest Point (ICP) [12] and Iterative Corresponding Contours (ICC) [1], utilize vessel centerline geometries for alignment. Recent approaches, like Monte Carlo Tree Search (MCTS) [22], focus on matching the specific endpoints and bifurcations using tree search technology. These methods heavily depend on hand-crafted point features, such as distances and angles, and also the traditional rule-based graph matching technology, which limits their robustness in complex scenarios, including complex vessel structure, substantial branch missing (due to CTO or segmentation failures), and overlap caused by projection, as shown in Fig. 2. These conditions are both common and critical in clinical applications. Furthermore, traditional methods are often evaluated on limited case studies rather than large-scale datasets, raising concerns about their generalizability.

In recent years, deep learning has demonstrated powerful image processing capabilities, effectively addressing the limitations of manually designed features and traditional graph matching. Using neural networks to extract features and perform point matching has achieved success in many studies [4,10,14,16,17]. Among these, SuperPoint [4] detects key points and features efficiently using CNNs, while SuperGlue [16] applies an attention mechanism for matching point sets. LightGlue [14] offers a lightweight, real-time solution for point matching and predicting occlusions. However, due to the lack of 2D/3D gold-standard datasets and vascular topology priors across modalities, which has long been a challenge in medical image registration [6], none of these methods have been applied to rigid registration between CTA and DSA.

In this paper, we introduce the first deep learning-based point-matching method for rigid DSA-CTA registration in cardiac navigation. More concretely, we propose a vascular-topology-aware deep structure matching framework that integrates transformer-based techniques to enhance vascular registration. Building on advances in small object segmentation [3,15,19], we first segment vessels, then sample key points along 2D and 3D centerlines, extracting CNN-based descriptors, thereby eliminating the reliance on manually defined features. To train this framework, we design a simulation strategy to generate realistic datasets that emulate branch missing and vessel overlap scenarios.

The main contributions of our method include: 1) proposing a novel structure matching framework for rigid registration between 2D DSA and 3D CTA that effectively addresses vessel overlap and branch missing; 2) collecting a large

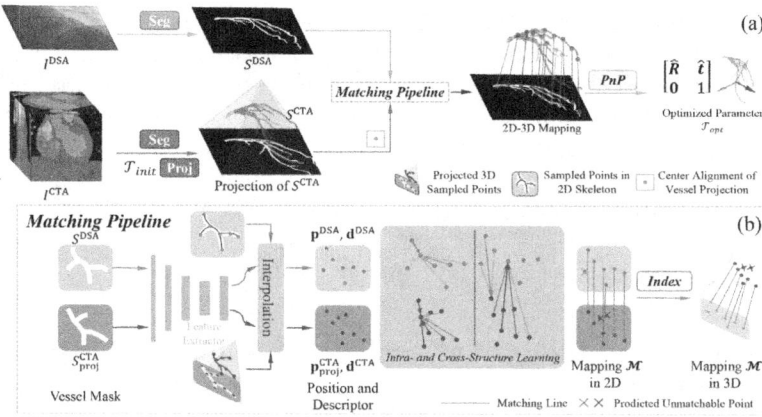

Fig. 3. Illustration of our proposed method. (a) depicts the segmentation and projection processes. (b) shows the key-point and descriptor extraction strategy, as well as the point-matching process.

clinical dataset and designing a simulation strategy to generate a large set of aired data for network training, replicating real-world challenges; and 3) demonstrating the effectiveness and robustness of our method on a large scale of both simulated and real-world datasets.

2 Problem Definition

Given a DSA image $I^{\text{DSA}} \in \mathbb{R}^{H \times W}$, CTA image $I^{\text{CTA}} \in \mathbb{R}^{H \times W \times D}$ and perspective projection process Π defined in DICOM file, we need to find an optimal transformation $\mathcal{T}^*$ of the 3D vessel to maximize the similarity function $\mathcal{D}$:

$$\mathcal{T}^* = \arg\max_{\mathcal{T} \in \Omega_{\mathcal{T}}} \mathcal{D}(\Pi \circ \mathcal{T}(I^{\text{CTA}}), I^{\text{DSA}}), \tag{1}$$

where $\Omega_{\mathcal{T}}$ is the domain of admissible transformation. For rigid registration, $\Omega_{\mathcal{T}}$ can be formulated as special Euclidean group SE(3) including three translations and three rotations:

$$\Omega_{\mathcal{T}} = \text{SE}(3) = \left\{ \begin{bmatrix} \mathbf{R} & \mathbf{t} \\ 0 & 1 \end{bmatrix} \;\middle|\; \mathbf{R} \in \text{SO}(3), \mathbf{t} \in \mathbb{R}^3 \right\}. \tag{2}$$

In feature-based methods, this can be reformulated into a key-point-matching problem. This task involves identifying two key-points sets, denoted as $\mathbf{p}^{\text{DSA}} \in \mathbb{R}^{K \times 2}$ and $\mathbf{p}^{\text{CTA}} \in \mathbb{R}^{M \times 3}$, which are indexed by $\Theta_{\text{DSA}} := \{1, \dots, k, \dots, K\}$ and $\Theta_{CTA} := \{1, \dots, m, \dots, M\}$ respectively, followed by establishing the mapping relationship $\mathcal{M} := \{(k,m)\} \subset \Theta^{\text{DSA}} \times \Theta^{\text{CTA}}$ between them. Finally, the Perspective-n-Points (PnP) algorithm is employed to compute the optimized rigid transformation:

$$\mathcal{T}_{\text{opt}} = \operatorname*{argmin}_{\mathcal{T} \in \Omega_{\mathcal{T}}} \sum_{(k,m) \in \mathcal{M}} \|\mathbf{p}_k^{\text{DSA}} - \Pi \circ \mathcal{T}(\mathbf{p}_m^{\text{CTA}})\|^2. \tag{3}$$

To ensure a robust rigid transformation, it is essential to have a sufficiently large $|\mathcal{M}|$, with these points well-distributed across the vascular structures in the DSA image, rather than being clustered in a localized region. Furthermore, achieving an accurate transformation requires establishing precise correspondences via $\mathcal{M}$, which need to overcome challenges such as vessel overlap in both modality, and correctly identifying unmatchable points index sets $\overline{\Theta^{\mathrm{DSA}}} \subset \Theta^{\mathrm{DSA}}$ and $\overline{\Theta^{\mathrm{CTA}}} \subset \Theta^{\mathrm{CTA}}$ caused by branch missing. To elaborate, let $\overline{\Theta^{\mathrm{DSA}}}$ be defined as follows:

$$\overline{\Theta^{\mathrm{DSA}}} := \Theta^{\mathrm{DSA}} \setminus \{k \mid \exists m \text{ such that } (k,m) \in \mathcal{M}\}. \tag{4}$$

3 Vascular-Topology-Aware Deep Structure Matching Framework

Our proposed framework is illustrated in Fig. 3. For given DSA and CTA images, the process begins with the key-point detection stage, where key-point sets $\mathbf{p}^{\mathrm{DSA}}$ and $\mathbf{p}^{\mathrm{CTA}}$ are extracted along with their corresponding descriptors, $\mathbf{d}^{\mathrm{DSA}}$ and $\mathbf{d}^{\mathrm{CTA}}$. This is followed by the key-point matching stage, where a 3D spatial-aware hybrid attention mechanism is employed to learn the inner and cross-structures of the point sets, obtaining the matching relationships $\mathcal{M}$ while explicitly predicting the matchability of points. Our network is trained on a simulated dataset that includes synthetic overlap and occlusion, as shown in Fig. 5. In the following sections, we will introduce the details of key-point extraction, key-point matching, and data simulation.

3.1 Key-Points Extraction and Point Matching

To minimize modality differences and ensure well-distributed key points across vascular structures, we segment the vessels from both modalities and extract key points along their centerlines. Specifically, two segmentation networks [9] are employed to extract vessel structures S^{DSA} and S^{CTA} from both DSA and CTA images. For DSA, centerlines are derived from the 2D S^{DSA} [20], and key points $\mathbf{p}^{\mathrm{DSA}}$ are sampled along the centerline. To improve the network's ability to handle overlapping projections in CTA, a special sampling strategy is used for CTA. Specifically, key points $\mathbf{p}^{\mathrm{CTA}}$ are sampled along the 3D coronary artery centerline [13]. Then, these 3D key points are then projected into the DSA coordinate system using initial rigid transformation parameters $\mathcal{T}_{\mathrm{init}}$, resulting in projected key points $\mathbf{p}^{\mathrm{CTA}}_{m,\mathrm{proj}} = \Pi \circ \mathcal{T}_{init}(\mathbf{p}^{\mathrm{CTA}}_m)$. With this projecting process, $\mathbf{p}^{\mathrm{CTA}}_{\mathrm{proj}}$ shared the same indexing set Θ^{CTA} with $\mathbf{p}^{\mathrm{CTA}}$. To capture the local geometric descriptors of each key point, S^{DSA} and projected CTA vessel segmentation $S^{\mathrm{CTA}}_{\mathrm{proj}}$, are fed into a same 2D CNN-based feature extraction network. This ensures consistency in feature representation across modalities. The network produces feature maps with dimensions $\mathbb{R}^{H/8 \times W/8 \times L}$. Bilinear interpolation is then applied to extract feature descriptors $\mathbf{d}^{\mathrm{DSA}}_k, \mathbf{d}^{\mathrm{CTA}}_m \in \mathbb{R}^L$ for each key-point $\mathbf{p}^{\mathrm{DSA}}_k, \mathbf{p}^{\mathrm{CTA}}_m$. Next, we establish correspondence $\mathcal{M}$ between the two sets of key points using self-attention and cross-attention mechanisms, as shown in Fig. 4. These mechanisms iteratively refine the descriptors **d** for each key-point, progressively aligning corresponding points across modalities by increasing the similarity of matched points while reducing the similarity of unmatched points. For clarity, we take the state update for points in DSA as an example (noting that the process for CTA is analogous). In each attention unit, the descriptor $\mathbf{d}_k^{\mathrm{DSA}}$ is updated using a multi-layer perceptron (MLP) to aggregate information from a source modality $M \in \{\mathrm{DSA}, \mathrm{CTA}\}$. The updating process is formulated as:

$$\mathbf{d}_k^{\mathrm{DSA}} \leftarrow \mathbf{d}_k^{\mathrm{DSA}} + \mathrm{MLP}([\mathbf{d}_k^{\mathrm{DSA}} | \Delta \mathbf{d}_k^{\mathrm{DSA} \leftarrow M}]), \qquad (5)$$

where $[\cdot|\cdot]$ denotes the concatenation of vectors and the update value $\Delta \mathbf{d}_k$ for a key-point k aggregating contextual information from M. When M refers to CTA, we employ the cross-attention to compute $\Delta \mathbf{d}^{\mathrm{DSA} \leftarrow \mathrm{CTA}}$ defined as:

$$\Delta \mathbf{d}_k^{\mathrm{DSA} \leftarrow \mathrm{CTA}} = \mathrm{Softmax}(\frac{\mathbf{Q}_k^{\mathrm{DSA}} (\mathbf{K}^{\mathrm{CTA}})^T}{\sqrt{C}}) \mathbf{V}^{\mathrm{CTA}}, \qquad (6)$$

where $\sqrt{C}$ is the scaling factor for the dot-product attention. Query, key and value matrices $\mathbf{Q}_k^M, \mathbf{K}_k^M$ and $\mathbf{V}_k^M$ are calculated as follows:

$$\mathbf{Q}^M, \mathbf{K}^M, \mathbf{V}^M = \mathbf{W}^Q \mathbf{d}^M, \mathbf{W}^K \mathbf{d}^M, \mathbf{W}^V \mathbf{d}^M. \qquad (7)$$

When M refers to DSA itself, we employ the self-attention mechanism combined with relative spatial encoding to compute $\Delta \mathbf{d}_k^{\mathrm{DSA} \leftarrow \mathrm{DSA}}$, which updates the feature descriptors of DSA key-points based on intra-modality interactions. The update is defined as:

$$\Delta \mathbf{d}_k^{\mathrm{DSA} \leftarrow \mathrm{DSA}} = \mathrm{Softmax}(\frac{\mathbf{E}(\mathbf{Q}_k^{\mathrm{DSA}}, \mathbf{p}_k^{\mathrm{DSA}}) \mathbf{E}(\mathbf{K}^{\mathrm{DSA}}, \mathbf{p}^{\mathrm{DSA}})^T}{\sqrt{C}}) \mathbf{V}^{\mathrm{DSA}}, \qquad (8)$$

where $\mathbf{E}(\mathbf{Q}_k^{\mathrm{DSA}}, \mathbf{p}_k^{\mathrm{DSA}})$ and $\mathbf{E}(\mathbf{K}^{\mathrm{DSA}}, \mathbf{p}^{\mathrm{DSA}})$ represent the query and key matrices after incorporating the relative positional encoding of the position $\mathbf{p}^{\mathrm{DSA}}$. For key points in CTA modality, $\Delta \mathbf{d}_k^{\mathrm{CTA} \leftarrow \mathrm{CTA}}$ and $\Delta \mathbf{d}_k^{\mathrm{CTA} \leftarrow \mathrm{DSA}}$ are calculated similarily. However, there is one important distinction. To better differentiate points $\mathbf{p}_{m,\mathrm{proj}}^{\mathrm{CTA}}$ and $\mathbf{p}_{n,\mathrm{proj}}^{\mathrm{CTA}}$ that belong to different branches but project onto overlap regions shown in Fig. 2, when applying position encoding $\mathbf{E}$, we use 3D spatial information after initial transformation $\mathbf{p}_{m,\mathcal{T}_{\mathrm{init}}}^{\mathrm{CTA}}, \mathbf{p}_{n,\mathcal{T}_{\mathrm{init}}}^{\mathrm{CTA}} \in \mathbb{R}^3$ to take advantage of larger distance differences:

$$\|\mathbf{p}_{m,\mathcal{T}_{\mathrm{init}}}^{\mathrm{CTA}} - \mathbf{p}_{n,\mathcal{T}_{\mathrm{init}}}^{\mathrm{CTA}}\| > \|\mathbf{p}_{m,\mathrm{proj}}^{\mathrm{CTA}} - \mathbf{p}_{n,\mathrm{proj}}^{\mathrm{CTA}}\| \approx 0. \qquad (9)$$

After obtaining the updated descriptor, the matching between key-points is determined by computing a pairwise similarity score $\mathbf{S}_{km} = \mathbf{d}_k^T \mathbf{d}_m$ between each corresponding point in the two modalities. To address the branch missing issue that frequently occurs in CTO cases, and following the approach in [14], we explicitly calculate the matchability σ for each key point, which is decoded with a learned linear transformation $\mathrm{Linear}(\cdot)$ with bias:

$$\sigma_k = \mathrm{Sigmoid}(\mathrm{Linear}(\mathbf{d}_k)) \in [0,1]. \qquad (10)$$

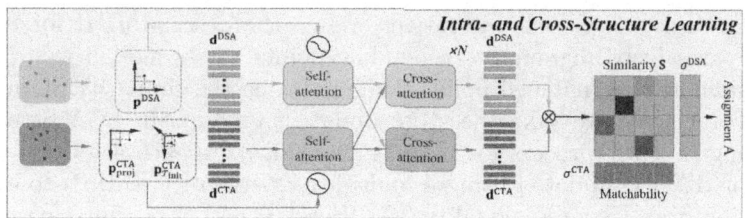

Fig. 4. The sketch of intra- and cross-structure learning.

To obtain $\mathcal{M}$, we apply the $\arg\max$ function to $\mathbf{S}$ to identify the matching key points $\mathbf{p}_k^{\text{DSA}}$ and $\mathbf{p}_m^{\text{CTA}}$, while ensuring that σ exceeds a threshold t_σ and $\mathbf{S}_{km}$ exceeds t_S. For training, we calculate a soft partial assignment matrix $\mathbf{A} \in [0,1]^{K \times M}$.

$$\mathbf{A}_{km} = \sigma_k^{\text{DSA}} \cdot \underset{i \in \Theta^{\text{DSA}}}{\text{Softmax}} (\mathbf{S}_{im})_k \cdot \sigma_m^{\text{CTA}} \cdot \underset{i \in \Theta^{\text{CTA}}}{\text{Softmax}} (\mathbf{S}_{ki})_m . \tag{11}$$

Then, given the ground truth matches $\mathcal{M}^{\text{GT}}$, along with unmatchable labels $\overline{\Theta^{\text{DSA}}}$, we can minimize the log-likelihood of the predicted assignment:

$$\begin{aligned} loss = & -\frac{1}{|\mathcal{M}^{\text{GT}}|} \sum_{(k,m) \in \mathcal{M}^{\text{GT}}} \log \mathbf{A}_{km} - \frac{1}{2|\overline{\Theta^{\text{DSA}}}|} \sum_{k \in \overline{\Theta^{\text{DSA}}}} \log \left(1 - \sigma_k^{\text{DSA}}\right) \\ & - \frac{1}{2|\overline{\Theta^{\text{CTA}}}|} \sum_{m \in \overline{\Theta^{\text{CTA}}}} \log \left(1 - \sigma_m^{\text{CTA}}\right) \end{aligned} \tag{12}$$

However, acquiring $\mathcal{M}^{\text{GT}}$, along with $\overline{\Theta^{\text{DSA}}}$ and $\overline{\Theta^{\text{CTA}}}$ for real clinical data is highly challenging. Therefore, in the following section, we propose a novel strategy for constructing a simulated dataset to address this challenge.

3.2 Branch-Missing-Aware Simulation Module

In this simulation module, we employ various rigid transformations and branch-trimming strategies to simulate DSA/CTA datasets and generate gold-standard matches, as illustrated in Fig. 5. To begin with, we apply various rigid transformation parameters $\mathcal{T}^{\text{DSA}}, \mathcal{T}^{\text{CTA}}$ to the CTA data to simulate the initial angular errors typically seen in DSA acquisition, as depicted in Fig. 5(a). As the range of applied transformations expands, the occurrence of overlapping issues in both DSA and CTA datasets also increases. Given that branch missing occurs in both CTA and DSA images, we apply a series of coronary branch trimming strategies to both modalities, as depicted in Fig. 5(b). Three trimming strategies are designed as follows: 1) Percentage trimming: simulating general chronic total occlusions, under-segmentation, and limited contrast flow by trimming vessels based on a preset percentage; 2) Ostiaum trimming: simulating full proximal occlusions, which has significant clinical relevance as it might cause perforations; 3) Root branch trimming: simulating root branch under-segmentation,

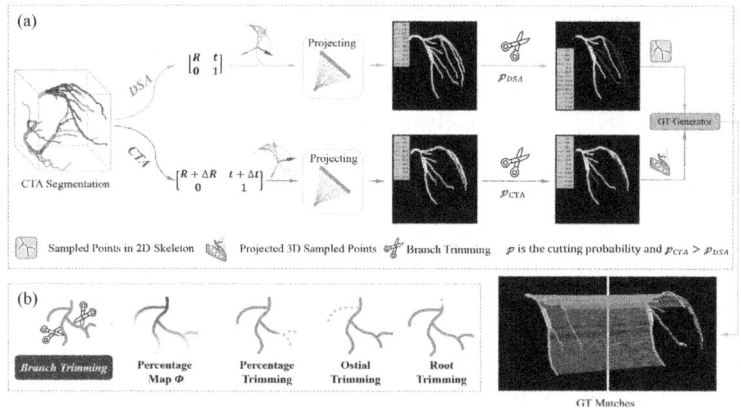

Fig. 5. Illustration of the simulation module.

often observed when contrast agents are injected at deeper vessel levels rather than at the coronary ostium. To control the extent of trimming, we first generate a percentage map Φ using the centerline of the CTA data, expressed as:

$$\Phi(x) = \frac{\ell_{\text{subtree}}(x)}{\ell_{\text{total}}}, \quad x \in \mathcal{C}, \tag{13}$$

where $\ell_{\text{subtree}}(x)$ is the total length of the branch at point x, including the length of the current branch and any sub-branches descending from it, ℓ_{total} represents the total length of coronary tree, and $\mathcal{C}$ denotes the set of points along the coronary artery centerline. By utilizing this control function, we can regulate the trimming proportion, preventing excessive trimming of branches, which could hinder network training. Using these trimming strategies, we generate simulated S^{DSA} and corresponding $S^{\text{CTA}}_{\text{proj}}$. Similar to the previous section, we sample key points along 2D and 3D centerline. Finally, since both segmentations derive from the same 3D coronary artery model, we can establish gold-standard point matches $\mathcal{M}^{\text{GT}}$ and unmatchable labels $\overline{\Theta^{\text{DSA}}}, \overline{\Theta^{\text{CTA}}}$ simply by finding the nearest-neighbor correspondences, as shown by the green match lines generated by the GT Generator in the bottom-right corner of Fig. 5.

4 Experiments

To evaluate the effectiveness of our method, we compare it with four approaches: (1) ICP-BP [12], an extension of ICP for 2D/3D applications; (2) TP-ICC [2], a tree topology matching method; (3) LG, the LightGlue [14] pipeline without fine-tuning; and (4) LG(2D/2D), LightGlue trained by simulated datasets with key points in CTA and DSA both sampled in 2D coordinate. The differences in key-point extraction strategies among LG, LG(2D/2D), and our method are clearly demonstrated in the zoomed areas of Fig. 7. These methods are evaluated on both simulated and real-world datasets alongside our proposed method.

4.1 Experimental Setting

Datasets: The dataset includes 3,500 CTA scans for simulation-based training and 700 for testing, collectively referred to as **SimuPaired**. For each scan, 250 simulated data samples are generated as detailed in Sect. 3.2. Additionally, the dataset contains 5,477 real-world paired DSA and CTA datasets from 783 sequence videos, referred to as **RealPaired**. These paired datasets are rigidly aligned through expert manual annotation. The RealPaired dataset is divided into two subsets: 4,461 pairs for training the DSA segmentation model and 1,016 pairs for evaluating both segmentation and registration algorithms.

Metrics: The evaluation metrics include the Projected Distance (PD) between 3D and 2D centerline points, angular distance d_R between $\mathbf{R}_{\text{opt}}$ and $\mathbf{R}_{\text{GT}} \in SO(3)$. The d_R is formulated as:

$$d_R(\mathbf{R}_{\text{opt}}, \mathbf{R}_{\text{GT}}) = \arccos\left(\frac{\text{trace}(\mathbf{R}_{\text{opt}}^T \mathbf{R}_{\text{GT}}) - 1}{2}\right). \tag{14}$$

Implementation Details: Our experiments are trained on four NVIDIA L40 GPUs, with testing conducted on an NVIDIA RTX A6000 and an Xeon Silver 4210 CPU. For network parameters, the similarity threshold t_S is set to 0.1 on the simulated dataset and 0.005 on the real dataset, while the matchability threshold t_σ is set to 0.95. For both DSA and CTA, we sample 1024 points during both training and testing. DSA segmentation and CTA projection are resized to (224, 224) to improve registration speed.

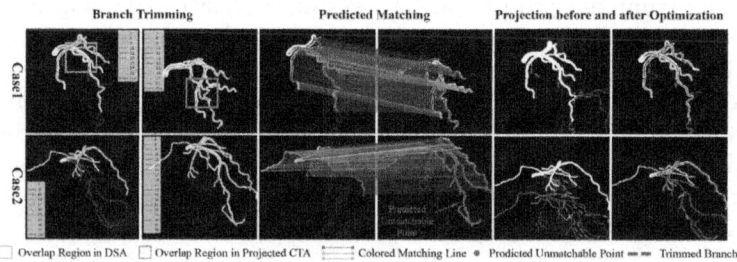

Fig. 6. Illustration of branch trimming, predicted matching, and projection before and after optimization on two selected cases from SimuPaired. Case 1 shows severe overlap, while Case 2 demonstrates a severe CTO. Since our method projects key points of CTA directly from 3D, matching lines are visualized by label colors of respective coronary branches.

Table 1. Robustness to Rotation and Branch Missing in **SimuPaired**.

Method	PD@1mm ↑	
	$R(10°/20°/30°)$	✂(20%/40%/Severe)
ICP-BP [12]	24.5/22.5/17.7	22.5/13.5/4.83
TP-ICC [2]	20.2/17.8/17.0	17.8/10.5/6.84
LG [14]	30.2/17.7/10.0	17.7/14.5/13.3
LG(2D/2D)	89.7/64.1/53.4	64.1/60.0/56.3
Ours	**99.4/98.9/95.9**	**98.9/96.2/93.2**

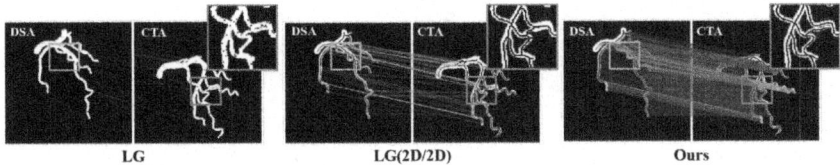

Fig. 7. Matching results for three different sampling methods. The zoomed-in areas highlight the corresponding sampled key points in the overlap region.

4.2 Performance in Simulated Datasets

To evaluate the robustness of our methods, we design two experiments focusing on robustness to branch missing and rotation noise. The ability to handle overlap issues is demonstrated through case evaluation under different rotations. **Robustness to Rotation Noise:** In this experiment, we fix the branch trimming ratio at 20% and applied rotational noise with variances of 10°, 20°, and 30° around all axes. We calculate the percentage of cases with PD below 1mm (PD@1mm) to measure the robustness of the model. Fig. 6 demonstrates the predicted matching and final optimized projection results on the simulated data. In Case 1 of Fig. 6, the simulated CTA and DSA exhibit large rotational differences, causing overlap regions to shift between the two modalities, as indicated by the green and blue boxes. Our method still achieve high-quality matching results in these areas, as indicated by the consistent colored matching lines, and produced projection results that closely aligned with the DSA segmentation. For this case, Fig. 7 clearly demonstrates that our method produces more abundant matching point pairs compared to other sampling methods, achieving more accurate matching predictions in overlapping regions. This is crucial for obtaining robust rigid matching results. **Robustness to Branch Missing:** This experiment evaluates the robustness of methods to an increase in missing branches. The rotation variance is fixed at 20°, while the branch trimming ratio increased from 20% to 40%. A severe CTO scenario is also introduced, where the longest segment and its child branches in the CTA are removed to simulate significant vessel occlusion. As shown in Table 1, compared to other methods, our method maintain the highest PD@1mm even as the trimming ratio increases. Case 2 in

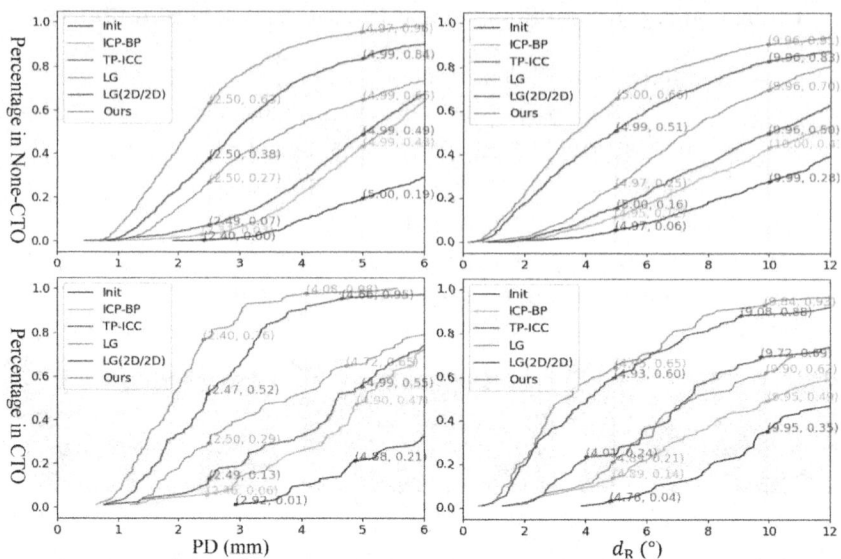

Fig. 8. Comparison of our method with four other methods on RealPaired.

Fig. 6 illustrates a severe CTO scenario. Our method achieve accurate matching results on the non-CTO branches and precisely predicted the CTO branches (indicated by red dots). In the final projection, the visible CTO branches in the CTA aligned accurately with the trimmed branch in DSA (indicated by the gray dashed line).

4.3 Performance in Real Clinical Datasets

Compare with Other Methods: We evaluate the performance of our method on real clinical datasets using cumulative error curves, which illustrate the percentage of cases with errors below a given threshold for PD and d_R. Results are provided for both None-CTO and CTO cases. Our method consistently outperforms others, achieving higher success rates across different thresholds, as shown in Fig. 8. In Fig. 9, our method demonstrates superior performance in handling overlapping regions, as highlighted by the green and blue boxes, and accurately predicts non-shared branches, indicated by the red points. Furthermore, in the final optimized projection results, the projected centerlines produced by our method closely aligned with the DSA angiographic vessels, with only minor deviations observed in small regions with significant non-rigid deformation. **Ablation Study:** Based on 2D/3D sampling strategy, we investigate the impact of two key components on matching performance: 1) using matchability σ to explicitly identify non-shared branches and 2) incorporating 3D attention α_{3D} from CTA. Ablation tests on real clinical datasets show significant improvements

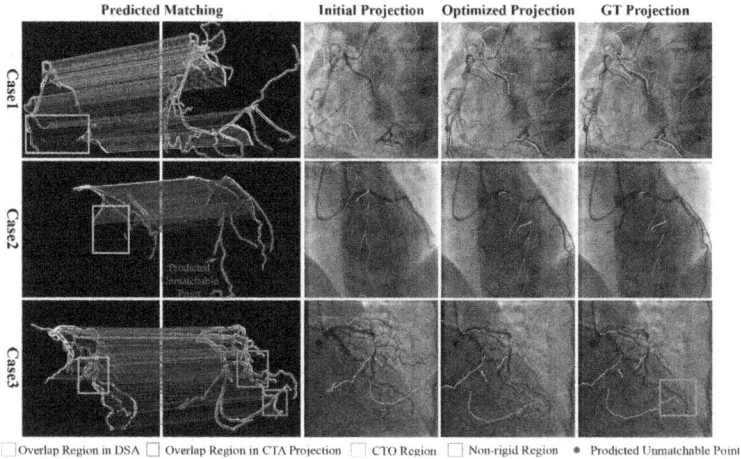

Fig. 9. Visualization of performance of our method on real clinical data: Case 1 (CTO in RCA), Case 2 (CTO in LCA), and Case 3 (Severe Overlap in LCA).

in PD@2.5mm, PD@5mm, d_R@5°, and d_R@10°, which represent the percentage of cases under given thresholds, as detailed in Table 2.

Table 2. Evaluation of σ and 3D Coordinate Integration in **RealRaired**.

Configuration	α_{3D}	σ	Metric			
			PD@2.5mm ↑	PD@5mm ↑	d_R@5° ↑	d_R@10° ↑
Ours(2D/3D)	✔		59.3	92.6	61.3	87.2
		✔	60.0	93.3	62.3	88.8
	✔	✔	**64.5**	**96.7**	**66.5**	**91.2**

5 Conclusion

In this paper, we develop a novel vascular-topology-aware deep structure matching framework for 2D/3D rigid registration between DSA and CTA images, aiming to improve cardiac navigation in PCI. By integrating a hybrid attention mechanism and also vascular topology awareness, our method is able to tackle challenges such as vessel overlap and branch missing. Experiments on both simulated and real datasets show that our method achieves high accuracy and strong generalization, potentially advancing PCI procedure precision and safety.

Acknowledgments. This work was supported in part by the National Natural Science Foundation of China (grant numbers 82441023, U23A20295, 62131015, 82394432,

82471982, and 62403380), the China Ministry of Science and Technology (grant numbers S20240085, STI2030-Major Projects-2022ZD0209000, and STI2030-Major Projects-2022ZD0213100), the Key R&D Program of Guangdong Province, China (grant number 2023B0303040001), the Shanghai Municipal Central Guided Local Science and Technology Development Fund (No.YDZX20233100001001) and the Medical and Health Science and Technology Project of Zhejiang Province (No.2023RC210). We also gratefully acknowledge the computational resources provided by the HPC cluster at ShanghaiTech University, as well as the support from United Imaging Intelligence.

References

1. Benseghir, T., Malandain, G., Vaillant, R.: Iterative closest curve: a framework for curvilinear structure registration application to 2D/3D coronary arteries registration. In: Mori, K., Sakuma, I., Sato, Y., Barillot, C., Navab, N. (eds.) MICCAI 2013. LNCS, vol. 8149, pp. 179–186. Springer, Heidelberg (2013). https://doi.org/10.1007/978-3-642-40811-3_23
2. Benseghir, T., Malandain, G., Vaillant, R.: A tree-topology preserving pairing for 3D/2D registration. Int. J. Comput. Assist. Radiol. Surg. **10**, 913–923 (2015)
3. Cui, Z., et al.: TsegNet: an efficient and accurate tooth segmentation network on 3D dental model. Med. Image Anal. **69**, 101949 (2021)
4. DeTone, D., Malisiewicz, T., Rabinovich, A.: Superpoint: self-supervised interest point detection and description. In: Proceedings of the IEEE Conference on Computer Vision and Pattern Recognition Workshops, pp. 224–236 (2018)
5. Doenst, T., et al.: PCI and CABG for treating stable coronary artery disease: JACC review topic of the week. J. Am. Coll. Cardiol. **73**, 964–976 (2019)
6. Fan, J., Cao, X., Wang, Q., Yap, P.T., Shen, D.: Adversarial learning for mono-or multi-modal registration. Med. Image Anal. **58**, 101545 (2019)
7. Ferraz, L., Binefa, X., Moreno-Noguer, F.: Very fast solution to the PnP problem with algebraic outlier rejection. In: Proceedings of the IEEE Conference on Computer Vision and Pattern Recognition, pp. 501–508 (2014)
8. Gopalakrishnan, V., Dey, N., Golland, P.: Intraoperative 2D/3D image registration via differentiable X-ray rendering. In: Proceedings of the IEEE/CVF Conference on Computer Vision and Pattern Recognition, pp. 11662–11672 (2024)
9. Isensee, F., Jaeger, P.F., Kohl, S.A., Petersen, J., Maier-Hein, K.H.: nnU-Net: a self-configuring method for deep learning-based biomedical image segmentation. Nat. Methods **18**, 203–211 (2021)
10. Jiang, H., Karpur, A., Cao, B., Huang, Q., Araujo, A.: OmniGlue: generalizable feature matching with foundation model guidance. In: Proceedings of the IEEE/CVF Conference on Computer Vision and Pattern Recognition, pp. 19865–19875 (2024)
11. Kendall, A., Grimes, M., Cipolla, R.: PoseNet: a convolutional network for real-time 6-DoF camera relocalization. In: Proceedings of the IEEE International Conference on Computer Vision, pp. 2938–2946 (2015)
12. Khoo, Y., Kapoor, A.: Non-iterative rigid 2D/3D point-set registration using semidefinite programming. IEEE Trans. Image Process. **25**, 2956–2970 (2016)
13. Lee, T.C., Kashyap, R.L., Chu, C.N.: Building skeleton models via 3D medial surface axis thinning algorithms. CVGIP: Graphical Models Image Process. **56**, 462–478 (1994)

14. Lindenberger, P., Sarlin, P.E., Pollefeys, M.: Lightglue: local feature matching at light speed. In: Proceedings of the IEEE/CVF International Conference on Computer Vision, pp. 17627–17638 (2023)
15. Ren, X., et al.: Interleaved 3D-CNN s for joint segmentation of small-volume structures in head and neck CT images. Med. Phys. **45**, 2063–2075 (2018)
16. Sarlin, P.E., DeTone, D., Malisiewicz, T., Rabinovich, A.: Superglue: learning feature matching with graph neural networks. In: Proceedings of the IEEE/CVF Conference on Computer Vision and Pattern Recognition, pp. 4938–4947 (2020)
17. Sun, J., Shen, Z., Wang, Y., Bao, H., Zhou, X.: LoFTR: detector-free local feature matching with transformers. In: Proceedings of the IEEE/CVF Conference on Computer Vision and Pattern Recognition, pp. 8922–8931 (2021)
18. Tajti, P., et al.: Update in the percutaneous management of coronary chronic total occlusions. JACC: Cardiovascular Interventions **11**, 615–625 (2018)
19. Xiong, X., et al.: Discontinuity-aware coronary artery segmentation on CCTA image. In: 2024 IEEE International Symposium on Biomedical Imaging (ISBI), pp. 1–5 (2024)
20. Zhang, T.Y., Suen, C.Y.: A fast parallel algorithm for thinning digital patterns. Commun. ACM **27**, 236–239 (1984)
21. Zhang, X., et al.: SPN-Net: structural points based registration for coronary arteries across systolic and diastolic phases. In: International Conference on Medical Image Computing and Computer-Assisted Intervention, pp. 791–801 (2023)
22. Zhu, J., et al.: 3D/2D vessel registration based on monte Carlo tree search and manifold regularization. IEEE Trans. Med. Imaging (2023)

Unsupervised Deformable Image Registration with Structural Nonparametric Smoothing

Hang Zhang[1], Renjiu Hu[1], Xiang Chen[2], Min Liu[2], Yaonan Wang[2], Rongguang Wang[3], Jinwei Zhang[4], Gaolei Li[5], Xinxing Cheng[6], and Jinming Duan[6,7(✉)]

[1] Cornell University, Ithaca, NY, USA
[2] Hunan University, Changsha, China
[3] University of Pennsylvania, Philadelphia, PA, USA
[4] Johns Hopkins University, Baltimore, MD, USA
[5] Shanghai Jiao Tong University, Shanghai, China
[6] University of Birmingham, Birmingham, UK
Jinming.Duan@manchester.ac.uk
[7] University of Manchester, Manchester, UK

Abstract. Learning-based deformable image registration (DIR) accelerates alignment by amortizing traditional optimization via neural networks. Label supervision further enhances accuracy, enabling efficient and precise nonlinear alignment of unseen scans. However, images with sparse features amid large smooth regions, such as retinal vessels, introduce aperture and large-displacement challenges that unsupervised DIR methods struggle to address. This limitation occurs because neural networks predict deformation fields in a single forward pass, leaving fields unconstrained post-training and shifting the regularization burden entirely to network weights. To address these issues, we introduce **SmoothProper**, a plug-and-play neural module enforcing smoothness and promoting message passing within the network's forward pass. By integrating a duality-based optimization layer with tailored interaction terms, SmoothProper efficiently propagates flow signals across spatial locations, enforces smoothness, and preserves structural consistency. It is model-agnostic, seamlessly integrates into existing registration frameworks with minimal parameter overhead, and eliminates regularizer hyperparameter tuning. Preliminary results on a retinal vessel dataset exhibiting aperture and large-displacement challenges demonstrate our method reduces registration error to 1.88 pixels on 2912 × 2912 images, marking the first unsupervised DIR approach to effectively address both challenges. The source code will be available at https://github.com/tinymilky/SmoothProper.

Keywords: Nonparametric smoothing · Deformable image registration · Neural networks · Plug-and-play · Aperture problem · Large displacement

1 Introduction

In recent years, learning-based deformable image registration (DIR) methods, fueled by progress in neural networks [16,27,67], have become mainstream for various registration tasks. Compared to classical iterative optimization methods [3,4,29,47,68], learning-based approaches [5,10,11,48,50] achieve faster performance through amortized optimization and offer potentially higher accuracy with the integration of label supervision (e.g., segmentation, keypoints).

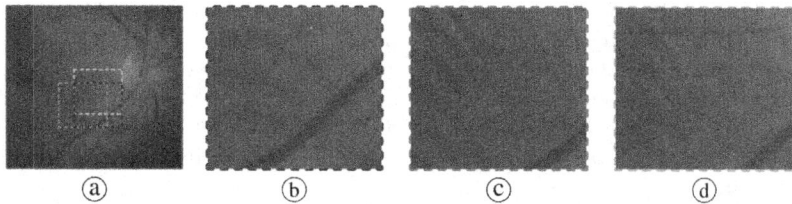

Fig. 1. Visual illustration of aperture and large displacement challenges. (a) shows a retinal vessel patch from the FIRE dataset, with blue, green, and yellow dashed boxes highlighting local vessel structures at different positions. The green and yellow boxes are horizontal and vertical translations of the blue box, respectively. (b), (c), and (d) are zoomed-in views of the blue, green, and yellow boxes, demonstrating that local information alone is insufficient to reconstruct their displacements. Furthermore, the displacements in (b)-(c) and (b)-(d) exceed the vessel diameter. (Color figure online)

Despite promising performance, unsupervised DIR methods struggle with tasks involving both the *aperture problem* and the *large displacement problem*. The aperture problem arises in homogeneous or textureless regions, where limited local evidence within the network's *effective receptive field (ERF)* [45] hinders accurate displacement estimation. The large displacement problem occurs when structural displacement between image pairs exceeds the structure's size. While many registration tasks involve either issue separately, retinal vessel datasets often present both simultaneously (see Fig. 1), significantly complicating unsupervised DIR. Current methods primarily rely on descriptor matching approaches [42,43], and to our knowledge, no existing unsupervised DIR methods have successfully handle both challenges in retinal vessel registration.

We hypothesize that the impact of the pairwise regularization term in the loss function has been underestimated. As demonstrated by the Horn-Schunck optical flow equation [34], this regularizer not only smooths the flow field but also facilitates message passing [25,26,28,75], propagating flow information from regions with strong signals (e.g., boundaries, texture-rich areas) to those with ambiguous or missing signals (e.g., large homogeneous areas). This mechanism is essential for addressing the aperture and large displacement problems. However, most unsupervised DIR methods incorporate this regularizer only as a loss term, shifting the burden for smoothness and signal propagation entirely to the network weights. See Fig. 2 for a visual illustration.

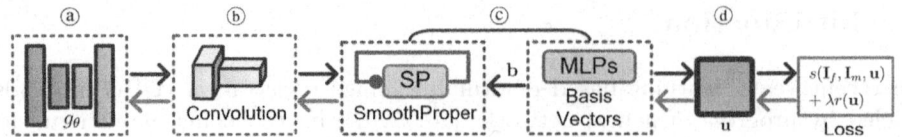

Fig. 2. Illustration of the registration framework and integration of SmoothProper into learning-based frameworks. Black arrows represent the forward pass, while yellow arrows indicate the backward pass. Dashed components are optional: removing (a), (b), and (c) reverts to a classic iterative approach where loss gradients directly update the flow field **u**, similar to ConvexAdam [59], which follows a modern learning framework but aligns with this strategy. Removing (c) leads to a conventional learning-based approach, where the $r(\mathbf{u})$ affects the flow field only through network weights via amortized optimization. Most learning-based frameworks, including pyramid or multi-scale models (except unrolling methods [28,38,54]), use a backbone network followed by convolution (b) to generate the flow field. Removing (b) allows SmoothProper with a basis vector generator to replace this convolution layer, adding minimal parameters. (Color figure online)

To address these challenges, we introduce **SmoothProper**, a plug-and-play unrolled neural network layer seamlessly integrating into existing frameworks with minimal parameter overhead (see Fig. 2, removing (b) or (c) for illustration). Our approach extends the Horn-Schunck variational formulation [34], employing a duality-based optimization framework [8,57,73] with quadratic relaxation [60] to decouple data fidelity from smoothness regularization. Unlike prior unrolling methods [7,35,38,54] emphasizing raw data consistency or neural network-based denoising, SmoothProper introduces adaptive, conditional basis vectors and tailored interaction terms in each unrolled iteration. This structure enables message passing, preserves local structural consistency, and eliminates the need for tuning regularization hyperparameters. SmoothProper resembles MoDL's unrolled architecture [1] and incorporates mechanisms analogous to loopy belief propagation [21] and mean-field message passing [40]. The main contributions of this work are as follows:

- We introduce SmoothProper, an unrolled neural network layer that enforces smoothness and message passing with local structural consistency during the network forward pass, addressing the aperture and large-displacement challenges in unsupervised DIR.
- SmoothProper is lightweight, model-agnostic, integrates seamlessly into unsupervised DIR frameworks with minimal parameter overhead, and removes the need for regularizer hyperparameter tuning.
- We have achieved the first unsupervised learning-based retinal vessel registration, reducing the average error to 1.88 pixels on 2912×2912 images.

2 Related Work

Preliminaries on Learning-Based DIR Methods: Classical iterative methods calculate a flow field by solving a *pairwise* variational optimization problem [34]. This allows direct updates to the flow field $\mathbf{u}$ using loss gradients (see Fig. 2 for a visual example). In contrast, learning-based methods employ *amortized optimization* over network parameters θ across a cohort of image pairs D:

$$\min_{\theta} \; \mathbb{E}_{(\mathbf{I}_f, \mathbf{I}_m) \sim D}[s(\mathbf{I}_f, \mathbf{I}_m, \mathbf{u}) + \lambda r(\mathbf{u})], \quad \text{s.t.} \quad \mathbf{u} = g_\theta(\mathbf{I}_f, \mathbf{I}_m). \tag{1}$$

Here, $\mathbf{I}_f$ and $\mathbf{I}_m$ denote the fixed and moving images, respectively; $\mathbf{u}$ is the displacement field output by a neural network g parameterized by θ; λ is a hyperparameter; s is the dissimilarity function; and r is typically a pairwise diffusive regularizer. Training can be accomplished by accumulating gradient backflows with respect to θ:

$$\frac{\partial \mathcal{L}}{\partial \theta} = \sum_{f,m} \frac{\partial s}{\partial g_\theta} \frac{\partial g_\theta}{\partial \theta} + \lambda \sum_{f,m} \frac{\partial r}{\partial g_\theta} \frac{\partial g_\theta}{\partial \theta} \tag{2}$$

Here, $\mathcal{L}$ denotes the total loss, with the Jacobians mapping flow gradients to network weights. As noted by Hansen and Heinrich [26], and shown in recent work [36], unsupervised DIR methods relying on amortized optimization yield limited gains in label matching (e.g., Dice score), since the first Jacobian in Eq. (2) contains no more information than raw intensity-label mutual information. We hypothesize that while the first Jacobian fails to improve label matching, the second Jacobian, beyond enforcing smoothness, can function as an effective message passing mechanism [21,26] if properly utilized. This potential has been largely overlooked, and our SmoothProper addresses it via a cost-efficient, easily integrated architecture grounded in modern neural network design.

Learning-Based and Deep Unrolling DIR Methods: To address aperture and large-displacement challenges, increasing the network ERF is common, involving methods like transformers [10,58,67], large-kernel convolutions [15,37], Laplacian pyramids [52,70,74], and coarse-to-fine designs [46,48]. However, relying solely on the loss term and network weights for message passing may risk the network acting as a low-pass filter [55,72], leading to oversmoothing that obscures structural details. To overcome these limitations, deep unrolling [32] integrates model-based methods [1,76–78] and iterative algorithms [12,23,63] directly into neural network architectures. Recent studies [7,28,35,38,54] have applied deep unrolling to image registration, enforcing both dissimilarity and regularity functions during the network forward pass. Among these works, most focus on either raw data consistency or neural network denoisers as regularizers, while PDD-Net [28] remains the only one that explicitly enforces smoothness, though it relies on a costly 6D dissimilarity map and underutilizes neural network capacity for message passing. GraDIRN [54] uses unrolled gradient descent for data consistency, PRIVATE [35] employs a plug-and-play denoiser with instance-wise optimization, and SUITS [7] and VR-Net [38] rely on separate feature extraction or alternating data consistency and denoising steps. In contrast, SmoothProper directly integrates smoothness regularization into the network forward pass, enabling efficient message passing without label supervision or heavy computational overhead.

3 Method

This section introduces the bi-level optimization framework, defines SmoothProper and its optimization, presents a conditional automatic adjustment for λ, and outlines the overall framework integrating SmoothProper.

3.1 The SmoothProper

Bi-level Optimization framework: The proposed SmoothProper adds a sub-optimization problem as a constraint to Eq. (1):

$$\min_{\theta} \mathbb{E}_{(\mathbf{I}_f, \mathbf{I}_m) \sim D}[\mathcal{L}(\mathbf{I}_f, \mathbf{I}_m, \mathbf{u}^*(\theta))], \tag{3a}$$

$$\text{s.t. } \mathbf{u}^*(\theta) = \arg\min_{\mathbf{u}} \mathcal{S}(\mathbf{I}_f, \mathbf{I}_m, \mathbf{u}, \theta). \tag{3b}$$

Equation (3b) constrains the expected loss in Eq. (3a), optimizing the displacement field by minimizing the SmoothProper function $\mathcal{S}$ based on input images and network parameters. The optimization $\mathcal{S}$ is embedded in the registration network by unrolling its iterative steps into the forward pass, with trainable parameters θ. Given the solution $\mathbf{u}^*$ for each image pair in the training set D, Eq. (3a) minimizes the expected loss $\mathcal{L}(\mathbf{I}_f, \mathbf{I}_m, \mathbf{u}^*(\theta))$ to optimize θ. Unlike GraDIRN [54], which unrolls its energy function to enforce data consistency, our approach unrolls $\mathcal{S}$ to shift the regularization burden away from network weights.

Quadratic Relaxation: Following prior works [8,60,68,73], we adopt a quadratic relaxation of Eq. (1) to decouple the data consistency term and the regularizer:

$$\min_{\mathbf{u}} \int_{\Omega} \left(\rho(\mathbf{v}) + \frac{1}{2\alpha} \|\mathbf{v} - \mathbf{u}\|^2 + \beta r(\mathbf{u}) \right) d\Omega. \tag{4}$$

Here, $\rho(\mathbf{v})$ measures the cost of applying the flow field $\mathbf{v}$ to the input image pair, while $\beta r(\mathbf{u})$ represents the smoothness regularizer with strength β. It has been shown [8] that as $\alpha \to 0$, minimizing Eq. (4) with the same data and regularization terms is equivalent to minimizing it without the interaction term. Although non-quadratic relaxation terms exist [6], we empirically found the sub-optimization problem easier to solve, whereas using an L1 term requires proximal gradients, adding complexity.

Many gradient-descent methods rely on first-order Taylor expansions to linearize $\rho(\mathbf{v})$, limiting them to small steps and requiring numerous iterations, which hinders handling large deformations. Deep unrolled DIR methods [7,38,54], built on gradient-based optimization, inherit these limitations.

In contrast, discrete optimization approaches [29,30,59] rely on predefined cost volumes, allowing each subproblem for **v** and **u** to be convex and optimally solvable, thus better accommodating large deformations. However, their computational cost rises sharply when expanding the discrete space to cover larger displacements.

Objective Function: SmoothProper leverages neural networks to adaptively linearize local intensities using a basis-constrained formulation. We discretize Eq. (4) as a sum of voxel- and pair-wise energies, with the basis matrix $\mathbf{b} \in \mathbb{R}^{m \times d}$, where $\mathbf{b} = [\mathbf{b}_1, \mathbf{b}_2, \ldots, \mathbf{b}_m]^\top$ and each $\mathbf{b}_i \in \mathbb{R}^d$ is a basis vector ($d = 2$ for 2D images). Given the neural network output $\mathbf{p} = g_\theta(\mathbf{I}_f, \mathbf{I}_m)$, where $\mathbf{p}(x) \in \mathbb{R}^m$ is a non-negative coefficient vector (from ReLU), representing the flow signal strength at $x \in \Omega$, the basis vectors are linearly combined to produce the displacement fields. This leads to the following energy optimization:

$$\min_{\mathbf{q},\mathbf{v}} \sum_{x \in \Omega} \|\mathbf{p}(x) - \mathbf{q}(x)\|^2 + \sum_{x \in \Omega} \sum_{i=1}^{m} \frac{1}{2\alpha} \mathbf{q}_i(x) \|\mathbf{v}(x) - \mathbf{b}_i\|^2$$
$$+ \sum_{x \in \Omega} \frac{1}{2\alpha} \|\mathbf{q}(x)\mathbf{b} - \mathbf{v}(x)\|^2 + \beta \mathbf{r}(\mathbf{v}). \quad (5)$$

Here, $\mathbf{q}(x)\mathbf{b}$ computes the displacement at location x, while $\mathbf{v}(x)$ represents the auxiliary displacement vector at x. The first term aligns $\mathbf{q}(x)$ with the network's initial prediction $\mathbf{p}(x)$, while the fourth term $\mathbf{r}(\mathbf{v}) = \|\nabla \mathbf{v}\|^2$ is a diffusive regularizer. The second term, unique to this work, introduces a directional bias between $\mathbf{q}(x)$ and $\mathbf{b}_i$, ensuring the displacement can be closely represented by a linear combination of the basis vectors $\mathbf{b}_i$, thereby guiding the reconstruction of displacement field from $\mathbf{q}$. The third coupling term pulls $\mathbf{v}(x)$ toward the linear combination $\mathbf{q}(x)\mathbf{b}$. This creates a feedback loop that enables two key properties:

- **Message Passing**: The term $\|\nabla \mathbf{v}\|^2$ encourages gradual changes in $\mathbf{v}$, smoothing $\mathbf{q}$ as $\mathbf{q}\mathbf{b}$ aligns with $\mathbf{v}$. This results in $\mathbf{q}$ serving as a smooth approximation of $\mathbf{p}$, effectively passing flow signals across regions.
- **Structural Consistency**: As $\mathbf{q}(x)$ and $\mathbf{v}(x)$ iteratively align with $\mathbf{p}(x)$ and the basis $\mathbf{b}_i$, the equation ensures that regions with strong flow signals (large $\mathbf{q}(x)$) remain anchored to their representative basis patterns $\mathbf{b}_i$.

This basis-constrained formulation employs a **"smooth-reinforce"** interplay: The smoothing step diffuses signals, slightly diminishing strong ones and enhancing weaker ones. The reinforcement step then **"locks in"** strong signals: when $\mathbf{p}(x)$ is large, the second term keeps $\mathbf{v}(x)$ close to the corresponding basis pattern, preventing dilution. Meanwhile, weak signals, associated with smaller $\mathbf{q}(x)$, can be influenced by stronger signals spreading through message passing of $\mathbf{p}(x)$. This iterative smooth-reinforce cycle promotes signal propagation across regions, preserves strong signals, and effectively extends them into weaker areas, addressing the aperture problem and mitigating oversmoothing.

Unlike previous methods that rely on manual linearization [7,38] or omit it entirely [35,54], our proposed Eq. (5) adaptively linearizes image intensities within the network's effective receptive field (ERF) into learned coefficients. In contrast to PDD-Net [28], which uses fixed displacement vectors and pre-computed cost volumes, SmoothProper jointly learns both basis vectors and coefficients during training. These learned basis vectors serve as "dictionary atoms" [61,74] or clustering centers [19,66], providing an adaptive representation that directly influences field smoothness. This approach achieves higher accuracy without computing a full cost volume, instead encoding disparity context directly within the ERF. Figure 3b illustrates the recursive SmoothProper with its feedback loop.

Unrolling the Recursive Networks. To solve Eq. (5), we employ coordinate descent, alternating between $\mathbf{q}^{(k)}$ and $\mathbf{v}^{(k)}$ while progressively reducing α to achieve $\mathbf{q}^{(K)}\mathbf{b} \approx \mathbf{v}^{(K)}$, thus approximating the solution of Eq. (5). Here, K is the total number of iterations, and k denotes the k_{th} iteration. Following prior work [30], we set $\alpha = \{150, 50, 15, 5, 1.5, 0.5\}$ for $K = 6$ alternating iterations. Let the directional bias and coupling terms be represented by $\mathcal{C}(\mathbf{q}, \mathbf{v}, \mathbf{b})$, we decompose the optimization of Eq. (5) into two subproblems as follows:

$$\mathbf{q}^{(k)} = \arg\min_{\mathbf{q}} \frac{1}{2\alpha}\mathcal{C}(\mathbf{q}, \mathbf{v}^{(k-1)}, \mathbf{b}) + \sum_{x \in \Omega} \|\mathbf{p}(x) - \mathbf{q}(x)\|^2;$$

$$\mathbf{v}^{(k)} = \arg\min_{\mathbf{v}} \frac{1}{2\alpha}\mathcal{C}(\mathbf{q}^{(k)}, \mathbf{v}, \mathbf{b}) + \beta\|\nabla\mathbf{v}\|^2. \tag{6}$$

The initial $\mathbf{q}^{(0)}$ is set to $\mathbf{p} = g_\theta(\mathbf{I}_f, \mathbf{I}_m)$, the output from the backbone network g_θ. The $\mathbf{q}$ subproblem can be solved optimally and pointwise across $x \in \Omega$ in

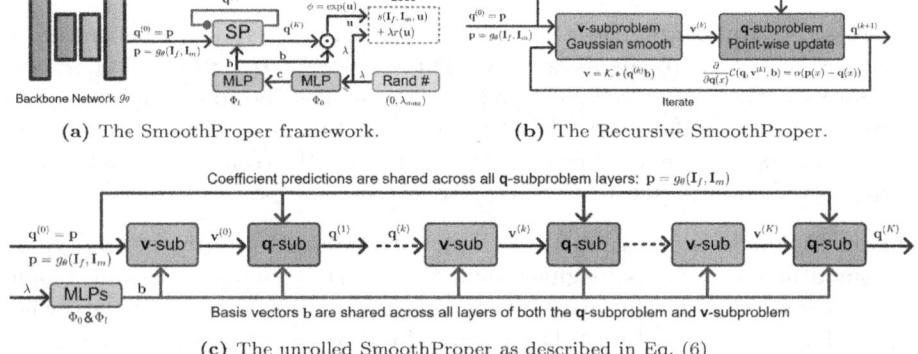

Fig. 3. (a) is the schematic of the SmoothProper (SP) framework. The yellow box represents SP optimization, where $\mathbf{p} = g_\theta(\mathbf{I}_f, \mathbf{I}_m)$ and $\mathbf{b} = \Phi_l(\Phi_0(\lambda))$, with λ randomly sampled from $(0, \lambda_{max})$. SP iteratively updates $\mathbf{q}^k$ to $\mathbf{q}^{k+1}$ until the maximum iteration count K is reached. (b) and (c) are visualizations of the SmoothProper module architecture in recursive and unrolled forms. Basis vectors $\mathbf{b}$ and α are omitted for brevity (details in Sect. 3.1). (Color figure online)

closed form by taking the derivative of $\mathcal{C}(\mathbf{q}, \mathbf{v}^{(k-1)}, \mathbf{b})$ with respect to $\mathbf{q}(x)$ and setting it to zero:

$$\frac{\partial}{\partial \mathbf{q}(x)}\mathcal{C}(\mathbf{q}, \mathbf{v}^{(k-1)}, \mathbf{b}) = \alpha\big(\mathbf{p}(x) - \mathbf{q}(x)\big). \qquad (7)$$

The $\mathbf{v}$ subproblem is solved through fixed-point iteration [8], involving neighboring values of $\mathbf{v}(x)$ and $\mathbf{q}(x)\mathbf{b}$, yielding $\mathbf{v}$ as a smoothed version of $\mathbf{qb}$. This can be approximated by applying Gaussian blurring to $\mathbf{qb}$ as $\mathbf{v}^{(k)} = \mathcal{K} * \mathbf{q}^{(k)}\mathbf{b}$ across spatial locations, with the blurring strength determined by β. A similar approach for optimizing $\mathbf{v}$ is discussed in Diffeomorphic Demons [68]. The unrolled SmoothProper architecture is illustrated in Fig. 3c.

3.2 Conditional Learning of Basis Vectors

SmoothProper serves as the constraint in Eq. (3), providing a smoothed version of $\mathbf{p} = g_\theta(\mathbf{I}_f, \mathbf{I}_m)$ with $\mathbf{q}$. The learnable basis vectors $\mathbf{b}$ link the smoothness of $\mathbf{v}$ to $\mathbf{q}$, adjusting the weight of $\mathbf{q}$ based on its proximity to $\mathbf{v}$. This enables adaptive regularization strength in the outer loop of Eq. (3a) through $\mathbf{b}$. We use two MLPs, Φ_0 and Φ_l: $\mathbf{c} \leftarrow \Phi_0(\lambda)$ generates a latent code $\mathbf{c}$, and $\mathbf{b} \leftarrow \Phi_l(\mathbf{c})$ produces conditional basis vectors. Using two MLPs instead of one provides model-agnostic flexibility, allowing $\mathbf{c}$ to be shared across all pyramid or scale levels if the model is multi-scale-based, while each level has its own Φ_l.

During training, we randomly sample $\lambda \in (0, \lambda_{max})$ at each forward pass, enabling $\mathbf{b}$ and the from-matching strength λ to adapt via $\Phi_l(\Phi_0(\lambda))$. This eliminates the need for manual λ-tuning and allows smoothness strength adjustments after amortized optimization. Prior works (e.g., HyperMorph [33] and cLapIRN [51]) also use hypernetworks [24], but they rely on inefficient, non-scalable designs: HyperMorph conditions the entire network, and cLapIRN requires extensive architectural changes. In contrast, our method is lightweight and efficient, requiring only a single convolution layer replacement with a SmoothProper layer at the backbone's end, leaving the backbone unchanged (Fig. 2).

3.3 Overall Framework

Figure 3a shows the SmoothProper framework, which consists of three components: a backbone network g_θ for initial coefficient estimation, MLPs Φ_0 and Φ_l for latent code and conditional basis vector generation, and unrolled SmoothProper layer. The SmoothProper module processes the network-predicted coefficients $\mathbf{p}$ and the MLP-predicted basis vectors $\mathbf{b}$ through K iterations to produce $\mathbf{q}^{(K)}$, a smooth and structurally consistent version of $\mathbf{p}$. The flow field $\mathbf{u}$ is obtained by linearly combining $\mathbf{q}^{(K)}$ with $\mathbf{b}$ and is integrated using scaling and squaring [2] to generate a diffeomorphic deformation field ϕ. As in prior work [13], the number of integration steps is fixed to 7. The deformation field ϕ is used for the dissimilarity term, while $\mathbf{u}$ is used for the regularization term in the loss function. In summary, SmoothProper not only smooths the displacement field $\mathbf{u}$ through the diffusive regularizer but also enables message passing, where flow signals (represented by coefficient vectors **p**) propagate with local structural consistency preserved via the feedback loop formed by the interaction term $\mathcal{C}(\mathbf{q}, \mathbf{v}, \mathbf{b})$.

4 Experiments and Results

4.1 Datasets and Evaluation Metrics

We evaluated our method on the public Fundus Image Registration (FIRE) dataset [31], which contains 134 image pairs from 39 subjects. Captured at 2912 × 2912 resolution with a 45° field of view using a Nidek AFC-210 fundus camera, the images were collected at Papageorgiou Hospital, Aristotle University of Thessaloniki. Each pair includes 10 annotated landmarks. To enhance details, we extract the green channel (richest in structural information due to red-free photography), apply Contrast Limited Adaptive Histogram Equalization (CLAHE) for local contrast enhancement, and use gamma correction ($\gamma = 1.2$) for better visibility. The dataset is divided into three categories: $\mathcal{S}$ (71 pairs, > 75% overlap, no anatomical changes), $\mathcal{P}$ (49 pairs, < 75% overlap, no anatomical changes), and $\mathcal{A}$ (14 pairs, > 75% overlap, anatomical changes). It was stratified by category and split into train, val, and test sets in a 7:1:2 ratio.

We evaluated model performance using Target Registration Error (TRE) and Area Under the Curve (AUC) for TRE thresholds. TRE measures the L2 distance between corresponding points in the fixed and warped images, with lower values indicating better alignment. AUC (AUC@15, AUC@25, AUC@50) represents the percentage of TRE values below specific thresholds, providing a normalized measure of registration success. We also report multiply-add operations (G) and parameter size (MB) to assess model complexity.

Table 1. Quantitative comparison on the FIRE dataset. Best results are in bold. ↓ indicates lower is better, and ↑ indicates higher is better. The initial alignment method for learning-based approaches is highlighted in gray.

Category	Methods	TRE ↓	AUC ↑			Group-wise AUC@25 ↑		
			mAUC@15	mAUC@25	mAUC@50	mAUC@$\mathcal{A}$	mAUC@$\mathcal{P}$	mAUC@$\mathcal{S}$
Detector-based Methods	XFeat [CVPR'24] [53]	10.858	0.560	0.637	0.794	0.853	0.102	0.915
	R2D2 [NeurIPS'19] [56]	7.926	0.553	0.701	0.850	0.813	0.333	0.899
	LightGlue [ICCV'23] [41]	7.802	0.575	0.710	0.855	0.853	0.338	0.904
	SuperPoint [CVPR'18] [14]	6.641	0.612	0.757	0.879	0.813	0.453	0.928
	Glampoints [ICCV'19] [64]	6.608	0.595	0.757	0.879	0.733	0.560	0.880
	SuperRetina [ECCV'22] [43]	6.382	0.622	0.767	0.884	0.813	0.516	0.909
	RetinaIPA [MICCAI'24] [71]	5.750	0.657	0.774	0.885	0.799	0.599	0.886
Detector-free Methods	LoFTR [CVPR'21] [62]	7.638	0.526	0.716	0.858	0.693	0.564	0.811
	DKM [CVPR'23] [17]	6.493	0.610	0.760	0.880	0.800	0.529	0.891
	Aspanformer [ECCV'22] [9]	6.415	0.602	0.761	0.881	0.800	0.556	0.877
	RoMa [CVPR'24] [18]	6.388	0.605	0.763	0.881	0.800	0.565	0.875
	GeoFormer [ICCV'23] [42]	6.201	0.625	0.770	0.887	0.813	0.587	0.925
Learning-based Methods w/o Deep Unrolling	KeyMorph [MIDL'22] [20]	5.947	0.640	0.784	0.892	0.813	0.551	0.917
	C2FViT [CVPR'22] [49]	5.842	0.642	0.786	0.893	0.827	0.547	0.920
	GAMorph [HBM'24] [44]	3.081	0.825	0.895	0.945	0.906	0.800	0.949
Learning-based Methods w/ Deep Unrolling	GraDIRN [MICCAI'22] [54]	6.344	0.657	0.774	0.885	0.799	0.599	0.886
	PDD-Net [MICCAI'19] [28]	5.765	0.688	0.792	0.893	0.819	0.598	0.915
	VR-Net [TMI'21] [38]	4.974	0.705	0.823	0.911	0.827	0.660	0.931
	SmoothProper (Ours)	**1.879**	**0.920**	**0.951**	**0.974**	**0.937**	**0.920**	**0.974**

4.2 Baseline Methods and Implementation Details

For retinal vessel datasets, most studies focus on descriptor matching methods using homography transformations. We benchmarked both detector-based methods, including SuperPoint [14], Glampoints [64], R2D2 [56], SuperRetina [43], LightGlue [41] with DISK [65], SiLK [22], XFeat [53], and RetinaIPA [71], and detector-free methods, such as LoFTR [62], DKM [17], AspanFormer [9], RoMa [18], and GeoFormer [42]. We also evaluated learning-based methods, including KeyMorph [69] and C2FViT [49], which claim to address large misalignments and arbitrary rotations ($[-180°, 180°]$). Additionally, we included deep unrolling methods, such as VR-Net [38], GraDIRN [54], and PDD-Net [28], as well as GAMorph [44], the only known method co-training descriptor matching with deformable registration, which shows promising results.

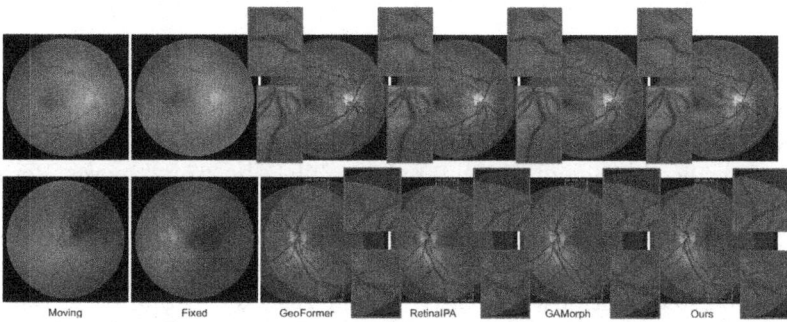

Fig. 4. Qualitative comparison. Fixed images are shown with the 'inferno' colormap, and moving images are overlaid with the 'viridis' colormap for clarity. Highlighted shadow regions show challenging cases with local anatomical changes, geometric shifts, or image shadowing.

All experiments, including SmoothProper and baseline models, were conducted on a machine with an NVIDIA A100 GPU (80GB), a 16-core CPU, and 32GB RAM, using Python 3.7 and PyTorch 1.9.0. We used the Adam optimizer [39] with a polynomially decayed learning rate starting at 4×10^{-4}, a batch size of 1, 100 epochs, and a regularization strength $\lambda = 1$ in Eq. (1) for all learning-based methods across both datasets. Local normalized cross-correlation served as the dissimilarity metric, with $r(\mathbf{u}) = \|\nabla \mathbf{u}\|^2$ as the regularirizer for learning based methods as well as outer loop regularizer for SmoothProper. For descriptor matching methods, only GeoFormer [42], SuperRetina [43], and RetinaIPA [71], designed specifically for retinal vessel registration, were trained from scratch on the FIRE dataset. For other models, fine-tuning on FIRE resulted in poorer performance compared to their pre-trained weights, likely due to the dataset's small size. For all learning-based methods, we initialized alignment using the best-performing descriptor matching method. This initial alignment is crucial for unsupervised DIR, as extreme conditions (e.g., 180° rotation) may surpass the capability of deformable-based methods.

While SmoothProper is model-agnostic, we employ a simple three-level Laplacian image pyramid as its backbone network. It uses three 2D convoloution, instance normalization and ReLU activations blocks (Conv-Block) as encoder to extract moving and fixed feature maps, followed by bilinear downsampling to form a 3-level image pyramid. At each pyramid level, the fixed feature map along with the moving feature map wapred by the previous level's deformation field (if any) concatenated through channels, followed by three 2D Conv-Blocks for coefficient vector extraction, and finalized with our SmoothProper module to produce the final field. The result of SmoothProper in Table 1 uses the parameters $m = 4 \times 3^2$, $K = 6$, and an input image size of 1024×1024. The effects of these parameters are discussed in Sect. 4.3 and shown in Fig. 6.

4.3 Results and Analysis

Quantitative Results: Table 1 presents the quantitative results on the FIRE dataset, where the proposed SmoothProper outperforms all other methods across all metrics. This marks the first successful application of unsupervised DIR for retinal vessel registration, achieving a TRE of 1.88 pixels on 2912×2912 images. Specifically, SmoothProper reduces TRE by 67.32% over RetinaIPA (detector-based), 69.68% over GeoFormer (detector-free), 39.01% over GAMorph (learning-based w/o unrolling), and 62.20% over VR-Net (learning-based w/ unrolling). SmoothProper also achieves the highest mAUC@25, the key metric for validating registrations with TRE within 25 pixels. Notably, in the breakdown of mAUC@25 into categories $\mathcal{A}$, $\mathcal{P}$, and $\mathcal{S}$, learning-based methods, especially GAMorph and SmoothProper, outperform descriptor-based methods significantly. This reflects the limitations of homography-based descriptor matching in longitudinal retinal imaging, where local deformations occur over time. In addition, most learning-based DIR methods, except GAMorph and VR-Net, perform worse than their initial alignment from descriptor matching method RetinalIPA. This likely occurs because the vessel regions occupy a small portion of the image, causing loss-term-based smooth regularization or message passing to dilute vessel displacements within large, smooth, signal-less regions.

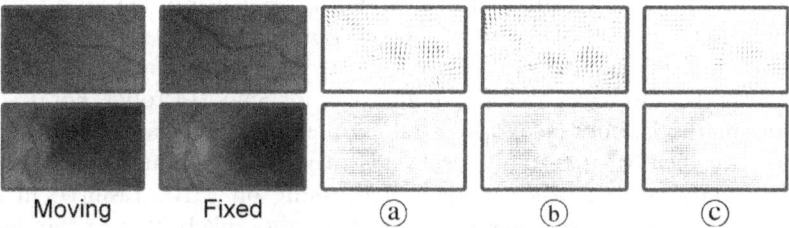

Fig. 5. Qualitative comparison of deformation fields: (a) Original SmoothProper, (b) SmoothProper without conditional basis vectors, and (c) GAMorph. Darker displacement vectors indicate larger magnitudes.

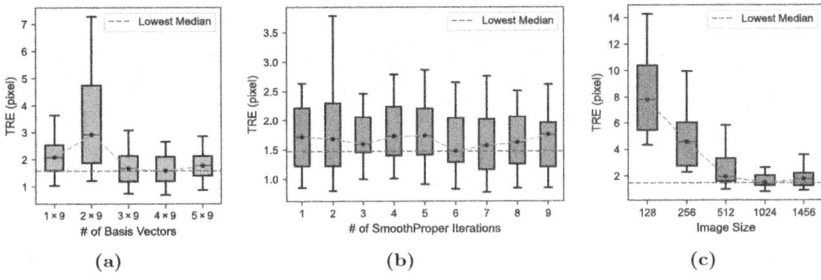

Fig. 6. Ablation analysis: boxplots and line plots showing the effect of basis vector count (a), inner-loop iterations (b), and input image size (c) on TRE (pixels). Red dashed lines indicate the lowest median TRE in each plot. (Color figure online)

Qualitative Results: Figure 4 shows the discrepancy between warped moving images and fixed images for exemplar methods in each category. Descriptor matching methods struggle with local deformations, and GAMorph shows limited adaptability. In contrast, SmoothProper best adapts to local deformations and is less affected by shadow artifacts. Figure 5 presents local image patches and their displacement fields. Comparing (a) and (c), SmoothProper better preserves local vessel structures, avoids shadowing artifacts, and maintains tissue integrity with a regular field that prevents oversmoothing or over-deemphasis.

Complexity and Smoothness Analysis: The SmoothProper (SP) counterpart without SP and CBV has 0.32 MB parameters and 31.84 G multi-adds. With SP (no CBV), it uses 0.32 MB and 31.95 G multi-adds. With SP and CBV, it increases to 41.01 MB and 50.18 G multi-adds. Despite the parameter-heavy MLPs for CBV, the SP module remains highly efficient. Readers can decide if the MLPs are worth using based on specific applications. As shown in Fig. 5 (a) and (b), CBV produces a more regular and plausible displacement field without tuning the λ-parameter, improving TRE slightly from 1.93 to 1.88. All baseline descriptor matching and learning-based methods, including initial alignment runtime, processed each scan pair in under one second.

Ablation Analysis on the Number of Basis Vectors m: In Fig. 6a, we vary m from 1×3^2 to 5×3^2, where 3^2 mimics the local 3×3 pixel neighborhood in discrete optimization. This neighborhood does not reflect actual displacements but corresponds to learned displacement atoms based on the local disparity context. Larger m allows for richer information querying and better local structural consistency. However, increasing m too much can lead to limitations due to the available training samples. As shown, in FIRE, TRE decreases with m initially, reaching a minimum at 4×3^2 before rising again.

Ablation Analysis on Iterations K: We vary K from 1 to 9, as shown in Fig. 6b, and find that the model achieves optimal performance at $K = 6$. This result aligns with prior work [79] using mean-field inference [40] with 5 iterations, where further increasing K yields limited improvement.

Ablation Analysis on Image Size: We varied the squared input image size from 128 to 1024, doubling at each step, and included 1456 (half of the original size). As shown in Fig. 6c, 128 yields worse performance than the initial alignment. TRE decreases steadily from 256 to 1024 but increases at 1456. Smaller sizes improve efficiency but reduce accuracy, likely due to vessels becoming less visible. Larger sizes can introduce local noise, limiting accuracy gains.

5 Conclusions

We introduce SmoothProper, a plug-and-play neural network module that enforces smoothness and message passing within the forward pass. By integrating a duality-based optimization layer with tailored interaction terms, it efficiently propagates flow signals while preserving structural consistency. SmoothProper addresses aperture and large displacement challenges in retinal vessel datasets, achieving a TRE of 1.88 pixels in unsupervised DIR for the first time. Although this study focuses on the 2D problem, extending it to 3D is as simple as replacing the 2D convolution with a 3D one. We hope our findings on utilizing the regularization term inspire new milestones and reignite interest in unsupervised DIR techniques.

References

1. Aggarwal, H.K., Mani, M.P., Jacob, M.: MODL: model-based deep learning architecture for inverse problems. IEEE Trans. Med. Imaging **38**(2), 394–405 (2018)
2. Arsigny, V., Commowick, O., Pennec, X., Ayache, N.: A log-euclidean framework for statistics on diffeomorphisms. In: Larsen, R., Nielsen, M., Sporring, J. (eds.) MICCAI 2006. LNCS, vol. 4190, pp. 924–931. Springer, Heidelberg (2006). https://doi.org/10.1007/11866565_113
3. Ashburner, J.: A fast diffeomorphic image registration algorithm. Neuroimage **38**(1), 95–113 (2007)
4. Avants, B.B., Tustison, N.J., Song, G., Cook, P.A., Klein, A., Gee, J.C.: A reproducible evaluation of ANTs similarity metric performance in brain image registration. Neuroimage **54**(3), 2033–2044 (2011)
5. Balakrishnan, G., Zhao, A., Sabuncu, M.R., Guttag, J., Dalca, A.V.: Voxelmorph: a learning framework for deformable medical image registration. IEEE Trans. Med. Imaging **38**(8), 1788–1800 (2019)
6. Black, M.J., Anandan, P.: A framework for the robust estimation of optical flow. In: 1993 (4th) International Conference on Computer Vision, pp. 231–236. IEEE (1993)
7. Blendowski, M., Hansen, L., Heinrich, M.P.: Weakly-supervised learning of multimodal features for regularised iterative descent in 3D image registration. Med. Image Anal. **67**, 101822 (2021)
8. Chambolle, A.: An algorithm for total variation minimization and applications. J. Math. Imaging Vis. **20**, 89–97 (2004)
9. Chen, H., et al.: Aspanformer: detector-free image matching with adaptive span transformer. In: European Conference on Computer Vision, pp. 20–36. Springer (2022)

10. Chen, J., Frey, E.C., He, Y., Segars, W.P., Li, Y., Du, Y.: Transmorph: transformer for unsupervised medical image registration. Med. Image Anal. **82**, 102615 (2022)
11. Chen, X., et al.: Spatially covariant image registration with text prompts. IEEE Trans. Neural Networks Learn. Syst. 1–11 (2024)
12. Chen, X., Liu, J., Wang, Z., Yin, W.: Theoretical linear convergence of unfolded ISTA and its practical weights and thresholds. In: Advances in Neural Information Processing Systems, vol. 31 (2018)
13. Dalca, A.V., Balakrishnan, G., Guttag, J., Sabuncu, M.R.: Unsupervised learning of probabilistic diffeomorphic registration for images and surfaces. Med. Image Anal. **57**, 226–236 (2019)
14. DeTone, D., Malisiewicz, T., Rabinovich, A.: Superpoint: self-supervised interest point detection and description. In: Proceedings of the IEEE Conference on Computer Vision and Pattern Recognition Workshops, pp. 224–236 (2018)
15. Ding, X., Zhang, X., Ma, N., Han, J., Ding, G., Sun, J.: RepVGG: making VGG-style convnets great again. In: Proceedings of the IEEE/CVF Conference on Computer Vision and Pattern Recognition, pp. 13733–13742 (2021)
16. Dosovitskiy, A., et al.: An image is worth 16x16 words: transformers for image recognition at scale. In: International Conference on Learning Representations
17. Edstedt, J., Athanasiadis, I., Wadenbäck, M., Felsberg, M.: DKM: dense kernelized feature matching for geometry estimation. In: Proceedings of the IEEE/CVF Conference on Computer Vision and Pattern Recognition, pp. 17765–17775 (2023)
18. Edstedt, J., Sun, Q., Bökman, G., Wadenbäck, M., Felsberg, M.: Roma: revisiting robust losses for dense feature matching. arXiv preprint arXiv:2305.15404 (2023)
19. Esser, P., Rombach, R., Ommer, B.: Taming transformers for high-resolution image synthesis. In: Proceedings of the IEEE/CVF Conference on Computer Vision and Pattern Recognition, pp. 12873–12883 (2021)
20. Evan, M.Y., Wang, A.Q., Dalca, A.V., Sabuncu, M.R.: Keymorph: robust multimodal affine registration via unsupervised keypoint detection. In: Medical Imaging with Deep Learning (2022)
21. Felzenszwalb, P.F., Huttenlocher, D.P.: Efficient belief propagation for early vision. Int. J. Comput. Vision **70**, 41–54 (2006)
22. Gleize, P., Wang, W., Feiszli, M.: Silk: simple learned keypoints. In: Proceedings of the IEEE/CVF International Conference on Computer Vision, pp. 22499–22508 (2023)
23. Gregor, K., LeCun, Y.: Learning fast approximations of sparse coding. In: Proceedings of the 27th International Conference on International Conference on Machine Learning, pp. 399–406 (2010)
24. Ha, D., Dai, A.M., Le, Q.V.: Hypernetworks. In: International Conference on Learning Representations (2017). https://openreview.net/forum?id=rkpACe1lx
25. Hansen, L., Heinrich, M.P.: Deep learning based geometric registration for medical images: how accurate can we get without visual features? In: Feragen, A., Sommer, S., Schnabel, J., Nielsen, M. (eds.) IPMI 2021. LNCS, vol. 12729, pp. 18–30. Springer, Cham (2021). https://doi.org/10.1007/978-3-030-78191-0_2
26. Hansen, L., Heinrich, M.P.: Revisiting iterative highly efficient optimisation schemes in medical image registration. In: de Bruijne, M., et al. (eds.) MICCAI 2021. LNCS, vol. 12904, pp. 203–212. Springer, Cham (2021). https://doi.org/10.1007/978-3-030-87202-1_20
27. He, K., Zhang, X., Ren, S., Sun, J.: Deep residual learning for image recognition. In: Proceedings of the IEEE Conference on Computer Vision and Pattern Recognition, pp. 770–778 (2016)

28. Heinrich, M.P.: Closing the Gap between deep and conventional image registration using probabilistic dense displacement networks. In: Shen, D., Liu, T., Peters, T.M., Staib, L.H., Essert, C., Zhou, S., Yap, P.-T., Khan, A. (eds.) MICCAI 2019. LNCS, vol. 11769, pp. 50–58. Springer, Cham (2019). https://doi.org/10.1007/978-3-030-32226-7_6
29. Heinrich, M.P., Jenkinson, M., Brady, M., Schnabel, J.A.: MRF-based deformable registration and ventilation estimation of lung CT. IEEE Trans. Med. Imaging **32**(7), 1239–1248 (2013)
30. Heinrich, M.P., Papież, B.W., Schnabel, J.A., Handels, H.: Non-parametric discrete registration with convex optimisation. In: Ourselin, S., Modat, M. (eds.) WBIR 2014. LNCS, vol. 8545, pp. 51–61. Springer, Cham (2014). https://doi.org/10.1007/978-3-319-08554-8_6
31. Hernandez-Matas, C., Zabulis, X., Triantafyllou, A., Anyfanti, P., Douma, S., Argyros, A.A.: Fire: fundus image registration dataset. Modeling Artif. Intell. Ophthalmol. **1**(4), 16–28 (2017)
32. Hershey, J.R., Roux, J.L., Weninger, F.: Deep unfolding: model-based inspiration of novel deep architectures. arXiv preprint arXiv:1409.2574 (2014)
33. Hoopes, A., Hoffmann, M., Fischl, B., Guttag, J., Dalca, A.V.: HyperMorph: amortized hyperparameter learning for image registration. In: Feragen, A., Sommer, S., Schnabel, J., Nielsen, M. (eds.) IPMI 2021. LNCS, vol. 12729, pp. 3–17. Springer, Cham (2021). https://doi.org/10.1007/978-3-030-78191-0_1
34. Horn, B.K., Schunck, B.G.: Determining optical flow. Artif. Intell. **17**(1–3), 185–203 (1981)
35. Hu, J., Gan, W., Sun, Z., An, H., Kamilov, U.: A plug-and-play image registration network. In: The Twelfth International Conference on Learning Representations (2024). https://openreview.net/forum?id=DGez4B2a6Y
36. Jena, R., Sethi, D., Chaudhari, P., Gee, J.C.: Deep learning in medical image registration: magic or mirage? arXiv preprint arXiv:2408.05839 (2024)
37. Jia, X., Bartlett, J., Zhang, T., Lu, W., Qiu, Z., Duan, J.: U-net vs transformer: is u-net outdated in medical image registration? In: International Workshop on Machine Learning in Medical Imaging, pp. 151–160. Springer (2022)
38. Jia, X., et al.: Learning a model-driven variational network for deformable image registration. IEEE Trans. Med. Imaging **41**(1), 199–212 (2021)
39. Kingma, D.P., Ba, J.: Adam: a method for stochastic optimization. arXiv preprint arXiv:1412.6980 (2014)
40. Krähenbühl, P., Koltun, V.: Efficient inference in fully connected CRFs with gaussian edge potentials. In: Advances in Neural Information Processing Systems, vol. 24 (2011)
41. Lindenberger, P., Sarlin, P.E., Pollefeys, M.: Lightglue: Local feature matching at light speed. In: Proceedings of the IEEE/CVF International Conference on Computer Vision, pp. 17627–17638 (2023)
42. Liu, J., Li, X.: Geometrized transformer for self-supervised homography estimation. In: Proceedings of the IEEE/CVF International Conference on Computer Vision, pp. 9556–9565 (2023)
43. Liu, J., Li, X., Wei, Q., Xu, J., Ding, D.: Semi-supervised keypoint detector and descriptor for retinal image matching. In: European Conference on Computer Vision, pp. 593–609. Springer (2022)
44. Liu, Y., et al.: Progressive retinal image registration via global and local deformable transformations. arXiv preprint arXiv:2409.01068 (2024)

45. Luo, W., Li, Y., Urtasun, R., Zemel, R.: Understanding the effective receptive field in deep convolutional neural networks. In: Advances in Neural Information Processing Systems, vol. 29 (2016)
46. Ma, T., Zhang, S., Li, J., Wen, Y.: Iirp-net: iterative inference residual pyramid network for enhanced image registration. In: Proceedings of the IEEE/CVF Conference on Computer Vision and Pattern Recognition, pp. 11546–11555 (2024)
47. Marstal, K., Berendsen, F., Staring, M., Klein, S.: SimpleElastix: a user-friendly, multi-lingual library for medical image registration. In: Proceedings of the IEEE Conference on Computer Vision and Pattern Recognition Workshops, pp. 134–142 (2016)
48. Meng, M., Feng, D., Bi, L., Kim, J.: Correlation-aware coarse-to-fine MLPs for deformable medical image registration. In: Proceedings of the IEEE/CVF Conference on Computer Vision and Pattern Recognition, pp. 9645–9654 (2024)
49. Mok, T.C., Chung, A.: Affine medical image registration with coarse-to-fine vision transformer. In: Proceedings of the IEEE/CVF Conference on Computer Vision and Pattern Recognition, pp. 20835–20844 (2022)
50. Mok, T., Chung, A.: Large deformation diffeomorphic image registration with laplacian pyramid networks. In: Martel, A.L., Abolmaesumi, P., Stoyanov, D., Mateus, D., Zuluaga, M.A., Zhou, S.K., Racoceanu, D., Joskowicz, L. (eds.) MICCAI 2020. LNCS, vol. 12263, pp. 211–221. Springer, Cham (2020). https://doi.org/10.1007/978-3-030-59716-0_21
51. Mok, T., Chung, A.: Conditional deformable image registration with convolutional neural network. In: de Bruijne, M., et al. (eds.) MICCAI 2021. LNCS, vol. 12904, pp. 35–45. Springer, Cham (2021). https://doi.org/10.1007/978-3-030-87202-1_4
52. Mok, T.C., Chung, A.C.: Large deformation image registration with anatomy-aware laplacian pyramid networks. In: Segmentation, Classification, and Registration of Multi-modality Medical Imaging Data: MICCAI 2020 Challenges, ABCs 2020, L2R 2020, TN-SCUI 2020, Held in Conjunction with MICCAI 2020, Lima, Peru, October 4–8, 2020, Proceedings 23, pp. 61–67. Springer (2021)
53. Potje, G., Cadar, F., Araujo, A., Martins, R., Nascimento, E.R.: Xfeat: accelerated features for lightweight image matching. In: Proceedings of the IEEE/CVF Conference on Computer Vision and Pattern Recognition, pp. 2682–2691 (2024)
54. Qiu, H., Hammernik, K., Qin, C., Chen, C., Rueckert, D.: Embedding gradient-based optimization in image registration networks. In: International Conference on Medical Image Computing and Computer-Assisted Intervention, pp. 56–65. Springer (2022)
55. Rahaman, N., et al.: On the spectral bias of neural networks. In: International Conference on Machine Learning, pp. 5301–5310. PMLR (2019)
56. Revaud, J., De Souza, C., Humenberger, M., Weinzaepfel, P.: R2d2: reliable and repeatable detector and descriptor. In: Advances in Neural Information Processing Systems, vol. 32 (2019)
57. Rudin, L.I., Osher, S., Fatemi, E.: Nonlinear total variation based noise removal algorithms. Physica D **60**(1–4), 259–268 (1992)
58. Shi, J., He, Y., Kong, Y., Coatrieux, J.L., Shu, H., Yang, G., Li, S.: Xmorpher: full transformer for deformable medical image registration via cross attention. In: International Conference on Medical Image Computing and Computer-Assisted Intervention, pp. 217–226. Springer (2022)
59. Siebert, H., Großbröhmer, C., Hansen, L., Heinrich, M.P.: Convexadam: self-configuring dual-optimisation-based 3d multitask medical image registration. IEEE Trans. Med. Imaging (2024)

60. Steinbrücker, F., Pock, T., Cremers, D.: Large displacement optical flow computation without warping. In: 2009 IEEE 12th International Conference on Computer Vision, pp. 1609–1614. IEEE (2009)
61. Sukhbaatar, S., Weston, J., Fergus, R., et al.: End-to-end memory networks. In: Advances in Neural Information Processing Systems, vol. 28 (2015)
62. Sun, J., Shen, Z., Wang, Y., Bao, H., Zhou, X.: LOFTR: detector-free local feature matching with transformers. In: Proceedings of the IEEE/CVF Conference on Computer Vision and Pattern Recognition, pp. 8922–8931 (2021)
63. Sun, J., Li, H., Xu, Z., et al.: Deep ADMM-net for compressive sensing MRI. In: Advances in Neural Information Processing Systems, vol. 29 (2016)
64. Truong, P., Apostolopoulos, S., Mosinska, A., Stucky, S., Ciller, C., Zanet, S.D.: Glampoints: Greedily learned accurate match points. In: Proceedings of the IEEE/CVF International Conference on Computer Vision, pp. 10732–10741 (2019)
65. Tyszkiewicz, M., Fua, P., Trulls, E.: Disk: Learning local features with policy gradient. Adv. Neural. Inf. Process. Syst. **33**, 14254–14265 (2020)
66. Van Den Oord, A., Vinyals, O., et al.: Neural discrete representation learning. In: Advances in Neural Information Processing Systems, vol. 30 (2017)
67. Vaswani, A., et al.: Attention is all you need. In: Advances in Neural Information Processing Systems, vol. 30 (2017)
68. Vercauteren, T., Pennec, X., Perchant, A., Ayache, N.: Diffeomorphic demons: efficient non-parametric image registration. Neuroimage **45**(1), S61–S72 (2009)
69. Wang, A.Q., Evan, M.Y., Dalca, A.V., Sabuncu, M.R.: A robust and interpretable deep learning framework for multi-modal registration via keypoints. Med. Image Anal. **90**, 102962 (2023)
70. Wang, H., Ni, D., Wang, Y.: Recursive deformable pyramid network for unsupervised medical image registration. IEEE Trans. Med. Imaging (2024)
71. Wang, J., et al.: Retinal IPA: iterative keypoints alignment for multimodal retinal imaging. arXiv preprint arXiv:2407.18362 (2024)
72. Wang, P., Zheng, W., Chen, T., Wang, Z.: Anti-oversmoothing in deep vision transformers via the fourier domain analysis: from theory to practice. arXiv preprint arXiv:2203.05962 (2022)
73. Zach, C., Pock, T., Bischof, H.: A duality based approach for realtime tv-l 1 optical flow. In: Pattern Recognition: 29th DAGM Symposium, Heidelberg, Germany, September 12-14, 2007. Proceedings 29, pp. 214–223. Springer (2007)
74. Zhang, H., Chen, X., Hu, R., Liu, D., Li, G., Wang, R.: Memwarp: discontinuity-preserving cardiac registration with memorized anatomical filters. In: Linguraru, M.G., et al. (eds.) Medical Image Computing and Computer Assisted Intervention - MICCAI 2024, pp. 671–681. Springer, Cham (2024)
75. Zhang, H., Chen, X., Wang, R., Hu, R., Liu, D., Li, G.: Slicer networks. arXiv preprint arXiv:2401.09833 (2024)
76. Zhang, H., Wang, R., Hu, R., Zhang, J., Li, J.: DEDA: deep directed accumulator. In: International Conference on Medical Image Computing and Computer-Assisted Intervention, pp. 765–775. Springer (2023)
77. Zhang, J., Liu, Z., Zhang, S., Zhang, H., Spincemaille, P., Nguyen, T.D., Sabuncu, M.R., Wang, Y.: Fidelity imposed network edit (fine) for solving ill-posed image reconstruction. Neuroimage **211**, 116579 (2020)
78. Zhang, J., et al.: Laro: learned acquisition and reconstruction optimization to accelerate quantitative susceptibility mapping. NeuroImage 119886 (2023)
79. Zheng, S., et al.: Conditional random fields as recurrent neural networks. In: Proceedings of the IEEE International Conference on Computer Vision, pp. 1529–1537 (2015)

Reconstruction

Unsupervised Accelerated MRI Reconstruction via Ground-Truth-Free Flow Matching

Xinzhe Luo[1], Yingzhen Li[2], and Chen Qin[1]()

[1] Department of Electrical and Electronic Engineering and I-X, Imperial College London, London, UK
{x.luo,c.qin15}@imperial.ac.uk
[2] Department of Computing, Imperial College London, London, UK
yingzhen.li@imperial.ac.uk

Abstract. Accelerated magnetic resonance imaging involves reconstructing fully sampled images from undersampled k-space measurements. Current state-of-the-art approaches have mainly focused on either end-to-end supervised training inspired by compressed sensing formulations, or posterior sampling methods built on modern generative models. However, their efficacy heavily relies on large datasets of fully sampled images, which may not always be available in practice. To address this issue, we propose an unsupervised MRI reconstruction method based on ground-truth-free flow matching (GTF^2M). Particularly, the GTF^2M learns a prior denoising process of fully sampled ground-truth images using only undersampled data. Based on that, an efficient cyclic reconstruction algorithm is further proposed to perform forward and backward integration in the dual space of image-space signal and k-space measurement. We compared our method with state-of-the-art learning-based baselines on the fastMRI database of both single-coil knee and multi-coil brain MRIs. The results show that our proposed unsupervised method can significantly outperform existing unsupervised approaches, and achieve performance comparable to most supervised end-to-end and prior learning baselines trained on fully sampled MRI, while offering greater efficiency than the compared generative model-based approaches.

1 Introduction

Magnetic resonance imaging (MRI) provides a non-invasive diagnostic modality that generates high-resolution depictions of anatomical structures and physiological processes within the human body. However, the sequential acquisition of measurements in the frequency domain (i.e., k-space) results in prolonged MRI acquisition time, imposing substantial demands on patients and making the process costly and less accessible. One common strategy to accelerate this process is to undersample the data in k-space and then reconstruct the fully-sampled signals, which naturally leads to an ill-posed inverse problem. Conventionally, the reconstruction can be achieved by compressed sensing (CS)-based optimisation procedures, via solving a variational problem with sparsity constraints in some transform domain [3,13,17,18]. Inspired by that, learning-based methods have been proposed to incorporate network-induced prior and unroll the optimisation steps of the variational problem into an end-to-end training framework [1,9,22,23,31]. This can be typically achieved by adopting the half quadratic splitting technique that alternates between a de-aliasing step using a regularisation network and a proximal mapping update for measurement consistency.

More recently, accelerated MRI has been approached through the lens of Bayesian inference. To solve the general inverse problems, these methods build upon generative models that learn the prior distribution of the ground-truth signals (i.e., fully-sampled images), followed by adapting the reverse sampling process to generate samples from the posterior distribution [5,20,24–26,30]. For instance, Song et al. [25], Chung et al. [6] and Wang et al. [30] proposed to incorporate data consistency projections into the reverse sampling process of an unconditional score-based diffusion models [26]. Further, DPS [5] and ΠGDM [24] devised strategies to approximate the intractable posterior score for posterior sampling of the ground truth. However, both end-to-end and prior learning approaches require large datasets of fully-sampled images, which can be hard to access in real-world scenarios. In addition, the high number of neural function evaluations (NFEs) (typically 100–2000) of the sampling process makes diffusion model-based MRI reconstruction less appealing in practice.

To overcome the above limitation, unsupervised learning methods for accelerated MRI have also recently raised research attention and been explored in previous work. Particularly, the robust equivariant imaging (REI) framework [4] proposed to exploit the group invariance property of the signal space, and employed Stein's Risk Estimator (SURE) [27] to enforce measurement consistency. The ENsemble SURE (ENSURE) framework [2] further proposed an unbiased estimate of the true mean squared error (MSE) between the prediction and the ground truth, which was then leveraged to train a model-based neural network for unsupervised MRI reconstruction.

In this work, we propose an unsupervised framework for MRI reconstruction, alleviating the demand for fully sampled ground-truth data. Inspired by ENSURE, we introduce a ground-truth-free learning framework to learn a flow-based denoiser network of fully sampled MR images, based on an induced forward model over the dual-space conditional vector fields. Moreover, a novel cyclic reconstruction algorithm is proposed that decouples data consistency updates from the training of the denoiser. The contributions of this work are summarised as:

– We introduce a ground-truth-free flow matching framework that learns a prior denoising process for fully sampled MR images using only undersampled k-space measurements.
– We propose an efficient cyclic reconstruction algorithm that forward integrates the induced k-space flow to obtain the hidden variable generating the aliased image, and then backward integrates the image-space flow to obtain the target image in a decoupled continuous de-aliasing process.
– We evaluated our method on both single-coil and multi-coil accelerated MRI reconstruction. Empirical results demonstrate that the proposed method

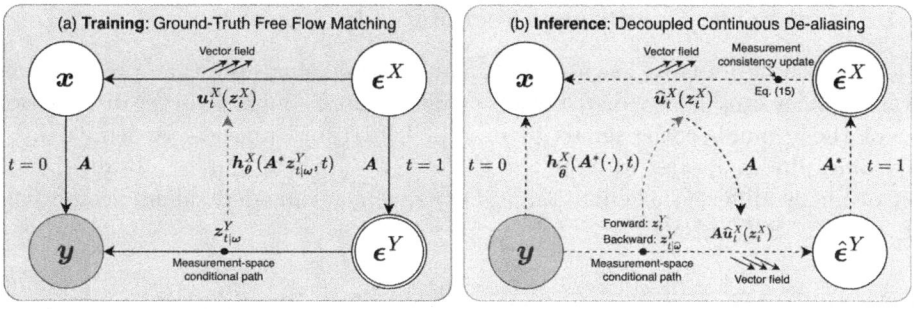

Fig. 1. Illustration of the proposed unsupervised reconstruction framework. **(a)** Graphical representation of the training framework via ground-truth-free flow matching. **(b)** Commutative diagram of the inference steps using cyclic reconstruction to form a decoupled continuous de-aliasing process. Random variables are in circles, deterministic variables are in double circles, and observed variables are shaded. Solid and dashed arrows indicate the generation and inference steps, respectively.

outperformed existing unsupervised approaches and achieved comparable performance to most state-of-the-art end-to-end and prior learning baselines trained with fully sampled images, whereas with significantly lower NFEs than other generative model-based approaches.

2 Methodology

In MRI reconstruction, we denote $x \in \mathbb{C}^D$ as the image-space *signal*, and $y \in \mathbb{C}^d$ as the undersampled k-space *measurement*. The forward model is defined as $y = Ax + e$, where noise e follows a complex Gaussian distribution $\mathcal{CN}(0, 2\sigma^2 I_d)$, and $A \in \mathbb{C}^{d \times D}$ is the forward operator combining the undersampling mask $M = [I_d \mid 0]T \in \mathbb{R}^{d \times D}$ for some permutation matrix $T \in \mathbb{R}^{D \times D}$, the Fourier transform matrix $F \in \mathbb{C}^{D \times D}$, and for the multi-coil case the sensitivity map $S \in \mathbb{C}^{D \times D}$. To reconstruct the fully-sampled signal, we propose a two-step strategy in our proposed method: **(a)** a denoising process of the signal is learnt using flow matching to model the prior of the fully sampled images, with undersampled measurement only; this is achieved by formulating an induced forward model on the dual-space conditional vector fields and optimising a ground-truth-free unbiased estimator of the conditional flow matching objective; **(b)** a decoupled continuous de-aliasing process with a novel cyclic integration algorithm is then performed to reconstruct the signal based on the learnt denoiser; this involves forward integration to the hidden variable generating the aliased undersampled image, and backward measurement-consistent integration to the target image. Figure 1 presents an overview of the proposed framework.

2.1 Dual-Space Conditional Vector Fields

Flow matching aims to learn a continuous normalising flow that progressively transforms a simple base distribution p_1 into a more complex target distribution p_0 of the ground-truth signals [8,14,15]. The transformation is defined by a diffeomorphic mapping $\phi_t : [0,1] \times \mathbb{R}^D \to \mathbb{R}^D$, which can be obtained with an ordinary differential equation (ODE) given a time-dependent vector field $\boldsymbol{u}_t : [0,1] \times \mathbb{R}^D \to \mathbb{R}^D$,

$$z_t \triangleq \phi_t(z_1), \quad dz_t = \boldsymbol{u}_t(z_t)\,dt, \quad \forall z_1 \sim p_1, \tag{1}$$

with the boundary condition $\phi_1 = \text{id}$. The base distribution p_1 usually follows a standard Gaussian, and thus $\phi_t(\cdot)$ can be viewed as a continuous denoising function with $\boldsymbol{u}_t(\cdot)$ as its infinitesimal residual.

The (marginal) vector field $\boldsymbol{u}_t(\cdot)$ generating the probability path $p_t = [\phi_t]_\sharp p_1$ (the push-forward distribution of p_1 by ϕ_t) can be constructed via the conditional path and vector field. Let $z^X_{t|\omega} \triangleq \psi^X_t(\boldsymbol{\omega}^X) = a_t \boldsymbol{x} + b_t \boldsymbol{\epsilon}^X$ be the conditional path in the ground-truth signal space X, where $\boldsymbol{x} \in \mathbb{C}^D \sim p_0$, $\boldsymbol{\epsilon}^X \sim p_1 = \mathcal{CN}(\boldsymbol{0}, 2\boldsymbol{I}_D)$, and $\boldsymbol{\omega}^X \triangleq (\boldsymbol{x}, \boldsymbol{\epsilon}^X) \sim p^X_\omega \in \mathcal{C}(p_0, p_1)$ which is the coupling of p_0 and p_1. We define the conditional vector field $\boldsymbol{u}^X_{t|\omega}(\cdot)$

$$\boldsymbol{u}^X_{t|\omega}(\cdot) \triangleq \frac{d\psi^X_t(\cdot)}{dt} = a'_t \boldsymbol{x} + b'_t \boldsymbol{\epsilon}^X, \quad \forall t \in [0,1], \tag{2}$$

where $a'_t = \frac{da_t}{dt}$ and $b'_t = \frac{db_t}{dt}$, such that $a_0 = b_1 = 1$ and $a_1 = b_0 = 0$. One common instantiation is the independent coupling $p^X_\omega = p_0 \times p_1$ and linear interpolation $\psi^X_t(\boldsymbol{\omega}^X) = (1-t)\boldsymbol{x} + t\boldsymbol{\epsilon}^X$ with $p_t(\cdot \mid \boldsymbol{\omega}^X) = \delta_{(1-t)\boldsymbol{x}+t\boldsymbol{\epsilon}^X}(\cdot)$, where δ is the Dirac delta function. Then, the marginal vector field $\boldsymbol{u}_t$ that generates p_t can be constructed by

$$\boldsymbol{u}^X_t(z_t) \triangleq \mathbb{E}_{\boldsymbol{\omega}^X \sim p_t(\cdot|z_t)}\left[\boldsymbol{u}^X_{t|\omega}(z_t)\right] = \mathbb{E}_{\boldsymbol{\omega}^X \sim p_0 \times p_1}\left[\boldsymbol{u}^X_{t|\omega}(z_t)\frac{p_t(z_t \mid \boldsymbol{\omega}^X)}{p_t(z_t)}\right], \tag{3}$$

which satisfies the continuity equation [29].

Given the above probability path defined with $\boldsymbol{u}^X_t(\cdot)$, flow matching learns the generative mapping between p_1 and p_0 by regressing the vector field with a predictor network $\boldsymbol{v}^X_\theta(\cdot)$ using the conditional flow matching objective

$$\mathcal{L}_{\text{CFM}}(\boldsymbol{\theta}) = \mathbb{E}_{t \sim p_T, \boldsymbol{\omega}^X \sim p_0 \times p_1, z^X_{t|\omega} \sim p_t(\cdot|\boldsymbol{\omega}^X)} \left\| \boldsymbol{v}^X_\theta(z^X_{t|\omega}, t) - \boldsymbol{u}^X_{t|\omega}(z^X_{t|\omega}) \right\|^2_2. \tag{4}$$

However, the scarcity of the ground-truth (i.e., fully-sampled) signal $\boldsymbol{x}$ in MRI makes the signal-space conditional path inaccessible. To address this, we instead define a measurement-space (Y) conditional path $z^Y_{t|\omega} \triangleq \psi^Y_t(\boldsymbol{y}, \boldsymbol{\epsilon}^Y) = a_t \boldsymbol{y} + b_t \boldsymbol{\epsilon}^Y$, where $\boldsymbol{y} = \boldsymbol{A}\boldsymbol{x} + \boldsymbol{e}$ and $\boldsymbol{\epsilon}^Y \triangleq \boldsymbol{A}\boldsymbol{\epsilon}^X$. Therefore, the signal- and measurement-space conditional vector fields take the form of

$$\boldsymbol{u}^X_{t|\omega}(z^X_{t|\omega}) \triangleq a'_t \boldsymbol{x} + b'_t \boldsymbol{\epsilon}^X, \text{ and } \boldsymbol{u}^Y_{t|\omega}(z^Y_{t|\omega}) \triangleq a'_t \boldsymbol{y} + b'_t \boldsymbol{\epsilon}^Y. \tag{5}$$

Substituting $y = Ax + e$ and $\epsilon^Y = A\epsilon^X$ into Eq. 5 yields

$$u_{t|\omega}^Y(z_{t|\omega}^Y) = Au_{t|\omega}^X(z_{t|\omega}^X) + a_t'e. \tag{6}$$

This implies that the original signal-measurement forward model $y = Ax + e$ induces another forward model over the dual-space conditional vector fields, which we can exploit to optimise the signal-space conditional flow matching objective in an unsupervised ground-truth-free manner.

2.2 Ground-Truth-Free Flow Matching (GTF²M)

In contrast to Eq. 4 which aims to predict the signal-space conditional vector field based on the *signal-space* conditional path that requires the knowledge of ground-truth signals, here we propose an unsupervised objective to regress the signal-space conditional vector field from the *measurement-space* conditional path with a predictor network $h_\theta^X(\cdot)$, namely,

$$\mathcal{L}_{\text{GTF}^2\text{M}}(\theta) \triangleq \mathbb{E}_{p_T, p_0 \times p_1, p_t^X(z_{t|\omega}^X | \omega^X), p_t(z_{t|\omega}^Y | z_{t|\omega}^X)} \left\| h_\theta^X(z_{t|\omega}^Y, t) - u_{t|\omega}^X(z_{t|\omega}^X) \right\|_2^2, \tag{7}$$

where $p_t(z_{t|\omega}^Y \mid z_{t|\omega}^X)$ is induced from $z_{t|\omega}^Y = Az_{t|\omega}^X + a_t e$. In addition, similar to Theorem 2 in [14], we can prove that

$$\mathcal{L}_{\text{GTF}^2\text{M}}(\theta) = \mathbb{E}_{p_T, p_t^X(z_t^X), p_t(z_t^Y|z_t^X)} \left\| h_\theta^X(z_t^Y, t) - u_t^X(z_t^X) \right\|_2^2 + \text{const.}, \tag{8}$$

which means we are essentially predicting the signal-space marginal vector field.

Meanwhile, noting from Eq. 5 that $z_{t|\omega}^Y = a_t y + b_t \epsilon^Y = \frac{a_t}{a_t'} u_{t|\omega}^Y(z_{t|\omega}^Y) - b_t'\left(\frac{a_t}{a_t'} - \frac{b_t}{b_t'}\right)\epsilon^Y$, we can write $h_\theta^X(z_{t|\omega}^Y, t) = h_\theta^X\left(u_{t|\omega}^Y(z_{t|\omega}^Y), t\right)$. Taking into account the randomness in the forward operator A_s indexed by some random variable $s \sim p(s)$ (i.e., a random permutation matrix T_s encoding the positions of k-space acquisition), the ENSURE framework [2] implies that the GTF²M objective has an unbiased estimate up to a constant, which does not require the fully sampled signal, namely,

$$\mathbb{E}_{s,t,\omega^X, z_{t|\omega,s}^X, z_{t|\omega,s}^Y | z_{t|\omega,s}^X} \left\| h_\theta^X(z_{t|\omega,s}^Y, t) - u_{t|\omega,s}^X(z_{t|\omega,s}^X) \right\|_2^2$$

$$= \mathbb{E}_{s,t,\omega^X, z_{t|\omega,s}^X, z_{t|\omega,s}^Y | z_{t|\omega,s}^X} \left\| R_s \left[h_\theta^X\left(u_{t|\omega,s}^Y(z_{t|\omega,s}^Y), t\right) - u_{t|\omega,s}^X(z_{t|\omega,s}^X) \right] \right\|_2^2$$

$$= \mathbb{E}_{s,t,\omega^Y, z_{t|\omega,s}^Y} \left[\left\| R_s \left[h_\theta^X(\mu_{t|\omega,s}^X, t) - \hat{u}_{t|\omega,s,\text{ML}}^X \right] \right\|_2^2 + 2\nabla_{\mu_{t|\omega,s}^X} \cdot R_s^* R_s h_\theta^X(\mu_{t|\omega,s}^X, t) \right] + \text{const.} \tag{9}$$

where $\mu_{t|\omega,s}^X \triangleq A_s^* C_t^{-1} u_{t|\omega,s}^Y(z_{t|\omega,s}^Y)$ is a sufficient statistic for $u_{t|\omega,s}^X(z_{t|\omega,s}^X)$ with $C_t = (a_t'\sigma)^2 I_d$, and $\hat{u}_{t|\omega,s,\text{ML}}^X \triangleq (A_s^* C_t^{-1} A_s)^\dagger A_s^* C_t^{-1} u_{t|\omega,s}^Y(z_{t|\omega,s}^Y)$ is the maximum likelihood solution of the forward model in Eq. 6. The projection operator is defined as $R_s \triangleq WP_s$ with $P_s = A_s^\dagger A_s$ and $W \triangleq \mathbb{E}_s[P_s]^{-1/2}$. For single-coil reconstruction where $A_s = M_s F$, R_s has a closed-form expression as $R_s = F^* T_s^T \text{diag}(p_1^{-0.5}, \ldots, p_d^{-0.5}, 0, \ldots, 0) T_s F$, where p_i denotes the probability that the i-th k-space measurement is acquired.

By ignoring the constant and rewriting $h_\theta^X(\mu_{t|\omega,s}^X, t) = h_\theta^X(A_s^* z_{t|\omega,s}^Y, t)$, we derive the GTF^{2}M objective $\mathcal{L}_{\text{GTF}^2\text{M}}(\theta)$ equivalent to

$$\mathbb{E}_{s,t,\omega Y, z_{t|\omega,s}^Y}\left[\left\|R_s\left[h_\theta^X(A_s^* z_{t|\omega,s}^Y, t) - \hat{u}_{t,s,\text{ML}}^X\right]\right\|_2^2 + 2a_t' a_t \sigma^2 \nabla_{A_s^* z_{t|\omega,s}^Y} \cdot R_s^* R_s h_\theta^X(A_s^* z_{t|\omega,s}^Y, t)\right], \tag{10}$$

which is essentially *ground-truth-free*. That is, the objective learns the ground-truth signal-space marginal vector field without fully sampled MRI data.

In practice, the divergence term can be computed using the Hutchinson trace estimator [11], namely,

$$\nabla_{A_s^* z_{t|\omega,s}^Y} \cdot R_s^* R_s h_\theta^X(A_s^* z_{t|\omega,s}^Y, t) = \mathbb{E}_{b \sim \mathcal{N}(0, R_s^* R_s)}\left[b^\mathsf{T} \nabla_{A_s^* z_{t|\omega,s}^Y} h_\theta^X(A_s^* z_{t|\omega,s}^Y, t) b\right], \tag{11}$$

where the expectation can be approximated by Monte-Carlo sampling.

2.3 Reconstruction as Decoupled Continuous De-Aliasing

To reconstruct the fully sampled image x from the observed undersampled k-space measurement y, we further propose a cyclic reconstruction algorithm based on the vector field learnt by the GTF^{2}M objective. Given the signal-space vector field $u_t^X(\cdot)$ and a random noise $z_1^X = \epsilon^X \sim p_1$, we can denoise it via *backward* integration of the ODE, namely

$$z_t^X = z_1^X + \int_1^t u_t^X(z_t^X)\,\mathrm{d}t, \tag{12}$$

and particularly at $t = 0$ we have $z_0^X = \epsilon^X + \int_1^0 u_t^X(z_t^X)\,\mathrm{d}t \sim p_0$. The marginal vector field $u_t^X(z_t^X)$ is estimated from $h_\theta^X(A^* z_{t|\omega}^Y)$ learnt by the GTF^{2}M objective, where $z_{t|\omega}^Y = a_t y + b_t A \epsilon^X$ given the observed measurement y.

However, instead of directly using a random ϵ^X, we propose starting from the aliased initial point $A^* A \hat{\epsilon}^X$ corresponding to the aliased image $A^* y$, where $\hat{\epsilon}^X$ is the unknown hidden variable generating the ground truth x via the learnt vector field $\hat{u}_t^X(z_t^X)$. This is analogous to unrolling-based learning methods that take as input the aliased image $A^* y$. Specifically, applying the MRI forward model to both sides of the *forward* ODE $z_t^X = x + \int_0^t u_t^X(z_t^X)\,\mathrm{d}t$, we can obtain

$$z_t^Y \triangleq A z_t^X + e = y + \int_0^t A u_t^X(z_t^X)\,\mathrm{d}t. \tag{13}$$

Therefore, we propose to estimate a better initial point $z_1^X \triangleq A^* \hat{\epsilon}^Y$ by

$$\hat{\epsilon}^Y \triangleq A \hat{\epsilon}^X \approx A \hat{\epsilon}^X + e = y + \int_0^1 A \hat{u}_t^X(z_t^X)\,\mathrm{d}t, \tag{14}$$

where the approximation comes from the fact that $\|A\hat{\epsilon}^X\|$ follows a Chi distribution whose expectation is proportional to the standard deviation of the i.i.d. Gaussian variables of $\hat{\epsilon}^X$, and we can always increase its variance by design.

Since we do not have access to the true trajectory z_t^X of the ground truth x, the above estimation is intractable. However, if the estimated signal-space flow is straight enough [15], we can integrate the following ODE as an approximation to Eq. 13:

$$z_t^Y = y + \int_0^t Ah_\theta^X(A^* z_t^Y) \, dt, \qquad (15)$$

which gives an estimate $\widehat{\epsilon}^Y = z_1^Y$ and then we set $z_1^X \triangleq A^* \widehat{\epsilon}^Y$.

Then, starting from the new initial point $z_1^X = A^* \widehat{\epsilon}^Y$, we backward integrate Eq. 12 by injecting measurement consistency updates, giving rise to a continuous de-aliasing process as the number of discrete Euler steps goes to infinity. Specifically, given K uniform time points $t = 1/K, \ldots, 1$ and, Eq. 12 can be approximated by discrete Euler steps

$$z_{t-1/K}^X \leftarrow z_t^X - \frac{1}{K} h_\theta^X(A^* z_{t|\omega}^Y). \qquad (16)$$

To align Az_t^X with $z_{t|\omega}^Y$ for ensuring data consistency, we propose to perform proximal mapping to enforce measurement consistency via the optimisation problem

$$z_t^X \leftarrow \arg\min_z \left\{ \gamma_t \left\| Az - z_{t|\omega}^Y \right\|_2^2 + \left\| z - z_t^X \right\|_2^2 \right\}, \qquad (17)$$

where the regularisation coefficient $\gamma_t = \frac{\zeta \sigma_t}{(a_t \sigma)^2} = \frac{\zeta}{b_t^2 \sigma^2}$ [33], with $\sigma_t \triangleq a_t^2/b_t^2$ the signal-to-noise ratio (SNR) at time t and ζ a hyper-parameter. In single-coil setting, a closed-form solution of Eq. 17 can be dervied as

$$z_t^X \leftarrow z_t^X - \lambda_t A^*(A z_t^X - z_{t|\omega}^Y), \qquad (18)$$

where the step size $\lambda_t \triangleq \frac{\gamma_t}{1+\gamma_t} = \left(1 + b_t^2 \sigma^2/\zeta\right)^{-1}$. Thus, unlike previous learning-based reconstruction methods that unroll conventional optimisation schemes, the proposed strategy decouples data consistency updates from the training of the denoiser network, leading to improved training efficiency and flexibility.

Furthermore, the forward and backward integration formulates a cyclic path, which represents a commutative diagram for inferring the ground-truth image as shown in Fig. 1 (b).

For multi-coil reconstruction, coil-wise reconstruction can be similarly performed following the aforementioned steps, yielding $z_{0,c}^X$ for $c = 1, \ldots, C$, followed by sensitivity combined reconstruction [21], i.e., $\widehat{x} = (\sum_c S_c^* S_c)^{-1} \sum_c S_c^* z_{0,c}^X$, where S_c is a diagonal matrix encoding the sensitivity map of the c-th coil. Algorithm 1 summarises the proposed reconstruction process.

Algorithm 1: Decoupled Continuous Dealiasing via Cyclic Integration

Input: k-space measurement $\boldsymbol{y} = (\boldsymbol{y}_c)_{c=1}^C$, pretrained flow predictor $\boldsymbol{h}_{\hat{\theta}}^X(\cdot, t)$, forward steps L, backward steps K, regularisation parameter ζ
Output: Reconstructed image $\hat{\boldsymbol{x}}$ of $\boldsymbol{y}$

1 $z_0^Y := \boldsymbol{y}$;
2 for $t = 0, \ldots, {}^{(L-1)}\!/\!_L$ do
3 $\quad z_{t+1/L}^Y \leftarrow z_t^Y + \frac{1}{L} \boldsymbol{A} \boldsymbol{h}_{\hat{\theta}}^X(\boldsymbol{A}^* z_t^Y, t)$; // forward integration
4 $\hat{\boldsymbol{\epsilon}}^Y \triangleq z_1^Y$, $z_1^X \triangleq \boldsymbol{A}^* \hat{\boldsymbol{\epsilon}}^Y$;
5 for $t \in \{1, \ldots, 1/K\}$ do
6 $\quad z_{t|\hat{\omega}}^Y = a_t \boldsymbol{y} + b_t \hat{\boldsymbol{\epsilon}}^Y$;
7 $\quad z_{t-1/K}^X \leftarrow z_t^X - \frac{1}{K} \boldsymbol{h}_{\hat{\theta}}^X(\boldsymbol{A}^* z_{t|\hat{\omega}}^Y, t)$; // backward integration
8 $\quad z_{t-1/K}^X \leftarrow z_{t-1/K}^X - \lambda_{t-1/K} \boldsymbol{A}^* (\boldsymbol{A} z_{t-1/K}^X - z_{t|\hat{\omega}}^Y)$; // measurement consistency update
9 $\hat{\boldsymbol{x}} = z_0^X$ or $\left(\sum_{c=1}^C \boldsymbol{S}_c^* \boldsymbol{S}_c \right)^{-1} \sum_{c=1}^C \boldsymbol{S}_c^* z_{0,c}^X$; // single- or multi-coil
10 return $\hat{\boldsymbol{x}}$.

3 Experiments and Results

3.1 Experimental Setups

Datasets and preprocessing. We used the NYU fastMRI Initiative database[1] [12,32] to evaluate the performance of our model for accelerated MRI reconstruction on both the single-coil knee and multi-coil brain datasets. A selection of 484/56/44 single-coil proton density (PD) weighted knee MRI volumes without fat suppression and 700/100/200 multi-coil T2 weighted brain MRI volumes were used for training/validation/testing. Each slice was cropped to the size of 320 × 320 and normalised to a constant norm. We simulated Cartesian subsampling masks for each volume following the setup in [32], which include 8% and 4% fully-sampled low-frequency k-space lines for acceleration factors of 4 and 8, respectively. The other lines were sampled random-uniformly and equidistantly respectively for knee and brain volumes. For knee data, we used the provided emulated single-coil (ESC) data as the ground truth, while for brain MRI we used the SENSE reconstruction [21], where a spatially invariant noise distribution was assumed. The coil sensitivity maps were estimated by ESPIRiT [28], following the implementation of [10]. Finally, reconstruction performance is measured by calculating the structural similarity index measure (SSIM) and the peak signal-to-noise ratio (PSNR) between prediction and ground truth images.

Implementation Details. For the conditional path coefficients, we adopt linear interpolation $a_t = 1 - t$ and $b_t = t$. The noise level is considered as $\sigma = 0.01$. We use the ADM (ablated diffusion model) network architecture [7] for the flow predictor $\boldsymbol{h}_{\hat{\theta}}^X(\cdot, t)$, which consists of a U-Net with adaptive group normalisation

---

[1] https://fastmri.med.nyu.edu/.

after each intermediate convolutional block. The parameters of group normalisation are given by linear projection of the positional embeddings of time points. In addition, multi-resolution attention and dropout with probability 0.1 are applied at the lowest three resolutions of the network. We trained the network using the AdamW optimiser [16] with a learning rate of 1×10^{-4} and a weight decay coefficient of 0.1. Exponential moving average of the network parameters was performed every 100 training steps with rate 0.99. The hyperparameter ζ is set to 1 by default. The code was implemented in PyTorch [19] with training and inference performed on NVIDIA L40S and A100 GPUs.

Compared Baselines. Three types of baseline methods were compared with our approach: (a) supervised end-to-end learning on paired samples of fully sampled images and their corresponding measurements, for which we compared MoDL [1], an architecture applicable to both single- and multi-coil reconstruction; (b) generative model-based methods that learn a prior distribution of fully sampled MRI and incorporate measurements into posterior sampling steps, for which we compared DDNM$^+$ [30], ΠGDM [24] and FlowPS [20]; in particular, DDNM$^+$ used range-null decomposition for measurement consistency, while ΠGDM and FlowPS relied on approximation of the posterior score function; (c) unsupervised approaches without prior learning using only undersampled k-space data, including REI [4] and ENSURE [2]. Note that except for MoDL and ENSURE that are directly applicable to MRI reconstruction, the other baselines were only developed for either inverse problems of natural images or single-coil MRI. We adapt them to both single- and multi-coil MRI reconstruction in our experiments.

3.2 Single-Coil Knee MRI Reconstruction

Comparison Study. Table 1 presents the reconstruction accuracy on the single-coil knee MRI dataset. Notably, our proposed approach performed significantly better than the unsupervised baselines. Compared with existing supervised and prior learning approaches that rely on fully sampled images, our method also achieved comparable performance in majority cases. Moreover, our method achieved superior efficiency than other generative model-based ones during inference, through a substantial reduction in the number of neural function evaluations (NFEs) required. Figure 2 visualises reconstruction and its error map on a sample single-coil knee MRI using various compared methods. In particular, the error map generated by our method is comparable to that produced by MoDL.

Ablation and parameter study. Table 2 presents the ablation study on the effect of applying forward integration steps to initialise noise in the proposed decoupled continuous de-aliasing algorithm. It is evident that the incorporation of forward integration markedly enhances the reconstruction accuracy. This suggests that a good initialisation, which aligns with the observed measurements, is crucial for the effectiveness of the de-aliasing process. Figure 3 presents the reconstruction accuracy as a function of the number of integration steps in the

Table 1. Qualitative results of 4× and 8× accelerated single-coil MRI reconstruction using various reconstruction methods on the fastMRI knee data with the random uniform undersampling pattern. Statistical significant difference ($p < 0.05$) between ours and the other methods indicated by two-sided paired t-tests was marked with asterisks (*). Best results among ground-truth-dependant and -free methods are highlighted in cyan and magenta, respectively.

Method	SSIM ↑		PSNR ↑		NFEs ↓
	4×	8×	4×	8×	
Zero-filled	0.684 ± 0.086*	0.556 ± 0.106*	27.60 ± 2.78*	23.92 ± 2.75*	N/A
(a) Supervised methods using fully sampled images					
MoDL [1]	0.786 ± 0.069*	0.692 ± 0.107*	30.72 ± 3.07*	28.58 ± 2.96*	1
(b) Prior learning methods using fully sampled images					
DDNM+ [30]	0.791 ± 0.076*	0.681 ± 0.108*	31.73 ± 3.29*	28.00 ± 3.20	100
ΠGDM [24]	0.728 ± 0.098*	0.581 ± 0.114*	30.27 ± 3.34*	25.83 ± 3.13*	100
FlowPS [20]	0.763 ± 0.077*	0.631 ± 0.101*	30.66 ± 2.73*	26.30 ± 2.55*	100
(c) Unsupervised methods w/o prior learning					
REI [4]	0.740 ± 0.087*	0.591 ± 0.110*	29.96 ± 2.87*	25.04 ± 2.97*	1
ENSURE [2]	0.684 ± 0.086*	0.556 ± 0.106*	27.65 ± 2.79*	23.91 ± 2.75*	1
(d) Unsupervised methods w/ prior learning					
Ours	0.801 ± 0.073	0.688 ± 0.095	31.64 ± 2.95	27.95 ± 2.64	20

Table 2. Ablation study on the effect of forward integration for initialisation in the proposed algorithm on single-coil knee MRI reconstruction. Statistical significant difference ($p < 0.05$) indicated by two-sided paired t-tests was marked with asterisks (*).

Method	SSIM ↑		PSNR ↑		NFEs ↓
	4×	8×	4×	8×	
w/o forward int.	0.755 ± 0.077*	0.648 ± 0.098*	30.29 ± 2.69*	27.02 ± 2.55*	10
w/ forward int.	0.801 ± 0.073	0.688 ± 0.095	31.64 ± 2.95	27.95 ± 2.64	20

proposed cyclic reconstruction algorithm. It can be observed that the enhancement in accuracy begins to plateau when the number of forward and backward steps reaches 10. This justifies our default configuration of $L = K = 10$ as an optimal balance between accuracy and efficiency.

3.3 Multi-coil Brain MRI Reconstruction

Comparison Study. Table 3 presents the reconstruction accuracy on the multi-coil brain MRI dataset. Our method performed the best among unsupervised approaches, and better than the flow-based method FlowPS that used fully

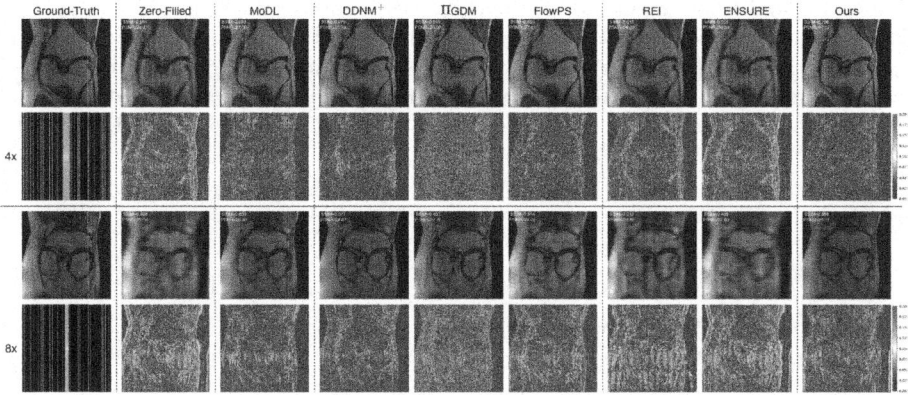

Fig. 2. Visualisation of reconstruction on a sample single-coil knee MRI using various compared methods. The k-space images are presented in log-scale absolute values. The error maps are presented in values relative to the peak intensity in the ground-truth image. It can be observed that our approach can achieve comparable performance to that obtained from supervised training using MoDL.

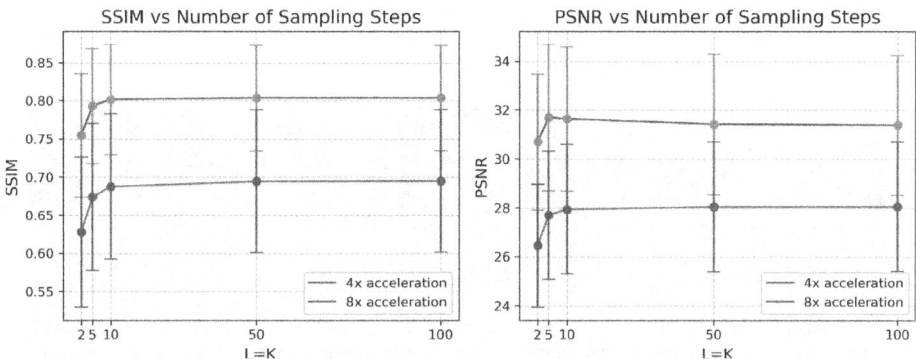

Fig. 3. Reconstruction accuracy on single-coil knee MRI as a function of the number of de-aliasing steps in the proposed algorithm.

sampled images for prior learning. However, there is still a performance gap between ours and the other methods that use fully sampled images. This could be attributed to the fact that our method is trained by the coil-wise images, which may be less efficient to capture the prior of the coil-combined ground-truth images. Figure 4 visualises reconstruction and its error map on a sample multi-coil brain MRI using the compared methods.

Table 3. Qualitative results of 4× and 8× accelerated multi-coil MRI reconstruction using various reconstruction methods on the fastMRI brain data with the uniform equidistant undersampling pattern. Statistical significant difference ($p < 0.05$) between ours and the other methods indicated by two-sided paired t-tests was marked with asterisks. Best results among ground-truth-dependant and -free methods are highlighted in cyan and magenta, respectively.

Method	SSIM ↑		PSNR ↑		NFEs ↓
	4×	8×	4×	8×	
Zero-filled	0.800 ± 0.089*	0.716 ± 0.117*	27.66 ± 3.78*	24.10 ± 3.97*	N/A
(a) Supervised methods using fully sampled images					
MoDL [1]	0.948 ± 0.044 *	0.820 ± 0.051*	38.28 ± 3.37*	30.18 ± 3.04*	1
(b) Prior learning methods using fully sampled images					
DDNM+ [30]	0.929 ± 0.045*	0.887 ± 0.048 *	40.61 ± 3.43 *	34.13 ± 2.92 *	100
FlowPS [20]	0.855 ± 0.060*	0.748 ± 0.069*	33.10 ± 2.73*	26.56 ± 3.50*	100
(c) Unsupervised methods w/o prior learning					
ENSURE [2]	0.825 ± 0.053*	0.739 ± 0.108*	31.75 ± 3.99*	25.56 ± 3.50*	1
(d) Unsupervised methods w/ prior learning					
Ours	0.920 ± 0.060	0.859 ± 0.054	34.65 ± 2.32	28.72 ± 2.92	20

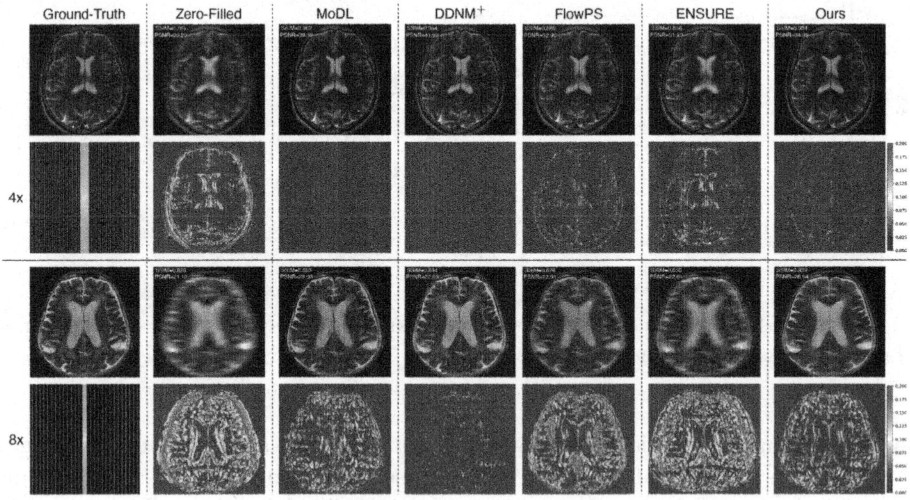

Fig. 4. Visualisation of reconstruction on a sample multi-coil brain MRI using various compared methods. The k-space images are presented as the log-scale root sum square of the multi-coil undersampled k-space maps. The error maps are presented in values relative to the peak intensity in the ground-truth image.

4 Conclusion

In this work, we have presented an unsupervised framework for accelerated MRI reconstruction based on flow matching. The proposed model adapts the original flow matching objective into a ground-truth-free variant based on an induced forward model between the dual-space conditional vector fields. A decoupled continuous de-aliasing process is further proposed through a cyclic reconstruction algorithm, where the forward integration estimates a better initialisation for backward integration, leading to improved reconstruction quality. The proposed method achieves the best performance among all the unsupervised approaches compared. However, its reconstruction accuracy may be outperformed by methods that leverage fully sampled images during training, especially in the multi-coil case. Future work will investigate methodological advancement that can mitigate this performance gap. For example, we can train the denoiser based on the combined forward operator of multi-coil MRI, which may entail a numerical solver for the projection operator in the proposed GTF^2M objective, whereas it might cause instability in the training process.

Acknowledgement. This work was supported by the Engineering and Physical Sciences Research Council (EPSRC) UK grants TrustMRI[EP/X039277/1] and [EP/Y002016/1].

References

1. Aggarwal, H.K., Mani, M.P., Jacob, M.: MODL: model-based deep learning architecture for inverse problems. IEEE Trans. Med. Imaging **38**(2), 394–405 (2018)
2. Aggarwal, H.K., Pramanik, A., John, M., Jacob, M.: Ensure: a general approach for unsupervised training of deep image reconstruction algorithms. IEEE Trans. Med. Imaging **42**(4), 1133–1144 (2022)
3. Candès, E.J., Romberg, J., Tao, T.: Robust uncertainty principles: exact signal reconstruction from highly incomplete frequency information. IEEE Trans. Inf. Theory **52**(2), 489–509 (2006)
4. Chen, D., Tachella, J., Davies, M.E.: Robust equivariant imaging: a fully unsupervised framework for learning to image from noisy and partial measurements. In: Proceedings of the IEEE/CVF Conference on Computer Vision and Pattern Recognition, pp. 5647–5656 (2022)
5. Chung, H., Kim, J., Mccann, M.T., Klasky, M.L., Ye, J.C.: Diffusion posterior sampling for general noisy inverse problems. In: International Conference on Learning Representations (2023)
6. Chung, H., Ye, J.C.: Score-based diffusion models for accelerated MRI. Med. Image Anal. **80**, 102479 (2022)
7. Dhariwal, P., Nichol, A.: Diffusion models beat GANs on image synthesis. Adv. Neural. Inf. Process. Syst. **34**, 8780–8794 (2021)
8. Esser, P., et al.: Scaling rectified flow transformers for high-resolution image synthesis. In: International Conference on Machine Learning (2024)
9. Hammernik, K., et al.: Learning a variational network for reconstruction of accelerated MRI data. Magn. Reson. Med. **79**(6), 3055–3071 (2018)

10. Hammernik, K., Schlemper, J., Qin, C., Duan, J., Summers, R.M., Rueckert, D.: Systematic evaluation of iterative deep neural networks for fast parallel MRI reconstruction with sensitivity-weighted coil combination. Magn. Reson. Med. **86**(4), 1859–1872 (2021)
11. Hutchinson, M.F.: A stochastic estimator of the trace of the influence matrix for Laplacian smoothing splines. Commun. Stat.-Simul. Comput. **18**(3), 1059–1076 (1989)
12. Knoll, F., et al.: fastMRI: a publicly available raw k-space and dicom dataset of knee images for accelerated MR image reconstruction using machine learning. Radiol. Artif. Intell **2**(1), e190007 (2020)
13. Liang, D., Liu, B., Wang, J., Ying, L.: Accelerating sense using compressed sensing. Magn. Reson. Med. **62**(6), 1574–1584 (2009)
14. Lipman, Y., Chen, R.T., Ben-Hamu, H., Nickel, M., Le, M.: Flow matching for generative modeling. In: International Conference on Learning Representations (2023)
15. Liu, X., Gong, C., Liu, Q.: Flow straight and fast: learning to generate and transfer data with rectified flow. In: International Conference on Learning Representations (2023)
16. Loshchilov, I., Hutter, F.: Decoupled weight decay regularization. In: International Conference on Learning Representations (2019)
17. Lustig, M., Donoho, D., Pauly, J.M.: Sparse MRI: the application of compressed sensing for rapid MR imaging. Magn. Reson. Med. **58**(6), 1182–1195 (2007)
18. Lustig, M., Donoho, D.L., Santos, J.M., Pauly, J.M.: Compressed sensing MRI. IEEE Signal Process. Mag. **25**(2), 72–82 (2008)
19. Paszke, A., et al.: Pytorch: an imperative style, high-performance deep learning library. In: Advances in Neural Information Processing Systems, vol. 32 (2019)
20. Pokle, A., Muckley, M.J., Chen, R.T.Q., Karrer, B.: Training-free linear image inverses via flows. Trans. Mach. Learn. Res. (2024)
21. Pruessmann, K.P., Weiger, M., Scheidegger, M.B., Boesiger, P.: Sense: sensitivity encoding for fast MRI. Magn. Reson. Med. **42**(5), 952–962 (1999)
22. Qin, C., Schlemper, J., Caballero, J., Price, A.N., Hajnal, J.V., Rueckert, D.: Convolutional recurrent neural networks for dynamic MR image reconstruction. IEEE Trans. Med. Imaging **38**(1), 280–290 (2019)
23. Schlemper, J., Caballero, J., Hajnal, J.V., Price, A.N., Rueckert, D.: A deep cascade of convolutional neural networks for dynamic MR image reconstruction. IEEE Trans. Med. Imaging **37**(2), 491–503 (2017)
24. Song, J., Vahdat, A., Mardani, M., Kautz, J.: Pseudoinverse-guided diffusion models for inverse problems. In: International Conference on Learning Representations (2023)
25. Song, Y., Shen, L., Xing, L., Ermon, S.: Solving inverse problems in medical imaging with score-based generative models. In: International Conference on Learning Representations (2022)
26. Song, Y., Sohl-Dickstein, J., Kingma, D.P., Kumar, A., Ermon, S., Poole, B.: Score-based generative modeling through stochastic differential equations. In: International Conference on Learning Representations (2021)
27. Stein, C.M.: Estimation of the mean of a multivariate normal distribution. Ann. Stat. 1135–1151 (1981)
28. Uecker, M., et al.: Espirit—an eigenvalue approach to autocalibrating parallel MRI: where sense meets grappa. Magn. Reson. Med. **71**(3), 990–1001 (2014)
29. Villani, C., et al.: Optimal Transport: Old and New, vol. 338. Springer (2009)

30. Wang, Y., Yu, J., Zhang, J.: Zero-shot image restoration using denoising diffusion null-space model. In: International Conference on Learning Representations (2023)
31. Yang, Y., Sun, J., Li, H., Xu, Z.: ADMM-CSNet: a deep learning approach for image compressive sensing. IEEE Trans. Pattern Anal. Mach. Intell. **42**(3), 521–538 (2018)
32. Zbontar, J., et al.: fastMRI: an open dataset and benchmarks for accelerated MRI. arXiv preprint arXiv:1811.08839 (2018)
33. Zhu, Y., et al.: Denoising diffusion models for plug-and-play image restoration. In: Proceedings of the IEEE/CVF Conference on Computer Vision and Pattern Recognition, pp. 1219–1229 (2023)

Optimization of Acquisition Schemes Towards a Better Estimation of Microstructure Parameters in Multidimensional Diffusion MRI

Constance Bocquillon[✉], Isabelle Corouge, and Emmanuel Caruyer

Univ Rennes, Inria, CNRS, Inserm, IRISA UMR 6074, Empenn ERL 1228,
Campus de Beaulieu, Rennes 35042, France
`constance.bocquillon@irisa.fr`

Abstract. Unlike traditional measurements by diffusion tensor imaging, multidimensional diffusion MRI allows the estimation of additional microstructural parameters such as anisotropy, kurtosis and orientation dispersion. To properly take advantage of this imaging modality and capturing microstructure parameters accurately and efficiently, it is crucial to use a dedicated acquisition scheme. Several models and acquisition representations can be used towards this goal.

In this paper, we focused on the q-space trajectory imaging, using b-tensor acquisition encoding and the diffusion tensor distribution (DTD) modeling. More specifically, our goal is to develop a framework for the optimization of acquisition scheme based on their ability to properly estimate microstructural parameters of interest.

We generated an extensive collection of b-tensor shapes with a fixed number of directions each, from which we efficiently selected an optimized acquisition scheme. In the spirit of fingerprinting, we proposed a dictionary-based approach. The dictionary columns were carefully adapted to the achievable resolution in the parameter space, the parameters of interest being the microscopic anisotropy, the tensor size variance and the orientation parameter. To solve the combinatorial optimization problem of selecting the best subset of b-tensor shapes, we implemented two approximation algorithms: a greedy approach based and a permutation strategy.

To assess the performance of our optimization procedure, we computed the estimation error for each parameter. The signal generated from our scheme yielded lower or comparable errors to those of a reference scheme proposed in the literature and designed for this purpose.

Keywords: Acquisition · Diffusion MRI · Diffusion tensor distribution · Q-space trajectory imaging · Microstructure · Optimal experimental design · Parameter estimation

1 Introduction

Multidimensional diffusion MRI captures detailed properties such as anisotropy, kurtosis and orientation dispersion. This allows for the differentiation of microstructural features like fiber crossings, cellularity, and membrane permeability. Multidimensional dMRI has emerged as a powerful tool for studying complex biological tissues, offering valuable insights into neural connectivity, tumor microenvironments, and other structural properties in both research and clinical settings. Specifically, q-space trajectory imaging (QTI) [17] aims at reconstructing parameters of diffusion tensor distribution (DTD) using b-tensor encoded acquisitions.

Acquisition optimization aims at improving the accuracy, precision, and efficiency of microstructural parameter estimation. By carefully selecting the MRI acquisition parameters—such as b-tensors and diffusion encoding directions—we can enhance the sensitivity of the measured signals to the underlying tissue features. The Cramer-Rao lower bound has been used as a measure of reference to evaluate the estimation accuracy when optimizing acquisition schemes [1,16]. This strategy has been recently extended to multidimensional dMRI [11].

One drawback of the Cramer-Rao lower bound is that it is computed locally, with respect to one point in the parameter space. In this paper, we propose an alternative method inspired by fingerprinting [5,14]. This method is based on a dictionary of signals where each column corresponds to a DTD and each line to a b-tensor. From a predefined superset of acquisitions, our objective is to select the subset of b-tensors that best preserves the separability of the columns of the dictionary. We used simulations to assess the ability of the selected subset to yield accurate estimations of the microstructural parameters of interest: the microscopic anisotropy, the tensor size variance and the orientation parameter.

2 Theory

In this paper, we represent scalar value by a lowercase (e.g. b), vector/first-order tensor by a bold lowercase letter (e.g. $\mathbf{q}$), second-order tensor by a bold capital letter (e.g. $\mathbf{B}$) and fourth-order tensor by a blackboard capital letter (e.g. $\mathbb{C}$).

2.1 B-Tensor Encoded Acquisitions

In [6,17], the authors generalized the concept of b-tensors using q-space trajectories. In its general form, it can be defined as:

$$\mathbf{B} = \int_0^\tau \mathbf{q}(t)^{\otimes 2} dt \quad (1)$$

where τ is the echo time, $\mathbf{q}(t)^{\otimes 2}$ is the outer product of $\mathbf{q}(t)$ with itself and $\mathbf{q}(t)$ is a time dependent trajectory in q-space associated with a gradient waveform $\mathbf{g}(t)$ as:

$$\mathbf{q}(t) = \gamma \int_0^\tau \mathbf{g}(t') dt' \quad (2)$$

where γ is the gyromagnetic ratio.

In this work, we focused on cylindrical b-tensors. They can be described by their diagonal form in their principal axis system by:

$$\mathbf{B} = \begin{pmatrix} b_{\|} & 0 & 0 \\ 0 & b_{\perp} & 0 \\ 0 & 0 & b_{\perp} \end{pmatrix} \quad (3)$$

where $b_{\|}$ is the axial eigenvalue of the b-tensor and $b_{\perp}$ is the radial eigenvalue. Cylindrical b-tensors shapes are described as planar, oblate, spherical, prolate or linear depending on the relative values of $b_{\|}$ and $b_{\perp}$ as shown in Fig. 1. The b-value is defined as the trace of the b-tensor and is given by:

$$b_{val} = b_{\|} + 2b_{\perp} \quad (4)$$

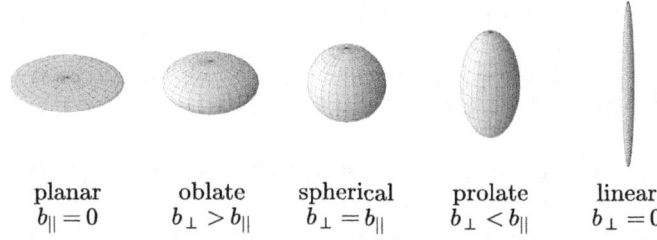

| planar | oblate | spherical | prolate | linear |
| $b_{\|}=0$ | $b_{\perp} > b_{\|}$ | $b_{\perp} = b_{\|}$ | $b_{\perp} < b_{\|}$ | $b_{\perp}=0$ |

Fig. 1. Cylindrical b-tensor shapes based on their eigenvalues.

2.2 Q-Space Trajectory Imaging

Q-space trajectory imaging [17] considers that the signal originates from a collection of microenvironments in which diffusion is assumed to be Gaussian.

$$s(\mathbf{B}) = s_0 \int p(\mathbf{D}) \exp(-\mathbf{B} : \mathbf{D}) d\mathbf{D} = s_0 \langle \exp(- < \mathbf{B} : \mathbf{D} >) \rangle \quad (5)$$

where s_0 is the non-diffusion weighted signal, $p(\mathbf{D})$ is the diffusion tensor distribution, $\mathbf{B} : \mathbf{D}$ is the inner product of $\mathbf{B}$ with $\mathbf{D}$, $\mathbf{D}$ is the diffusion tensor and $\langle . \rangle$ represents the averaging over the distribution.

The cumulant expansion of s gives:

$$s(\mathbf{B}) \approx s_0 \exp(- < \mathbf{B} : \langle \mathbf{D} \rangle > + \frac{1}{2} < \mathbb{B} : \mathbb{C} >) \quad (6)$$

where $\mathbb{B}$ is the outer product of $\mathbf{B}$ with itself, $\mathbb{C}$ is the covariance matrix of $\mathbf{D}$ given by:

$$\mathbb{C} = \langle \mathbf{D}^{\otimes 2} \rangle - \langle \mathbf{D} \rangle^{\otimes 2} \quad (7)$$

This framework was further improved by adding positivity constrains on $\mathbf{D}$ and $\mathbb{C}$ in [8].

2.3 The Diffusion Tensor Distribution (DTD) Parameters

The DTD introduced in Eq. 5 gives several microstructure parameters that can be defined through the moments of the distribution [17]. Here, we focused on the microscopic anisotropy, the tensor size variance and the orientation parameter. Their effect is illustrated in Fig. 2.

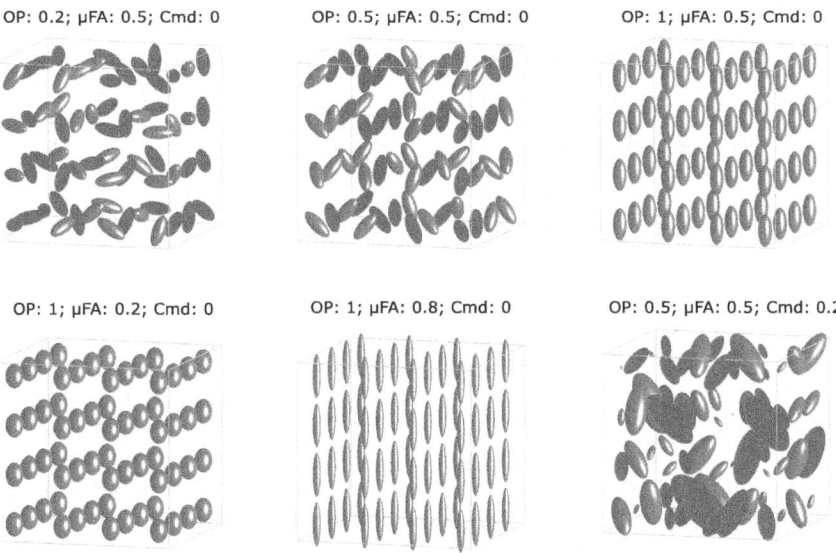

Fig. 2. Collection of DTDs with several values for each parameter of interest.

Tensor-size variance (C_{MD}). The tensor size variance describes the variance of the mean diffusivity of the microscopic tensors. It starts at 0 where all the microscopic tensors have the same size and increases with the size disparity. It can be computed by:

$$C_{MD} = \frac{<\mathbb{C}, \mathbb{E}_{bulk}>}{<\langle \mathbf{D}^{\otimes 2}\rangle, \mathbb{E}_{bulk}>} \quad (8)$$

Microscopic Anisotropy (μFA). The microscopic anisotropy ranges from 0 to 1 where 0 corresponds to spherical microscopic tensors, and 1 corresponds to a distribution of sticks-like tensors. It can be computed by:

$$C_\mu = \frac{<\langle \mathbf{D}^{\otimes 2}\rangle, \mathbb{E}_{shear}>}{<\langle \mathbf{D}^{\otimes 2}\rangle, \mathbb{E}_{iso}>} = \mu FA^2 \quad (9)$$

Orientation Parameter (OP). The orientation parameter ranges from 0 to 1 where 0 corresponds to fully randomly oriented tensors and 1 to fully aligned ones. It can be computed by:

$$OP^2 = \frac{<\langle \mathbf{D} \rangle^{\otimes 2}, \mathbb{E}_{shear}>}{<\langle \mathbf{D}^{\otimes 2} \rangle, \mathbb{E}_{shear}>} \tag{10}$$

To compute these parameters, the isotropic 4th order isotropic tensor $\mathbb{E}_{iso}$ and the bulk modulus $\mathbb{E}_{bulk}$ are defined as

$$\mathbb{E}_{iso} = \frac{1}{3}\begin{pmatrix} 1 & 0 & 0 & 0 & 0 & 0 \\ 0 & 1 & 0 & 0 & 0 & 0 \\ 0 & 0 & 1 & 0 & 0 & 0 \\ 0 & 0 & 0 & 1 & 0 & 0 \\ 0 & 0 & 0 & 0 & 1 & 0 \\ 0 & 0 & 0 & 0 & 0 & 1 \end{pmatrix}; \mathbb{E}_{bulk} = \mathbb{E}_{iso}^{\otimes 2} = \frac{1}{9}\begin{pmatrix} 1 & 1 & 1 & 0 & 0 & 0 \\ 1 & 1 & 1 & 0 & 0 & 0 \\ 1 & 1 & 1 & 0 & 0 & 0 \\ 0 & 0 & 0 & 0 & 0 & 0 \\ 0 & 0 & 0 & 0 & 0 & 0 \\ 0 & 0 & 0 & 0 & 0 & 0 \end{pmatrix} \tag{11}$$

and the shear modulus as

$$\mathbb{E}_{shear} = \mathbb{E}_{iso} - \mathbb{E}_{bulk} \tag{12}$$

3 Methods

In order to optimize the sensitivity of the acquisition to microstructural parameters of interest, we designed an approach in several steps. First, we generated a superset of acquisitions from which we will select the best subset of b-tensors. Then, we generated a collection of DTD parameterized by OP, µFA and C_{MD} which will become the columns of our dictionary. Next, we optimized the choice of a subset of acquisitions to maximize the separability of the columns in our dictionary. Finally, we evaluated the accuracy of the reconstructed parameters from the acquisitions associated with the selected subset.

3.1 Acquisition Superset

We constructed a superset of b-tensors, $\mathcal{S}$, containing every axisymmetric b-tensor shape with $b_{\parallel}$ and $b_{\perp}$ in the range [0, 2000] s/mm^2 divided into 60 intervals. We kept only those with a b-value under 2000 s/mm^2 and ensuring a step of 100 in b-values. The principal axis of each b-tensor shape was rotated into 30 directions generated from [9]. This results in a superset, $\mathcal{S}$, containing $n = 331$ unique b-tensor shapes, each rotated into 30 directions.

3.2 Fingerprinting the Parameter Space

We generated a dictionary of signals, inspired by fingerprinting methods [14], where each column corresponds to a DTD and each row corresponds to a b-tensor $\mathbf{B} \in \mathcal{S}$. Each DTD can be parameterized by three parameters: µFA, OP and C_{MD}. The range of values is [0, 1] for μFA and OP, and [0, 0.3] for C_{MD}. We restricted ourselves to DTD solely containing axisymmetric micro-tensors. We sampled 10000 micro-tensors for each DTD; the shape, size and orientation of each micro-tensor are sampled independently. The corresponding signal is simply computed according to Eq. 5.

Instead of dividing the parameter space on a regular grid, such as commonly done in fingerprinting [5,14], we locally adapted the granularity of the subdivision to the resolution permitted by the superset $\mathcal{S}$. More precisely, in a construction inspired by octrees, we started from the initial cube representing the whole space of parameters. At each step, and for each parameter, we tried to divide the cube into two half-cubes; the division was only kept if it could discriminate the pair of newly created DTDs using our superset of acquisitions, $\mathcal{S}$.

For a given acquisition scheme, $\mathcal{S}$, we consider that two DTDs can be discriminated if:

$$\max_{\mathbf{B} \in \mathcal{S}}(|s_{DTD1}(\mathbf{B}) - s_{DTD2}(\mathbf{B})|) \geq \frac{1}{SNR} \quad (13)$$

where $\mathbf{B}$ is a b-tensor in the acquisition scheme, $s_{DTD1}(\mathbf{B})$ (resp. $s_{DTD2}(\mathbf{B})$) is the signal generated from DTD 1 (resp. DTD 2) for the b-tensor $\mathbf{B}$.

3.3 Selection of Acquisition Schemes

Among the superset $\mathcal{S}$, the objective is to select a small subset of k b-tensor shapes, such that the number of pairs of DTD that can be distinguished is maximized. The number of possible subsets, C_n^k, grows very quickly with n and k which renders the brute-force approach impractical.

The criterion in Eq. 13 ensures that if we take a subset $\mathcal{S}_\infty \subset \mathcal{S}$, the number of pairs of DTD that can be discriminated is at most the same. To compare the shapes, we assigned a score to each. This score is the number of pairs of DTD that can be discriminated using a scheme with this b-tensor shape on every direction considered. This way, the stronger the shape, the more DTD can be discriminated, the weaker the shape, the less DTD can be discriminated. We focused on two approaches to select the acquisition subsets: a greedy algorithm (see Algorithm 1) and a permutation algorithm (see Algorithm 2).

We chose the number of selected b-tensor shapes so as to be comparable to the reference scheme, Q3. The scheme Q3, described in Table 1, is optimized using the Cramér-Rao lower bound trace of the mean diffusivity, μFA, the isotropic kurtosis and the anisotropic kurtosis described in [11]. Since Q3 uses 120 acquisitions, we selected 4 shapes of b-tensor acquired on 30 directions each. The 30 directions were generated analytically following Koay's method [10]. The upper b-value in Q3 is $2000\,\mathrm{s/mm^2}$, which is the same as what we retained for our superset.

3.4 Parameter Estimation

We evaluated the accuracy of the parameters of interest, using signal simulated following each of the optimized acquisition design. In order to do that, we esti-

Algorithm 1: Greedy algorithm

Input: Distance matrix, nb_shape
Output: Acquisition subset

subset ← [];
scores ← Array of size nb_shape, containing the number of pair of DTDs that can be discriminated by each shape;
for $i \leftarrow 1$ **to** nb_shape **do**
    best_shape ← Best shape available outside of subset;
    subset ← subset+best_shape;
    scores ← Number of new pairs that can be discriminated using each shape;
end
return subset

Algorithm 2: Permutation algorithm

Input: Distance matrix, nb_shape, nb_rep
Output: Best acquisition subset generated

best_subset_seen ← [];
best_nb_dtds ← 0;
for $i \leftarrow 1$ **to** nb_rep **do**
    subset ← random selection of nb_shape shapes;
    changed ← True;
    while *changed* **do**
        weakest_shape ← Weakest shape of subset;
        nb_weakest ← number of pairs for the subset with the weakest shape;
        next_best_shape ← Best shape available outsite of the subset;
        nb_next_best ← number of pairs for the subset with the next_best_shape instead;
        if nb_weakest < nb_next_best **then**
            Switching the weakest shape with the next best one in subset;
        end
        else
            changed ← False
        end
    end
    nb_dtds_subset ← number of pairs of DTDs that can be discriminated;
    if nb_dtds_subset > best_nb_dtds **then**
        best_nb_dtds *gets* nb_dtds_subset;
        best_subset_seen ← subset;
    end
end
return best_subset_seen

mated each parameter using QTI+ constraints [8] and its implementation in Dipy [7].

The acquisition scheme that we use as the reference is Q3 [11]. We focused more specifically on the estimation of OP and the estimation of μFA. For a given

OP (resp. μFA), we generated 11 DTDs with 10000 tensors and a C_{MD} at 0.1. The 11 DTDs were generated using the full range of the varying parameter: 0 to 1 with a step of 0.1 between values.

Then, for each of these DTDs, we generated the signals from each acquisition scheme that we did or did not corrupt with Rician noise. In the case of a corruption with noise, we fixed the SNR at 30 and generated 100 signals corrupted with noise for each DTD.

4 Results

The hierarchical division of the parameter space yields a total of 124 cubes when we set $SNR = 25$, out of a total of $2^{3 \times 6} = 262144$ possible cubes, since we used 6 recursive levels. On the resulting subdivision, on Fig. 3, we can see that the C_{MD} value has no significant impact on signals as only one division was made within the range. Also, we clearly have very few subdivisions in the parameter space when the μFA is small. This can be explained as it is hardly possible to find direction information within sphere-like tensors.

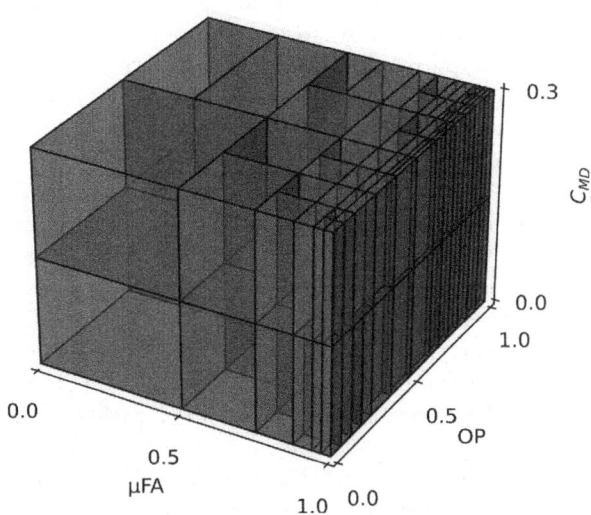

Fig. 3. Division of the parameter space in boxes of DTD with generated signal according to the achievable resolution.

Table 1 details the repartition of the b-tensor encoded acquisitions for the three acquisition schemes we compared. The first one is Q3 [11] and includes planar, spherical and linear b-tensors. The second one is the scheme generated from the greedy algorithm. The b-values selected are between 1000 and 2000 s/mm² and includes planar, oblate and linear b-tensors but no spherical. The last one

Table 1. Repartition of the b-tensor encoded acquisitions. Q3 is the reference scheme, PTE corresponds to planar b-tensors, OTE to oblate b-tensors, STE to spherical b-tensors and LTE to linear b-tensors.

Acquisition scheme	Q3				Greedy algorithm				Permutation algorithm			
b (s/mm^2)	PTE	OTE	STE	LTE	PTE	OTE	STE	LTE	PTE	OTE	STE	LTE
100	7	–	–	9	–	–	–	–	–	–	–	–
800	9	–	–	50	–	–	–	–	–	–	–	–
1000	–	–	–	–	30	–	–	–	–	–	–	–
1100	–	–	–	–	–	–	–	–	30	–	–	–
1200	–	–	–	–	–	30	–	–	30	–	–	30
1300	–	–	–	–	–	–	30	–	–	–	–	30
2000	–	–	30	15	–	–	–	30	–	–	–	–

is the scheme generated from the permutation algorithm with 100 random initializations. Selected b-values are 1000, 1100 and 1200 s/mm^2 with some planar, oblate and linear b-tensor selected.

Figure 4 shows the error on the estimation of OP over its full range for two values of µFA, without noise or with Rician noise with an SNR of 30. On the top left, we have the comparison of the estimation error for our three schemes at SNR 30 with a µFA of 0.6. In this case, similar results were achieved for the reference scheme Q3 and the scheme selected through the greedy algorithm. The permutation algorithm gives a more important error. It is interesting to note that the error is fairly high for every scheme considered here. When we look at the top right figure, we keep the same value of the noise but we increase the µFA. While the scheme from the permutation algorithm still gives the least interesting results, the scheme generated from the greedy algorithm gives an error that is decreased and better than the reference scheme. This pattern is also present in the bottom row where the signal was not corrupted with noise.

Figure 5 shows the error on the estimation of µFA over its full range for two values of OP, with a Rician noise at an SNR of 30 or without noise. The top row presents the estimation with Rician noise and the bottom row the estimation without noise. The first column gives the estimation for an OP of 0.4 while the second gives it for an OP of 0.8. Overall, the estimation error is smaller than it is for the estimation of OP. In all four cases, Q3 and the greedy algorithm scheme perform similarly. In presence of noise, the permutation algorithm generated the most errors. However, without noise, it generates the best results.

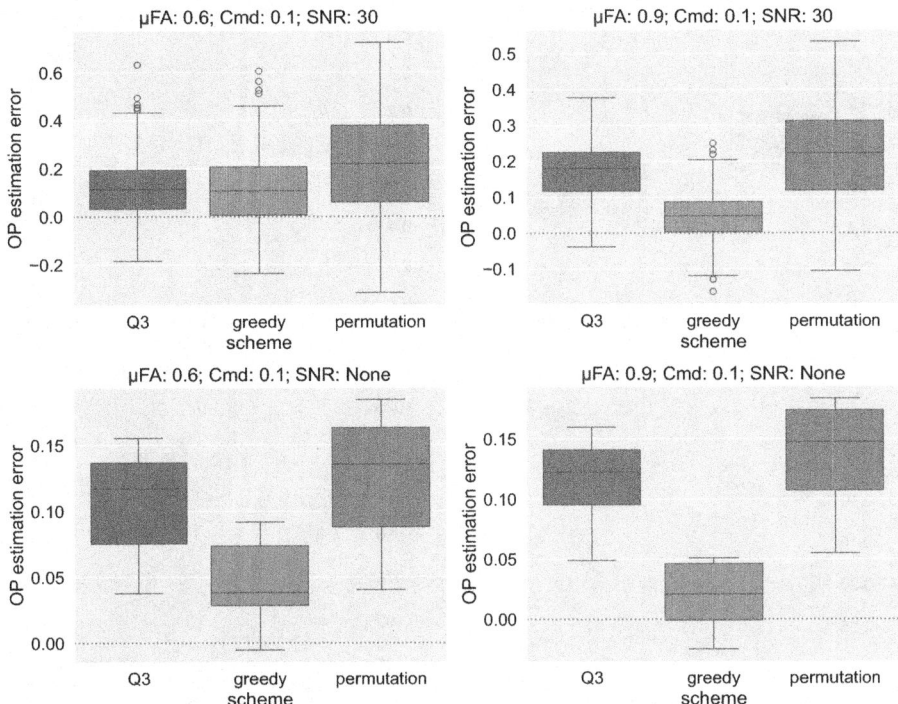

Fig. 4. OP estimation error on its full range. In blue, the reference scheme; in orange, the greedy algorithm result; in green, the permutation algorithm result. (top row) The signal was corrupted with a Rician noise at an SNR 30. (bottom row) The signal was not corrupted with any noise. (first column) μFA was set at a medium value. (second colomn) μFA was set at a high value. (Color figure online)

5 Discussion

We proposed a framework aiming to improve the accuracy of DTD parameters estimation. We found that the b-tensor encoded acquisition scheme generated by the greedy approach provided the best results in terms of estimation compared to the previously proposed scheme we used as a reference.

Our approach is promising and can be further improved in several ways. One improvement that needs to be implemented is to adapt the number of directions based on the b-value and the shape of the b-tensor. That can be achieved by computing the band limit in spherical harmonic for each b-tensor shape [4]. Then, we can select a number of direction based on the band-limit using spherical codes [12]. Another direction for improvement is to take into consideration the impact of the shape of a b-tensor and its b-value on the echo time (TE), directly related to the waveform [15]. That way, we can not only select an upper bound in the number of acquisitions selected, but also guide the choice towards b-tensor shapes with shorter TE, leading to an overall improvement in SNR.

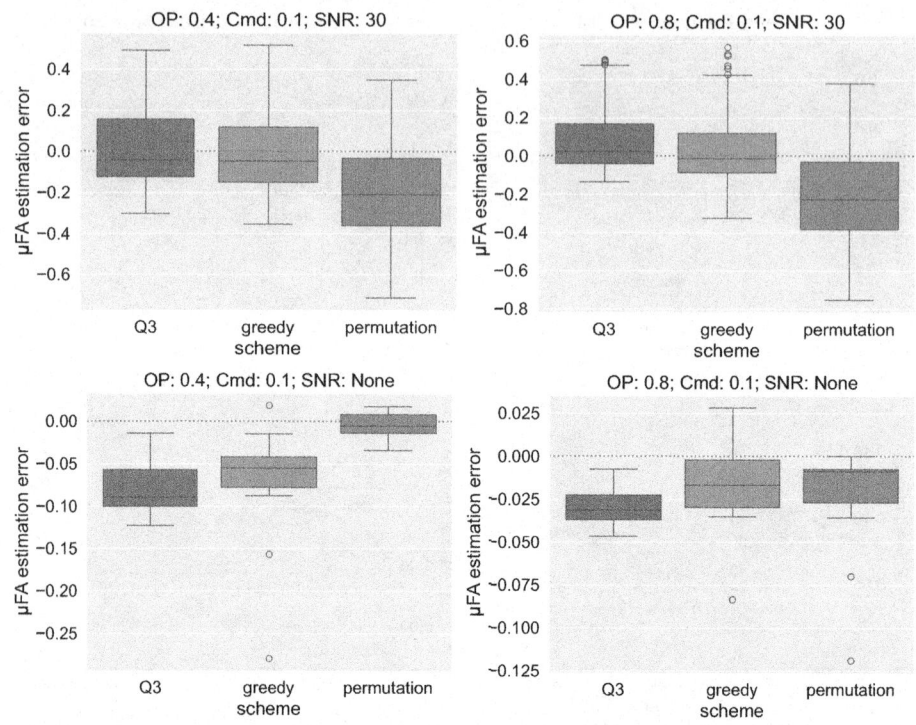

Fig. 5. μFA estimation error on its full range. In blue, the reference scheme; in orange, the greedy algorithm result; in green, the permutation algorithm result. (top row) The signal was corrupted with a Rician noise at an SNR 30. (bottom row) The signal was not corrupted with any noise. (first column) OP was set at a medium value. (second colomn) OP was set at a high value. (Color figure online)

In this work, we focused on the DTD framework but this could easily be adapted to work with biophysical models such as AxCaliber [3], CHARMED [2], NODDI [18] or SANDI [13] as they have a limited number of parameters of interest.

6 Conclusion

The estimation of microstructure parameter is greatly influenced by the acquisition scheme. Our approach allows us to generate acquisition schemes optimized towards the reconstruction of selected parameters.

While the current results can be further improved, we generated acquisition schemes with similar or better estimation errors while ensuring its feasibility on clinical MR scanner.

Our approach could also be implemented for other models with a reasonable number of parameters of interest in order to generate acquisition schemes tailored specifically for their reconstruction.

References

1. Alexander, D.C.: A general framework for experiment design in diffusion MRI and its application in measuring direct tissue-microstructure features. Magn. Reson. Med. **60**(2), 439–448 (2008)
2. Assaf, Y., Basser, P.J.: Composite hindered and restricted model of diffusion (charmed) MR imaging of the human brain. Neuroimage **27**(1), 48–58 (2005)
3. Assaf, Y., Blumenfeld-Katzir, T., Yovel, Y., Basser, P.J.: Axcaliber: a method for measuring axon diameter distribution from diffusion MRI. Magn. Reson. Med. **59**(6), 1347–1354 (2008)
4. Bates, A., Daducci, A., Sadeghi, P., Caruyer, E.: A 4D basis and sampling scheme for the tensor encoded multi-dimensional diffusion MRI signal. IEEE Signal Process. Lett. **27**, 790–794 (2020). https://doi.org/10.1109/LSP.2020.2991832, https://inserm.hal.science/inserm-02881980
5. Daducci, A., Canales-Rodríguez, E.J., Zhang, H., Dyrby, T.B., Alexander, D.C., Thiran, J.P.: Accelerated microstructure imaging via convex optimization (AMICO) from diffusion MRI data. Neuroimage **105**, 32–44 (2015)
6. Eriksson, S., Lasič, S., Nilsson, M., Westin, C.F., Topgaard, D.: NMR diffusion-encoding with axial symmetry and variable anisotropy: Distinguishing between prolate and oblate microscopic diffusion tensors with unknown orientation distribution. J. Chem. Phys. **142**(10), 104201 (2015)
7. Garyfallidis, E., et al.: Dipy, a library for the analysis of diffusion MRI data. Front. Neuroinform. **8** (2014)
8. Herberthson, M., Boito, D., Haije, T.D., Feragen, A., Westin, C.F., Özarslan, E.: Q-space trajectory imaging with positivity constraints (qti+). Neuroimage **238**, 118198 (2021)
9. Jones, D.K., Horsfield, M.A., Simmons, A.: Optimal strategies for measuring diffusion in anisotropic systems by magnetic resonance imaging. Magn. Reson. Med.: Official J. Int. Soc. Magn. Reson. Med. **42**(3), 515–525 (1999)
10. Koay, C.G.: A simple scheme for generating nearly uniform distribution of antipodally symmetric points on the unit sphere. J. Comput. Sci. **2**(4), 377–381 (2011)
11. Morez, J., Szczepankiewicz, F., den Dekker, A.J., Vanhevel, F., Sijbers, J., Jeurissen, B.: Optimal experimental design and estimation for q-space trajectory imaging. Hum. Brain Mapp. **44**(4), 1793–1809 (2023)
12. Sloane, N.J.A., with the collaboration of R. H. Hardin, W.D.S., et al.: Tables of spherical codes, NeilSloane.com/packings/
13. Palombo, M., et al.: Sandi: a compartment-based model for non-invasive apparent soma and neurite imaging by diffusion MRI. Neuroimage **215**, 116835 (2020)
14. Rensonnet, G., et al.: Towards microstructure fingerprinting: estimation of tissue properties from a dictionary of monte Carlo diffusion MRI simulations. Neuroimage **184**, 964–980 (2019)
15. Sjölund, J., Szczepankiewicz, F., Nilsson, M., Topgaard, D., Westin, C.F., Knutsson, H.: Constrained optimization of gradient waveforms for generalized diffusion encoding. J. Magn. Reson. **261**, 157–168 (2015)
16. Truffet, R., et al.: An evolutionary framework for microstructure-sensitive generalized diffusion gradient waveforms (2020)
17. Westin, C.F., et al.: Q-space trajectory imaging for multidimensional diffusion MRI of the human brain. Neuroimage **135**, 345–362 (2016)
18. Zhang, H., Schneider, T., Wheeler-Kingshott, C.A., Alexander, D.C.: NODDI: practical in vivo neurite orientation dispersion and density imaging of the human brain. Neuroimage **61**(4), 1000–1016 (2012)

Bilinear Projector: Mitigating Discretization Artifacts in Model Based Iterative Reconstruction for X-Ray CT

Ke Chen[✉] and Alireza Entezari

University of Florida, Florida, USA
`ke.chen@uft.edu`

Abstract. Pixel and voxel-basis are commonly used for designing X-ray CT projectors that underly existing projection and back-projection methods that are widely used in practice. We introduce a bilinear projector for modeling X-ray optics and provide experimental evidence showing its impact in reducing discretization artifacts in Model-Based Iterative Reconstruction (MBIR). We demonstrate the advantages of using this bilinear model over the conventional pixel-based projectors (piecewise constant) for MBIR in realistic simulations in presence of nonlinearity of the Lambert-Beer law. We derive efficient piecewise polynomial forms for footprint and detector blur computations for this new projector and demonstrate its efficient GPU implementation for high resolution fan-beam geometry.

Keywords: X-ray CT · Iterative Reconstruction · Discretization

1 Introduction

Tomographic reconstruction algorithms enable reconstruction of images from projection data, which is a prevalent problem in biomedical imaging. Prominent medical imaging applications are transmission X-ray computed tomography (CT), and emission tomography (PET and SPECT). The flexibility of X-ray imaging hardware has led to the adoption of this imaging modality in a growing list of applications not only in diagnostic but also in image-guided radiation therapy and surgery.

Although analytical methods, most notably the Filtered Back Projection (FBP), are commonly used in commercial CT scanners [13], their imaging quality heavily depends on availability of a large number of high-dose X-ray views. Model-Based Iterative Reconstruction (MBIR) enables imaging from low-dose or limited/sparse view projection data by integrating a statistical model for X-ray physics (i.e., the forward model) together with a discrete representation of the attenuation map over the Field of View (FOV) into an optimization framework for image reconstruction [5,12]. This optimization framework allows for integrating priors for images such as statistical (e.g., Markov random fields [1]), sparsity promoting (e.g., total variation [15]), and more recently deep learning [14,16] priors that allow for improving quality of reconstruction.

A key step in design of MBIR algorithms is the discretization of the attenuation map over the FOV into a number of pixels/voxels that is necessary to formulating the image reconstruction problem as a finite-dimensional optimization problem. In this optimization framework, the forward model (projector) provides a simulation of the acquisition process relating the pixel intensities (CT numbers) in the image domain to the detector measurements in the sinogram domain. The projector uses a piecewise-constant (over the extent of pixel/voxel) model in the image domain and by integrating the line integrals (rays) over a detector cell, provides an linearized approximation to the LambertâĂŞBeer law. While this approximation introduces a number of well-known artifacts [3, 6] such as the exponential edge gradient effect [7], it underlies existing projection/back-projection methods that are widely used in practice [9]. To reduce these artifacts in MBIR, one can increase the resolution, by finer discretization in the image domain. However, not only this leads to significant increase in computational cost of forward and back-projection during iterations, but also the worsening condition of the inverse problem that leads to longer iterations necessary for convergence as well as numerical instabilities [12].

Contribution. We introduce a bilinear projector for modeling X-ray optics in forward/back-projection algorithms and provide experimental evidence showing its impact in reducing discretization artifacts in MBIR without increasing resolution. We demonstrate the advantages of using this bilinear model over the conventional pixel-based projectors (piecewise constant) for MBIR in realistic simulations in presence of nonlinearity of the Lambert-Beer law. We derive efficient piecewise polynomial forms for footprint and detector blur computations for this new projector and demonstrate its efficient GPU implementation.

2   Discretization of Forward Model

The X-ray transform relates the attenuation function to its 1-D projections parametrized by directions: The line-integration of a function $f(\boldsymbol{x})$, $\boldsymbol{x} \in \mathbb{R}^d$ along various directions $\boldsymbol{u} \in S^{d-1}$ ($d = 2, 3$ for 2-D and 3-D problems respectively). Let $\mathbf{P}_{\boldsymbol{u}}$ be a matrix whose columns span the hyperplane that is orthogonal to the direction $\boldsymbol{u}$. Then $\mathbb{R}^d$ can be re-parameterized along $\boldsymbol{u}$ and orthogonal to it as $\boldsymbol{x} = \boldsymbol{u}t + \mathbf{P}_{\boldsymbol{u}}^T \boldsymbol{s}$ where $\boldsymbol{s} \in \mathbb{R}^{d-1}$ parameterizes the projection (sinogram) domain and t parametrizes the direction of integration. Then the X-ray transform of f with respect to the direction $\boldsymbol{u}$ (angle θ) is:

$$g(\boldsymbol{u}, \boldsymbol{s}) = (\mathcal{P}_{\boldsymbol{u}} f)(\boldsymbol{s}) = \int_{\mathbb{R}} f(t\boldsymbol{u} + \mathbf{P}_{\boldsymbol{u}}^T \boldsymbol{s}) \mathrm{d}t. \tag{1}$$

In 2-D the X-ray and Radon transforms coincide; hence, the terms are sometimes interchangeably used although they describe different transforms in 3-D. For a viewing angle θ_i, identified by the direction $\boldsymbol{u}_i$, the *line integral* of f along a ray connecting the source $\boldsymbol{p}$ to a point $\boldsymbol{s}_j$ is a direct sampling of the *footprint* of f

by $g(\boldsymbol{u}_i, \boldsymbol{s}_j)$. In practice, however, the detector measurements are modeled by integrating g across the detector cell, a process that introduces a *detector blur*:

$$b_m = \{g(\boldsymbol{u}_i, \cdot), h_j\} = \int_{\mathbb{R}^{d-1}} g(\boldsymbol{u}_i, \boldsymbol{s}) h_j(\boldsymbol{s}) d\boldsymbol{s} \quad \text{with} \quad m = ij. \tag{2}$$

Here b_m denotes the measurement from the jth detector cell with the detector blur function h_j which is positioned on the ith viewing direction $\boldsymbol{u}_i$. This measurement b_m is contaminated with noise which is often modeled by a (compound) Poisson distribution [8]. Given a discretization model, f_N, for the image f, one can set up an inverse problem relating continuous-domain X-ray optics, measured by the detectors, to the image coefficients (Fig. 1).

For an N-dimensional model (i.e., with N pixels/voxels), a continuous-domain basis function φ provides a *discrete-continuous model* (Fig. 2),

$$f_N(\boldsymbol{x}) = \sum_{n=1}^{N} y_n \varphi(\boldsymbol{x} - \boldsymbol{p}_n) \tag{3}$$

that allows for formulating CT reconstruction as an inverse problem: the system matrix $\mathbf{A}$, that relates $\boldsymbol{y}$, the discretized image (with coefficients y_n), to the logarithmically-transformed detector measurements b_m from (2) provides a *forward model*:

$$\mathbf{A}\boldsymbol{y} = \boldsymbol{b}, \tag{4}$$

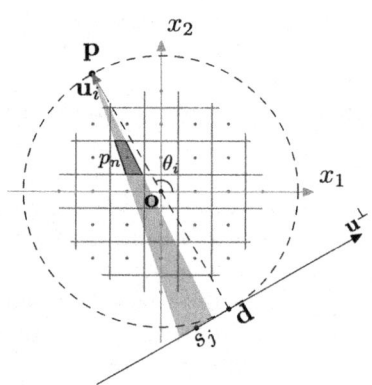

Fig. 1. Discretization in the forward model.

where $\boldsymbol{y}$ and $\boldsymbol{b}$ denote the vector of approximation coefficients and the measurements respectively. The most common choice of basis function is the pixel basis: $\varphi(\boldsymbol{x}) = \varphi(x_1, x_2) = B_0(x_1) B_0(x_2)$ where $B_0(x) = (x+1/2)_+^0 - (x-1/2)_+^0$ with $(x)_+ = x$ if $x > 0$ and 0 otherwise that can be viewed as first-order B-spline [4]. Kaiser-Bessel functions (aka blobs) have also been proposed for CT imaging [10]. The (m, n) entry of $\mathbf{A}$ is computed from the contribution of the nth basis function $\varphi_n(\boldsymbol{x}) := \varphi(\boldsymbol{x} - \boldsymbol{p}_n)$ to the mth observation in (2). Specifically the contribution of φ_n is computed by (1) integrating it along incident rays to form its *footprint*: $\mathcal{P}_{\boldsymbol{u}_i} \varphi_n$ and (2) integrating these footprints across a detector cell in the projection plane (the detector blur) $\{\mathcal{P}_{\boldsymbol{u}_i} \varphi_n, h_j\}$. These integral calculations are the most commonly-used model of the X-ray optics in the imaging system [18]:

$$\begin{aligned} \text{Footprint} \quad & \mathcal{P}_{\boldsymbol{u}_i} \varphi_n(\boldsymbol{s}) = \int_{\mathbb{R}} \varphi_n(t\boldsymbol{u}_i + \mathbf{P}_{\boldsymbol{u}_i}^T \boldsymbol{s}) dt \\ \text{Detector blur} \quad & \{\mathcal{P}_{\boldsymbol{u}_i} \varphi_n, h_j\} = \int_{\mathbb{R}^{d-1}} h_j(\boldsymbol{s}) \mathcal{P}_{\boldsymbol{u}_i} \varphi_n(\boldsymbol{s}) d\boldsymbol{s}. \end{aligned} \tag{5}$$

Given any N, representing the total number of pixels/voxels in an image (i.e., resolution), the MBIR reconstruction is often formulated as an optimization problem for finding the approximation coefficients in (3) by solving the inverse problem (4) by an iterative method.

The discretization artifacts [12] are introduced by inconsistencies in projection data arising from attenuation gradients occurring within the detector cell that are exacerbated due to the logarithmic relationship between attenuation line integrals and the photon counts measured via detector blur. Therefore, with a finite pixel/voxel size and finite detector sizes discretization artifacts are unavoidable. Even in ideal situations where each voxel represents the average attenuation of the real object, the projections of the voxels will be mismatched with the measured projections [19]. In contrast, the back-projection result of detector data will also deviate from the true data. Since the MBIR utilizes the forward and backward model iteratively, the error will be accumulated and causing aliasing artifacts around sharp material boundaries between tissues and other object.

Increasing the resolution (i.e., finer discretization) in MBIR has been explored as a potential solution to mitigate those artifacts. However, this approach introduces significant increase in computational cost of forward and back-projection often resulting in impractical reconstruction times for resolutions relevant to clinical settings. Additionally, increasing resolution leads to degradation of conditioning of the inverse problem which can exacerbate numerical instability in the iterative optimization process, slowing convergence and increasing susceptibility to noise. These drawbacks necessitate novel strategies to balance computational efficiency, image quality, and artifact suppression [5,12].

3 Bilinear Projector in Fanbeam Geometry

While pixel and voxel-basis are the most common basis set [9] used in CT imaging, alternative basis functions have been considered. Most prominently spherically symmetric Kaiser-Bessel functions (as known as blobs) have been explored in iterative reconstruction [10]. Despite the easy calculation of blob footprints [23], their integration for detector blur is not straightforward. More importantly Kaiser-Bessel functions do not form a partition of unity (even for first order of approximation) and require expensive filtering operations [11] to minimize discretization errors. It has been observed that pixel-based reconstruction often outperforms blob-based reconstruction in terms of edge artifact reduction [20].

Another class of basis sets that have been proposed for CT imaging are box splines [4] that were originally developed in parallel geometry and have been recently extended to divergent geometry [21,22]. This class of basis functions, include pixel/voxel-basis and tensor product higher-order B-splines as special cases but also include non-separable functions.

The convolutional framework presented in [21] provides a closed-form expression for efficient computation of footprint and detector blur for pixel and voxel basis functions. We leverage this convolutional framework to derive an efficient projector for bilinear basis: $\varphi(\mathbf{x}) = \varphi(x_1, x_2) = B_1(x_1)B_1(x_2)$ with the second order B-spline defined as $B_1(x) = (x+1)_+ - 2(x)_+ + (x-1)_+$. (recalling the notation $(x)_+ = x$ if $x > 0$ and 0 otherwise. Specifically, the proposed bilinear projector we develop is obtained by efficient computation of footprint and detector blur in (5) for the bilinear basis.

In the convolutional framework, the bilinear basis corresponds to a box spline M_Ξ corresponding to four directions defined as: $\Xi = [\xi_1, \xi_2, \xi_3, \xi_4] = \begin{bmatrix} 1 & -1 & 0 & 0 \\ 0 & 0 & 1 & -1 \end{bmatrix}$.

Using the convolution properties of box splines [4], the convolution of two pairs of vectors of equal length and opposite direction can be decomposed into two one-dimensional convolutions. This implies the four-direction box splines is simply a tensor-product of two triangle functions.

With the box spline decomposition of the bilinear basis function and the forward projection algorithm using box splines in fan-beam geometry [21], we can project the bilinear basis function onto the detector plane. Let R_u denote the parallel X-ray projection for a viewing direction given by the vector $\mathbf{u}$ (corresponding to viewing direction θ). Then:

$$R_u\{\varphi(\mathbf{p})\}(s) = M_Z(s) = (M_{u_\perp \xi_1} * M_{u_\perp \xi_2} * M_{u_\perp \xi_3} * M_{u_\perp \xi_4})(x), \qquad (6)$$

where $u_\perp$ is the transformation matrix projecting the vectors onto the detector plane. This transformation preserves the same lengths and opposite-direction relationships of the four vectors for each basis φ. Leveraging the concept of effective blur [21], approximating the fan-beam X-ray projection with detector blur involves convolving the parallel X-ray projection function with the detector blur vector $\mathbf{v}_s$. Thus, the intensity at each detector bin can be computed by the convolution of five vectors.

Let $a = \max(\|u_\perp \xi_1\|_2, \|u_\perp \xi_3\|)$ and $b = \min(\|u_\perp \xi_1\|_2, \|u_\perp \xi_3\|)$. Then, the parallel projection onto the detector plane can be written as:

$$R_u\{\varphi(\mathbf{p})\}(s) = M_{[a,-a,b,-b]}(s) = \frac{1}{a}\Lambda\left(\frac{s}{a}\right) * \frac{1}{b}\Lambda\left(\frac{s}{b}\right) \qquad (7)$$

where $*$ denotes convolution.

Next, we include detector blur. Suppose the detector blur is a rectangular function $\Pi(x)$. Convolving $R_u\{\varphi(\mathbf{p})\}(s)$ with $\Pi(x)$ yields the final measured intensity $P_\theta(s)$. To simplify notation and computation, we introduce a cumulative integral function $M_C(x)$, which transforms the piecewise complexity into manageable closed-form expressions. Detailed definitions of the piecewise function are in Appendix A.

Given X, Y indicating the left and right margin of the rectangular function $\Pi(x)$, the intensity at detector cell centered at s is:

$$P_\theta(s) = \begin{cases} |M_C(|X|) - M_C(|Y|)| & \text{if } XY > 0, \\ M_C(|X|) + M_C(|Y|) & \text{if } XY < 0. \end{cases}$$

Intuitively, the final detector intensity $P_\theta(s)$ can be interpreted as the integrated area under two triangle functions, scaled by geometry, and "windowed" by the detector blur (rectangle). This analytic approach avoids costly numerical integrations for each detector bin, improving computational efficiency.

By expanding and simplifying the explicit form of the convolution, we can calculate the bilinear pixel projection with detector blur fast and accurately, as shown in the experiment section.

4 Experiments and Results

To validate the effectiveness of the proposed bilinear basis forward and backprojection operators, we conducted comparisons with a number of pixel-based projectors including the Separable Footprints (SF) [9] and CNSF [21]. While we observed similar results with other projectors, due to space constraints we report comparisons against CNSF which is the most accurate in modeling of X-ray optics in pixel-basis [17].

As for the projector implementation, we utilized the code structure of the SF projection for 3D flat detectors from the MCR Toolkit [2], which incorporates ILP, data coalescing, and a pre-computed projection matrix to accelerate forward and backward projections. This served as a reference framework for implementing our forward and backward models across all methods. Further optimizations were performed on the original code with the assistance of the NVIDIA Nsight Compute Profiler to enhance performance.

We note that the objective of our experiments is to quantify the accuracy of X-ray optics in (5) for pixel projectors and the proposed bilinear projector. Hence, our study examines an analytical benchmark, designed for this purpose, where line integrals, for simulating Beers-Lambert law, detector integration and Poisson noise can be calculated exactly.

4.1 Analytical Benchmark

We employed a modified FORBILD phantom as the benchmark dataset as shown in the left of Fig. 3. This phantom retained the original primary elliptical components and the right ear structure but replaced the original resolution dots with three parallel lines for quantitatively assessing resolution loss due to discretization. These lines, spaced 10 mm apart and 1 mm in width, enabled controlled testing. Additionally, we introduced a star-shaped pattern consisting of six lines oriented at 0°, 30°, 60°, 90°, 120°, and 150°. The first three lines spanned 1 mm vertically (along the y-axis), while the remaining three spanned 1 mm horizontally. This addition allowed resolution evaluation in multiple directions. The density of the added lines was set to 1.8 g/cm^3, compared to the background ellipse's density of 1.05 g/cm^3.

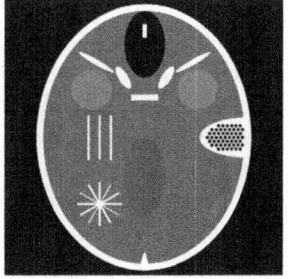

Fig. 2. FORBILD analytical benchmark.

For CT projection data simulation, we projected the FORBILD phantom onto 1800 evenly distributed flat detectors over a 360° rotation. The detector bins were spaced by $\Delta = 25/512$ mm, with a total of 1710 bins. The distance between the X-ray source and the detector center was $D_{sd} = 78.125$ mm, while the distance between the source and the object center was $D_{so} = 39.0625$ mm. The X-ray fan beam's isocenter was precisely aligned with the detector center.

To generate sinograms for reconstruction, we computed the line integral function $P_u\{f\}$ from the X-ray source to specific detector points, performing numerical integration over each detector bin with absolute and relative error tolerances of $E_{abs} = E_{rel} = 1e^{-10}$. To account for the edge-gradient effect induced by the logarithmic attenuation-intensity relationship, we calculated intensity values using Function. We can sample from the 1800 different angles and combine adjacent detector bins to generate intensity maps with varying resolutions and perspectives.

We conducted experiments on both noisy and noiseless projection data. To simulate Poisson noise in the projections, we assumed an initial photon count of $N_0 = 2 \times 10^4$ per detector bin. The attenuation coefficient of the material was proportional to its density, with a mass attenuation coefficient of $\mu/\rho = 0.1\,\mathrm{cm}^2/\mathrm{g}$. The numerical intergral won't work with noisy data. Thus we sampled 100 points evenly in each detector bin to calculate the intensity value.

4.2 Reconstruction Performance

To evaluate the performance of projectors, we conducted Model-Based Iterative Reconstruction (MBIR) without a regularization term on projection data at varying resolutions. Reconstruction experiments were performed on 512^2 pixel grids using sinograms of size 900×855 obtained from the dataset. The optimization was carried out using a simple iterative least-squares solver from the Python package "PyLops," with the number of iterations ranging from 2 to 64.

The metrics used for performance evaluation included Root Mean Square Error (RMSE), Peak Signal-to-Noise Ratio (PSNR), Structural Similarity Index Measure (SSIM), and Full Width at Half Maximum (FWHM). These metrics were chosen for their ability to assess image quality. RMSE provides a direct indication of overall fidelity by measuring the average reconstruction error. PSNR, as a logarithmic transformation of RMSE, highlights differences in image quality more distinctly across models. SSIM emphasizes perceptual quality by evaluating luminance, contrast, and structural similarity between the reconstructed and reference images. FWHM assesses spatial resolution by measuring the width of intensity profiles at half their maximum value, offering a critical measure of sharpness.

In our experiments, FWHM was measured using intensity profiles extracted from 8 horizontal lines sampled across the three vertical bars in the phantom image, as well as from 16 horizontal and vertical lines sampled around the star pattern. The half maximum value used was $1.425\,\mathrm{g/cm}^3$, which represents the midpoint between the density of the background and the density of the resolution

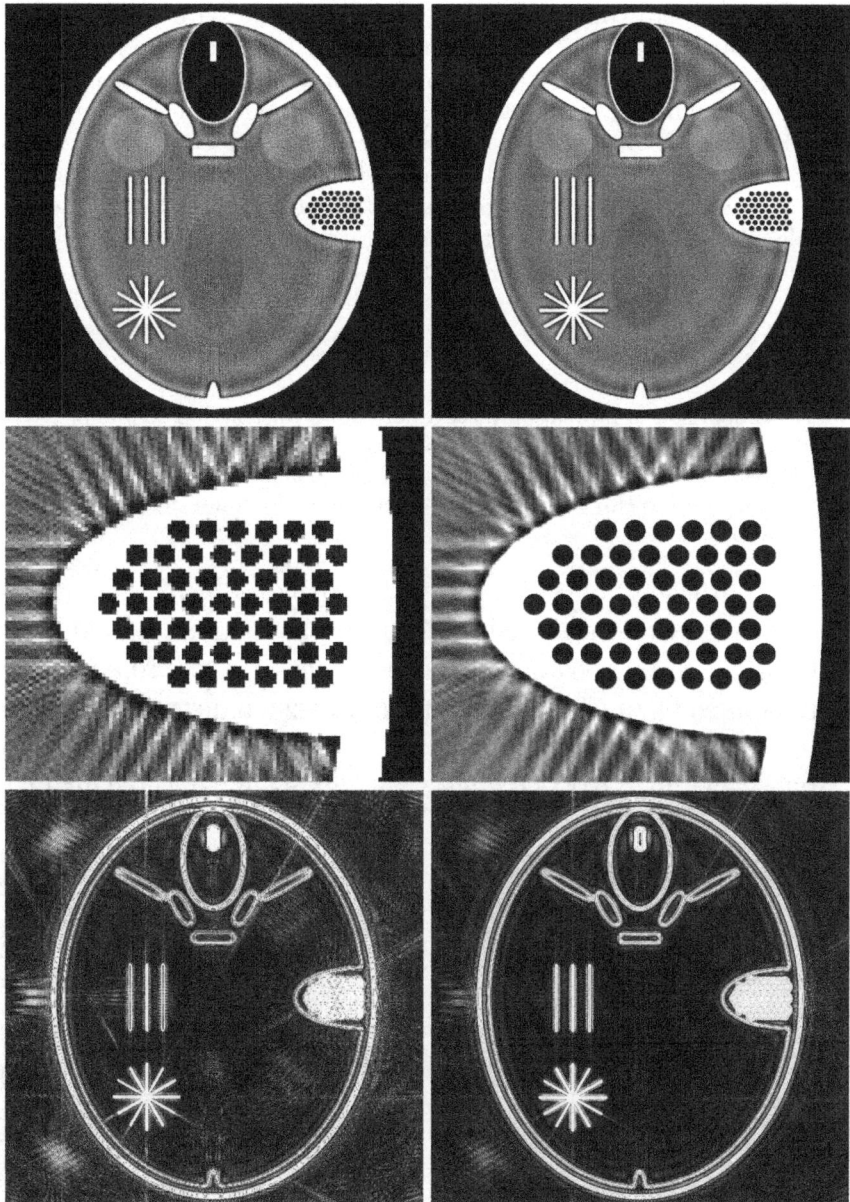

Fig. 3. Reconstruction of the modified FORBILD phantom with pixel-basis (left) and bilinear basis (right). The display range for the images is from $1.0\,\text{g/cm}^3$ to $1.1\,\text{g/cm}^3$. The second row show a close up view and the third row shows the error maps.

bars. This methodology ensured a robust evaluation of spatial resolution across various features in the dataset.

Experiments were conducted on datasets with and without Poisson noise to examine robustness under different noise conditions. Since no regularization terms were applied, the least-squares algorithm exhibited overfitting behavior after a certain number of iterations. To ensure meaningful comparisons, we selected iterations that achieved the highest PSNR (or equivalently the lowest RMSE) and the highest FWHM values for each projector. This approach enabled us to identify the optimal reconstruction performance for each projector while balancing fidelity and resolution metrics.

Table 1 presents the quantitative results of the evaluation metrics. In both the iterations with the highest SSIM and those with the highest PSNR, the CNSF-based pixel projector provided a more accurate approximation of forward projec-

Table 1. Comparison of evaluation metrics for the reconstruction results of the modified FORBILD phantom from a 900×855 sinogram to 512×512 pixels, with and without Poisson noise. The unit of RMSE is mm^{-1} and the unit of FWHM is mm. Empty slots in the FWHM column indicate that FWHM could not be measured for certain iterations.

Iteration		Highest SSIM			Highest PSNR		
Metrics		PSNR/RMSE	SSIM	FWHM	PSNR/RMSE	SSIM	FWHM
Clean	Pixel	26.14/0.00973	0.8493	1.3002	26.14/0.00973	**0.8493**	1.2989
	Bilinear	**26.98/0.00800**	**0.8838**	**1.2463**	**27.24/0.00754**	0.8457	**1.1392**
Noisy	Pixel	22.87/0.02065	0.7361	/	25.07/0.01244	0.5874	**1.2936**
	Bilinear	**24.20/0.01518**	**0.7668**	/	**25.88/0.01031**	**0.6495**	1.2926

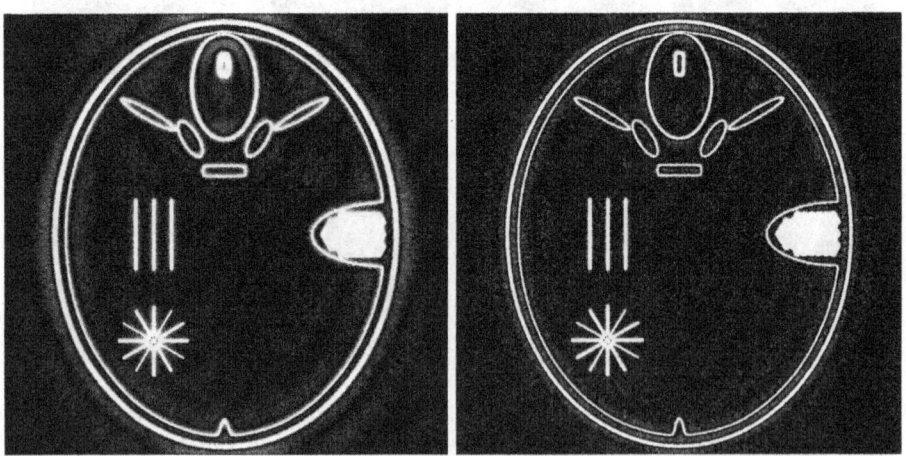

Fig. 4. Error map in reconstruction of the modified FORBILD phantom with Poisson noise with pixel-basis (left) and bilinear basis (right).

tion and demonstrated superior performance compared to traditional methods. However, the bilinear projector outperformed the CNSF method, achieving significantly higher quality on both the clean dataset and the dataset with Poisson noise. Notably, the proposed bilinear projector exhibited substantial improvements in reconstruction quality, as evidenced by higher PSNR, lower RMSE, improved SSIM metrics, and greater resolution as measured by the FWHM metric.

Figure 3 illustrates the reconstruction results of the projectors at the iterations with the highest SSIM. The images show that the reconstruction produced by the bilinear projector exhibits significantly fewer edge-gradient artifacts compared to those produced by the pixel basis projector. Moreover, the bilinear projector inherently supports high-resolution oversampling without requiring post-processing mechanisms such as interpolation. This capability allows it to recover more detailed features than traditional pixel-based projectors without prior knowledge. Under conditions with insufficient intensity data from the CT scanner—such as in limited-angle or fewer-shot scenarios—merely increasing the resolution of the reconstruction grid does not enhance resolution quality. However, the bilinear basis projector effectively addresses this limitation, providing improved resolution and detail recovery.

Figure 4 shows the error map in reconstruction results of sinogram simulated with Poisson noise, the bilinear projector also shows improved performance.

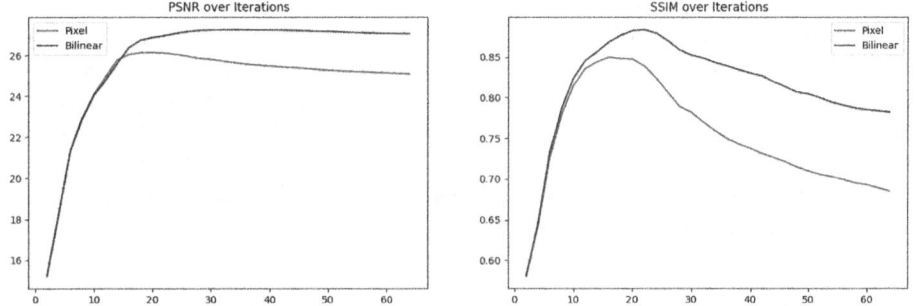

Fig. 5. Change in PSNR and SSIM over iterations (2 to 64) for the dataset without Poisson noise, corresponding to the results in Table 1.

To further analyze the iterative reconstruction mechanism, we plotted the progression of the PSNR and SSIM over iterations. The results (Fig. 5 indicate that the bilinear basis projector reaches its peak PSNR and SSIM values later than the other projectors. This behavior can be attributed to two main reasons. First, the bilinear basis projector provides a more accurate data projection, delaying overfitting compared to the pixel-based projectors. Second, the projection matrix A of the bilinear projector contains a higher number of non-zero elements, which enhances stability during iterative reconstruction and contributes to its superior performance over time.

4.3 Speed Comparison

Table 2. Running times of the pixel basis and bilinear basis projectors under varying sinogram and reconstruction grid resolutions. The values represent the average running time across 10 runs for each scenario. All experiments are performed on a single NVIDIA GTX 3090 GPU and CUDA v12.4.

Sinogram Resolution		360 × 855		900 × 855		1800 × 1710	
Reconstruction Grid Size		512 × 512	1024 × 1024	512 × 512	1024 × 1024	512 × 512	1024 × 1024
Pixel	Forward	44.3 ms	72.1 ms	58.2 ms	285.1 ms	304.3 ms	613.4 ms
	Backward	26.6 ms	71.9 ms	37.4 ms	285.0 ms	280.9 ms	619.2 ms
Bilinear	Forward	51.9 ms	84.3 ms	125.1 ms	347.2 ms	384.5 ms	807.5 ms
	Backward	32.7 ms	85.2 ms	83.6 ms	348.0 ms	385.3 ms	812.1 ms

Beyond reconstruction quality, the efficiency of the projector is a critical factor in practical applications. While MBIR methods are significantly slower than analytical reconstruction methods like FBP, recent deep learning-based approaches also depend heavily on the projector's speed during training and inference. Our proposed method achieves a speed comparable to the pixel basis projector through the efficient implementation of vector convolution on GPUs by explicit form extraction in Eq. 3.

To evaluate performance under different conditions, we measured the running time of the projector in three sinogram resolutions and two reconstruction grid sizes, for both forward and backward projections. As shown in Table 2, in the most computationally intensive scenario (1800×1710 → 1024×1024), the proposed method is approximately 1.3 times slower than the CNSF projector. However, this slight performance tradeoff is acceptable given the substantial improvements in reconstruction quality.

5 Conclusion

This paper demonstrates the advantages of bilinear projector for modeling X-ray optics in forward and back-projection steps of MBIR algorithms in fan-beam geometry. The experiments suggest significant improvements in reduction of discretization artifacts without the need for increasing resolution. Our future research will plan to expand the bilinear model to a 3-D in cone-beam geometry version and assess the observed improvements in surface and volume estimation problems in medical image analysis.

Acknowledgements. This study was funded in part by National Science Foundation grant CCF-2210866.

Disclosure of Interests. Authors have no conflicts of interest to report.

A Appendix: Detailed Piecewise Formulas

Below we provide the explicit piecewise definitions for $M_C(x)$ and $P_\theta(s)$.

Definitions

For $a, b > 0$, define:

$$M_C(x) = \int_0^x \left(\frac{1}{a} \Lambda\left(\frac{s}{a}\right) * \frac{1}{b} \Lambda\left(\frac{s}{b}\right) \right) ds.$$

Case: $a > 2b$

$$M_C(x) = \begin{cases} \frac{x(12ab^2 - 4b^3 - 4bx^2 + x^3)}{12a^2 b^2} & 0 \leq x < a-b \\ \frac{a^4 - 4a^3(b+x) + 6a^2(b+x)^2 - 4a(b^3 - 3b^2 x + 3bx^2 + x^3) + b^4 - 4b^3 x + 6b^2 x^2 - 4bx^3 + 3x^4}{24a^2 b^2} & a-b \leq x < b \\ \frac{a^4 - 4a^3(b+x) + 6a^2(b+x)^2 - 4a(b^3 - 3b^2 x + 3bx^2 + x^3) - b^4 + 4b^3 x - 6b^2 x^2 + 4bx^3 + x^4}{24a^2 b^2} & b \leq x < a \\ -\frac{a^4 + 4a^3(b-x) - 6a^2(b^2 + 2bx - x^2) + 4a(b-x)^3 + (b-x)^4}{24a^2 b^2} & a \leq x < a+b \\ \frac{1}{2} & x \geq a+b \end{cases}$$

Case: $a \leq 2b$

$$M_C(x) = \begin{cases} \frac{x(12ab^2 - 4b^3 - 4bx^2 + x^3)}{12a^2 b^2} & 0 \leq x < b \\ -\frac{6x(x-2a) + b^2}{12a^2} & b \leq x < a-b \\ \frac{a^4 - 4a^3(b+x) + 6a^2(b+x)^2 - 4a(b^3 - 3b^2 x + 3bx^2 + x^3) - b^4 + 4b^3 x - 6b^2 x^2 + 4bx^3 + x^4}{24a^2 b^2} & b \leq x < a \\ -\frac{a^4 + 4a^3(b-x) - 6a^2(b^2 + 2bx - x^2) + 4a(b-x)^3 + (b-x)^4}{24a^2 b^2} & a \leq x < a+b \\ \frac{1}{2} & x \geq a+b \end{cases}$$

References

1. Bouman, C., Sauer, K.: A generalized gaussian image model for edge-preserving map estimation. IEEE Trans. Image Process. **2**(3), 296–310 (1993)
2. Clark, D.P., Badea, C.T.: Mcr toolkit: a gpu-based toolkit for multi-channel reconstruction of preclinical and clinical x-ray ct data. Med. Phys. **50**(8), 4775–4796 (2023). https://doi.org/10.1002/mp.16532, https://aapm.onlinelibrary.wiley.com/doi/abs/10.1002/mp.16532
3. De Man, B., Nuyts, J., Dupont, P., Marchal, G., Suetens, P.: Metal streak artifacts in x-ray computed tomography: a simulation study. In: 1998 IEEE Nuclear Science Symposium Conference Record. 1998 IEEE Nuclear Science Symposium and Medical Imaging Conference (Cat. No. 98CH36255), vol. 3, pp. 1860–1865. IEEE (1998)

4. Entezari, A., Nilchian, M., Unser, M.: A box spline calculus for the discretization of computed tomography reconstruction problems. IEEE Trans. Med. Imaging **31**(8), 1532–1541 (2012)
5. Fessler, J.A.: Comprehensive Biomedical Physics, Vol. 2, chap. Fundamentals of CT reconstruction in 2D and 3D, pp. 263–95. Elsevier (2014)
6. Glover, G.H., Pelc, N.J.: Nonlinear partial volume artifacts in xray computed tomography. Med. Phys. **7**(3), 238–248 (1980). https://doi.org/10.1118/1.594678, https://aapm.onlinelibrary.wiley.com/doi/abs/10.1118/1.594678
7. Joseph, P.M., Spital, R.: The exponential edge-gradient effect in x-ray computed tomography. Phys. Med. Biol. **26**(3), 473 (1981)
8. Lasio, G.M., Whiting, B.R., Williamson, J.F.: Statistical reconstruction for x-ray computed tomography using energy-integrating detectors. Phys. Med. Biol. **52**(8), 2247 (2007)
9. Long, Y., Fessler, J.A., Balter, J.M.: 3d forward and back-projection for x-ray ct using separable footprints. IEEE Trans. Med. Imaging **29**(11), 1839–1850 (2010)
10. Matej, S., Lewitt, R.M.: Efficient 3d grids for image reconstruction using spherically-symmetric volume elements. IEEE Trans. Nucl. Sci. **42**(4), 1361–1370 (1995)
11. Nilchian, M., Ward, J.P., Vonesch, C., Unser, M.: Optimized kaiser-bessel window functions for computed tomography. IEEE Trans. Image Process. **24**(11), 3826–3833 (2015)
12. Nuyts, J., Man, B.D., Fessler, J.A., Zbijewski, W., Beekman, F.J.: Modelling the physics in the iterative reconstruction for transmission computed tomography. Phys. Med. Biol. **58**(12), R63 (2013). https://doi.org/10.1088/0031-9155/58/12/R63, https://dx.doi.org/10.1088/0031-9155/58/12/R63
13. Pan, X., Sidky, E.Y., Vannier, M.: Why do commercial ct scanners still employ traditional, filtered back-projection for image reconstruction? Inverse Prob. **25**(12), 123009 (2009)
14. Shu, Z., Entezari, A.: Rbp-dip: residual back projection with deep image prior for ill-posed ct reconstruction. Neural Netw. **180**, 106740 (2024)
15. Sidky, E.Y., Pan, X.: Image reconstruction in circular cone-beam computed tomography by constrained, total-variation minimization. Phys. Med. Biol. **53**(17), 4777 (2008)
16. Wang, G., Ye, J.C., Mueller, K., Fessler, J.A.: Image reconstruction is a new frontier of machine learning. IEEE Trans. Med. Imaging **37**(6), 1289–1296 (2018)
17. Xie, S., Zhang, K., Entezari, A.: An evaluation of state-of-the-art projectors in the presence of noise and nonlinearity in the beer-lambert law. In: International Conference on Medical Image Computing and Computer-Assisted Intervention, pp. 78–87. Springer (2024)
18. Yu, H., Wang, G.: Finite detector based projection model for high spatial resolution. J. Xray Sci. Technol. **20**(2), 229–238 (2012)
19. Zbijewski, W., Beekman, F.J.: Characterization and suppression of edge and aliasing artefacts in iterative x-ray ct reconstruction. Phys. Med. Biol. **49**(1), 145 (2003). https://doi.org/10.1088/0031-9155/49/1/010, https://dx.doi.org/10.1088/0031-9155/49/1/010
20. Zbijewski, W., Beekman, F.J.: Comparison of methods for suppressing edge and aliasing artefacts in iterative x-ray ct reconstruction. Phys. Med. Biol. **51**(7), 1877 (2006)
21. Zhang, K., Entezari, A.: A convolutional forward and back-projection model for fan-beam geometry (2019). https://arxiv.org/abs/1907.10526

22. Zhang, K., Entezari, A.: Convolutional forward models for x-ray computed tomography. SIAM J. Imaging Sci. **16**(4), 1953–1977 (2023). https://doi.org/10.1137/21M1464191
23. Ziegler, A., Köhler, T., Nielsen, T., Proksa, R.: Efficient projection and backprojection scheme for spherically symmetric basis functions in divergent beam geometry. Med. Phys. **33**(12), 4653–4663 (2006)

Subspace Implicit Neural Representations for Real-Time Cardiac Cine MR Imaging

Wenqi Huang[1]([✉]), Veronika Spieker[1,2], Siying Xu[3], Gastao Cruz[4], Claudia Prieto[5,6], Julia A. Schnabel[1,2,6], Kerstin Hammernik[1], Thomas Kuestner[3], and Daniel Rueckert[1,7]

[1] Technical University of Munich, Munich, Germany
wenqi.huang@tum.de
[2] Helmholtz Munich, Munich, Germany
[3] University of Tübingen, Tuebingen, Germany
[4] University of Michigan, Ann Arbor, USA
[5] Pontificia Universidad Católica de Chile, Santiago, Chile
[6] King's College London, London, UK
[7] Imperial College London, London, UK

Abstract. Conventional cardiac cine MRI methods rely on retrospective gating, which limits temporal resolution and the ability to capture continuous cardiac dynamics, particularly in patients with arrhythmias and beat-to-beat variations. To address these challenges, we propose a reconstruction framework based on subspace implicit neural representations for real-time cardiac cine MRI of continuously sampled radial data. This approach employs two multilayer perceptrons to learn spatial and temporal subspace bases, leveraging the low-rank properties of cardiac cine MRI. Initialized with low-resolution reconstructions, the networks are fine-tuned using spoke-specific loss functions to recover spatial details and temporal fidelity. Our method directly utilizes the continuously sampled radial k-space spokes during training, thereby eliminating the need for binning and non-uniform FFT. This approach achieves superior spatial and temporal image quality compared to conventional binned methods at the acceleration rate of 10 and 20, demonstrating potential for high-resolution imaging of dynamic cardiac events and enhancing diagnostic capability (Code available: https://github.com/wenqihuang/SubspaceINR-CMR).

Keywords: Image Reconstruction · Non-Cartesian Sampling · Cardiac MRI · Implicit Neural Representations · Deep Learning · Low-rank

1 Introduction

Cardiac cine magnetic resonance imaging (MRI) plays an important role in accurate clinical evaluation of cardiovascular function and morphology, offering a non-invasive way to capture the beating heart. However, the presence of both cardiac and respiratory motion makes it challenging to acquire and reconstruct high-quality images, as achieving high spatial and temporal resolution simultaneously remains difficult [39]. Therefore, fast data acquisition with high undersampling rates in the MR signal space, also called k-space, has attracted great interest in the field of reconstruction.

Reconstruction of images from undersampled k-space data has been significantly advanced by the introduction of parallel imaging (PI) and compressed sensing (CS) techniques in MRI. These approaches have driven substantial improvements in acquisition efficiency, leveraging both hardware and algorithmic innovations. PI utilizes multiple receiver coils in MRI scanners, exploiting the redundancy among coils to enable higher undersampling rates during data acquisition. In contrast, CS incorporates sparse priors as regularization terms to constrain the solution space and mitigate the risk of local minima in the reconstruction process. Building upon these foundations, the development of low-rank and subspace methods has opened up new avenues for dynamic cardiac MRI reconstruction [7,12,23,28]. By explicitly leveraging the low-rank structure across temporal phases, these approaches model the cardiac motion as lying close to a lower-dimensional manifold, which can be spanned by just a few temporal basis functions. This effective dimensionality reduction not only simplifies the reconstruction problem but also enables the use of fewer measurements, ultimately improving scan efficiency. As a result, both spatial and temporal information are faithfully recovered, facilitating more accurate representations of the underlying cardiac motion. More recently, deep learning-based approaches have demonstrated remarkable potential in leveraging acquisition physics and raw k-space data for improved reconstruction [14,16,19,29,30,32,35]. Despite their success, most of these methods are supervised and require large, fully-sampled training datasets, and consequently prone to hallucinations when there are domain shifts. Several approaches that do not require fully sampled data have been proposed [2,8,9,15,20,45], and while recent work has begun to address radial sampling [3,24], the majority of these methods have been developed primarily for Cartesian sampling trajectories.

Non-Cartesian sampling techniques, such as radial sampling, have become increasingly popular. Unlike Cartesian sampling, non-Cartesian approaches vary both phase encoding directions simultaneously, which helps to distribute motion- and undersampling-related noise throughout the image, leading to more incoherent and less prominent artifacts [44]. This property makes non-Cartesian sampling particularly well-suited for cardiac cine MRI, especially in patients with irregular heart rhythms. However, non-Cartesian sampling introduces new challenges. Reconstruction from non-Cartesian data requires non-uniform fast Fourier transforms (NUFFT), which can be computationally expensive and affected by density compensation and interpolation errors [31,44]. Due to the moving nature of the heart and the limitations of MR physics, only a limited number of data points can be rapidly sampled for each motion state in cardiac cine MRI. Existing methods typically rely on binning k-space samples along the temporal dimension into a fixed number of motion states to achieve high image quality for each cardiac frame [1,18,34,43]. Binning of real-time data simplifies data representation and optimization, but results in a loss of temporal information.

In recent developments, Implicit Neural Representations (INRs) have shown impressive potential to represent continuous signals through coordinate-value mappings using multi-layer perceptrons (MLPs) [6,25,26,38]. The flexibility of INRs to sample at arbitrary locations makes them ideal to reconstruct images from non-Cartesian sampling trajectories without the need for NUFFT, and the ability of continuous representation is ideal for learning the continuous spatiotemporal information in cardiac cine MRI. Previous studies show the effectiveness of INRs for both Cartesian [10,21,37] and non-Cartesian sampling [5], but cardiac cine MRI methods still either rely on binned data [5] or attempt to directly learn the entire spatiotemporal space [13,17,40,41], which is particularly challenging under sparse sampling schemes.

Despite significant progress in cardiac cine MRI reconstruction, achieving high-quality images for real-time cardiac cine MRI under sparse sampling conditions remains a challenging task. Traditional binning-based approaches sacrifice temporal resolution by grouping data into discrete motion states, while non-Cartesian sampling often relies on computationally expensive and error-prone NUFFT operations. Moreover, directly modeling the full spatiotemporal domain of the beating heart can overwhelm reconstruction algorithms, particularly when acquired data are limited. To overcome these hurdles, we propose a novel approach that exploits the inherent low-rank structure of cardiac cine MRI through the integration of subspace learning and INRs. By avoiding binning and NUFFT altogether, our approach enables efficient, high-quality reconstruction that preserves crucial cardiac dynamics. Our main contributions are as follows:

1. We introduce a reconstruction pipeline that is tailored for radially sampled real-time cardiac cine MRI that removes the need for temporal binning and avoids the limitations of NUFFT.
2. We propose a subspace learning strategy that directly exploits the low-rank properties of cardiac cine data. By representing the data as a combination of spatial and temporal bases, we effectively reduce the dimensionality of the representation space. This reduction enables our approach to employ INRs to accurately learn these representations, thereby capturing intricate details of cardiac motion.
3. Our framework achieves superior reconstruction of both spatial details and temporal motion compared to conventional binning-based methods, delivering improved image quality and more faithful depictions of cardiac function.

2 Methods

This section presents our proposed reconstruction framework for real-time cardiac cine MRI, integrating INRs and subspace learning. By leveraging the low-rank properties of cardiac cine MRI and bypassing the need for binning and NUFFT, our method achieves superior spatiotemporal fidelity while directly utilizing continuous k-space sampling. The following subsections detail the problem

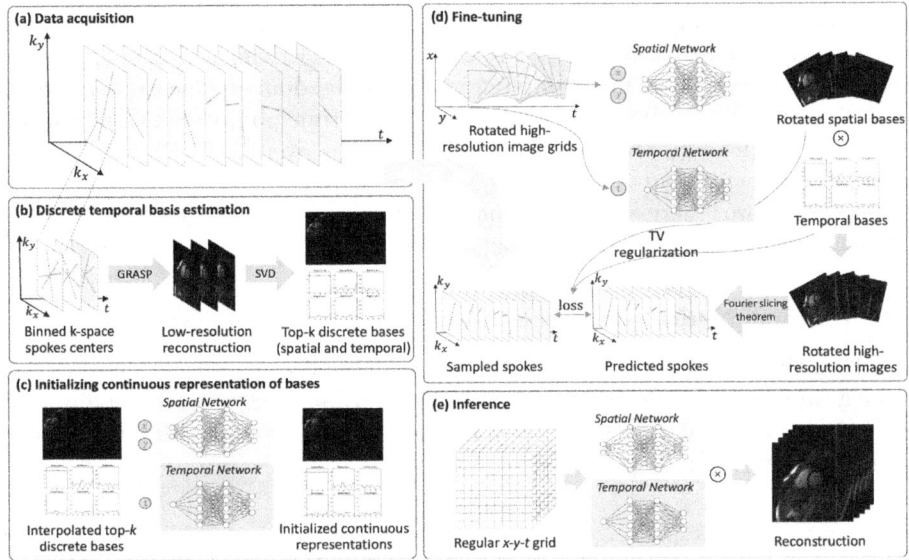

Fig. 1. Method overview: a) Continuous k-space sampling with tiny golden angle radial trajectory; b) Binning spoke centers to reconstruct a low-resolution image with GRASP, decomposing into spatial and temporal bases via SVD, retaining top-k components; c) Interpolating and fitting low-resolution bases to spatial and temporal networks; d) Inputting rotated spatial and accurate temporal coordinates into networks to obtain spatial and temporal bases, whose product forms rotated images, networks optimized with Eq. 5; e) Final reconstruction by inputting a regular x-y-t grid.

formulation, the INR-based reconstruction strategy, and the integration of low-rank properties for efficient representation.

2.1 Cardiac Cine MRI with Continuous Radial Sampling

The raw signal of cardiac cine MRI with continuous radial sampling trajectories is acquired for each spoke in the k-space (frequency domain). The goal of MRI reconstruction is to recover the optimal image $\mathbf{x}^*$ by solving:

$$\mathbf{x}^* = \arg\min_{\mathbf{x}} DC(\mathbf{x}, \mathbf{y}) + R(\mathbf{x}). \tag{1}$$

Here $DC(\mathbf{x}, \mathbf{y})$ enforces data consistency between the reconstructed image $\mathbf{x}$ and the k-space measurements $\mathbf{y}$, and $R(\mathbf{x})$ is a regularization term. Each spoke $\mathbf{y}_i, i = 1, 2, \cdots, N$ (N for total number of spokes), sampled at time intervals defined by the repetition time (TR), represents k-space information from a single moment in time. However, it is impossible to reconstruct an image frame from a single spoke due to the extreme sparsity of information. Conventionally, neighboring k-space spokes (in time) are grouped into segments using a binning operator B, and the data consistency term is defined as:

$$DC(\mathbf{x}, \mathbf{y}) = ||A\mathbf{x} - B\mathbf{y}||_2^2, \tag{2}$$

where $A = MFS$, representing the radial sampling trajectories M, NUFFT operator F, and coil sensitivity maps S. To counter the lack of measurements per time point and the resulting undersampling in the spatial domain, the binning operator B aggregates neighboring spokes from various time points (see various colors in 1a). For larger bin sizes, more cardiac motion is averaged within one frame, which in turn can lead to an increased motion blurring in the reconstructed image. Therefore, there will be an unavoidable trade-off between temporal and spatial resolution when using binning.

2.2 Implicit Neural Representations for MRI Reconstruction

To fully utilize the temporal information encoded in the acquired k-space spokes, an ideal approach would reconstruct N image frames, each corresponding to a single spoke, eliminating the temporal information loss from the binning operator B. This modifies the data consistency term to:

$$DC(\mathbf{x}, \mathbf{y}) = ||A\mathbf{x} - \mathbf{y}||_2^2. \tag{3}$$

However, reconstructing image individually from each single spoke using NUFFT leads to severe undersampling artifacts and does not effectively exploit the temporal correlation between spokes. To avoid that, we represent the image $\mathbf{x}$ as a continuous function using an INR network, $\mathbf{x} = G_\theta(\mathbf{c})$, where $\mathbf{c}$ denotes coordinates in the spatiotemporal space, and $G_\theta(\cdot) : \mathbb{R}^3 \to \mathbb{C}$ is the INR network mapping from the spatiotemporal coordinates to the complex image voxel value. The optimal reconstruction is then obtained by optimizing the network parameters θ:

$$\theta^* = \arg\min_\theta DC(G_\theta(\mathbf{c}), \mathbf{y}) + R(G_\theta(\mathbf{c})),$$
$$\mathbf{x}^* = G_{\theta^*}(\mathbf{c}_{\text{grid}}), \tag{4}$$

where θ^* represents the optimal parameters of the INR network, $\mathbf{c}_{\text{grid}}$ represents the coordinates of a target MRI image on a Cartesian grid.

The sparsity of k-space, as well as the high-dimensional x-y-t image space, poses a challenge for learning a well-represented INR. To address this, we leverage the inherent temporal redundancies of cardiac cine MRI by integrating a compact subspace into the INR training. Inspired by the low-rank representations of cardiac cine MRI, where the dynamic image can be approximated by the product of two low-rank matrices (e.g., top-k components)—spatial bases and temporal bases - we decompose the dynamic image into these two components. The image is decomposed into spatial and temporal bases, modeled by two networks: $G_{\theta_s}(\mathbf{s})$ for spatial representation and $G_{\theta_t}(\mathbf{t})$ for temporal representation. The optimization problem is reformulated as:

$$\theta_s^*, \theta_t^* = \arg\min_{\theta_s, \theta_t} DC(G_{\theta_s}(\mathbf{s}) \cdot G_{\theta_t}(\mathbf{t}), \mathbf{y}) + R_s(G_{\theta_s}(\mathbf{s})) + R_t(G_{\theta_t}(\mathbf{t})),$$
$$\mathbf{x}^* = G_{\theta_s^*}(\mathbf{s}) \cdot G_{\theta_t^*}(\mathbf{t}). \tag{5}$$

Here θ_s^* and θ_t^* represent the optimal parameters for the spatial and temporal model, respectively, while R_s and R_t denote their networks' corresponding regularization terms; **s** are the spatial coordinates and **t** represents the temporal coordinates. The spatial model $G_{\theta_s}(\cdot) : \mathbb{R}^2 \to \mathbb{C}^k$ maps the spatial coordinates to a k-dimensional output for spatial bases, and the temporal model $G_{\theta_t}(\cdot) : \mathbb{R} \to \mathbb{C}^k$ maps the temporal coordinates to a k-dimension output. This formulation reduces the parameter space while preserving essential spatiotemporal information, enabling efficient and accurate reconstruction.

Leveraging the INR's ability to sample arbitrary coordinates, we use the Fourier slice theorem [4,27] to establish a direct relationship between the image and its k-space spokes. The theorem states that the 1D Fourier transform of a line integral (projection) $p_v(t)$ of a 2D image $f(x,y)$ along a certain direction v is equivalent to a corresponding spoke in k-space:

$$\mathcal{F}_1\{p_v(t)\}(\omega) = \mathcal{F}_2\{f(x,y)\}(\omega_x, \omega_y). \tag{6}$$

Here (ω_x, ω_y) lies along the spoke's direction (orthogonal to projection direction v). This means that, in our case, each spoke corresponds to the 1D Fourier transform of the image's projection along the spoke's orthogonal direction,

$$\mathcal{F}_1 P_v(S\mathbf{x}) = \mathbf{y}, \tag{7}$$

where P_v is the projection operator for the image **x** along the orthogonal direction of each spoke, and S represents the coil sensitivities. This theorem establishes a direct relationship between radial spokes in k-space and the image space without requiring NUFFT. Furthermore, when combined with INR, it eliminates the need for image interpolation when applying the projection P_v to handle rotated grid coordinates in the problem formulation (Eq. 4), as INRs can directly sample points on a rotated image grid. The data consistency term is reformulated as:

$$DC(G_{\theta_s}(\mathbf{s}) \cdot G_{\theta_t}(\mathbf{t}), \mathbf{y}) = w(\mathcal{F}_1 P_v(S \cdot G_{\theta_s}(\mathbf{s}) \cdot G_{\theta_t}(\mathbf{t})) - \mathbf{y}) \tag{8}$$

Here w is a ramp weight, which is the distance to the spoke center of each data points, compensating the value range of low and high frequency area in k-space spokes. By sampling on a rotated image grid according to the rotation angle of each spoke, the projection simplifies to a summation along the orthogonal direction to each spoke. This allows for direct utilization of continuous k-space spokes without binning, as inspired by [5].

2.3 Network Initialization and Optimization

Proper initialization is critical for fast convergence. Inspired by [12,33], we first reconstruct a low-resolution image using GRASP [11], aggregating the central spokes of low spatial and temporal resolution. This low-resolution image is decomposed via singular value decomposition (SVD), retaining the top-k components (Fig. 1b), which are interpolated to full resolution using linear interpolation. The spatial and temporal networks are initialized by minimizing the mean squared error (MSE) between their outputs and these interpolated bases

(Fig. 1c). Subsequently, the network parameters are optimized by minimizing Eq. 5. Once trained, the reconstructed image is obtained by inputting the x-y-t grid $\mathbf{c}_{\text{grid}}$ into the networks. Figure 1d and e illustrate the fine-tuning and inference steps of the proposed method.

3 Experimental Setup

Dataset. Experiments were conducted on 17 healthy subjects using a 1.5T scanner (Ingenia, Philips, Best, The Netherlands) with 28 MR receiver coils. To reduce training time, we used data from 6 selected coils sensitive to the heart region. All experiments were IRB-approved with written informed consent. The following sequence parameters were used: 7–8 short axis slices; FOV = 256×256 mm^2 (considering inherent 1.6× frequency encoding oversampling of radial trajectories); 8 mm slice thickness; resolution = 2×2 mm^2; TE/TR = 1.16/2.3 ms; b-SSFP readout; radial tiny golden angle of $\sim$23.6°; flip angle 60°; 8960 acquired radial spokes (800 spokes were used, covering 2–3 cardiac cycles) without ECG gating; nominal scan time $\sim$20 s ($\sim$1.8 s for 800 spokes); breath-hold acquisition. No fully-sampled ground truth images are available in this real-time acquisition setting.

Model Details. The proposed INR framework consists of a spatial network and a temporal network. Both networks are based on multi-layer perceptrons (MLPs) with hash-grid encoding [26]. The hash-grid encoding uses a hashmap of size 20 with 16 levels, 2 features per level, a base resolution of 16, and a per-level scale of 1.26. The MLPs have 2 hidden layers, each containing 64 nodes. The number of main components k is set to 6 for both spatial and temporal bases. The networks generate an output dimension of 12, which represents the concatenation of the real and imaginary parts of the basis components.

Initialization Settings. The low-resolution image (1/2.56 of the original resolution) was reconstructed using GRASP [11,42] with 100 iterations on binned spoke centers. Binning was performed with 20 spokes per phase and a total variation regularization weight of 0.025. By cropping the center of the radial k-space for the low-resolution reconstruction, sparsity towards the outside k-space can be reduced. Thus, the effective acceleration factor is reduced, resulting in low-resolution reconstructions with minimized undersampling artefacts [10]. SVD was applied to the image, decomposing it into spatial bases and temporal bases. Then we retained the top 6 components for further processing. The temporal and spatial bases were interpolated to resolutions of 2.56 times and 20 times their original sizes, respectively, to match the full spatial resolution and maximum temporal resolution (TR). The spatial and temporal networks were trained on these interpolated bases for 1000 steps with a learning rate of 0.01, using MSE loss.

Fine-Tuning Settings. Training on non-binned spokes was conducted using the Adam optimizer for 150 iterations on all sampled spokes at a reduced learning rate of 3×10^{-5} for this phase. During the first 10 iterations, the temporal model was frozen to accelerate the optimization of the spatial model. The entire initiaslization and fine-tuning process for each slice took approximately 3 min on an NVIDIA RTX A6000 GPU with 48GB of memory.

Evaluation Settings. We compared our proposed method quantitatively and qualitatively to NUFFT and GRASP. Since NUFFT and GRASP require data binning, we configured them to use 20 and 40 spokes per bin (R=20 and R=10) to balance spatial and temporal resolution. For the proposed method, we sampled on the same image grid as the 20 spokes/bin configuration for a fair comparison. GRASP implementation was performed using the BART toolbox [42]. As ground truth images were unavailable, image quality was assessed quantitatively using edge sharpness (ES) and estimated signal-to-noise ratio (SNR) in the end-systolic and end-diastolic cardiac phases. ES is defined as the inverse distance between the positions corresponding to 20% and 80% of the maximum intensity along the line profile [22,36]. ES was measured by extracting 12 line profiles between the left ventricle blood pool and the endocardium (6 lines for each phase). A higher ES value indicates sharper edges. SNR was estimated by extracting patches from the background region (noise region) and the left ventricle blood pool (signal region). SNR was calculated as $10 \cdot \log_{10}(P_s/P_n)$, where P_s and P_n are the mean values of the signal and noise regions, respectively. Only the 2nd to 6th short-axis slices of each subject were considered in the ES and SNR calculations, as the other slices typically lacked sufficient information about the ventricular blood pool.

4 Results

Table 1 summarizes the mean and standard deviation of the quantitative results for systole and diastole phases. For SNR, the proposed method outperformed the comparison methods by a considerable margin in both systolic and diastolic phases, reflecting its ability to maintain a higher SNR during dynamic cardiac motion. For ES, the proposed method showed a notable advantage in the systolic phase and comparable performance in the diastolic phase, indicating its capability to preserve edge sharpness under different motion conditions.

Table 1. SNR and Edge Sharpness (ES) of comparison methods

Method	SNR (mean ± std, dB)		ES (mean ± std, $1/mm$)	
	Systolic	Diastolic	Systolic	Diastolic
NUFFT	6.85 ± 2.45	7.62 ± 2.27	0.177 ± 0.024	0.243 ± 0.0287
GRASP	13.57 ± 2.79	13.92 ± 2.79	0.201 ± 0.030	**0.316 ± 0.043**
Proposed	**20.21 ± 6.88**	**20.89 ± 6.96**	**0.227 ± 0.034**	0.316 ± 0.052

Figure 2a and b present qualitative comparisons between NUFFT, GRASP, and the proposed method for two representative cases. Higher temporal resolutions (fewer spokes per frame) are compared to higher spatial resolutions (more spokes per frame). Qualitative results demonstrate that the proposed method preserves finer spatial details and richer temporal fidelity compared to binned approaches, with negligible motion blur. In the region of interest, our method retains intricate details that are blurred in the reconstructions of binned methods (NUFFT and GRASP) marked by yellow arrows. This difference can be attributed to the averaging of neighboring time frames in binned methods, which introduces spatial and temporal blurring. The x-t profiles further illustrate that binning leads to noticeable temporal blurring, while the proposed reconstruction framework effectively preserves temporal continuity, offering clear visualization of dynamic cardiac motion.

Figures 3a and b illustrate the first four components of temporal and spatial bases learned by the networks under three different settings: low-resolution initialization only, fine-tuned results based on low-resolution initialization, and directly learned bases without low-resolution initialization. By comparing the results of directly learned and fine-tuned bases, we observe that without initialization, the INRs struggle to learn the spatiotemporal information, particularly failing to capture the temporal details. This highlights that initialization is crucial for achieving fast convergence and effective representation learning in our proposed method. Comparing the initialized and fine-tuned bases, we notice that the spatial bases initially appear blurry and lack fine details, as they are generated from low-resolution reconstructions during the initialization phase. After fine-tuning on the non-binned k-space spokes, the spatial bases exhibit significantly enhanced clarity, and the temporal bases recover fine details that were previously lost. These results validate our hypothesis that the INR framework, combined with fine-tuning, effectively recovers high-frequency spatial and temporal information. This comparison further demonstrates the effectiveness of leveraging the low-rank property of cardiac cine MRI within the INR framework.

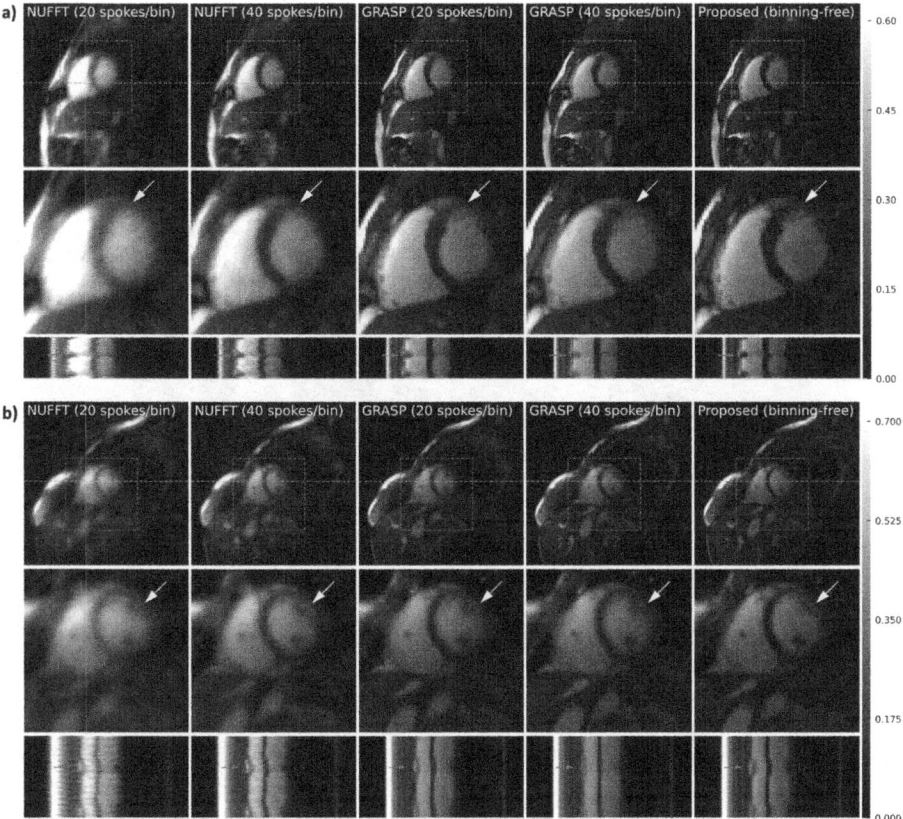

Fig. 2. Comparison of reconstruction results. a) and b) show two subjects. The first row of each subfigure shows $x-y$ images at a selected cardiac phase. The second row zooms into a region of interest (ROI). The third row presents $x-t$ profiles along a chosen y-coordinate. Our method preserves details in the ROI (yellow arrow) and minimizes temporal blurring seen in binned methods (red arrow). The color bars indicate intensity ranges. (Color figure online)

In summary, both qualitative and quantitative evaluations indicate that the proposed method achieves superior reconstruction quality compared to conventional NUFFT and GRASP methods. By preserving both spatial and temporal details, the proposed framework demonstrates significant advantages for the reconstruction of real-time cardiac cine MRI. The results suggest that this approach is particularly well-suited for capturing dynamic cardiac events, providing high-quality reconstructions without the need for binning or NUFFT.

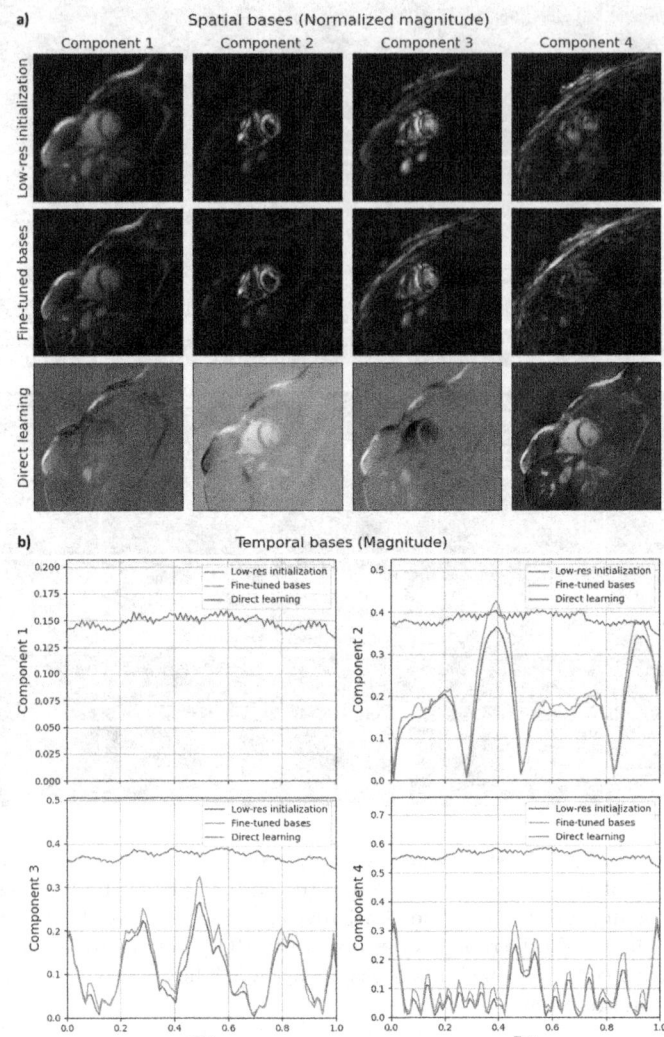

Fig. 3. The first four components of the spatial and temporal bases for an example case. a) The spatial bases are shown across four columns, where the first row displays the initialized spatial bases derived from low-resolution reconstruction, the second row shows the fine-tuned representations obtained by training based on the low-resolution initialization, and the third row presents the bases directly learned without initialization. b) The temporal bases correspond to the spatial components in a), with orange lines representing the low-resolution initialization, green lines showing the fine-tuned bases, and blue lines depicting the directly learned components without initialization. The results demonstrate that with low-resolution initialization, the models successfully capture finer spatial structures and temporal details after training on non-binned k-space spokes. In contrast, without initialization, the models fail to converge effectively. (Color figure online)

5 Discussion and Conclusion

In this work, we presented a novel subspace INR framework for the reconstruction of real-time cardiac cine MRI. Conventional approaches often rely on binning k-space data to handle motion. Similarily, they often rely on non-uniform FFT (NUFFT) to handle non-Cartesian sampling trajectories. While effective in certain scenarios, these methods face inherent limitations, such as losing temporal resolution due to binning and reconstruction artifacts arising from NUFFT density compensation and interpolation errors. Our framework addresses these challenges by leveraging subspace learning with INRs, eliminating the need for both binning and NUFFT. Our proposed method exploits low-rank properties of cardiac cine MRI, enabling accurate and data-efficient reconstruction from sparse k-space samples without compromising spatial or temporal resolution.

The foundation of our method lies in decomposing the spatiotemporal image into separate spatial and temporal bases represented by two distinct INR networks. This decomposition reduces the complexity of the reconstruction problem by projecting the high-dimensional spatiotemporal space into a lower-dimensional subspace, significantly improving data efficiency. The networks are initialized using low-resolution reconstructions obtained via GRASP, ensuring stable and reliable starting points. Fine-tuning networks with spoke-specific losses recovers fine details lost at initialization, resulting in superior image quality. Our experiments show that the proposed method outperforms traditional binning-based approaches, better preserving temporal dynamics and finer spatial details. Additionally, the flexibility of INR-based representations allows the reconstruction of continuous cardiac motion, effectively minimizing motion artifacts that are common in conventional methods.

One of the key contributions of this work is a paradigm shift in the preprocessing of k-space data. Instead of treating groups of k-space measurements, such as binned spokes, as the minimal unit, we explored treating individual spokes as the minimal unit. This approach unlocks the potential to achieve a temporal resolution equal to the repetition time (TR). While similar concepts have been explored for implicit neural representations in k-space [17], our work extends the representation to image space while retaining the spoke-wise minimal unit through the use of the Fourier slicing theorem, showcasing its utility for real-time cardiac cine MRI. Furthermore, this concept is not limited to cardiac imaging, but can be generalized to multi-contrast imaging and other quantitative MRI applications, highlighting its broad potential impact.

Despite its very promising results, this study still has some limitations. First, while training and inference times have been reduced to approximately 15 min, this remains a bottleneck for clinical routine use. Second, the method requires training and additional optimization for each new scan, which prevents direct inference on unseen data. Although this scan-specific approach minimizes the risk of inpainting artifacts from training data, it introduces potential issues with overfitting. The current stopping criteria for the number of iterations are empirically determined and may not always be optimal. Lastly, our evaluation focused on imaging quality metrics; further studies are needed to assess the method's performance on more clinical parameters. These limitations underscore the need for continued refinement and validation.

Future work will focus on optimizing the spatial and temporal network architectures to improve computational efficiency and scalability. This is particularly important for extending the approach to higher-dimensional problems, such as 4D MRI, where the trade-off between data acquisition time and resolution is even more pronounced. Expanding the framework to other applications and imaging modalities, such as functional MRI or positron emission tomography (PET), could further demonstrate its versatility and applicability across the broader field of medical imaging. Finally, rigorous clinical validation, including evaluations on pathological cases and diverse multi-institution datasets, will be critical to ensure its practical utility in real-world clinical settings.

In conclusion, the proposed subspace INR framework represents a significant step forward in real-time cardiac cine MRI reconstruction. By overcoming the limitations of binning and NUFFT, our method preserves both spatial and temporal fidelity, enabling high-quality imaging of dynamic cardiac events. With further development and clinical validation, this approach has the potential to transform real-time cardiac cine MRI and expand its applications across a range of imaging scenarios.

Acknowledgements. This work was supported in part by the European Research Council (Deep4MI project, Grant Agreement Nr. 884622). We would also like to thank Dr. Jing Cheng, Dr. Yuanyuan Liu and Dr. Zhilang Qiu for the insightful discussion during ISMRM 2024.

References

1. Ahmad, R., Xue, H., Giri, S., Ding, Y., Craft, J., Simonetti, O.P.: Variable density incoherent spatiotemporal acquisition (vista) for highly accelerated cardiac mri. Magn. Reson. Med. **74**(5), 1266–1278 (2015)
2. Akçakaya, M., Moeller, S., Weingärtner, S., Uğurbil, K.: Scan-specific robust artificial-neural-networks for k-space interpolation (RAKI) reconstruction: database-free deep learning for fast imaging. Magn. Reson. Med. **81**(1), 439–453 (2019)
3. Blumenthal, M., Fantinato, C., Unterberg-Buchwald, C., Haltmeier, M., Wang, X., Uecker, M.: Self-supervised learning for improved calibrationless radial mri with nlinv-net. Magn. Reson. Med. **92**(6), 2447–2463 (2024)
4. Bracewell, R.N.: Strip integration in radio astronomy. Aust. J. Phys. **9**(2), 198–217 (1956)
5. Catalán, T., Courdurier, M., Osses, A., Botnar, R., Costabal, F.S., Prieto, C.: Unsupervised reconstruction of accelerated cardiac cine mri using neural fields. arXiv preprint arXiv:2307.14363 (2023)
6. Chibane, J., Alldieck, T., Pons-Moll, G.: Implicit functions in feature space for 3D shape reconstruction and completion. In: Proceedings of the IEEE/CVF Conference on Computer Vision and Pattern Recognition, pp. 6970–6981 (2020)

7. Christodoulou, A.G., Hitchens, T.K., Wu, Y.L., Ho, C., Liang, Z.P.: Improved subspace estimation for low-rank model-based accelerated cardiac imaging. IEEE Trans. Biomed. Eng. **61**(9), 2451–2457 (2014)
8. Cui, Z.X., et al.: Self-score: self-supervised learning on score-based models for mri reconstruction. arXiv preprint arXiv:2209.00835 (2022)
9. Desai, A.D., et al.: Noise2recon: enabling snr-robust mri reconstruction with semi-supervised and self-supervised learning. Magn. Reson. Med. **90**(5), 2052–2070 (2023)
10. Feng, J., Feng, R., Wu, Q., Zhang, Z., Zhang, Y., Wei, H.: Spatiotemporal implicit neural representation for unsupervised dynamic mri reconstruction. arXiv preprint arXiv:2301.00127 (2022)
11. Feng, L., et al.: Golden-angle radial sparse parallel mri: combination of compressed sensing, parallel imaging, and golden-angle radial sampling for fast and flexible dynamic volumetric mri. Magn. Reson. Med. **72**(3), 707–717 (2014)
12. Feng, L., Wen, Q., Huang, C., Tong, A., Liu, F., Chandarana, H.: Grasp-pro: improving grasp dce-mri through self-calibrating subspace-modeling and contrast phase automation. Magn. Reson. Med. **83**(1), 94–108 (2020)
13. Haft, P.T., Huang, W., Cruz, G., Rueckert, D., Zimmer, V.A., Hammernik, K.: Neural implicit k-space with trainable periodic activation functions for cardiac mr imaging. In: BVM Workshop, pp. 82–87. Springer (2024)
14. Hammernik, K., Klatzer, T., Kobler, E., Recht, M.P., Sodickson, D.K., Pock, T., Knoll, F.: Learning a variational network for reconstruction of accelerated MRI data. Magn. Reson. Med. **79**(6), 3055–3071 (2018)
15. Hu, C., Li, C., Wang, H., Liu, Q., Zheng, H., Wang, S.: Self-supervised learning for MRI reconstruction with a parallel network training framework. In: de Bruijne, M., Cattin, P.C., Cotin, S., Padoy, N., Speidel, S., Zheng, Y., Essert, C. (eds.) MICCAI 2021. LNCS, vol. 12906, pp. 382–391. Springer, Cham (2021). https://doi.org/10.1007/978-3-030-87231-1_37
16. Huang, W., Ke, Z., Cui, Z.X., Cheng, J., Qiu, Z., Jia, S., Ying, L., Zhu, Y., Liang, D.: Deep low-rank plus sparse network for dynamic MR imaging. Med. Image Anal. **73**, 102190 (2021)
17. Huang, W., Li, H.B., Pan, J., Cruz, G., Rueckert, D., Hammernik, K.: Neural implicit k-space for binning-free non-cartesian cardiac mr imaging. In: International Conference on Information Processing in Medical Imaging, pp. 548–560. Springer (2023)
18. Jung, H., Park, J., Yoo, J., Ye, J.C.: Radial k-t focuss for high-resolution cardiac cine mri. Magnetic Resonance in Medicine: An Official Journal of the International Society for Magnetic Resonance in Medicine **63**(1), 68–78 (2010)
19. Ke, Z., Huang, W., Cui, Z.X., Cheng, J., Jia, S., Wang, H., Liu, X., Zheng, H., Ying, L., Zhu, Y., et al.: Learned low-rank priors in dynamic mr imaging. IEEE Trans. Med. Imaging **40**(12), 3698–3710 (2021)
20. Korkmaz, Y., Cukur, T., Patel, V.M.: Self-supervised mri reconstruction with unrolled diffusion models. In: International Conference on Medical Image Computing and Computer-Assisted Intervention, pp. 491–501. Springer (2023)
21. Kunz, J.F., Ruschke, S., Heckel, R.: Implicit neural networks with fourier-feature inputs for free-breathing cardiac mri reconstruction. IEEE Trans. Comput. Imaging (2024)
22. Küstner, T., Bustin, A., Jaubert, O., Hajhosseiny, R., Masci, P.G., Neji, R., Botnar, R., Prieto, C.: Isotropic 3d cartesian single breath-hold cine mri with multi-bin patch-based low-rank reconstruction. Magn. Reson. Med. **84**(4), 2018–2033 (2020)

23. Lingala, S.G., Hu, Y., DiBella, E., Jacob, M.: Accelerated dynamic mri exploiting sparsity and low-rank structure: kt slr. IEEE Trans. Med. Imaging **30**(5), 1042–1054 (2011)
24. Mancu, A., Huang, W., da Cruz, G.L., Rueckert, D., Hammernik, K.: Self-supervised low-rank plus sparse network for radial mri reconstruction. In: NeurIPS 2023 Workshop on Deep Learning and Inverse Problems (2023)
25. Mildenhall, B., Srinivasan, P.P., Tancik, M., Barron, J.T., Ramamoorthi, R., Ng, R.: NeRF: representing scenes as neural radiance fields for view synthesis. Commun. ACM **65**(1), 99–106 (2021)
26. Müller, T., Evans, A., Schied, C., Keller, A.: Instant neural graphics primitives with a multiresolution hash encoding. ACM Trans. Graph. (TOG) **41**(4), 1–15 (2022)
27. Nishimura, D.G.: Principles of Magnetic Resonance Imaging. Lulu.com (2010)
28. Otazo, R., Candes, E., Sodickson, D.K.: Low-rank plus sparse matrix decomposition for accelerated dynamic MRI with separation of background and dynamic components. Magn. Reson. Med. **73**(3), 1125–1136 (2015)
29. Pan, J., Hamdi, M., Huang, W., Hammernik, K., Kuestner, T., Rueckert, D.: Unrolled and rapid motion-compensated reconstruction for cardiac cine mri. Med. Image Anal. **91**, 103017 (2024)
30. Pan, J., Huang, W., Rueckert, D., Küstner, T., Hammernik, K.: Reconstruction-driven motion estimation for motion-compensated mr cine imaging. IEEE Trans. Med. Imaging (2024)
31. Pipe, J.G., Menon, P.: Sampling density compensation in MRI: rationale and an iterative numerical solution. Magnetic Resonance in Medicine: An Official Journal of the International Society for Magnetic Resonance in Medicine **41**(1), 179–186 (1999)
32. Qin, C., Schlemper, J., Caballero, J., Price, A.N., Hajnal, J.V., Rueckert, D.: Convolutional recurrent neural networks for dynamic MR image reconstruction. IEEE Trans. Med. Imaging **38**(1), 280–290 (2018)
33. Qiu, Z., Hu, S., Zhao, W., Sakaie, K., Sun, J.E., Griswold, M.A., Jones, D.K., Ma, D.: Self-calibrated subspace reconstruction for multidimensional mr fingerprinting for simultaneous relaxation and diffusion quantification. Magn. Reson. Med. **91**(5), 1978–1993 (2024)
34. Rajiah, P.S., François, C.J., Leiner, T.: Cardiac mri: state of the art. Radiology **307**(3), e223008 (2023)
35. Schlemper, J., Caballero, J., Hajnal, J.V., Price, A., Rueckert, D.: A deep cascade of convolutional neural networks for MR image reconstruction. In: International Conference on Information Processing in Medical Imaging, pp. 647–658. Springer (2017)
36. Shea, S.M., Kroeker, R.M., Deshpande, V., Laub, G., Zheng, J., Finn, J.P., Li, D.: Coronary artery imaging: 3d segmented k-space data acquisition with multiple breath-holds and real-time slab following. J. Magnet. Resonance Imaging Official J. Int. Soc. Magnet. Resonance Medicine **13**(2), 301–307 (2001)
37. Shen, L., Pauly, J., Xing, L.: Nerp: implicit neural representation learning with prior embedding for sparsely sampled image reconstruction. IEEE Trans. Neural Networks Learn. Syst. **35**(1), 770–782 (2022)
38. Sitzmann, V., Martel, J., Bergman, A., Lindell, D., Wetzstein, G.: Implicit neural representations with periodic activation functions. Adv. Neural. Inf. Process. Syst. **33**, 7462–7473 (2020)
39. Spieker, V., et al.: Deep learning for retrospective motion correction in mri: a comprehensive review. IEEE Trans. Med. Imaging (2023)

40. Spieker, V., et al.: Self-supervised k-space regularization for motion-resolved abdominal mri using neural implicit k-space representations. In: International Conference on Medical Image Computing and Computer-Assisted Intervention, pp. 614–624. Springer (2024)
41. Spieker, V., et al.: Iconik: generating respiratory-resolved abdominal mr reconstructions using neural implicit representations in k-space. In: International Conference on Medical Image Computing and Computer-Assisted Intervention, pp. 183–192. Springer (2023)
42. Uecker, M., Tamir, J.I., Ong, F., Lustig, M.: The bart toolbox for computational magnetic resonance imaging. In: Proc Intl Soc Magn Reson Med., vol. 24, p. 1 (2016)
43. Wang, X., Uecker, M., Feng, L.: Fast real-time cardiac mri: a review of current techniques and future directions. Investigative Magnetic Resonance Imaging **25**(4), 252–265 (2021)
44. Wright, K.L., Hamilton, J.I., Griswold, M.A., Gulani, V., Seiberlich, N.: Non-Cartesian parallel imaging reconstruction. J. Magn. Reson. Imaging **40**(5), 1022–1040 (2014)
45. Yaman, B., Hosseini, S., Moeller, S., Ellermann, J., Uğurbil, K., Akçakaya, M.: Self-supervised learning of physics-guided reconstruction neural networks without fully sampled reference data. Magn. Reson. Med. **84**(6), 3172–3191 (2020)

Image Synthesis

3D Shape-to-Image Brownian Bridge Diffusion for Brain MRI Synthesis from Cortical Surfaces

Fabian Bongratz[1,2(✉)], Yitong Li[1,2], Sama Elbaroudy[1], and Christian Wachinger[1,2]

[1] Lab for AI in Medical Imaging, Technical University of Munich, Munich, Germany
fabi.bongratz@tum.de
[2] Munich Center for Machine Learning, Munich, Germany

Abstract. Despite recent advances in medical image generation, existing methods struggle to produce anatomically plausible 3D structures. In synthetic brain magnetic resonance images (MRIs), characteristic fissures are often missing, and reconstructed cortical surfaces appear scattered rather than densely convoluted. To address this issue, we introduce Cor2Vox, the first diffusion model-based method that translates continuous cortical shape priors to synthetic brain MRIs. To achieve this, we leverage a Brownian bridge process which allows for direct structured mapping between shape contours and medical images. Specifically, we adapt the concept of the Brownian bridge diffusion model to 3D and extend it to embrace various complementary shape representations. Our experiments demonstrate significant improvements in the geometric accuracy of reconstructed structures compared to previous voxel-based approaches. Moreover, Cor2Vox excels in image quality and diversity, yielding high variation in non-target structures like the skull. Finally, we highlight the capability of our approach to simulate cortical atrophy at the sub-voxel level. Our code is available at https://github.com/ai-med/Cor2Vox.

1 Introduction

Generating synthetic medical images has demonstrated numerous benefits, including improved model robustness [14], enhanced fairness [2,24], reduced manual annotation burdens [31,46], and support in individualized simulations and treatment planning [49]. Recent progress in diffusion models has enabled the generation of realistic-looking brain MRIs [22,36,38] as in Fig. 1(a). However, examining beyond 2D views to analyze 3D cortical surface reconstructions reveals significant limitations, as most cases exhibit implausible folding patterns as shown in Fig. 1(b). We investigated 20 randomly generated MRIs and reconstructed their corresponding cortical surfaces of the left hemisphere. Severe irregularities were observed in 18 out of 20 reconstructions, including missing central gyri/sulci, unrealistic grooves, and scattered surface structures. While initially

---
F. Bongratz and Y. Li—Equal contribution.

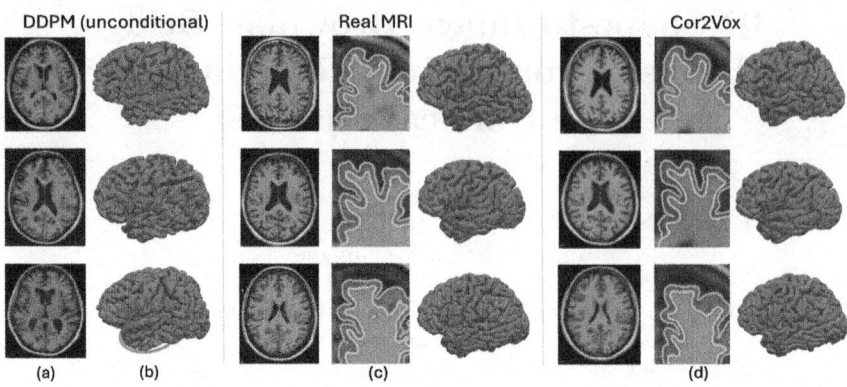

Fig. 1. Conditioning on realistic cortical surfaces is crucial to avoid anatomical implausibilities in synthetic MRIs. (a) Generated MRIs from an unconditional 3D diffusion model. (b) Anatomical implausibilities; from top to bottom: missing central sulcus, unrealistic wide groove, scattered surface structure. (c) Real MRIs and corresponding cortical surfaces; we highlighted the characteristic central sulcus in red. (d) Our approach (Cor2Vox) incorporates real cortical surfaces to guide the generative process. (Color figure online)

surprising, this issue aligns with the well-established challenge diffusion models face in generating anatomically accurate hands [33], a problem that is further amplified given the geometric complexity of the cerebral cortex. This issue has likely received limited attention because traditional image quality metrics, such as Structural Similarity Index Measure (SSIM) and Peak Signal-to-Noise-Ratio (PSNR), primarily evaluate visual similarity and fail to capture structural inconsistencies and critical neuroanatomical inaccuracies.

The problem of anatomical implausibility highlights a core challenge in image generation: achieving effective *control* over the generative process. While recent methods [32,47,48] have introduced localized conditions like 2D pose skeletons or edge maps to guide image synthesis, they are insufficient for the generation of brain MRIs, as controlling the intricate 3D geometry of the highly folded cortex demands more sophisticated representations. Existing continuous shape representations, such as triangular meshes [6,29,35] and signed distance fields (SDF) [5,13,16], excel at modeling the dense convolutions of the cortex and are instrumental in simulating sub-millimeter-level brain atrophy [3,40], aiding the benchmarking of segmentation and registration algorithms [4,20,25,41]. Incorporating such representations into generative models therefore holds promise for synthesizing anatomically realistic 3D brain scans.

However, integrating geometric representations into standard denoising diffusion probabilistic models (DDPM) [17,42] is suboptimal, as they assume a pure Gaussian noise as the prior, requiring anatomical constraints to be integrated externally. A recent method, Brownian Bridge diffusion models (BBDM) [27], offers a more flexible approach by modeling the transport between two arbitrary distributions, allowing for a direct mapping between geometric conditions and brain MRIs. Yet, current models [26,27] were developed for 2D images, and commonly operate in the latent space after a pre-trained autoencoder, making them unsuited for precisely generating intricate 3D structure of the cerebral cortex.

Thus, we introduce *Cor2Vox*, the first diffusion model-based method leveraging a 3D shape-to-image Brownian bridge diffusion model for medical image synthesis incorporating continuous shape priors. We combine inner white matter and outer pial cortical surfaces deliberately to represent detailed brain structures with 3D SDFs. By leveraging both surfaces, Cor2Vox is tailored to model changes in cortical thickness, offering novel possibilities to accurately simulate cortical thinning. With the cortical control, we can compute surface distances to precisely quantify the accuracy of the generated tissue boundaries, surpassing the indirect evaluation with Cohen's d [45]. We show a significant improvement in the accuracy of the generated images over previous methods, while maintaining the model's ability to introduce variability in other image regions, such as the skull. Finally, we used Cor2Vox to simulate cortical atrophy in MRIs, supporting the benchmarking of cortical thickness estimation methods.

2 Related Work

To date, generative adversarial networks (GAN) [12] and DDPMs [17,42] are among the most prevalent approaches for image generation. Both have been extended to incorporate geometric and semantic constraints, such as segmentation masks and edge maps, to achieve more controlled synthesis [19,34,39,44,47]. However, these approaches are typically tailored to 2D images, making their adaptation to 3D medical data inherently challenging. Current approaches for 3D medical image generation commonly either operate in a lower-dimensional latent space [22,37,38] or generate 3D volumes slice by slice [15,28,36]. Nonetheless, latent diffusion models have inherent limitations in achieving high precision [23,39], while slice-wise generative models introduce inter-slice inconsistencies which demand further post-processing [15,28]. Recent advancements, such as Med-DDPM [9], directly generate 3D medical scans conditioned on voxel segmentation masks, providing control over the location and shape of structures like tumors [9,22]. Yet, their level of detail is limited by the voxel size, which poses challenges when generating fine details for complex geometries, such as the cortex. While this discussion focuses on geometric condition-based image generation, it is worth noting that alternative approaches for synthesizing medical images exist as well, including methods based on Jacobian deformations [20,41] and biophysical models [3,4,21].

3 Methods

3.1 Cortical Shape Representations

Triangular surface meshes are a common *explicit* representation of cortical tissue boundaries [1,10], directly usable for visualization and rendering. Alternatively, surfaces can be represented *implicitly* through signed distance fields (SDF). Given a surface $\mathcal{S} \subset \mathbb{R}^3$, an SDF is defined as a function:

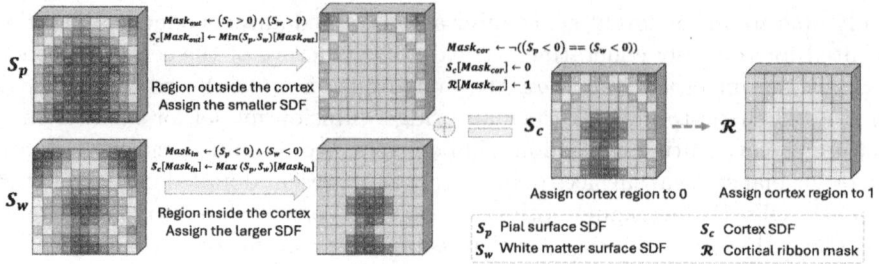

Fig. 2. Generation of cortex SDF ($\mathcal{S}_c$) and cortical ribbon mask ($\mathcal{R}$).

$$g : \Omega \subseteq \mathbb{R}^3 \to \mathbb{R}, \tag{1}$$

which maps each point $x \in \Omega$ to its orthogonal distance to the surface $\mathcal{S}$. The zero-level set of g corresponds to the surface $\mathcal{S}$, and the sign of g indicates whether x lies inside ($g(x) < 0$) or outside ($g(x) > 0$) the surface. With distance transforms [7], surface meshes can be converted into SDFs by computing the orthogonal distance from an input point to its closest triangle. Hence, SDFs can be represented as dense voxel grids, making them compatible with 3D image data. Additionally, meshes can be converted into binary edge maps by identifying the occupancy of voxels with mesh faces.

As shown in Fig. 1(c), the cerebral cortex has outer and inner surface boundaries, namely the pial and white matter surface, which are commonly represented as separate meshes $\mathcal{M}_p$ and $\mathcal{M}_w$, respectively. We aim to combine both surfaces into a single cortex-describing SDF, $\mathcal{S}_c$, which will be the source input to our framework. To achieve this, we first convert the cortical meshes into dense SDFs using a distance transform, yielding $\mathcal{S}_p$ and $\mathcal{S}_w$. Subsequently, we construct the cortex SDF $\mathcal{S}_c$ by combining $\mathcal{S}_p$ and $\mathcal{S}_w$ following the procedure in Fig. 2. We begin by initializing $\mathcal{S}_c$ as a tensor of the same size as $\mathcal{S}_p$ and $\mathcal{S}_w$, with all entries set to zero. The regions of interest are then identified and processed as follows. i) The region outside the cortex is identified where $\mathcal{S}_p$ and $\mathcal{S}_w$ have both positive values; within this region, we assign the lower value, i.e., the distance to the pial surface, to $\mathcal{S}_c$. ii) Next, the region inside the cortical ribbon is determined as $\mathcal{S}_p$ and $\mathcal{S}_w$ having both negative values; within this region, we assign the higher value, i.e., the distance to the white matter surface, to $\mathcal{S}_c$. iii) Lastly, the cortical ribbon is given by the region where $\mathcal{S}_p$ and $\mathcal{S}_w$ have opposite signs; we set $\mathcal{S}_c$ in this region to zero to ensure a clear cortical boundary. We also extract this cortical ribbon as a separate mask $\mathcal{R}$.

3.2 Shape-to-Image Brownian Bridge Diffusion

Our objective is to generate an image $\mathcal{I} = f_\theta(\mathcal{S}_c)$ that precisely matches the geometry given by a shape prior $\mathcal{S}_c$. Inspired by [27], we use a Brownian bridge process and adapt it for directly transforming the shape to the 3D image domain. We show an overview of our method, *Cor2Vox*, in Fig. 3. Cor2Vox uses paired data $(\mathcal{S}_c^i, \mathcal{I}^i)_{i=1}^N$ for training, comprising N pairs of 3D MRIs $\mathcal{I} \in \mathbb{R}^{H \times W \times D}$ and

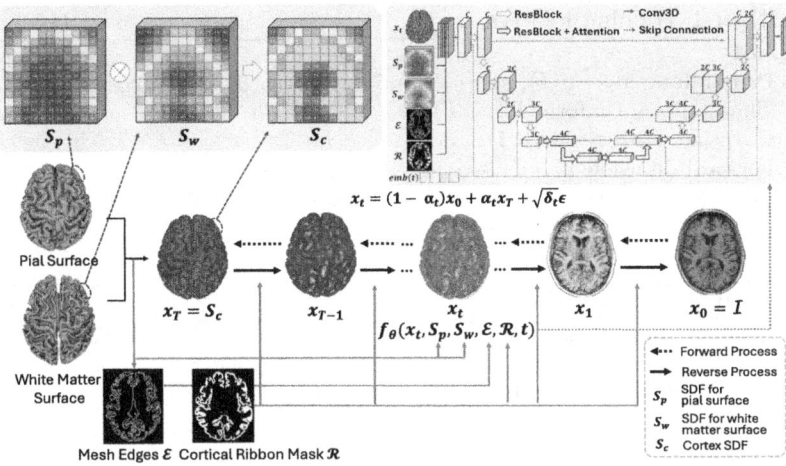

Fig. 3. Cor2Vox leverages a shape-to-image Brownian bridge diffusion process to learn a stochastic mapping f_θ between the shape prior $\mathcal{S}_c$ and the MRI domain $\mathcal{I}$. During the reverse diffusion process, additional shape conditions are incorporated to improve structural alignment in the generated MRIs. f_θ is modeled using a 3D UNet.

corresponding cortical shapes. We assume that shapes are given as dense SDFs $\mathcal{S}_c \in \mathbb{R}^{H \times W \times D}$ with the same dimensionality as the MRIs.

Forward Process. Unlike DDPM, which perturbs the input until pure Gaussian noise $\varepsilon \sim \mathcal{N}(0, \mathbf{I})$ at timestep T, the Brownian bridge diffusion process maps between data points from the joint distribution of structured source and target domains, i.e., $(\boldsymbol{x}_T, \boldsymbol{x}_0) \sim q_{data}(\mathcal{S}_c, \mathcal{I})$. Starting from an initial state $\boldsymbol{x}_0$ (i.e., the MRI volume $\mathcal{I}$), the forward process of the Brownian bridge connects it to the destination state $\boldsymbol{x}_T$ (i.e., the shape prior $\mathcal{S}_c$) by:

$$q(\boldsymbol{x}_t \mid \boldsymbol{x}_0, \boldsymbol{x}_T) = \mathcal{N}\big(\boldsymbol{x}_t; (1-\alpha_t)\boldsymbol{x}_0 + \alpha_t \boldsymbol{x}_T, \delta_t \mathbf{I}\big), \quad \text{where} \quad \mathcal{S}_c = \boldsymbol{x}_T, \quad (2)$$

$\alpha_t = t/T$, T the total number of steps in the diffusion process, and $\delta_t = 2(\alpha_t - \alpha_t^2)$ the variance term. The transition probability between two consecutive steps can be derived as [27]:

$$q_{BB}(\boldsymbol{x}_t \mid \boldsymbol{x}_{t-1}, \boldsymbol{x}_T) = \mathcal{N}\left(\boldsymbol{x}_t; \frac{1-\alpha_t}{1-\alpha_{t-1}} \boldsymbol{x}_{t-1} + \left(\alpha_t - \frac{1-\alpha_t}{1-\alpha_{t-1}} \alpha_{t-1}\right) \boldsymbol{x}_T, \delta_{t|t-1} \mathbf{I}\right),$$

$$\delta_{t|t-1} = \delta_t - \delta_{t-1} \frac{(1-\alpha_t)^2}{(1-\alpha_{t-1})^2}. \quad (3)$$

Thus, at the beginning of the diffusion process, i.e., $t = 0$, we have $\alpha_0 = 0$, starting from the mean value of $\boldsymbol{x}_0 = \mathcal{I}$ with probability 1 and variance $\delta_0 = 0$. As the diffusion progresses, the variance δ_t first increases to its maximum value $\delta_{max} = \delta_{T/2} = 1/2$ at the midpoint of the process, and then decreases until it reaches $\delta_T = 0$ at $\alpha_T = 1$. At this point, the diffusion concludes in the destination

Algorithm 1 Training Process

1: **repeat**
2:     Paired data: $\boldsymbol{x}_0 \sim q(\mathcal{I}), \boldsymbol{x}_T \sim q(\mathcal{S}_c), \mathcal{C} = (\mathcal{S}_p, \mathcal{S}_w, \mathcal{E}, \mathcal{R}) \sim q(\mathcal{S}_p, \mathcal{S}_w, \mathcal{E}, \mathcal{R})$
3:     Timestep $t \sim \text{Uniform}(1, \ldots, T)$
4:     Gaussian noise $\varepsilon \sim \mathcal{N}(0, \mathbf{I})$
5:     Forward diffusion $\boldsymbol{x}_t = (1 - \alpha_t)\boldsymbol{x}_0 + \alpha_t \boldsymbol{x}_T + \sqrt{\delta_t}\varepsilon$
6:     Take gradient descent step on $\nabla_\theta \| \alpha_t(\boldsymbol{x}_T - \boldsymbol{x}_0) + \sqrt{\delta_t}\varepsilon - f_\theta(\boldsymbol{x}_t, \mathcal{C}, t)\|_1$
7: **until** converged

Algorithm 2 Sampling Process

1: Sample conditional input $x_T = \mathcal{S}_c \sim q(\mathcal{S}_c)$, $\mathcal{C} = (\mathcal{S}_p, \mathcal{S}_w, \mathcal{E}, \mathcal{R}) \sim q(\mathcal{S}_p, \mathcal{S}_w, \mathcal{E}, \mathcal{R})$
2: **for** $t = T, \ldots, 1$ **do**
3:     **if** $t > 1$ **then** $\varepsilon \sim \mathcal{N}(0, \mathbf{I})$, **else** $\varepsilon = 0$
4:     $x_{t-1} = c_{xt}\boldsymbol{x}_t + c_{st}\mathcal{S}_c - c_{ft}f_\theta(\boldsymbol{x}_t, \mathcal{C}, t) + \sqrt{\tilde{\delta}_t}\varepsilon$
5: **end for**
6: **return** x_0

with a mean value equal to $\mathcal{S}_c$. Through this mechanism, the forward diffusion process establishes a stochastic mapping with fixed endpoints from the image domain $\mathcal{I}$ to the shape domain $\mathcal{S}_c$.

Reverse Process. The reverse process of BBDM is designed to predict $\boldsymbol{x}_{t-1}$ given $\boldsymbol{x}_t$, starting directly with the conditional input $\boldsymbol{x}_T = \mathcal{S}_c$, different from standard diffusion models that start from the pure Gaussian noise. To improve the structural alignment, we incorporate supplementary shape representations that can be obtained from the cortical meshes. This includes the SDF of the pial surface $\mathcal{S}_p$, the SDF of the white matter surface $\mathcal{S}_w$, a binary edge map $\mathcal{E}$ containing both surfaces, and the cortical ribbon mask $\mathcal{R}$, cf. Fig. 2. We collectively represent these conditions as $\mathcal{C} = (\mathcal{S}_p, \mathcal{S}_w, \mathcal{E}, \mathcal{R})$, and integrate them at each timestep t to aid the prediction:

$$p_\theta(\boldsymbol{x}_{t-1} \mid \boldsymbol{x}_t, \mathcal{C}, \mathcal{S}_c) = \mathcal{N}\left(\boldsymbol{x}_{t-1}; \boldsymbol{\mu}_\theta(\boldsymbol{x}_t, \mathcal{C}, t), \tilde{\delta}_t \mathbf{I}\right). \tag{4}$$

Here, $\boldsymbol{\mu}_\theta(\boldsymbol{x}_t, \mathcal{C}, t)$ represents the predicted mean value while $\tilde{\delta}_t$ denotes the variance of noise at each timestep. Following the reparameterization strategy used in DDPM [17], we can train a neural network $f_\theta(\cdot)$ to predict solely the noise instead of the mean $\boldsymbol{\mu}_\theta$. Namely, we reformulate $\boldsymbol{\mu}_\theta$ as a linear combination of $\boldsymbol{x}_t$, $\mathcal{S}_c$, and the estimated part f_θ:

$$\boldsymbol{\mu}_\theta(\boldsymbol{x}_t, \mathcal{C}, \mathcal{S}_c, t) = c_{xt}\boldsymbol{x}_t + c_{st}\mathcal{S}_c + c_{ft}f_\theta(\boldsymbol{x}_t, \mathcal{C}, t), \text{ where} \tag{5}$$

$$c_{xt} = \frac{\delta_{t-1}}{\delta_t}\frac{1-\alpha_t}{1-\alpha_{t-1}} + \frac{\delta_{t|t-1}}{\delta_t}(1-\alpha_{t-1}),$$

$$c_{st} = \alpha_{t-1} - \alpha_t \frac{1-\alpha_t}{1-\alpha_{t-1}}\frac{\delta_{t-1}}{\delta_t}, \text{ and}$$

$$c_{ft} = (1-\alpha_{t-1})\frac{\delta_{t|t-1}}{\delta_t}.$$

The parameters c_{xt}, c_{st}, and c_{ft} are non-trainable. Instead, they are derived from α_t, α_{t-1}, θ_t, and θ_{t-1}. The variance term $\tilde{\delta}_t$ does not need to be learned either; it can be derived in the analytical form as $\tilde{\delta}_t = \frac{\delta_{t|t-1} \cdot \delta_{t-1}}{\delta_t}$.

Training. We outline the training process in Algorithm 1. The training is designed to minimize the disparity between the joint distribution predicted by the model f_θ and the training data. This is achieved by optimizing the Evidence Lower Bound (ELBO) defined below:

$$\begin{aligned}\text{ELBO} = &-\mathbb{E}_q\big(\text{D}_{\text{KL}}(q_{\text{BB}}(\boldsymbol{x}_T|\,\boldsymbol{x}_0,\mathcal{S}_c)\,\|\,p(\boldsymbol{x}_T|\,\mathcal{C},\mathcal{S}_c))\\&+\sum_{t=2}^T \text{D}_{\text{KL}}(q_{\text{BB}}(\boldsymbol{x}_{t-1}|\,\boldsymbol{x}_t,\boldsymbol{x}_0,\mathcal{S}_c)\,\|\,p_\theta(\boldsymbol{x}_{t-1}|\,\boldsymbol{x}_t,\mathcal{C},\mathcal{S}_c))\\&-\log p_\theta(\boldsymbol{x}_0|\,\boldsymbol{x}_1,\mathcal{C},\mathcal{S}_c)\big).\end{aligned} \quad (6)$$

By combining the ELBO with Equations (3) to (5), the training objective is given by:

$$\mathcal{L}_{\text{Cor2Vox}} = \mathbb{E}_{\boldsymbol{x}_0,\mathcal{S}_c,\varepsilon\sim\mathcal{N}(0,\mathbf{I})}\left[\left\|\alpha_t(\mathcal{S}_c - \boldsymbol{x}_0) + \sqrt{\delta_t}\varepsilon - f_\theta(\boldsymbol{x}_t,\mathcal{C},t)\right\|_1\right]. \quad (7)$$

Sampling. We outline the sampling process in Algorithm 2. The sampling process can be derived as:

$$\boldsymbol{x}_{t-1} = c_{xt}\boldsymbol{x}_t + c_{st}\mathcal{S}_c - c_{ft}f_\theta(\boldsymbol{x}_t,\mathcal{C},t) + \sqrt{\tilde{\delta}_t}\varepsilon, \quad (8)$$

where $\varepsilon \sim \mathcal{N}(0,\mathbf{I})$ when $t > 1$, otherwise $\varepsilon = 0$. We accelerate the sampling process using the DDIM [43] strategy, which adopts a non-Markovian process with the same marginal distributions as Markovian inference.

3.3 Surface-Based Evaluation of Geometric Accuracy

Since we use continuous brain surfaces, $\mathcal{M}^{\text{Ref}}$, as the input to our model, we can evaluate the geometric accuracy of the generated images via surface-based distance metrics. This is not possible for voxel-based methods, where no such reference is available for comparison. Specifically, we propose to use V2C-Flow [1] to obtain cortical surfaces, $\mathcal{M}^{\text{Pred}}$, from the generated images. Based on these reconstructions, we compute the average symmetric surface distance (ASSD):

$$\text{ASSD}(\mathcal{M}^{\text{Pred}}, \mathcal{M}^{\text{Ref}}) = \frac{\sum_{p\in\mathcal{P}^{\text{Pred}}} d(p,\mathcal{M}^{\text{Ref}}) + \sum_{p\in\mathcal{P}^{\text{Ref}}} d(p,\mathcal{M}^{\text{Pred}})}{|\mathcal{P}^{\text{Pred}}| + |\mathcal{P}^{\text{Ref}}|}, \quad (9)$$

using $|\mathcal{P}^{\text{Pred}}| = |\mathcal{P}^{\text{Ref}}| = 100{,}000$ randomly sampled surface points. The distance $d(p,\mathcal{M})$ measures the orthogonal distance from a point $p \in \mathbb{R}^3$ to its closest triangle in the mesh $\mathcal{M}$.

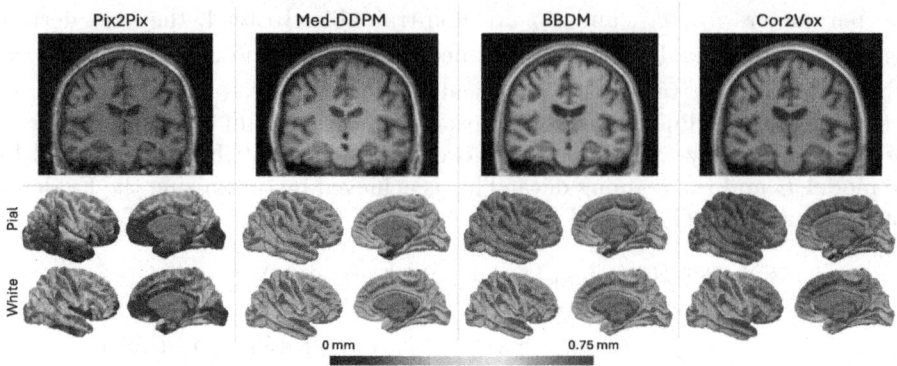

Fig. 4. Generated brain MRIs (top) and vertex-wise errors (bottom, in mm) comparing reconstructed cortical shapes to their original inputs. We show the mean error values on the cortical surface across the test set for the right hemisphere, separately for pial and white matter surfaces.

4 Results

4.1 Experimental Setting

Models and Hyperparameters. We employ a 3D UNet to model f_θ, which is a 3D adaptation of the ADM architecture [8]. We use channels of $[C, 2C, 3C, 4C]$ for each residual stage, where $C = 64$. Global attention is applied at downsampling factor 8, with 4 heads and 64 channels. We use adaptive group normalization to inject the timestep embedding into each residual block. The model is trained using the Adam optimizer with an initial learning rate of 1×10^{-4}, reduced by a factor of 0.5 on the plateau, a batch size of 2, and an exponential moving average (EMA) rate of 0.995, for 400 epochs with one NVIDIA H100 GPU. We set 1,000 timesteps for training, while 10 for inference with DDIM [43] sampling strategy for a balance between image quality and computational efficiency.

Data and Preprocessing. We used data from the Alzheimer's Disease Neuroimaging Initiative (ADNI, https://adni.loni.usc.edu), containing T1-weighted brain MRIs from cognitively normal subjects, subjects diagnosed with mild cognitive impairment (MCI) and Alzheimer's disease (AD). We registered all scans to MNI152 space via affine registration, implemented in NiftyReg (v1.5.69). To avoid subject bias, we used only baseline scans, with 1,155/169/323 samples for train/validation/test sets. We used the V2C-Flow [1] model, pre-trained on FreeSurfer (v7.2) [10] meshes, to provide surfaces as input to Cor2Vox for both training and sampling. From the surface meshes, we created white matter and pial SDFs, merging both hemispheres into one SDF, with a KDTree [30] implemented in Scipy (v1.10.0). Internally, our models use a resolution of 128^3 voxels but we rescaled all outputs to MNI space (1mm isotropic resolution, trilinear interpolation) for evaluation. To assess image quality, we used SynthStrip [18] to skull-strip both generated and original images (orig.mgz files from FreeSurfer),

Table 1. Quantitative comparison of implemented methods for 3D brain MRI generation. We report mean ± SD across all samples in our test set. Geometric accuracy of the cortex is in mm.

Model	Image Quality		Geometric Accuracy—ASSD↓			
			Left Hemisphere		Right Hemisphere	
	SSIM↑	PSNR↑	White	Pial	White	Pial
Pix2Pix$_\mathcal{E}$ [19]*	0.891±0.011	**24.76±1.96**	0.784±0.190	0.853±0.227	0.760±0.174	0.804±0.195
Med-DDPM$_\mathcal{E}$ [9]	0.894±0.015	22.38±2.63	0.506±0.042	0.437±0.027	0.516±0.043	0.436±0.026
BBDM$_\mathcal{E}$ [27]*	0.898±0.020	19.95±2.61	0.349±0.032	0.307±0.021	0.368±0.035	0.307±0.021
Pix2Pix$_\mathcal{R}$ [19]*	0.885±0.013	23.69±2.06	0.697±0.137	0.691±0.155	0.698±0.183	0.680±0.221
Med-DDPM$_\mathcal{R}$ [9]	0.899±0.016	22.30±2.80	0.359±0.043	0.347±0.034	0.373±0.106	0.352±0.101
BBDM$_\mathcal{R}$ [27]*	0.898±0.021	19.67±2.54	0.328±0.037	0.283±0.034	0.338±0.128	0.287±0.121
Cor2Vox/DDPM	0.902±0.015	23.13±2.71	0.335±0.038	0.300±0.033	0.348±0.048	0.305±0.037
Cor2Vox	**0.906±0.018**	21.10±2.74	**0.283±0.029**	**0.251±0.019**	**0.289±0.031**	**0.251±0.017**

* Method adapted for 3D image generation.

to avoid the scanner noise in the original data to confound the evaluation. Nonetheless, all models were trained to generate the entire MRI, i.e., including the skull and background, as shown in our visualizations.

Baselines Implementation. We compare Cor2Vox to state-of-the-art conditional image generation models, namely Pix2Pix [19], Med-DDPM [9], and BBDM [27]. We adapted Pix2Pix and BBDM for 3D data as both were originally developed for 2D images. For BBDM, we applied the Brownian diffusion process in the image space instead of the latent space, in analogy to Med-DDPM and Cor2Vox. We adapted Med-DDPM by switching the intensity scaling from slice-wise to volumetric and the denoising target from noise to denoised image, which considerably enhanced its performance. We provided the mesh edges ($\mathcal{E}$) and cortical ribbon masks ($\mathcal{R}$) as input conditions to these models, as they closely resemble the masks and edge maps used in the original works. In addition, we developed a variant of Cor2Vox by replacing the Brownian bridge process with a standard denoising diffusion process with the same input as Cor2Vox; we call this variant Cor2Vox/DDPM.

4.2 Image Quality and Accuracy

We visualize the generated brain MRIs from Cor2Vox and implemented baseline methods in Fig. 4. Additionally, we show local surface-based errors on the FsAverage template that serves as input to V2C-Flow. We report quantitative scores for all methods in Table 1. Note that the reconstruction errors shown in Fig. 4 are not equivalent to the ASSD; the ASSD is bi-directional and independent of vertices, whereas the surface plots show the average distance of vertices in the reconstructed surfaces to the original cortical boundaries.

From the qualitative inspection of the generated images, we observe that all methods are capable of generating the same cortical anatomy based on the provided shape condition. However, the quantitative evaluation in Table 1 reveals a clear improvement of Cor2Vox over other methods in geometric accuracy, confirmed by surface-based plots in Fig. 4. White matter surfaces are, on average, by approximately 0.04 mm (~10%) more accurate than those from the 3D BBDM. For the pial surfaces, the gain is similar with a reduced error of around 0.03 mm. At the same time, Cor2Vox does not sacrifice image quality and achieves the highest SSIM score, albeit only by a slight margin. The 3D Pix2Pix excels in terms of PSNR, however, with an ASSD of more than 0.5 mm on all surfaces, Pix2Pix is not competitive regarding reconstruction accuracy. Med-DDPM and the Cox2Vox/DDPM variant also achieve low errors on both white and pial surfaces but cannot keep up with the best Brownian bridge-based methods. Generally, the surface errors are slightly larger on the white matter than on the pial surfaces, especially in the precentral gyrus. This could be explained by the challenge of precisely synthesizing the subtle contrast between white and gray matter, compared to the more pronounced intensity difference between gray matter and cerebrospinal fluid. Yet, Cor2Vox is the only method achieving an ASSD below 0.3 mm for the white matter surface. Finally, paired Wilcoxon signed-rank tests between Cor2Vox and baseline methods indicated highly significant improvements with $p < 10^{-7}$ for the geometric accuracy on both surfaces and hemispheres.

4.3 Ablation Study

We evaluate the impact of various shape conditions, whether as the source domain for the generative process or as additional conditioning inputs in Cor2Vox, on the geometric accuracy. As reported in Table 2, we initially tested individual shape conditions—pial surface SDF $\mathcal{S}_p$, white matter surface SDF $\mathcal{S}_w$, edge map $\mathcal{E}$, cortical ribbon mask $\mathcal{R}$, cortex SDF $\mathcal{S}_c$—as single-source domains in the Brownian bridge diffusion process. The pial surface achieves the best accuracy when conditioned solely on $\mathcal{S}_p$, and similarly for the white matter surface with $\mathcal{S}_w$. The cortical ribbon mask ($\mathcal{R}$) helps with the accuracy of the white matter while falling short on the pial surface. Next, we explored using $\mathcal{R}$ as the source domain and its combination with other conditions. This approach performs better than either using no additional conditions or employing two parallel bridge processes for both $\mathcal{S}_p$ and $\mathcal{S}_w$. Finally, we adopt the cortex SDF $\mathcal{S}_c$ as the source domain and experiment with incorporating different sets of shape conditions as additional inputs during the reverse diffusion process. Results show that including more conditions generally improves performance, with the combination of all four shape conditions yielding the best accuracy on the white matter surface and comparable performance on the pial surface across hemispheres. This validates the effectiveness of Cor2Vox with $\mathcal{S}_c$ as the source domain and the importance of including extra shape information via conditioning.

* This model requires two parallel Brownian bridge processes.

Table 2. Ablation study of different source domains for the Brownian bridge process and extra input conditions provided to Cor2Vox. We report the mean±SD on the validation set. The geometric accuracy of the cortex is in mm. ($\mathcal{S}_p$: Pial SDF, $\mathcal{S}_w$: White matter SDF, $\mathcal{S}_c$: Cortex SDF, $\mathcal{E}$: Edge map, $\mathcal{R}$: Cortical ribbon mask)

Source	Condition	Geometric Accuracy—ASSD↓			
		Left Hemisphere		Right Hemisphere	
		White	Pial	White	Pial
$\mathcal{S}_p$	—	$0.375_{\pm 0.066}$	$\mathbf{0.233_{\pm 0.021}}$	$0.386_{\pm 0.068}$	$\mathbf{0.236_{\pm 0.022}}$
$\mathcal{S}_w$	—	$0.289_{\pm 0.066}$	$0.375_{\pm 0.031}$	$0.303_{\pm 0.067}$	$0.376_{\pm 0.034}$
$\mathcal{E}$	—	$0.351_{\pm 0.037}$	$0.307_{\pm 0.023}$	$0.369_{\pm 0.042}$	$0.307_{\pm 0.024}$
$\mathcal{R}$	—	$0.336_{\pm 0.034}$	$0.279_{\pm 0.021}$	$0.335_{\pm 0.037}$	$0.280_{\pm 0.022}$
$\mathcal{S}_c$	—	$0.381_{\pm 0.066}$	$0.245_{\pm 0.026}$	$0.390_{\pm 0.070}$	$0.247_{\pm 0.026}$
$\mathcal{S}_p, \mathcal{S}_w$*	—	$0.394_{\pm 0.074}$	$0.535_{\pm 0.064}$	$0.418_{\pm 0.079}$	$0.553_{\pm 0.071}$
$\mathcal{R}$	$\mathcal{S}_p, \mathcal{S}_w$	$0.293_{\pm 0.027}$	$0.262_{\pm 0.018}$	$0.305_{\pm 0.031}$	$0.261_{\pm 0.016}$
$\mathcal{R}$	$\mathcal{S}_p, \mathcal{S}_w, \mathcal{E}$	$0.301_{\pm 0.038}$	$0.263_{\pm 0.021}$	$0.305_{\pm 0.040}$	$0.264_{\pm 0.022}$
$\mathcal{S}_c$	$\mathcal{E}$	$0.326_{\pm 0.036}$	$0.241_{\pm 0.021}$	$0.339_{\pm 0.039}$	$0.243_{\pm 0.022}$
$\mathcal{S}_c$	$\mathcal{R}$	$0.305_{\pm 0.030}$	$0.261_{\pm 0.021}$	$0.312_{\pm 0.033}$	$0.262_{\pm 0.020}$
$\mathcal{S}_c$	$\mathcal{S}_p, \mathcal{S}_w$	$0.310_{\pm 0.060}$	$0.236_{\pm 0.020}$	$0.321_{\pm 0.064}$	$0.239_{\pm 0.022}$
$\mathcal{S}_c$	$\mathcal{E}, \mathcal{R}$	$0.289_{\pm 0.031}$	$0.252_{\pm 0.016}$	$0.294_{\pm 0.034}$	$0.254_{\pm 0.016}$
$\mathcal{S}_c$	$\mathcal{S}_p, \mathcal{S}_w, \mathcal{E}$	$0.294_{\pm 0.043}$	$0.236_{\pm 0.017}$	$0.311_{\pm 0.047}$	$0.238_{\pm 0.018}$
$\mathcal{S}_c$	$\mathcal{S}_p, \mathcal{S}_w, \mathcal{R}$	$0.305_{\pm 0.036}$	$0.255_{\pm 0.019}$	$0.308_{\pm 0.037}$	$0.255_{\pm 0.019}$
$\mathcal{S}_c$	$\mathcal{S}_p, \mathcal{S}_w, \mathcal{E}, \mathcal{R}$	$\mathbf{0.282_{\pm 0.032}}$	$0.250_{\pm 0.017}$	$\mathbf{0.287_{\pm 0.035}}$	$0.250_{\pm 0.017}$

* This model requires two parallel Brownian bridge processes.

4.4 Image Variability

While we aim to generate MRI scans that conform to the shape of given cortical surfaces, it is desirable for applications like dataset augmentation to have high diversity in the remaining image regions. We generated five synthetic MRIs per subject in the test set using different random seeds and calculated the average variance in individual voxels across the synthetic MRIs. Additionally, we computed the mean absolute difference between these synthetic scans and the corresponding real MRIs. As illustrated in Fig. 5, the largest variation can be expected in the skull regions, whereas the cortical regions exhibit minimal differences to the original data and also comparably low variance. This indicates that Cor2Vox effectively generates realistic MRI scans that accurately align with the cortical surfaces while promoting diversity in the remaining image parts.

4.5 Synthesizing Cortical Atrophy

The deliberate combination of white matter and pial surfaces in Cor2Vox enables the creation of synthetic datasets with simulated cortical thickness changes. Inspired by previous work on cortical atrophy simulation [40], we mimicked

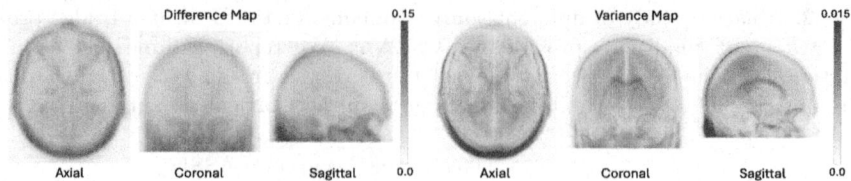

Fig. 5. We show the mean absolute difference between synthetic and original MRIs, and voxel-wise variance across synthetic MRIs across five random seeds among the whole test set, averaged along three anatomical axes, with darker colors indicating higher variability.

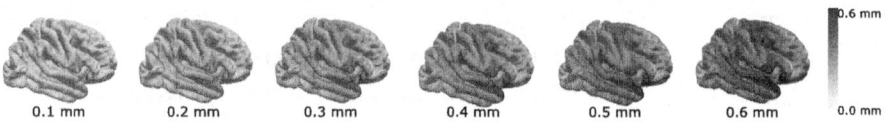

Fig. 6. Synthetic global atrophy in the range of 0.1–0.6 mm recovered from MRIs generated by Cor2Vox. We show vertex-wise mean values based on 124 cognitively normal test cases. The closer the recovered values to the introduced atrophy, the better.

the process of global cortical thinning in the range of 0.1–0.6 mm by deforming pial surfaces from cognitively normal subjects in our test set ($n = 124$) inside towards the white matter boundary. We used the Euclidean point-face distance between pial and white surfaces and uniform scaling. After generating matching MRIs from the altered surfaces (Cor2Vox) and reconstructing respective cortical surfaces (V2C-Flow), we compared the recovered change in cortical thickness (measured bi-directionally [11]) to the introduced atrophy, see Fig. 6. Except for a small occipital area, the introduced changes in the subvoxel range were well recovered, supporting the applicability of Cor2Vox for algorithm benchmarking.

5 Conclusion

We introduced Cor2Vox, a novel method for 3D brain MRI generation based on cortical surfaces. For the first time, we leveraged a shape-to-image Brownian bridge diffusion process for synthesizing anatomically plausible 3D medical scans. We extended the Brownian diffusion process by incorporating multiple complementary shape conditions to enhance anatomical alignment. By conditioning on realistic 3D shapes, Cor2Vox effectively addresses the problem of anatomical implausibilities commonly seen in synthetic brain MRIs. Our results demonstrated state-of-the-art image quality and significant improvements in geometric accuracy compared to existing methods. Moreover, we conducted a comprehensive ablation study and showcased the application of Cor2Vox for simulating individual cortical atrophy. Our findings highlight the critical role of realistic anatomical modeling in synthetic medical image generation and pave the way for advanced data augmentation and algorithm benchmarking.

Acknowledgements. This research was supported by the German Research Foundation (DFG) and the Munich Center for Machine Learning (MCML). We gratefully acknowledge the computational resources provided by the Leibniz Supercomputing Centre (http://www.lrz.de).

References

1. Bongratz, F., Rickmann, A.M., Wachinger, C.: Neural deformation fields for template-based reconstruction of cortical surfaces from mri. Med. Image Anal. **93**, 103093 (2024)
2. Burlina, P., Joshi, N., Paul, W., Pacheco, K.D., Bressler, N.M.: Addressing artificial intelligence bias in retinal diagnostics. Transl. Vis. Sci. Technol. **10**(2), 13–13 (2021)
3. Camara, O., Schweiger, M., Scahill, R., Crum, W., Sneller, B., Schnabel, J., Ridgway, G., Cash, D., Hill, D., Fox, N.: Phenomenological model of diffuse global and regional atrophy using finite-element methods. IEEE Trans. Med. Imaging **25**(11), 1417–1430 (2006)
4. Castellano Smith, A.D., Crum, W.R., Hill, D.L.G., Thacker, N.A., Bromiley, P.A.: Biomechanical simulation of atrophy in mr images. In: Sonka, M., Fitzpatrick, J.M. (eds.) Medical Imaging 2003: Image Processing, vol. 5032, p. 481. SPIE, May 2003
5. Cruz, R.S., Lebrat, L., Bourgeat, P., Fookes, C., Fripp, J., Salvado, O.: Deepcsr: a 3d deep learning approach for cortical surface reconstruction. In: 2021 IEEE Winter Conference on Applications of Computer Vision (WACV). IEEE, January 2021
6. Dale, A.M., Fischl, B., Sereno, M.I.: Cortical surface-based analysis. Neuroimage **9**(2), 179–194 (1999)
7. Danielsson, P.E.: Euclidean distance mapping. Comput. Graphics Image Process. **14**(3), 227–248 (1980)
8. Dhariwal, P., Nichol, A.: Diffusion models beat gans on image synthesis. In: NeurIPS (2021)
9. Dorjsembe, Z., Pao, H.K., Odonchimed, S., Xiao, F.: Conditional diffusion models for semantic 3d brain mri synthesis. IEEE J. Biomed. Health Inform. **28**(7), 4084–4093 (2024)
10. Fischl, B.: FreeSurfer. NeuroImage **62**(2), 774–781 (2012)
11. Fischl, B., Dale, A.M.: Measuring the thickness of the human cerebral cortex from magnetic resonance images. Proc. Natl. Acad. Sci. **97**(20), 11050–11055 (Sep 2000), publisher: Proceedings of the National Academy of Sciences
12. Goodfellow, I., et al.: Generative adversarial nets. In: Ghahramani, Z., Welling, M., Cortes, C., Lawrence, N., Weinberger, K. (eds.) Advances in Neural Information Processing Systems, vol. 27. Curran Associates, Inc. (2014)
13. Gopinath, K., Desrosiers, C., Lombaert, H.: Segrecon: Learning joint brain surface reconstruction and segmentation from images. In: Medical Image Computing and Computer Assisted Intervention – MICCAI 2021: 24th International Conference.

Strasbourg, France, September 27 – October 1, 2021, Proceedings, Part VII, pp. 650–659. Springer, Heidelberg (2021)
14. Gopinath, K., et al.: Synthetic data in generalizable, learning-based neuroimaging. Imaging Neuroscience (2024)
15. Han, K., et al.: MedGen3D: A Deep Generative Framework for Paired 3D Image and Mask Generation, pp. 759–769. Springer Nature Switzerland (2023)
16. Han, X., Pham, D.L., Tosun, D., Rettmann, M.E., Xu, C., Prince, J.L.: Cruise: cortical reconstruction using implicit surface evolution. Neuroimage **23**(3), 997–1012 (2004)
17. Ho, J., Jain, A., Abbeel, P.: Denoising diffusion probabilistic models. In: Advances in Neural Information Processing Systems, vol. 33, pp. 6840–6851. Curran Associates, Inc. (2020)
18. Hoopes, A., Mora, J.S., Dalca, A.V., Fischl, B., Hoffmann, M.: Synthstrip: skull-stripping for any brain image. Neuroimage **260**, 119474 (2022)
19. Isola, P., Zhu, J.Y., Zhou, T., Efros, A.A.: Image-to-image translation with conditional adversarial networks. CVPR (2017)
20. Karacali, B., Davatzikos, C.: Simulation of tissue atrophy using a topology preserving transformation model. IEEE Trans. Med. Imaging **25**(5), 649–652 (2006)
21. Khanal, B., Lorenzi, M., Ayache, N., Pennec, X.: A biophysical model of brain deformation to simulate and analyze longitudinal mris of patients with Alzheimer's disease. Neuroimage **134**, 35–52 (2016)
22. Kim, K., et al.: Controllable text-to-image synthesis for multi-modality mr images. In: 2024 IEEE/CVF Winter Conference on Applications of Computer Vision (WACV), pp. 7921–7930. IEEE, January 2024
23. Konz, N., Chen, Y., Dong, H., Mazurowski, M.A.: Anatomically-Controllable Medical Image Generation with Segmentation-Guided Diffusion Models, pp. 88–98. Springer Nature Switzerland (2024)
24. Ktena, I., et al.: Generative models improve fairness of medical classifiers under distribution shifts. Nat. Med. **30**(4), 1166–1173 (2024)
25. Larson, K.E., Oguz, I.: Synthetic atrophy for longitudinal cortical surface analyses. Frontiers in Neuroimaging **1**, June 2022
26. Lee, E., Jeong, S., Sohn, K.: EBDM: Exemplar-Guided Image Translation with Brownian-Bridge Diffusion Models, pp. 306–323. Springer Nature Switzerland, October 2024
27. Li, B., Xue, K., Liu, B., Lai, Y.K.: Bbdm: Image-to-image translation with brownian bridge diffusion models. In: Proceedings of the IEEE/CVF Conference on Computer Vision and Pattern Recognition, pp. 1952–1961 (2023)
28. Li, Y., Yakushev, I., Hedderich, D.M., Wachinger, C.: Pasta: Pathology-aware mri to pet cross-modal translation with diffusion models. In: Medical Image Computing and Computer Assisted Intervention – MICCAI 2024, pp. 529–540. Springer Nature Switzerland, Cham (2024)
29. MacDonald, D., Kabani, N., Avis, D., Evans, A.C.: Automated 3-d extraction of inner and outer surfaces of cerebral cortex from mri. Neuroimage **12**(3), 340–356 (2000)
30. Maneewongvatana, S., Mount, D.: Analysis of approximate nearest neighbor searching with clustered point sets, pp. 105–123. American Mathematical Society (Dec 2002)
31. Menten, M.J., Paetzold, J.C., Dima, A., Menze, B.H., Knier, B., Rueckert, D.: Physiology-Based Simulation of the Retinal Vasculature Enables Annotation-Free Segmentation of OCT Angiographs, pp. 330–340. Springer Nature Switzerland (2022)

32. Mou, C., Wang, X., Xie, L., Wu, Y., Zhang, J., Qi, Z., Shan, Y.: T2i-adapter: learning adapters to dig out more controllable ability for text-to-image diffusion models. In: Proceedings of the AAAI Conference on Artificial Intelligence, vol. 38, pp. 4296–4304 (2024)
33. Narasimhaswamy, S., Bhattacharya, U., Chen, X., Dasgupta, I., Mitra, S., Hoai, M.: Handiffuser: text-to-image generation with realistic hand appearances. In: Proceedings of the IEEE/CVF Conference on Computer Vision and Pattern Recognition, pp. 2468–2479 (2024)
34. Park, T., Liu, M.Y., Wang, T.C., Zhu, J.Y.: Semantic image synthesis with spatially-adaptive normalization. In: Proceedings of the IEEE Conference on Computer Vision and Pattern Recognition (2019)
35. Patenaude, B., Smith, S.M., Kennedy, D.N., Jenkinson, M.: A bayesian model of shape and appearance for subcortical brain segmentation. Neuroimage **56**(3), 907–922 (2011)
36. Peng, W., Adeli, E., Bosschieter, T., Park, S.H., Zhao, Q., Pohl, K.M.: Generating Realistic Brain MRIs via a Conditional Diffusion Probabilistic Model, pp. 14–24. Springer Nature Switzerland (2023)
37. Peng, W., Bosschieter, T., Ouyang, J., Paul, R., Sullivan, E.V., Pfefferbaum, A., Adeli, E., Zhao, Q., Pohl, K.M.: Metadata-conditioned generative models to synthesize anatomically-plausible 3d brain mris. Med. Image Anal. **98**, 103325 (2024)
38. Pinaya, W.H.L., et al.: Brain Imaging Generation with Latent Diffusion Models, p. 117–126. Springer Nature Switzerland (2022)
39. Rombach, R., Blattmann, A., Lorenz, D., Esser, P., Ommer, B.: High-resolution image synthesis with latent diffusion models. In: 2022 IEEE/CVF Conference on Computer Vision and Pattern Recognition (CVPR), pp. 10674–10685. IEEE, June 2022
40. Rusak, F., Santa Cruz, R., Lebrat, L., Hlinka, O., Fripp, J., Smith, E., Fookes, C., Bradley, A.P., Bourgeat, P.: Quantifiable brain atrophy synthesis for benchmarking of cortical thickness estimation methods. Med. Image Anal. **82**, 102576 (2022)
41. Sharma, S., Noblet, V., Rousseau, F., Heitz, F., Rumbach, L., Armspach, J.P.: Evaluation of brain atrophy estimation algorithms using simulated ground-truth data. Med. Image Anal. **14**(3), 373–389 (2010)
42. Sohl-Dickstein, J., Weiss, E.A., Maheswaranathan, N., Ganguli, S.: Deep unsupervised learning using nonequilibrium thermodynamics. In: Proceedings of the 32nd International Conference on International Conference on Machine Learning - Volume 37. pp. 2256–2265. ICML'15, JMLR.org (2015)
43. Song, J., Meng, C., Ermon, S.: Denoising diffusion implicit models. In: International Conference on Learning Representations (2021)
44. Wang, T.C., Liu, M.Y., Zhu, J.Y., Tao, A., Kautz, J., Catanzaro, B.: High-resolution image synthesis and semantic manipulation with conditional gans. In: 2018 IEEE/CVF Conference on Computer Vision and Pattern Recognition, pp. 8798–8807. IEEE, June 2018
45. Wu, J., Peng, W., Li, B., Zhang, Y., Pohl, K.M.: Evaluating the Quality of Brain MRI Generators, pp. 297–307. Springer Nature Switzerland (2024)
46. Xanthis, C.G., Filos, D., Haris, K., Aletras, A.H.: Simulator-generated training datasets as an alternative to using patient data for machine learning: An example in myocardial segmentation with mri. Comput. Methods Programs Biomed. **198**, 105817 (2021)
47. Zhang, L., Rao, A., Agrawala, M.: Adding conditional control to text-to-image diffusion models. In: Proceedings of the IEEE/CVF International Conference on Computer Vision (ICCV), pp. 3836–3847, October 2023

48. Zhao, S., Chen, D., Chen, Y.C., Bao, J., Hao, S., Yuan, L., Wong, K.Y.K.: Unicontrolnet: All-in-one control to text-to-image diffusion models. Advances in Neural Information Processing Systems **36** (2024)
49. Zhu, B., Gu, L., Zhang, J., Yan, Z., Pan, L., Zhao, Q.: Simulation of organ deformation using boundary element method and meshless shape matching. In: 2008 30th Annual International Conference of the IEEE Engineering in Medicine and Biology Society, pp. 3253–3256 (2008)

Cascaded Diffusion Model and Segment Anything Model for Medical Image Synthesis via Uncertainty-Guided Prompt Generation

Haowen Pang, Xiaoming Hong, Peng Zhang, and Chuyang Ye[✉]

School of Integrated Circuits and Electronics, Beijing Institute of Technology, Beijing, China
chuyang.ye@bit.edu.cn

Abstract. Multi-modal medical images of the same anatomical structure enhance the diversity of diagnostic information. However, limitations such as scanning time, economic costs, and radiation dose constraints can hinder the acquisition of certain modalities. In such cases, synthesizing missing images from available images offers a promising solution. In recent years, methods based on deep learning, particularly *Diffusion Models* (DMs), have shown significant success in medical image synthesis. However, these methods can still struggle with the presence of notable image anomalies caused by pathologies. In this work, to improve the synthesis robustness to anomalies, we propose a model cascading the DM and the *Segment Anything Model* (SAM) with uncertainty-guided prompt generation for medical image synthesis. SAM is originally a foundational model for image segmentation. We hypothesize that SAM (the medical variant) is beneficial to the synthesis tasks because 1) the SAM encoder trained on large, diverse datasets allows the model to grasp a deep understanding of complex anomaly patterns of pathologies and 2) its ability to take prompt inputs naturally allows the synthesis to pay special attention to abnormal regions that are hard to synthesize. To effectively integrate the DM and SAM, we propose the uncertainty-guided prompt generation framework, where DM synthesis results with higher uncertainty are considered regions potentially with worse synthesis quality and prompts are generated for SAM accordingly to improve the result. First, as the DM produces outputs based on randomly sampled noise, we propose to estimate the uncertainty of its synthesis output by repeated noise sampling and represent the output uncertainty by the prediction standard deviation. Then, prompts are generated based on the standard deviation and given to SAM, together with the DM input image and output. For effective interaction between the prompt, DM input, and DM output, we propose an *Uncertainty Guided Cross Attention* (UGCA) module, where the prompt serves as a query to guide the model to focus on relevant regions of the DM input and output. Finally, a synthesis decoder replaces the SAM decoder, and it is trained together with UGCA. Experimental results on two public datasets demonstrate that the proposed method outperforms existing state-of-the-art methods when images manifest notable anomalies.

Keywords: Medical image synthesis · Segment anything model · Diffusion model

1 Introduction

The acquisition of multi-modal medical images, such as multi-contrast *Magnetic Resonance Imaging* (MRI) and *Computed Tomography* (CT), of the same anatomical structure provides comprehensive information for patient condition evaluation and diagnostic insights [6,15]. However, practical challenges including prolonged scanning times, high costs, and concerns about radiation exposure often limit the acquisition of all desired modalities for a specific examination [22]. In such cases, synthesis of missing medical images from available modalities offers a promising solution to complementing the missing information [6,20,37].

The advances in *deep learning* (DL), particularly *Diffusion Models* (DMs) [10], have significantly enhanced the quality of medical image synthesis [26,37,42]. However, these methods are mostly applied to normal subjects, and the synthesis remains challenging for pathological regions that are crucial for diagnosis, prognosis, and treatment planning [34]. The pathological regions often exhibit various appearances, which are difficult to learn and hinder model generalization and accuracy in clinical settings. Therefore, further development of medical image synthesis methods that are robust to pathology continues to be a key focus of research.

The *Segment Anything Model* (SAM) [19], which is a foundation model for image segmentation, has also recently been adapted for medical image synthesis [12]. The SAM encoder (its medical variant [24]) is trained on diverse datasets, which empowers the model with a deep understanding of medical images, including the understanding of complex anomaly patterns of pathologies. Therefore, the use of the SAM encoder may benefit medical image synthesis, and in [12] SAM-I2I is proposed by borrowing the powerful SAM encoder for the synthesis task, where a decoder is trained based on the image features extracted by the SAM encoder. However, this method neglects another unique characteristic of SAM, the ability to allow prompt inputs to attend to specific regions. Such prompts, if generated correctly, can naturally allow the model to focus on abnormal regions that are hard to synthesize, and it is desirable to further exploit the prompting capability for adapting SAM to medical image synthesis.

In this work, we propose a medical image synthesis model that cascades the DM with SAM via uncertainty-guided prompt generation, where SAM further refines the synthesis results of the DM, especially for pathological regions with lower synthesis quality. To effectively integrate the DM and SAM, we generate automated prompts for SAM based on the DM synthesis output. As the DM synthesizes images with randomly sampled noise, it provides a natural way of estimating its synthesis uncertainty by repeated noise sampling. We assume that regions with higher uncertainty are likely those associated with poorer quality, and prompts are generated for SAM accordingly to improve the DM output. Specifically, we estimate the uncertainty of the DM output by computing the prediction standard deviation using different sampled noise. This uncertainty is then provided to SAM, along with the DM input image and output. To achieve effective interaction between the prompt, DM input, and DM output, we propose an *Uncertainty Guided Cross Attention* (UGCA) module. In this module, the prompt serves as the query to guide the attention mechanism, enabling the model to focus on contextually relevant regions of both the DM input and output. Finally, we replace the SAM decoder with a synthesis decoder, which is trained together with the UGCA module. The major contributions of this work are summarized as follows:

- We have proposed a new medical image synthesis model that cascades the DM with SAM via uncertainty-guided prompt generation. The DM-SAM cascade has not been explored before.
- Unlike the previous method that adapts SAM for synthesis by simply borrowing the SAM encoder and instead of naive cascading, we have proposed to exploit the prompting capability of SAM for cascading, where an automated method for prompt generation is developed based on the uncertainty of DM output.
- We have proposed the UGCA module that achieves effective interaction between the prompt, DM input, and DM output to produce the final synthesis result.
- Comprehensive experiments performed on two public datasets demonstrate the benefit of the proposed method over existing methods.

2 Related Works

2.1 Medical Image Synthesis

Conventional medical image synthesis methods often rely on models with handcrafted features to convert images between different modalities [16,28,29]. The development of DL techniques has led to more intelligent and effective synthesis methods [20]. In [32], pGAN, a *generative adversarial network* (GAN) method, is proposed for multi-contrast MRI synthesis. Dalmaz et al. [5] propose ResViT, a model for multi-modal medical image synthesis that combines the contextual sensitivity of vision transformers [7], the precision of convolutional operations, and the realism of adversarial learning. Cao et al. [4] propose an autoencoder-driven multi-modal collaborative learning framework for medical image synthesis. Liu et al. [23] formulate missing data synthesis as a sequence-to-sequence learning problem and propose a multi-contrast multi-scale Transformer for the problem. Atli et al. [2] propose I2I-Mamba for multi-modal medical image synthesis, where selective state space modeling (SSM) [9,45] is employed to efficiently capture long-range context while preserving local precision. Besides these models, DMs [10] have gained significant attention and success in medical image synthesis [17,25,38,44]. They excel in generating realistic image appearances compared to other models. However, these DM methods, as well as other synthesis models, often struggle to accurately synthesize pathological regions, such as lesions, tumors, and other abnormal structures, which are critical for the accurate assessment of medical conditions. Therefore, addressing the limitations of existing synthesis methods in capturing abnormalities is a key challenge in advancing medical image synthesis techniques [34].

2.2 SAM and Its Medical Variants

SAM has emerged as a powerful image segmentation model [19] for a wide range of computer vision applications [1,41]. Unlike conventional segmentation models, it is trained on a huge amount of diverse segmentation tasks, and it requires various prompt inputs, including points, boxes, and even textual descriptions, to indicate the object to segment [1,41]. SAM has also been adapted to medical image segmentation for its ability to handle various types of segmentation tasks efficiently and accurately [8,21]. In particular, MedSAM [24] is developed with a large-scale medical image dataset comprising over 1.5 million image-mask pairs across ten different imaging modalities. In addition to the segmentation task, SAM has been recently adapted to medical image synthesis. In [12], the SAM image encoder is used to extract multiscale semantic features from the source image, and a new decoder is attached to synthesize target modality images.

3 Method

3.1 Problem Formulation and Method Overview

Our goal is to train a synthesis model that generates a target modality image from a source modality image, where paired source and target images are acquired for model training. An overview of the proposed method is shown in Fig. 1, which aims to combine the strength of the DM and SAM via informative prompting. Our method first trains a DM that takes the source image and produces an initial estimate of the target image. Then, we estimate the uncertainty of the DM output with repeated noise sampling, where the resulting prediction standard deviation represents the uncertainty. SAM further refines the average DM output with special attention to regions with high uncertainty, where the uncertainty information serves as prompt, and both the source image and the DM output are provided to SAM for the refinement. These difference sources of information are fused in the proposed UGCA module, and the final target image is synthesized from the fused features.

3.2 DM Synthesis and Uncertainty-Guided Prompt Generation

The DM starts the synthesis process with random noise and a conditional image, i.e., the source image. It can produce diverse synthesis results if the sampled random noise is different. We have empirically observed that for regions that are

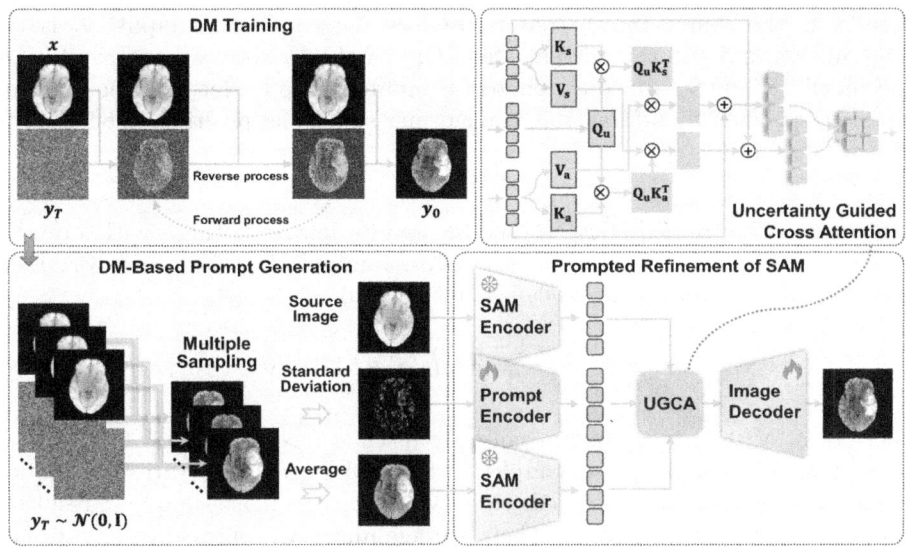

Fig. 1. An overview of our method. Our method cascades the DM with SAM, where the uncertainty of DM output serves as the prompt for SAM. The UGCA module allows interaction between the prompt, source image, and DM output for the final synthesis.

relatively easy to synthesize, e.g., normal tissue, the synthesis results across different random noise samples tend to be similar, whereas for pathological regions with high appearance variability, the DM output tends to differ from each other with different noise samples. Therefore, the DM provides a natural capability of quantifying the output uncertainty by repeatedly running the DM with multiple random noise samples [39], and this allows automated generation of prompts for SAM to highlight pathological regions for further refinement of the synthesis result. The detailed training and inference procedures of the DM in our work are presented below.

Training. The forward diffusion process in DM training is a Markovian process that gradually adds Gaussian noise to the target image y_0 over T iterations. With this, the noisy image $\hat{y}_t$ at the t-th step in the forward process can be written as

$$\hat{y}_t = \sqrt{\gamma_t} y_0 + \sqrt{1-\gamma_t}\epsilon, \tag{1}$$

where $\epsilon \sim \mathcal{N}(0, I)$ is a random noise term and $\gamma_t = \prod_{s=1}^{t} \alpha_s$ with α_s being the noise schedule parameters. DMs seeks to estimate the noise and thus gradually remove it during the reverse process to recover the target image y_0. Like [10], we use the PixelCNN++ [31] backbone, which is a U-Net structure [27] based on Wide ResNet [40], to estimate the noise with an L2 loss:

$$\mathcal{L} = \|f_\theta(\mathbf{x}, \hat{y}_t, t) - \epsilon\|_2^2. \tag{2}$$

Here, **x** is the source image that is used as the conditional input, $\hat{\mathbf{y}}_t$ is the noisy image, and t is the current step. $f_\theta(\cdot)$ is the U-Net parameterized by θ. Specifically, **x** and $\hat{\mathbf{y}}_t$ are concatenated as inputs to the U-Net, and the diffusion time **t** is specified by adding the Transformer sinusoidal position embedding to each residual block.

Inference. The trained DM can then be used for image synthesis with a reverse process. The reverse process begins with a random Gaussian noise $\mathbf{y}_T$. Each iteration of the reverse process can be formulated as

$$\mathbf{y}_{t-1} = \frac{1}{\sqrt{\alpha_t}}\left(\mathbf{y}_t - \frac{1-\alpha_t}{\sqrt{1-\gamma_t}} f_\theta(\mathbf{x}, \mathbf{y}_t, t)\right) + \sqrt{1-\alpha_t}\epsilon_t, \quad (3)$$

where $\epsilon_t \sim \mathcal{N}(0, I)$ is a random noise term when $t > 1$ and $\epsilon_t = 0$ otherwise. Since different random noise samples in the DM can lead to varied prediction results, we perform multiple predictions with repeated noise sampling, resulting in multiple synthesized images. Their average image $\mathbf{y}_0^{\text{avg}}$ provides the estimate of the target image and the standard deviation $\mathbf{y}_0^{\text{std}}$ of these synthesized images naturally quantifies the estimation uncertainty. Mathematically, they are calculated as

$$\mathbf{y}_0^{\text{avg}} = \frac{1}{N_p}\sum_{p=1}^{N_p}\mathbf{y}_0^p \quad \text{and} \quad \mathbf{y}_0^{\text{std}} = \sqrt{\frac{1}{N_p}\sum_{i=1}^{N}(\mathbf{y}_0^i - \mathbf{y}_0^{\text{avg}})^2}, \quad (4)$$

where $\mathbf{y}_0^p$ is the synthesized image associated with the p-th noise sample and N_p is the number of predictions. In this work, N_p is set to 8.

The uncertainty map highlights regions with higher prediction variability. These regions are likely difficult to synthesize with poorer synthesis quality. Therefore, they are used as prompt for subsequent refinement of the DM output. Figure 2 shows an example of the average DM synthesis result and uncertainty map, where a large brain tumor is present. Eight repeated predictions are produced, and they are also shown in Fig. 2 together with the source image and the gold standard target image. In the tumor core and its adjacent regions, the synthesis result is more different from the target image than in other areas, and these regions are highlighted by the uncertainty map.

3.3 Prompted Refinement of SAM

The average DM output $\mathbf{y}_0^{\text{avg}}$ is then refined by a SAM-based image synthesis model with the prompt $\mathbf{y}_0^{\text{std}}$ and the source image **x**. This model consists of a SAM image encoder, a prompt encoder, the UGCA module, and an image decoder. **x** and $\mathbf{y}_0^{\text{avg}}$ are fed into the SAM encoder, whereas $\mathbf{y}_0^{\text{std}}$ is fed into the prompt encoder. The encoded results are fused in the UGCA module. Finally, the image decoder synthesizes the target image based on the UGCA output.

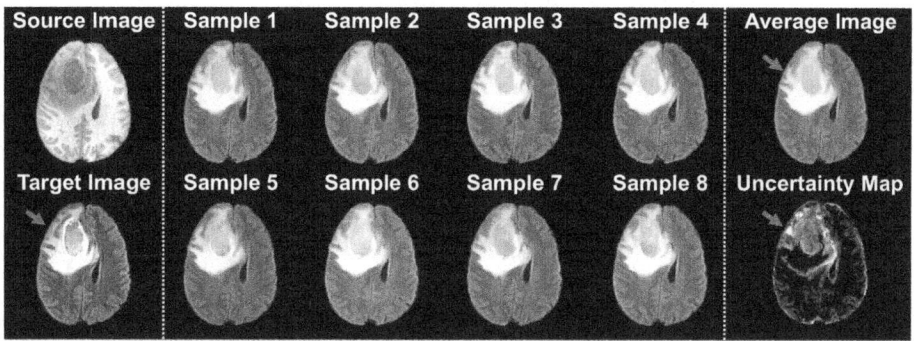

Fig. 2. Uncertainty-guided prompt generation based on DM output. Multiple synthesis results are produced by the DM with repeated noise sampling. The average result gives the estimate of the target image, whereas the prediction standard deviation represents uncertainty, which is used as prompt. The source image and gold standard target image are shown for reference.

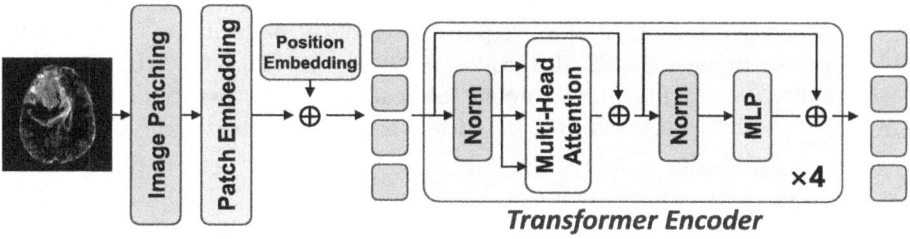

Fig. 3. The network architecture of the prompt encoder.

Image Encoder. The SAM image encoder is used solely for feature extraction, and its weights are frozen during the training process. This work uses the encoder of MedSAM [24], a SAM variant that is more suitable for medical images. $\mathbf{x}$ and $\mathbf{y}_0^{avg}$ are separately sent into the SAM encoder.

Prompt Encoder. The prompt encoder is constructed based on a four-layer vision Transformer [7]. Specifically, the prompt is first divided into non-overlapping patches. Each patch is flattened into a vector and then linearly projected to obtain the patch embeddings. Positional embeddings are subsequently added to these patch embeddings to incorporate spatial information. The resulting embeddings are passed through the four-layer Transformer encoder, which consists of Layer Normalization, *Multi-Head Self-Attention* (MHSA), and a *Multi-Layer Perceptron* (MLP) module. The structure of the prompt encoder is summarized in Fig. 3.

UGCA. We propose the UGCA module so that the prompt information can be effectively integrated with the image information. To encourage the model to focus on high uncertainty areas, in UGCA the uncertainty map queries the source image and DM output separately. Moreover, to allow the three sources of information to simultaneously interact, the prompt, source image, and DM output are mixed in the cross-attention. Mathematically, the query vector of the uncertainty map is denoted by $\mathbf{Q}_u$, the key and value vectors of the source image are denoted by $\mathbf{K}_s$ and $\mathbf{V}_s$, respectively, and the key and value vectors of the average DM output are denoted by $\mathbf{K}_a$ and $\mathbf{V}_a$, respectively. The output $\mathbf{F}$ of UGCA is computed as

$$\mathbf{F} = \text{Concat}\left(\text{Softmax}\left(\frac{\mathbf{Q}_u \mathbf{K}_s^T}{\sqrt{d_k}}\right)\mathbf{V}_a + \mathbf{E}_a, \text{Softmax}\left(\frac{\mathbf{Q}_u \mathbf{K}_a^T}{\sqrt{d_k}}\right)\mathbf{V}_s + \mathbf{E}_s\right), \quad (5)$$

where $\sqrt{d_k}$ represents the dimension of the key vectors, and $\mathbf{E}_a$ and $\mathbf{E}_s$ are the feature representations obtained from the image encoder of the source image and DM output, respectively. Equation (5) is expected to highlight the image information relevant to the uncertain areas, and its output fused feature is then fed into the image decoder.

Image Decoder. As the SAM decoder is trained for segmentation, it is not suitable for our synthesis task that requires prediction of continuous intensities. Therefore, we replace the SAM decoder by the decoder of VQ-VAE [36], which is originally developed to predict image intensities. The image decoder is trained together with the UGCA module based on the *mean squared error* (MSE) loss.

3.4 Implementation Details

All images are normalized by clipping the intensity values to the range between the 0.5th and 99.5th percentiles, followed by rescaling to the range of [0,1]. To meet the input requirements of SAM, 3D images are split into 2D slices and resized to 512 × 512, and the channels are repeated three times for input consistency. For the DM, the number of training iterations is set to 200,000. The Adam optimizer [18] is used with a batch size of 16, and the learning rate is set to 8×10^{-5}. For the prompted refinement of SAM, the number of training epochs is set to 50. The Adam optimizer [18] is used with a batch size of 16 and default hyperparameters. The codes of our method are available at https://github.com/PangHaowen-hub/DM-SAM.

4 Results

4.1 Data Description

We evaluated our method on two public datasets containing data from patients with structural abnormalities: BraSyn [22] and SynthRAD [11,35]. The BraSyn dataset includes brain MRI scans of 1,470 patients, comprising aligned T1w, T2w, FLAIR, and T1CE images. We used the T1w and FLAIR images for the T1-to-FLAIR synthesis task. The SynthRAD dataset includes brain images of 180 patients, comprising aligned MRI and CT scans. We used the MRI and CT scans for the MRI-to-CT synthesis task.

Table 1. The means and standard deviations of the PSNR and SSIM between synthesized and real images for the T1-to-FLAIR and MRI-to-CT synthesis tasks. The best results are highlighted in bold. Asterisks indicate statistically significant differences between the results of DM-SAM and each competing method based on the Wilcoxon signed-rank test: *** $p < 0.001$.

Model	T1-to-FLAIR		MRI-to-CT	
	PSNR	SSIM	PSNR	SSIM
Pix2Pix	$21.77^{***}_{\pm 2.34}$	$84.37^{***}_{\pm 4.08}$	$23.49^{***}_{\pm 3.33}$	$79.98^{***}_{\pm 9.78}$
ResViT	$21.91^{***}_{\pm 2.39}$	$84.75^{***}_{\pm 4.54}$	$25.03^{***}_{\pm 2.76}$	$84.67^{***}_{\pm 7.02}$
PTNet	$21.20^{***}_{\pm 2.37}$	$84.24^{***}_{\pm 4.40}$	$22.86^{***}_{\pm 2.80}$	$80.07^{***}_{\pm 8.37}$
Palette	$23.15^{***}_{\pm 2.98}$	$88.74^{***}_{\pm 3.72}$	$25.24^{***}_{\pm 2.73}$	$85.93^{***}_{\pm 6.05}$
Cold Diffusion	$22.12^{***}_{\pm 2.19}$	$85.72^{***}_{\pm 4.51}$	$21.57^{***}_{\pm 1.76}$	$78.60^{***}_{\pm 6.76}$
DM-SAM	**$24.14_{\pm 2.73}$**	**$89.98_{\pm 3.48}$**	**$25.87_{\pm 2.59}$**	**$86.65_{\pm 5.54}$**

Experiments were performed independently on BraSyn and SynthRAD, and for each dataset, the preprocessed images were split into training, validation, and test sets. For BraSyn, the training/validation/test set consisted of images of 1,251/109/110 patients, respectively; for SynthRAD, the training/validation/test set consisted of images of 108/36/36 patients, respectively.

4.2 Comparison with Existing Image Synthesis Methods

We compared our method with a number of existing image synthesis methods, including Pix2Pix [14], ResViT [5], PTNet [43], Palette [30], and Cold Diffusion [3]. Pix2Pix is a GAN-based image synthesis method using a convolutional neural network architecture. ResViT is a multimodal medical image synthesis model that combines vision Transformers with convolutional operators and adversarial learning. PTNet is a brain MRI synthesis method that consists of Transformer layers, skip connections, and a multi-scale pyramid representation. Palette is a simple yet effective image synthesis method based on a diffusion model. Cold Diffusion is a diffusion model that does not rely on typical Gaussian noise addition; instead, it operates with deterministic degradations to generate images. All these competing methods used the same training, validation, and test sets as our method.

We first qualitatively evaluated our method, and for convenience it is referred to as DM-SAM due to the cascade of the DM and SAM. Figure 4 presents four examples of T1-to-FLAIR and MRI-to-CT synthesis results. The synthesized images of DM-SAM are more consistent with the real images than those from the competing methods. In the abnormal regions (highlighted by red arrows), DM-SAM produces higher-quality results that agree with the real image, whereas the competing methods tend to produce distorted structures.

Next, we quantitatively compared DM-SAM with the competing methods by calculating the *Peak Signal-to-Noise Ratio* (PSNR) and *Structural Similarity*

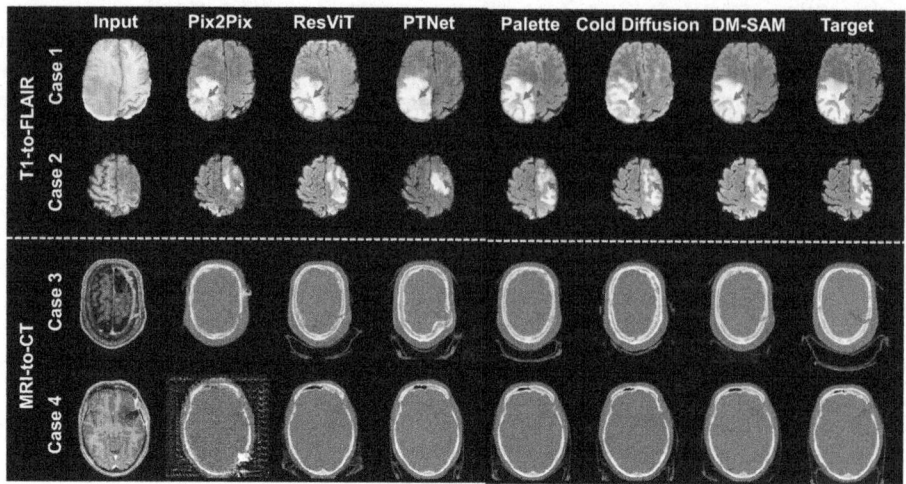

Fig. 4. Four examples of T1-to-FLAIR and MRI-to-CT synthesis results, shown together with the input image and real target image for reference. Note the pathological areas highlighted by red arrows for comparison. (Color figure online)

Table 2. The means and standard deviations of the PSNR and SSIM of the synthesis result within the tumor region for the T1-to-FLAIR synthesis task. The best results are highlighted in bold. Asterisks indicate statistically significant differences between the results of DM-SAM and each competing method based on the Wilcoxon signed-rank test: *** $p < 0.001$.

Model	PSNR	SSIM
Pix2Pix	$14.30^{***}_{\pm 3.69}$	$64.56^{***}_{\pm 16.61}$
ResViT	$16.08^{***}_{\pm 3.53}$	$69.95^{***}_{\pm 15.57}$
PTNet	$14.34^{***}_{\pm 3.72}$	$65.97^{***}_{\pm 16.12}$
Palette	$15.70^{***}_{\pm 4.06}$	$72.78^{***}_{\pm 14.18}$
Cold Diffusion	$15.81^{***}_{\pm 3.83}$	$71.40^{***}_{\pm 17.28}$
DM-SAM	$\mathbf{17.35}_{\pm 3.69}$	$\mathbf{77.77}_{\pm 13.06}$

Index Measure (SSIM) between the synthesized and real images. The Wilcoxon signed-rank test was also performed for the comparison. The results on the BraSyn and SynthRAD datasets are presented in Table 1. DM-SAM outperforms the other methods in all cases, achieving higher PSNR and SSIM. Moreover, in all cases the improvement of DM-SAM is highly statistically significant ($p < 0.001$).

As existing synthesis methods struggle with pathological regions, we further analyzed the synthesis results within such regions. Specifically, we segmented the tumor regions for the test set of BraSyn. As the training set of the BraSyn dataset provides tumor annotations, we trained a state-of-the-art segmentation model nnU-Net [13] for brain tumor segmentation based on the annotations. Note

Table 3. Ablation studies performed on the BraSyn dataset for investigating the individual benefit of each major component of DM-SAM. The means and standard deviations are presented and the best results are in bold. Asterisks indicate statistically significant differences between the results of DM-SAM and each setting based on the Wilcoxon signed-rank test: *** $p < 0.001$

Prompt generation	Repeated sampling	DM	SAM	PSNR	SSIM
✓	✓	✓	✓	$\mathbf{24.14_{\pm 2.73}}$	$\mathbf{89.98_{\pm 3.48}}$
✗	✓	✓	✓	$23.89^{***}_{\pm 2.69}$	$89.46^{***}_{\pm 3.50}$
✗	✗	✓	✓	$23.69^{***}_{\pm 2.66}$	$88.82^{***}_{\pm 3.64}$
✗	✗	✗	✓	$23.68^{***}_{\pm 2.60}$	$88.75^{***}_{\pm 3.61}$
✗	✓	✓	✗	$23.93^{***}_{\pm 2.87}$	$89.83^{***}_{\pm 3.54}$

that because it is challenging to define the boundary of abnormal regions for the MRI-to-CT task, the analysis within pathological regions was not applied for the SynthRAD dataset. The PSNR and SSIM within the tumor are summarized in Table 2 for each method. DM-SAM significantly outperforms the competing models in terms of both PSNR and SSIM. The margin between DM-SAM and the other methods is greater in the tumor regions than across the whole brain (compare Table 2 with Table 1). This further highlights the benefit of our method in handling the synthesis of abnormal regions.

4.3 Ablation Study

To confirm the individual benefit of each major component of DM-SAM, including the prompt generation, repeated sampling, DM, and SAM, we performed ablation studies on the BraSyn dataset. In the ablation studies, the overall pipeline of DM-SAM was revised and retrained using the same data split as described in Sect. 4.1. The detailed results are shown in Table 3 and described below.

Prompt Generation. First, we removed the prompt generation for SAM, where the source image and average DM output were separately fed into the SAM encoder, then concatenated, and finally sent into the decoder. Table 3 (second row) shows that the removal of prompt generation results in reduced PSNR and SSIM.

Repeated Sampling. In addition to the removal of prompt generation, we further removed the repeated sampling in DM. In this case, only one image was synthesized by the DM. The single DM output and the source image were again separately fed into the SAM encoder without prompt generation. The results (third row in Table 3) indicate further decline in the synthesis quality, which highlights the importance of repeated sampling.

DM. Based on the configuration above, we additionally removed the DM, where the source image is directly fed into the frozen SAM encoder to obtain image features, and the features are passed to the image decoder to synthesize the target image. Table 3 (fourth row) shows that the removal of the DM further results in slightly lower PSNR and SSIM, which indicates the utility of the DM in image synthesis.

SAM. Based on the configuration that removed the prompt generation, we additionally removed SAM to investigate its individual benefit, where only the DM was used for synthesis with repeated sampling. As shown in Table 3 (fifth row), the removal of SAM leads to decreased PSNR and SSIM compared with our method (first row). Interestingly, its result is better than the use of both DM and SAM without prompt generation (second row). This further demonstrates the importance of prompt generation, where simple addition of SAM may not lead to improved result; it is important to provide SAM with informative prompt to unleash its power.

5 Discussion and Conclusion

In this study, we have presented an approach to enhancing the robustness of medical image synthesis, particularly in the presence of anomalies such as lesions. Our method cascades a DM with SAM by developing an uncertainty-guided prompt generation framework, where the prompt allows attention to challenging regions and thus improves the synthesis quality. SAM refines the DM output based on the prompt and input image. To integrate the prompt with other image information, we have designed a UGCA module that facilitates meaningful interactions between the prompt, input image, and DM output. Experimental results on two publicly available datasets demonstrate that our approach outperforms existing state-of-the-art methods in synthesizing high-quality medical images, especially for regions with complex anomalies.

Several challenges remain that warrant future investigation. First, besides the two tasks considered in our experiments, there are other clinically relevant image synthesis tasks that involve anomalies, such as the synthesis of cerebral blood volume maps and diffusion-weighted MRI from standard MRI. Future research will explore these additional tasks. Second, in our method the SAM encoder is used directly without any tailoring. It is possible that image synthesis may require appropriate modification of the complete encoder for feature extraction, and this can be investigated in future work to optimize the use of SAM in medical image synthesis. Additionally, our method currently requires repeated sampling to generate uncertainty maps, resulting in relatively long inference times. Future work can address this problem by accelerating the DM process, such as using DDIM [33], thereby reducing computational costs.

Acknowledgement. This work is supported by the Beijing Municipal Natural Science Foundation (7242273).

References

1. Ali, M., et al.: A review of the Segment Anything Model (SAM) for medical image analysis: accomplishments and perspectives. Computerized Medical Imaging Graph., 102473 (2024)
2. Atli, O.F., Kabas, B., Arslan, F., Yurt, M., Dalmaz, O., Çukur, T.: I2I-Mamba: multi-modal medical image synthesis via selective state space modeling. arXiv preprint arXiv:2405.14022 (2024)
3. Bansal, A., et al.: Cold diffusion: Inverting arbitrary image transforms without noise. Advances in Neural Information Processing Systems 36 (2024)
4. Cao, B., Bi, Z., Hu, Q., Zhang, H., Wang, N., Gao, X., Shen, D.: Autoencoder-driven multimodal collaborative learning for medical image synthesis. Int. J. Comput. Vision **131**(8), 1995–2014 (2023)
5. Dalmaz, O., Yurt, M., Çukur, T.: ResViT: residual vision transformers for multimodal medical image synthesis. IEEE Trans. Med. Imaging **41**(10), 2598–2614 (2022)
6. Dayarathna, S., Islam, K.T., Uribe, S., Yang, G., Hayat, M., Chen, Z.: Deep learning based synthesis of MRI, CT and PET: Review and analysis. Med. Image Anal. **92**, 103046 (2024)
7. Dosovitskiy, A.: An image is worth 16x16 words: Transformers for image recognition at scale. arXiv preprint arXiv:2010.11929 (2020)
8. Gong, S., Zhong, Y., Ma, W., Li, J., Wang, Z., Zhang, J., Heng, P.A., Dou, Q.: 3DSAM-adapter: holistic adaptation of SAM from 2D to 3D for promptable tumor segmentation. Med. Image Anal. **98**, 103324 (2024)
9. Gu, A., Dao, T.: Mamba: Linear-time sequence modeling with selective state spaces. arXiv preprint arXiv:2312.00752 (2023)
10. Ho, J., Jain, A., Abbeel, P.: Denoising diffusion probabilistic models. Adv. Neural. Inf. Process. Syst. **33**, 6840–6851 (2020)
11. Huijben, E.M., et al.: Generating synthetic computed tomography for radiotherapy: SynthRAD2023 challenge report. Med. Image Anal. **97**, 103276 (2024)
12. Huo, J., Ourselin, S., Sparks, R.: SAM-I2I: unleash the power of segment anything model for medical image translation. arXiv preprint arXiv:2411.12755 (2024)
13. Isensee, F., Jaeger, P.F., Kohl, S.A., Petersen, J., Maier-Hein, K.H.: nnU-Net: a self-configuring method for deep learning-based biomedical image segmentation. Nat. Methods **18**(2), 203–211 (2021)
14. Isola, P., Zhu, J.Y., Zhou, T., Efros, A.A.: Image-to-image translation with conditional adversarial networks. In: Proceedings of the IEEE Conference on Computer Vision and Pattern Recognition, pp. 1125–1134 (2017)
15. James, A.P., Dasarathy, B.V.: Medical image fusion: a survey of the state of the art. Inf. Fusion **19**, 4–19 (2014)
16. Jog, A., Roy, S., Carass, A., Prince, J.L.: Magnetic resonance image synthesis through patch regression. In: 2013 IEEE 10th International Symposium on Biomedical Imaging, pp. 350–353. IEEE (2013)
17. Kazerouni, A., Aghdam, E.K., Heidari, M., Azad, R., Fayyaz, M., Hacihaliloglu, I., Merhof, D.: Diffusion models in medical imaging: a comprehensive survey. Med. Image Anal. **88**, 102846 (2023)
18. Kingma, D.P., Ba, J.: Adam: A method for stochastic optimization. arXiv preprint arXiv:1412.6980 (2014)
19. Kirillov, A., et al.: Segment anything. In: Proceedings of the IEEE/CVF International Conference on Computer Vision, pp. 4015–4026 (2023)

20. Kong, L., Lian, C., Huang, D., Hu, Y., Zhou, Q., et al.: Breaking the dilemma of medical image-to-image translation. Adv. Neural. Inf. Process. Syst. **34**, 1964–1978 (2021)
21. Lei, W., Xu, W., Li, K., Zhang, X., Zhang, S.: MedLSAM: localize and segment anything model for 3D CT images. Med. Image Anal. **99**, 103370 (2025)
22. Li, H.B., et al.: The brain tumor segmentation (BraTS) challenge 2023: Brain MR image synthesis for tumor segmentation (BraSyn). ArXiv (2023)
23. Liu, J., Pasumarthi, S., Duffy, B., Gong, E., Datta, K., Zaharchuk, G.: One model to synthesize them all: Multi-contrast multi-scale transformer for missing data imputation. IEEE Trans. Med. Imaging **42**(9), 2577–2591 (2023)
24. Ma, J., He, Y., Li, F., Han, L., You, C., Wang, B.: Segment anything in medical images. Nat. Commun. **15**(1), 654 (2024)
25. Meng, X., Sun, K., Xu, J., He, X., Shen, D.: Multi-modal modality-masked diffusion network for brain MRI synthesis with random modality missing. IEEE Trans. Med. Imaging (2024)
26. Pang, H., et al.: NCCT-CECT image synthesizers and their application to pulmonary vessel segmentation. Comput. Methods Programs Biomed. **231**, 107389 (2023)
27. Ronneberger, O., Fischer, P., Brox, T.: U-net: convolutional networks for biomedical image segmentation. In: International Conference on Medical Image Computing and Computer-Assisted Intervention, pp. 234–241. Springer (2015)
28. Roy, S., Carass, A., Prince, J.: A compressed sensing approach for MR tissue contrast synthesis. In: Székely, G., Hahn, H.K. (eds.) IPMI 2011. LNCS, vol. 6801, pp. 371–383. Springer, Heidelberg (2011). https://doi.org/10.1007/978-3-642-22092-0_31
29. Roy, S., Carass, A., Prince, J.L.: Magnetic resonance image example-based contrast synthesis. IEEE Trans. Med. Imaging **32**(12), 2348–2363 (2013)
30. Saharia, C., et al.: Palette: image-to-image diffusion models. In: ACM SIGGRAPH 2022 Conference Proceedings, pp. 1–10 (2022)
31. Salimans, T., Karpathy, A., Chen, X., Kingma, D.P.: Pixelcnn++: improving the pixelcnn with discretized logistic mixture likelihood and other modifications. arXiv preprint arXiv:1701.05517 (2017)
32. Shamsolmoali, P., Zareapoor, M., Granger, E., Zhou, H., Wang, R., Celebi, M.E., Yang, J.: Image synthesis with adversarial networks: a comprehensive survey and case studies. Inf. Fusion **72**, 126–146 (2021)
33. Song, J., Meng, C., Ermon, S.: Denoising diffusion implicit models. In: International Conference on Learning Representations (2021)
34. Sun, L., Wang, J., Huang, Y., Ding, X., Greenspan, H., Paisley, J.: An adversarial learning approach to medical image synthesis for lesion detection. IEEE J. Biomed. Health Inform. **24**(8), 2303–2314 (2020)
35. Thummerer, A., et al.: SynthRAD2023 grand challenge dataset: generating synthetic CT for radiotherapy. Med. Phys. **50**(7), 4664–4674 (2023)
36. Van Den Oord, A., Vinyals, O., et al.: Neural discrete representation learning. Advances in Neural Information Processing Systems 30 (2017)
37. Wang, T., et al.: A review on medical imaging synthesis using deep learning and its clinical applications. J. Appl. Clin. Med. Phys. **22**(1), 11–36 (2021)
38. Wang, Z., et al.: Mutual information guided diffusion for zero-shot cross-modality medical image translation. IEEE Trans. Med. Imaging (2024)
39. Wolleb, J., Sandkühler, R., Bieder, F., Valmaggia, P., Cattin, P.C.: Diffusion models for implicit image segmentation ensembles. In: International Conference on Medical Imaging with Deep Learning, pp. 1336–1348. PMLR (2022)

40. Zagoruyko, S.: Wide residual networks. arXiv preprint arXiv:1605.07146 (2016)
41. Zhang, L., Deng, X., Lu, Y.: Segment Anything Model (SAM) for medical image segmentation: a preliminary review. In: 2023 IEEE International Conference on Bioinformatics and Biomedicine (BIBM), pp. 4187–4194. IEEE (2023)
42. Zhang, T., et al.: BreathVisionNet: a Pulmonary-function-guided CNN-Transformer hybrid model for expiratory CT image synthesis. Computer Methods and Programs in Biomedicine, p. 108516 (2024)
43. Zhang, X., et al.: PTNet: a high-resolution infant MRI synthesizer based on transformer. arXiv preprint arXiv:2105.13993 (2021)
44. Zhou, Y., et al.: Cascaded multi-path shortcut diffusion model for medical image translation. Med. Image Anal. **98**, 103300 (2024)
45. Zhu, L., Liao, B., Zhang, Q., Wang, X., Liu, W., Wang, X.: Vision mamba: efficient visual representation learning with bidirectional state space model. arXiv preprint arXiv:2401.09417 (2024)

DIReCT: Domain-Informed Rectified Flow for Controllable Brain MRI to PET Translation

Tuo Liu[1,3], Haifeng Wang[1,3], Heng Chang[1,3], Fan Wang[2,3], Chunfeng Lian[1,3,4(✉)], and Jianhua Ma[2,3,4(✉)]

[1] School of Mathematics and Statistics, Xian Jiaotong University, Xi'an, China
[2] Key Laboratory of Biomedical Information Engineering of Ministry of Education, School of Life Science and Technology, Xi'an Jiaotong University, Xi'an, China
[3] Research Center for Intelligent Medical Equipment and Devices (IMED), Xi'an Jiaotong University, Xi'an, China
[4] Pazhou Lab (Huangpu), Guangzhou, China
{chunfeng.lian,jhma}@xjtu.edu.cn

Abstract. Recent advancements in generative learning have enabled PET image synthesis from relatively more accessible MRI scans, offering a safer, cost-effective, and scalable alternative to traditional PET imaging, e.g., for Alzheimer's disease (AD) diagnosis. However, current MRI-to-PET translation methods face limitations in controllability and fidelity, often failing to capture personalized metabolic activations and fine-grained structural details in critical regions. To address these challenges, we propose a novel controllable MRI-to-PET translation framework, termed DIReCT, which leverages rectified flow to generate high-fidelity PET images tailored to downstream diagnostic and analytical needs. By injecting cross-modal guidance from a pretrained vision-language model (BiomedCLIP), DIReCT incorporates both common imaging knowledge and individualized clinical information to enhance the personalization of PET synthesis. Extensive experiments on the ADNI dataset demonstrate that DIReCT significantly outperforms existing methods across various image quality metrics. Notably, the synthesized FDG-PET images by DIReCT achieve analytical performance comparable to real FDG-PET scans, excelling in capturing AD-related pathological features for reliable group comparisons and personalized diagnosis.

Keywords: Controllable Cross-Modal Synthesis · Domain-Knowledge Encoding · PET Imaging

1 Introduction

The diagnosis of brain diseases relies on various advanced imaging tools, with magnetic resonance imaging (MRI) and positron emission tomography (PET) being widely used [1]. Structural magnetic resonance imaging, with its excellent resolution, effectively detects anatomical changes, such as localized brain atrophy, while fluorodeoxyglucose PET (FDG-PET) dynamically tracks changes in glucose metabolism. In Alzheimer's disease (AD) and other related brain disorders, glucose uptake in specific brain regions significantly decreases. This metabolic feature renders PET highly sensitive and accurate for the early detection and differential diagnosis of the disease [3,12,13]. However, the widespread adoption of PET imaging is constrained by its high cost and radiation exposure, which limit its implementation in many medical centers worldwide [10]. In contrast, MRI is non-invasive and more accessible, though it still lags behind PET in observing functional brain activities [6]. In this context, generating PET images from MRI has emerged as a promising solution. However, existing MRI-to-PET conversion methods exhibit significant limitations. The diffusion model (DDPM), while generating target images through gradual diffusion and reverse denoising, suffers from low inference efficiency, as it requires hundreds to thousands of steps for the generation process [8]. Although the generated images are fine-grained, the fidelity of local features remains limited, making it challenging to accurately capture subtle metabolic changes. The improved DDIM model accelerates the generation process through a deterministic approach, but it reduces generation diversity and inadequately represents pathological features [21]. Additionally, the Swin-Unet method, which is based on the Swin Transformer and UNet architecture, performs well in capturing global contextual information, but its high computational complexity and significant hardware resource requirements remain notable drawbacks. Moreover, due to the substantial differences between MRI and PET modalities, the generated images still lack consistency in metabolic features [4]. More importantly, existing methods have failed to fully utilize personalized prior information from patients, leading to insufficient capability in capturing individual pathological features.

To address these challenges, this study proposes DIReCT, an image translation framework based on 3D rectified flow [23] combined with domain knowledge encoding (BiomedCLIP) [24]. DIReCT generates high-quality PET images that are both anatomically consistent and pathologically meaningful by incorporating encoded guidance from common domain knowledge (such as anatomical structures and functional area distributions) and personalized patient information (such as clinical features and medical scores), as shown in Fig. 1. The contributions of our DIReCT are as follows:

1. We propose a new end-to-end MRI-to-PET conversion framework based on the Flow model, enabling the efficient generation of high-quality PET images.
2. By leveraging the BiomedCLIP encoder, text information is transformed into guiding signals during the generation process. This integration of common domain knowledge and personalized patient information provides accurate and individualized guidance for image generation.
3. The implementation of single-step sampling significantly enhances generation efficiency, surpassing the speed of existing methods.
4. The reliability and practical value of the synthesized PET images in medical diagnosis are demonstrated through comprehensive validation using personalized classification and region-level group comparison tasks.

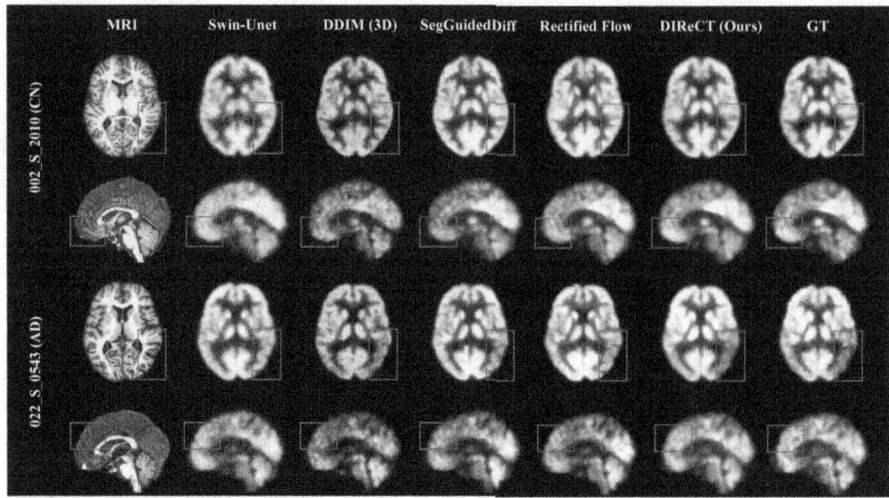

Fig. 1. By comparing the prediction results of Swin-Unet, DDIM (3D), SegGuidedDif, Rectified Flow, and DIReCT (Ours) with real PET images, it can be observed that DIReCT more accurately captures the pixel-level details of real PET images, regardless of whether the subject is cognitively normal (CN) or diagnosed with Alzheimer's disease (AD). This superiority is consistently demonstrated across axial and sagittal views, both in the overall structure and the regions of interest highlighted by the red boxes.

Related Work. Cross-modal MRI-to-PET conversion has evolved from generative adversarial networks (GANs), such as the sketch-to-refinement framework [22], GANDALF for Alzheimer's diagnosis [19], and BiGANs for 3D synthesis [9], to diffusion models that outperform GANs in capturing complex data distributions [16,17,20,25]. Recently, rectified flow [23] has emerged as a more efficient approach, which aligns better with true data manifolds. This progression highlights the shift from GANs to diffusion models and rectified flow as promising directions for MRI-to-PET generation.

2 Proposed Method

2.1 Preliminary

Rectified flow is an improved generative method based on diffusion models, designed to simplify the complex, step-by-step denoising process of traditional diffusion models [23]. By optimizing the generation pathway and leveraging geometric information, it optimizes sampling trajectories to better align with the true data manifold. Unlike traditional diffusion models, which require computationally expensive multi-step sampling, rectified flow significantly reduces the number of sampling steps while maintaining high-quality output. This approach not only accelerates the generation process but also effectively addresses the dual demands for efficiency and fidelity in medical image generation.

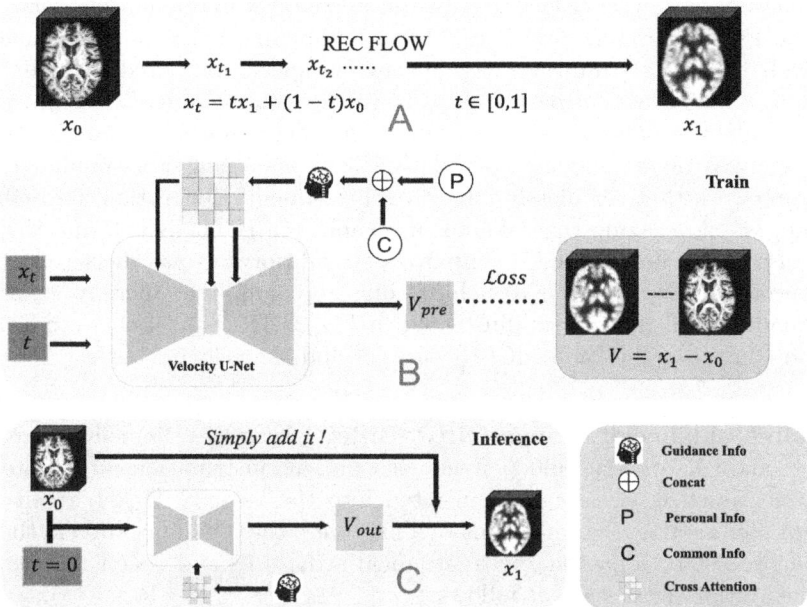

Fig. 2. Overview of DIReCT: (A) Linear interpolation between MRI and PET within the rectified flow framework. (B) The training architecture of the proposed DIReCT model. (C) One-step PET generation during inference through distillation.

BiomedCLIP is a multimodal pretraining model tailored for biomedical applications and is inspired by OpenAI's CLIP (Contrastive Language-Image Pretraining) [24]. It jointly trains biomedical images (e.g., MRI, X-rays, pathology slides) and textual descriptions (e.g., pathology reports, radiology annotations) to achieve semantic alignment between images and text. The core idea is to use a contrastive learning framework to project paired images and text into a shared semantic space, enabling efficient cross-modal retrieval and generation. Additionally, BiomedCLIP integrates personalized patient information (e.g., medical history and clinical scores) to enhance task-specific performance. By combining public-domain knowledge with individualized data, BiomedCLIP significantly enhances adaptability and generative capabilities for biomedical tasks.

2.2 DIReCT

The objective of cross-modal image translation is to establish a mapping from a source modality to a target modality, producing high-quality images that align with the real data while preserving pathological evidence. In the context of medical image generation, current methods face two primary challenges: (1) insufficient translation accuracy, which complicates the precise capture of pathological features; and (2) low generation efficiency, as traditional diffusion models (DMs) are hindered by slow inference speeds due to complex mathematical derivations and multi-step sampling processes, thereby impeding efficient deployment [8]. Although DDIM [21] improves the generation speed compared to DDPM, its ability to capture critical pathological features remains limited, failing to meet the high-fidelity requirements essential for the generation of medical images.

To address these challenges, our DIReCT forms a new cross-modal translation framework that seamlessly integrates high-fidelity generation with efficient sampling. By leveraging the guidance mechanism of BiomedCLIP and the principles of rectified flow, DIReCT employs a straightforward yet effective ordinary differential equation (ODE) to achieve one-step sampling, thereby generating high-quality PET images, as illustrated in Fig. 2. The core architecture of our method comprises the BiomedCLIP encoder and the velocity U-Net.

Domain Guidance of BiomedCLIP. DIReCT leverages BiomedCLIP to integrate domain knowledge and personalized patient information into contextual guidance, which is subsequently injected into the velocity U-Net architecture through a cross-attention mechanism. This contextual guidance directs the generation process to focus on key anatomical structures and lesion regions. The implementation details are as follows:

1. Domain Information Encoding: Common knowledge (e.g., anatomical structures and functional region distributions) and patient-specific data (e.g., clinical scale assessments including CDR, MMSE, GDS, and FAQ) are concatenated as textual input and encoded into high-dimensional vectors using BiomedCLIP. Notably, our approach does not rely on diagnostic labels such as AD or CN.
2. Contextual Integration: The output of BiomedCLIP functions as "context" and is integrated into the downsampling, intermediate, and upsampling modules of the velocity network to guide the generation process.

Rectified Flow. The core of DIReCT is a continuous motion system based on the concept of rectified flow, which constructs a "straight-as-possible" generative trajectory guided by domain-informed guidance. This approach aims to reduce sampling complexity while preserving the fidelity of the generated results. Given two distributions of medical data $\mathbf{x}_0 \sim \mathcal{M}$ (MRI) and $\mathbf{x}_1 \sim \mathcal{P}$ (PET), a rectified flow model is modeled with a learnable velocity field $V_\theta : \mathbb{R}^d \times \mathbb{R}^c \times [0,1] \to \mathbb{R}^d$, where θ represents its learnable parameters. The velocity field V_θ, which is our velocity U-Net, must satisfy the following ODE:

$$\frac{\mathrm{d}}{\mathrm{d}t}\mathbf{x}_t = V_\theta\left(\mathbf{x}_t, \mathbf{c}, t\right), \tag{1}$$

where $\mathbf{x}_t \sim \mathbb{R}^d$ and $\mathbf{c} \sim \mathbb{R}^c$ is the high-dimensional vectors form BiomedCLIP. This ODE governs the transition from the initial state $\mathbf{x}_0 \in \mathcal{M}$ to the final state $\mathbf{x}_1 \in \mathcal{P}$, ensuring that the velocity field $V_\theta\left(\mathbf{x}_t, \mathbf{c}, t\right)$ drives the flow along the desired path. To determine the velocity field V_θ, we need to solve a straightforward least squares regression problem:

$$\min_\theta \int_0^1 \mathbb{E} \left\| (\mathbf{x}_1 - \mathbf{x}_0) - V_\theta(\mathbf{x}_t, \mathbf{c}, t) \right\|^2 dt, \quad \text{with} \quad \mathbf{x}_t = t\mathbf{x}_1 + (1-t)\mathbf{x}_0, \quad (2)$$

where $\mathbf{x}_t$ represents the linear interpolation of $\mathbf{x}_0$ and $\mathbf{x}_1$. In order to train the velocity U-Net, we formulate the loss function based on Eq. 2 as follows:

$$\mathcal{L}(\theta) = \mathbb{E}_{t, \mathbf{x}_0 \sim \mathcal{M}, \mathbf{x}_1 \sim \mathcal{P}} \left\| (\mathbf{x}_1 - \mathbf{x}_0) - V_\theta(\mathbf{x}_t, \mathbf{c}, t) \right\|^2, \quad (3)$$

where $t \sim \mathcal{U}([0, 1])$ is the uniform distribution.

Distillation and One-Step Sampling. A key advantage of rectified flow is its capacity for one-step sampling, enabling DIReCT to generate target PET images directly from initial conditions in a single step. This is achieved through the straight paths learned by the ODE, allowing exact simulation without time discretization. This approach not only reduces computational complexity and improves efficiency but also balances accuracy and speed. Additionally, it excels in batch generation, maintaining consistent performance while processing multiple inputs in parallel, making it ideal for large-scale clinical applications.

To further enhance performance, we introduce a distillation process. After training, this process refines one-step sampling by training the neural network to approximate the relationship between the initial and target data, thereby enabling more accurate and efficient one-step generation. The one-step sampling can be mathematically represented as:

$$\mathbf{x}_1 \approx \mathbf{x}_0 + V_\theta(\mathbf{x}_0, \mathbf{c}, 0), \quad (4)$$

where V_θ is the velocity field learned by our velocity U-Net, $\mathbf{x}_0$ is the initial MRI image, and $\mathbf{x}_1$ is the generated target PET image, as shown in Fig. 2 C.

3 Data and Implementation Details

3.1 Dataset Description

This study utilized two subsets from the Alzheimer's Disease Neuroimaging Initiative (ADNI) project: ADNI-1 and ADNI-2 [14]. Subjects in these subsets were categorized into the following five groups: (1) Alzheimer's Disease (AD), (2) Cognitively Normal (CN), (3) Mild Cognitive Impairment progressing to AD within 36 months (pMCI), (4) Stable Mild Cognitive Impairment (sMCI), and (5) Mild Cognitive Impairment with an unclear progression path (MCI). After removing duplicate subjects across ADNI-1 and ADNI-2, a total of 1,159 paired MRI and PET images were obtained. Following the exclusion of severely defective PET images, the final dataset consisted of 1,129 paired MRI and PET images, including 290 CN, 230 AD, 418 sMCI, 83 pMCI, and 108 MCI.

3.2 Implementation Details

To ensure spatial consistency between MRI and PET images, data preprocessing was conducted as follows: (1) Skull stripping was performed on MRI images using FreeSurfer [5]; (2) each PET scan was linearly registered to its corresponding MRI scan; (3) MRI images were aligned with the standard MNI template using an affine transformation in SPM [11], and the same parameters were applied to the corresponding PET images; and (4) the MRI and PET images were cropped to $160 \times 192 \times 160$ with a voxel spacing of $1 \times 1 \times 1$, and their intensities were normalized to the [0, 1] range using min-max normalization. All models were trained on such images of size $160 \times 192 \times 160$. Additionally, brain regions were segmented into 98 labels using the SynthSeg tool [2] for further analysis.

To ensure a rigorous evaluation, we utilized a dataset comprising 1,129 paired MRI and PET images, which was randomly divided into two subsets: Subset 1 (564 pairs) and Subset 2 (565 pairs). To facilitate cross-validation, the two subsets were alternately used for training and testing. In the first setup, Subset 1 was used for training and validation, while Subset 2 was used for testing and generating synthetic PET (Syn-PET) images. In the second setup, the roles were reversed, with Subset 2 used for training and validation, and Subset 1 for testing and Syn-PET generation. All generated Syn-PET images were subsequently aggregated to form a comprehensive Syn-PET dataset, which was then utilized for downstream analysis.

Experiments were conducted on the PyTorch platform using an NVIDIA RTX A6000 GPU with the Adam optimizer and an initial learning rate of 2.5×10^{-5}. The model was trained for 140 epochs, with knowledge distillation initiated at the 100th epoch. The entire training process required approximately 47 GPU hours on the 48 GB NVIDIA RTX A6000 GPU.

4 Synthesis Results

4.1 Comparison Methods

To comprehensively evaluate the performance of DIReCT, we selected the following baseline methods for comparison: (1) **Swin-Unet** [4], which leverages the global characteristics of Transformers and the symmetric encoder-decoder structure of UNet, demonstrating excellent performance in fine-grained segmentation and high-resolution image generation; (2) **SegGuidedDif** [15], which represents high-quality generative capabilities and serves as a reference for the generation accuracy of DIReCT; (3) **DDIM** (3D), which optimizes generation speed through fast sampling and is compared with DIReCT in terms of efficiency [18]; and (4) **Rectified Flow**, a variant of DIReCT without the domain-informed guidance module, used to validate the role of BiomedCLIP. In all comparative experiments, the baseline methods were trained and evaluated on the same dataset as DIReCT, with hyperparameters adjusted according to their respective characteristics to ensure a fair comparison.

4.2 Qualitative Results

In the qualitative analysis, we compared the PET images generated by different methods with the real PET images in terms of anatomical structures and metabolic features. Our findings indicate that PET images generated by DIReCT exhibit higher fidelity and greater similarity to the ground truth (GT). Specifically, DIReCT accurately captured the metabolic features of key regions in samples 002_S_2010 (CN) and 022_S_0543 (AD), significantly outperforming other methods. In contrast, PET images generated by Swin-Unet deviated further from the GT and exhibited noticeable artifacts. While DDIM (3D) demonstrated fast sampling capabilities, it lacked fidelity in pixel value differences and pathological features. SegGuidedDif improved sampling accuracy compared to DDIM (3D) but suffered from slower inference speed, requiring approximately 5 min for 1,000 sampling steps. The ablation study on Rectified Flow (DIReCT without the domain-informed guidance module) showed that it could restore anatomical structures to a certain extent but failed to accurately capture pathological features. By leveraging domain knowledge from BiomedCLIP, DIReCT successfully focused on critical anatomical and pathological regions, achieving higher consistency with the GT. This validates the importance of domain guidance in improving the fidelity of generated images, as shown in Fig. 1.

Table 1. Quantitative comparison between baseline methods and our DIReCT in the task of FDG-PET translation from MRI.

Methods	SSIM (%) ↑	PSNR ↑	MSE (%) ↓	MAE (%) ↓	Inference Speed ↓
Swin-Unet [4]	73.92	19.82	1.27	5.81	0.46 s
DDIM (3D) [18]	73.34	22.09	0.96	4.92	15.91 s
SegGuidedDif [15]	84.25	23.16	0.61	3.73	318 s
Rectified Flow [23]	85.65	23.74	0.55	3.69	0.31 s
DIReCT (ours)	**87.62**	**24.80**	**0.45**	**3.18**	**0.31 s**

4.3 Quantitative Results

Quantitative analysis demonstrates the superior performance of DIReCT across multiple evaluation metrics, as shown in Table 1. In terms of generation quality, DIReCT achieves the best results across metrics such as Mean Absolute Error (MAE: 3.18%), Mean Squared Error (MSE: 0.45%), Peak Signal-to-Noise Ratio (PSNR: 24.80), and Structural Similarity Index Measure (SSIM: 87.62%). These results indicate that the PET images generated by DIReCT significantly outperform those produced by other methods in terms of detail fidelity and metabolic feature preservation. In terms of inference efficiency, DIReCT achieves an impressive inference time of 0.31 s through single-step sampling, substantially outperforming methods such as SegGuidedDif and DDIM (3D), which require multi-step sampling. Although SegGuidedDif achieves generation quality close to that of DIReCT, its slower inference speed limits its applicability in practical scenarios. Conversely, Swin-Unet exhibits poorer metabolic feature consistency, failing to meet the high-fidelity requirements of medical imaging. Ablation study results further validate the contribution of BiomedCLIP. Removing domain guidance from DIReCT results in a decline in metabolic feature fidelity, underscoring the critical role of domain knowledge in generating pathological features. In summary, the comprehensive advantages of DIReCT in both generation quality and inference efficiency firmly establish its practicality and reliability in the field of medical image generation.

5 Downstream Validation

To comprehensively assess the clinical applicability of synthetic PET (Syn-PET) generated by our DIReCT model, we designed two validation experiments focusing on two key aspects: regional consistency and overall task performance. The first downstream validation experiment focused on brain region segmentation and statistical analyses to evaluate whether metabolic differences between Syn-PET and real PET images were significant, particularly in critical functional regions. These analyses provide strong experimental evidence for the clinical utility of Syn-PET, demonstrating its potential for effective use in medical practice. The second downstream validation experiment involved a classification task to diagnose Alzheimer's disease and related disorders using Syn-PET. This was conducted to verify whether key pathological features relevant to the disease were preserved, ensuring that Syn-PET retains clinically significant markers necessary for accurate diagnosis.

5.1 Evaluation via Regional Significance Analysis

To further validate the anatomical and functional consistency of the generated PET images in local regions, we designed a **significance analysis experiment** based on brain region segmentation. Specifically, MRI images were first used to perform brain region segmentation, extracting key functional regions closely associated with the disease (e.g., hippocampus, temporoparietal junction, and posterior cingulate cortex). The segmented brain region masks were then applied to both real PET and Syn-PET images, and the metabolic values for each brain region were computed. For the AD and CN groups, a paired-sample t-test was conducted to statistically compare the metabolic values of the two types of images, assessing whether the generated Syn-PET accurately reflects the characteristics of real PET in these critical regions, as shown in Fig. 3.

Fluorodeoxyglucose PET (FDG-PET) imaging has demonstrated that reduced glucose metabolism in the temporoparietal regions, frontal lobes, hippocampus, and posterior cingulate cortex is a hallmark of Alzheimer's disease (AD) [14]. By analyzing the data distributions of real PET and Syn-PET in key

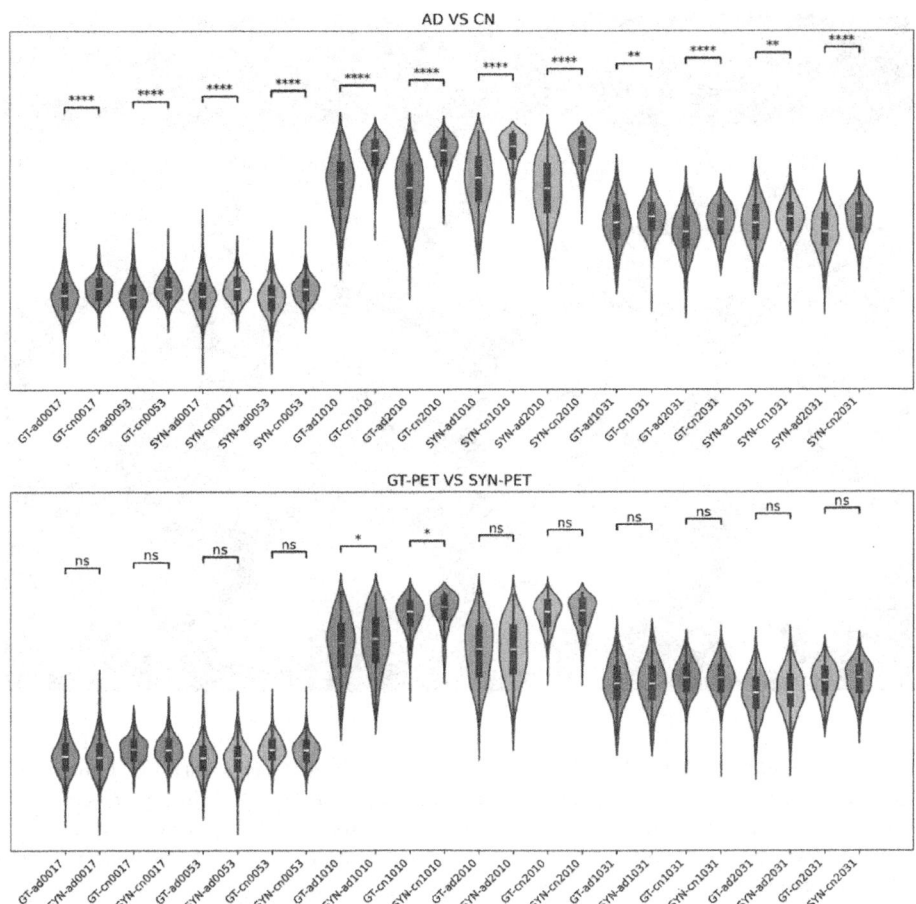

Fig. 3. In the upper panel, a comparison between the AD and CN groups reveals significant differences in specific brain regions for real PET, which are also effectively captured by Syn-PET. In the lower panel, however, no significant differences are observed between real PET and Syn-PET in the corresponding brain regions for either the AD or CN groups, except for a minor difference in the ctx-lh-isthmuscingulate region (1010), which remains within an acceptable range.

regions, such as the hippocampus (regions 0017 and 0053), posterior cingulate cortex (regions 1010 and 2010), and temporoparietal junction (regions 1031 and 2031), these regions were identified as the most significantly different between AD and cognitively normal (CN) groups. The results indicate that Syn-PET successfully captures the significant differences between AD and CN in these regions, achieving high consistency with real PET, as shown in Fig. 4. Moreover, within the same brain regions, the distributions of Syn-PET and real PET do not exhibit significant differences (regardless of the AD or CN group), further validating the ability of Syn-PET to accurately simulate regional characteristics.

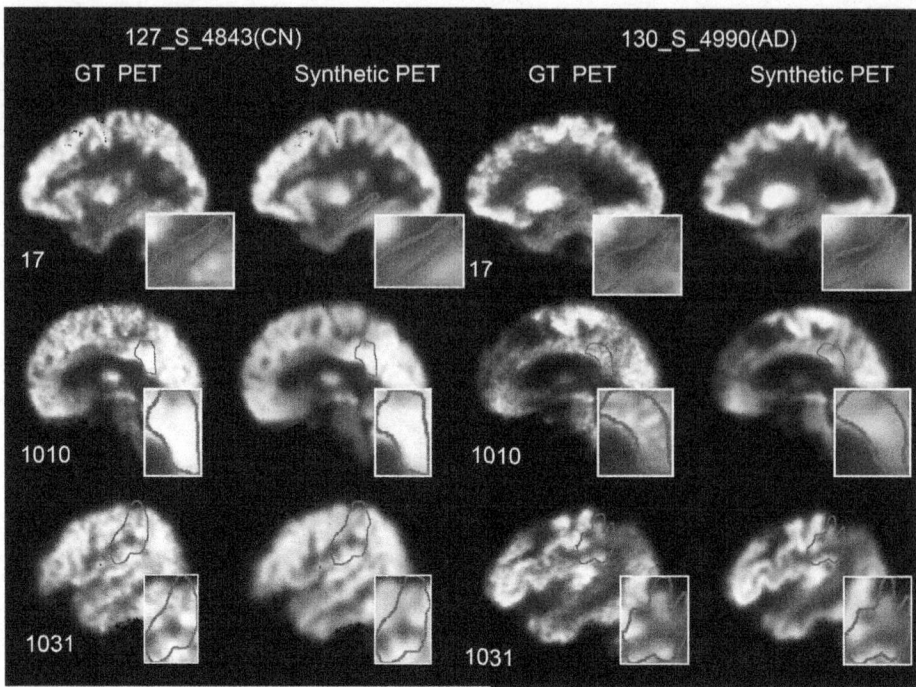

Fig. 4. The Syn-PET demonstrates remarkable accuracy in capturing the details of three brain regions—0017 (Left Hippocampus), 1010 (Left Cortex-Isthmus Cingulate), and 1031 (Left Cortex-Supramarginal)—in both CN and AD individuals, closely mirroring the observations from real PET images. Moreover, the significant differences identified between CN and AD in these regions from the real PET images are faithfully and effectively preserved in the Syn-PET.

Overall, Syn-PET not only replicates disease-related differentiating features but also demonstrates strong agreement with real PET in intra-region distribution consistency, highlighting its potential as a viable alternative.

5.2 Evaluation via Disease Diagnosis

To validate the effectiveness of Syn-PET in downstream tasks, we selected classification as the evaluation benchmark, as shown in Table 2. Specifically, classification experiments were conducted on Syn-PET to differentiate between Alzheimer's disease (AD) and cognitively normal (CN) subjects, as well as between progressive mild cognitive impairment (pMCI) and stable mild cognitive impairment (sMCI). These experiments aimed to assess whether Syn-PET preserved the pathological features relevant to disease diagnosis. To ensure the robustness and reliability of the classification results, we employed a 5-fold cross-validation approach. Additionally, a comprehensive comparison was conducted between the classification performance of Syn-PET and real PET, providing critical evidence of the quality and clinical relevance of the images generated by our model.

Table 2. Regardless of the prediction method employed, the classification metrics of the synthetic PET closely approximate those of the real PET when evaluated using the 3D DenseNet classifier.

Modality	AD vs CN					pMCI vs sMCI				
	AUC	ACC	SPE	SEN	FIS	AUC	ACC	SPE	SEN	FIS
MRI	91.88	87.69	93.04	85.00	84.21	77.26	87.65	94.12	53.85	58.33
PET	95.50	91.28	94.78	88.75	87.31	80.66	88.89	94.12	61.54	69.57
Syn-PET	94.80	91.76	95.65	86.25	87.09	79.30	86.42	94.42	60.85	66.67
MRI + PET	95.20	91.79	92.17	91.25	89.32	80.77	90.12	95.59	61.54	66.67
MRI + Syn-PET	95.07	90.77	91.17	88.75	88.14	79.95	88.89	94.65	69.23	63.64

In the classification tasks of AD vs. CN and pMCI vs. sMCI using 3D DenseNet [7], Syn-PET demonstrates performance highly comparable to real PET, with minimal differences across key metrics. For the AD vs. CN task, Syn-PET achieves an AUC within 0.7 of real PET, and the accuracy difference is less than 0.5% (Syn-PET: 91.76%, real PET: 91.28%). Furthermore, Syn-PET outperforms real PET in specificity (SPE) by 0.87%, while exhibiting slight disadvantages in sensitivity (SEN) and F1 score by 2.5% and 0.22%, respectively. In the more challenging pMCI vs. sMCI task, both modalities achieve identical specificity (94.12%), with Syn-PET trailing real PET in sensitivity and F1 score by only 0.69% and 2.9%, respectively. These results underscore the near-equivalent performance of Syn-PET to real PET, particularly in terms of specificity. Thus, Syn-PET presents a promising, cost-effective alternative for imaging analysis, particularly in scenarios where access to real PET is limited.

6 Conclusion

In this paper, we present DIReCT, a framework that integrates the rectified flow model with a pretrained vision-language model (BiomedCLIP) to generate high-fidelity PET images with anatomical and pathological consistency. Experiments on the ADNI dataset demonstrate DIReCT's superiority. Statistical analyses confirm that the differences between synthetic PET (Syn-PET) and real PET scans are within acceptable ranges, supporting their use as reliable substitutes. In disease diagnostic tasks, Syn-PET performs comparably to real PET, effectively capturing AD-related pathological features for group-level analysis and personalized diagnosis. DIReCT's ability to capture both population-level and patient-specific features underscores its potential for precision disease prediction and diagnosis. We plan to conduct comprehensive comparative studies with additional state-of-the-art methods to further validate and enhance the robustness of our framework in the future.

Acknowledgements. This work was supported in part by the National Natural Science Foundation of China (Grant No. 12326616), the Natural Science Basic Research Program of Shaanxi, China (Grant No. 2024JC-TBZC-09), and Tianyuan Fund for Mathematics of the National Natural Science Foundation of China (Grant No. 12426105).

Disclosure of Interests. The authors have no competing interests to declare.

References

1. Aisen, P., et al.: On the path to 2025: understanding the Alzheimer's disease continuum. Alzheimer's Res. Therapy **9**, 1–10 (2017)
2. Billot, C., et al.: SynthSeg: segmentation of brain MRI scans of any contrast and resolution without retraining. Med. Image Anal. **86**, 102789 (2023)
3. Bloudek, L.M., Spackman, D.E., Blankenburg, M., Sullivan, S.D.: Review and meta-analysis of biomarkers and diagnostic imaging in Alzheimer's disease. J. Alzheimer's Disease **26**(4), 627–645 (2011)
4. Cao, H., et al.: Swin-Unet: Unet-like Pure Transformer for Medical Image Segmentation. arXiv preprint arXiv:2105.05537 (2021)
5. Fischl, B.: Freesurfer. NeuroImage **62**(2), 774–781 (2012)
6. Frisoni, G.B., Bocchetta, M., Chételat, G., Rabinovici, G.D., et al.: Imaging markers for Alzheimer Disease: which vs how. Neurology **81**(5), 487–500 (2013)
7. Huang, G., et al.: Densely Connected Convolutional Networks. In: Proceedings of the IEEE Conference on Computer Vision and Pattern Recognition (CVPR), pp. 4700–4708 (2017)
8. Ho, J., Jain, A., Abbeel, P.: Denoising diffusion probabilistic models. In: Advances in Neural Information Processing Systems (NeurIPS), vol. 33, pp. 6840–6851 (2020)
9. Hu, S., Lei, B., Wang, S., Wang, Y., Feng, Z., Shen, Y.: Bidirectional mapping generative adversarial networks for brain MR to PET synthesis. IEEE Trans. Med. Imaging **41**(1), 145–157 (2021)
10. Keppler, J.S., Conti, P.S.: A cost analysis of positron emission tomography. Am. J. Roentgenol. **177**(1), 31–40 (2001)
11. Kurth, F., Gaser, C., Luders, E.: A 12-step user guide for analyzing voxel-wise gray matter asymmetries in statistical parametric mapping (SPM). Nat. Protoc. **10**, 293–304 (2015)
12. Kyrtata, N., Emsley, H., Sparasci, O., Parkes, L.M., Dickie, B.R.: A systematic review of glucose transport alterations in Alzheimer's Disease. Front. Neurosci. **15**, 626636 (2021)
13. Marcus, C., Mena, E., Subramaniam, R.M.: Brain PET in the diagnosis of Alzheimer's disease. Clin. Nucl. Med. **39**(10), E413 (2014)
14. Mueller, T.G., Weiner, M.W., Thal, L.J., Petersen, R.C., Jack, C.R., Jagust, W., et al.: The Alzheimer's disease neuroimaging initiative. Neuroimaging Clin. N. Am. **15**(4), 869–877 (2005)
15. Konz, N., Chen, Y., H.D.M.A.M.: Anatomically-controllable medical image generation with segmentation-guided diffusion models. In: Medical Image Computing and Computer-Assisted Intervention (MICCAI) (2024)
16. Ozbey, M., et al.: Unsupervised medical image translation with adversarial diffusion models. IEEE Trans. Med. Imaging **42**(12), 3524–3539 (2023)

17. Peng, W., Adeli, E., Bosschieter, T., Park, S.H., Zhao, Q., Pohl, K.M.: Generating realistic brain MRIs via a conditional diffusion probabilistic model. In: Medical Image Computing and Computer-Assisted Intervention (MICCAI) (2023)
18. Graf, R., et al.: Denoising diffusion-based MRI to CT image translation enables automated spinal segmentation. European Radiol. Exp. **7**(70) (2023)
19. Shin, H.C., Ihsani, A., Xu, Z., Mandava, S., Sreenivas, S.T., Forster, C., Cha, J.: Gandalf: generative adversarial networks with discriminator-adaptive loss fine-tuning for Alzheimer's Disease diagnosis from MRI. In: Medical Image Computing and Computer-Assisted Intervention (MICCAI) (2020)
20. Sikka, A., Virk, S.J., Bathula, D.R.: MRI to PET cross-modality translation using globally and locally aware GAN (GLA-GAN) for multi-modal diagnosis of Alzheimer's Disease. arXiv preprint arXiv:2108.02160 (2021)
21. Song, J., Meng, C., Ermon, S.: Denoising Diffusion Implicit Models. arXiv preprint arXiv:2010.02502 (2020)
22. Wei, W., et al.: Learning myelin content in multiple sclerosis from multimodal MRI through adversarial training. In: Medical Image Computing and Computer-Assisted Intervention (MICCAI) (2018)
23. Liu, X., Chengyue Gong, Q.L.: Flow Straight and Fast: Learning to Generate and Transfer Data with Rectified Flow. arXiv preprint arXiv:2209.03003 (2022)
24. Zhang, S., et al.: BiomedCLIP: A Multimodal Biomedical Foundation Model Pretrained from Fifteen Million Scientific Image-Text Pairs. arXiv preprint arXiv:2303.00915 (2023)
25. Zhu, L., et al.: Make-a-volume: leveraging latent diffusion models for cross-modality 3D brain MRI synthesis. In: Medical Image Computing and Computer-Assisted Intervention (MICCAI) (2023)

IGG: Image Generation Informed by Geodesic Dynamics in Deformation Spaces

Nian Wu[1]([✉]), Nivetha Jayakumar[1], Jiarui Xing[1], and Miaomiao Zhang[1,2]

[1] Department of Electrical and Computer Engineering, University of Virginia, Charlottesville, USA
bsw3ac@virginia.edu
[2] Department of Computer Science, University of Virginia, Charlottesville, USA

Abstract. Generative models have recently gained increasing attention in image generation and editing tasks. However, they often lack a direct connection to object geometry, which is crucial in sensitive domains such as computational anatomy, biology, and robotics. This paper presents a novel framework for Image Generation informed by Geodesic dynamics (IGG) in deformation spaces. Our IGG model comprises two key components: (i) an efficient autoencoder that explicitly learns the geodesic path of image transformations in the latent space; and (ii) a latent geodesic diffusion model that captures the distribution of latent representations of geodesic deformations conditioned on text instructions. By leveraging geodesic paths, our method ensures smooth, topology-preserving, and interpretable deformations, capturing complex variations in image structures while maintaining geometric consistency. We validate the proposed IGG on plant growth data and brain magnetic resonance imaging (MRI). Experimental results show that IGG outperforms the state-of-the-art image generation/editing models with superior performance in generating realistic, high-quality images with preserved object topology and reduced artifacts. Our code is publicly available at https://github.com/nellie689/IGG.

1 Introduction

Generative diffusion models have gained significant attention in recent years due to their ability to capture complex data distributions and produce high-quality and realistic samples [7,29]. Among these, language-image generative models [34,36] leverage text prompts to guide the diffusion process, allowing users to perform highly customizable and context-aware edits to images. In medical imaging, such models are widely utilized to create diverse datasets, improving downstream tasks such as image classification [18,45], segmentation [1,10], and object recognition [12]. Additionally, they achieve great performance in image-to-image translation, addressing key challenges in the synthesis of high-quality medical scans from low-quality data [9,25] and the simulation of missing modalities [31,50]. With their capability to potentially improve disease diagnostic accuracy and support personalized treatment planning, generative diffusion models hold great promise for advancing real-world clinical applications in healthcare.

Despite their success, existing diffusion models primarily focus on manipulating the intensity and texture features extracted from images, often neglecting direct connections to object geometry in the editing or generation process [13,20,24,30]. This limitation can lead to unrealistic or biologically implausible samples, posing risks to downstream tasks that require accurate geometric preservation [4]. Recent advances have introduced simple geometric constraints such as shape localization [11,32], boundary conditions [26], and 3D shape priors [17,48] in the modeling process. However, these approaches lack fine-grained structural preservation, which is highly desirable in sensitive domains, including computational anatomy, robotics, and medical imaging. Another important yet under-investigated limitation of current diffusion models is their inability to provide interpretable metrics or quantitative measures of topological integrity for image objects. Commonly used evaluation metrics include Fréchet Inception Distance (FID) for assessing distribution similarity between real and generated data [14], Inception Score (IS) for measuring diversity and quality [37], and Structural Similarity Index (SSIM) for evaluating pixel-level fidelity [41]. While these metrics effectively assess visual quality, they do not account for the preservation of object geometry and topology, raising questions about their reliability and trustworthiness in structure-sensitive applications.

To overcome these limitations, we introduce a novel framework, Image Generation informed by Geodesic dynamics (IGG), that for the first time generates images by deforming a given template along random geodesics in deformation spaces guided by text instructions. In contrast to current approaches [8,16,21], our IGG ensures topological consistency by treating each generated image as a deformed variation of a template or reference image through learned diffeomorphic transformations (i.e., a one-to-one, smooth, and invertible mapping). A dynamic geodesic, producing natural transformations and providing interpretable metrics to quantify topological changes, will be explicitly learned during the diffusion process. Our proposed IGG model comprises two key components: a latent representation learning framework of diffeomorphic transformations informed by geodesic dynamics, and a latent geometric diffusion model conditioned on user-defined text instructions. To summarize, the contributions of IGG are threefold:

(i) Develop a novel image generation model that integrates geometric principles through the geodesic learning of image deformations with a powerful conditional diffusion process.
(ii) Enable quantitative metrics to assess and ensure the topology consistency of generated samples.
(iii) Establish a new paradigm for producing anatomically intact and realistic image samples, potentially advancing applications in domains requiring structural fidelity and precision.

We demonstrate the effectiveness of IGG on a diverse set of real-world datasets, including Komatsuna plant growth data [38] and brain MRIs [23]. We compare the performance of the model with state-of-the-art text-instructed generative models. Experimental results show that IGG achieves significantly improved results in generating image samples with well-preserved topological structures.

2 Background: Geodesics in Deformation Spaces

In this section, we briefly review the basic concept of geodesics in deformation spaces, which provides a principled way to model smooth and natural deformations between images while preserving structural integrity [28]. With the underlying assumption that objects of a generic class (e.g., human brains, hearts, or lungs) are described as deformed variants of a given template, descriptors of that class arise naturally by deforming the template to other images along a **geodesic** - a shortest path with minimal energy of transformations [3,19]. In theory, every topological property of the deformed template can be preserved by enforcing the resulting transformations to be diffeomorphisms, i.e., differentiable, bijective mappings with differentiable inverses [2,5,28]. Violations of such constraints introduce image artifacts, such as tearing, crossing, or passing through itself.

Let $\text{Diff}^\infty(\Omega)$ denote the space of smooth diffeomorphisms on a d-dimensional torus domain $\Omega = \mathbb{R}^d/\mathbb{Z}^d$. The tangent space of diffeomorphisms is the space $V = \mathfrak{X}^\infty(T\Omega)$ of smooth vector fields on Ω. Considering a time-varying velocity field, $\{v_t\} : [0, \tau] \to V$, we can generate diffeomorphisms $\{\phi_t\}$ between pairwise images by solving

$$\frac{d\phi_t}{dt} = v_t(\phi_t), \ t \in [0, 1], \tag{1}$$

where ϕ_0 is the identity map and ϕ_1 is the target transformation.

The geodesic minimizes the functional $\int_0^1 (\mathcal{L}v_t, v_t)\, dt$ [5], where $\mathcal{L} : V \to V^*$ is a symmetric, positive-definite differential operator that maps a tangent vector $v(t) \in V$ into its dual space as a momentum vector $m(t) \in V^*$. We typically write $m(t) = \mathcal{L}v(t)$, or $v(t) = \mathcal{K}m(t)$, with $\mathcal{K}$ being an inverse operator of $\mathcal{L}$. In this paper, we adopt a commonly used Laplacian operator $\mathcal{L} = (-\alpha\Delta + \text{Id})^3$, where α is a weighting parameter that controls the smoothness of transformation fields and Id is an identity matrix. The $(\cdot, \cdot)$ is a dual paring, which is similar to an inner product between vectors. According to a well-known geodesic shooting algorithm [40], the minimum of the functional mentioned above is uniquely determined by solving a Euler-Poincaré differential (EPDiff) equation [40,47] with a given initial condition. That is, for $\forall v_0 \in V$, a geodesic path $t \mapsto \phi_t$ in the space of diffeomorphisms can be computed by forward shooting the EPDiff equation

$$\frac{\partial v_t}{\partial t} = -K\left[(Dv_t)^T m_t + Dm_t\, v_t + m_t \,\text{div}\, v_t\right], \tag{2}$$

where D denotes the Jacobian matrix and div is a divergence operator.

Derive Geodesics from Images. Consider a source image S and a target image T defined in the domain Ω $(S(x), T(x) : \Omega \to \mathbb{R})$. The optimization of geodesic transformations can be formulated as minimizing an energy function over the initial velocity, subject to the EPDiff equations, i.e.,

$$E(v_0) = (\mathcal{L}v_0, v_0) + \lambda \mathrm{Dist}(S(\phi_1), T) \text{ s.t. Equation (1) \& (2)}.$$

Here, $\mathrm{Dist}(\cdot, \cdot)$ is a distance function that measures the dissimilarity between images and λ is a positive weighting parameter. In this paper, we will use the commonly used sum-of-squared intensity differences [5,42,49].

3 Our Method: IGG

This section introduces our proposed model, IGG, that for the first time generates images in deformation spaces guided by text instructions. A dynamic geodesic, producing natural transformations and providing interpretable metrics to quantify topological changes, will be explicitly learned during the diffusion process. Our proposed IGG model consists of two main components: (i) an autoencoder-based deformable registration network that learns the latent representations of diffeomorphic transformations informed by geodesic dynamics; and (ii) a latent geodesic diffusion model that captures the latent distribution of sequential deformation features, conditioned on user-defined text inputs. An overview of our model is illustrated in Fig. 1.

3.1 Network Architecture

Learning Latent Representations of Geodesics in Deformation Space.
For a number of N template and target images associated with text instructions $\{s^n, f^n, \Lambda^n\}_{n=1}^N$, we employ an autoencoder-based deformable registration network to learn the latent representation of geodesic transformations, parameterized by velocity fields $\{v_t\}$, between a template s^n and a target f^n. The encoder of this network, $\mathcal{E}_{\theta_v}$, maps input data to the latent representation of a sequence of velocity fields, $\{z_{v_t}\}$. Inspired by the recent work NeurEPDiff [43], which develops a neural operator for learning the mapping functions of the EPDiff equation, we incorporate the NeurEPDiff module, $\mathcal{G}_{\theta_r}$, in our model IGG to learn geodesics in latent deformation spaces. More specifically, starting with a latent initial velocity z_{v_0}, the operator $\mathcal{G}_{\theta_r}$ iteratively propagates it along the geodesic path, $z_{v_0} \longmapsto \cdots z_{v_t} \longmapsto \cdots z_{v_1}$. At each time point t, a multilayer neural network is employed to simulate the geodesic mapping function from z_{v_t} to $z_{v_{t+1}}$. Each hidden layer combines a local linear transformation, W^j, and a global convolutional kernel, $\mathcal{H}^j$, to extract both global and local representations. Additionally, a smooth nonlinear activation function, $\sigma(\cdot)$, is used to encourage a smooth evolution of the geodesic path. This process is defined as

$$z_{v_{t+1}} := \sigma(W^J, \mathcal{H}^J) \circ \cdots \circ \sigma(W^2, \mathcal{H}^2) \circ \sigma(W^1, \mathcal{H}^1, z_{v_t}),$$

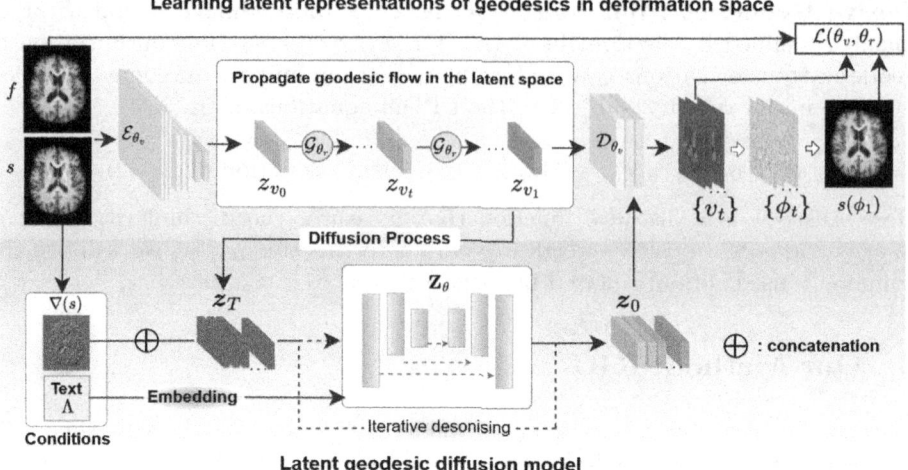

Fig. 1. An overview of the proposed IGG model.

where ∘ denotes the composition of network operations. The decoder, $\mathcal{D}_{\theta_v}$, then projects these latent representations back to the full-dimensional image space to produce a geodesic flow of velocity fields, $v_0 \mapsto \cdots v_t \mapsto \cdots v_1$, which are integrated to generate the corresponding deformations, $\phi_0 \mapsto \cdots \phi_t \mapsto \cdots \phi_1$ as described in Eq. 1.

The network loss combines contributions from an unsupervised registration loss, and a geodesic loss guided by numerical solutions of the EPDiff equation obtained via Euler integration. These solutions are represented as $\{\hat{v}_t^n\}, t \in (0, 1]$ with a simultaneously learned initial velocity $\hat{v}_0^n$. To simplify the notation, we define $\hat{\boldsymbol{v}}^n \triangleq \{\hat{v}_t^n\}$ before formulating the loss function as

$$\mathcal{L}(\theta_v, \theta_r) = \sum_{n=1}^{N} \lambda \, \|(s^n(\phi_1^n(\theta_v)) - f^n\|_2^2 + \frac{1}{2}(\mathcal{L}v_0^n(\theta_v), v_0^n(\theta_v)) \\ + \eta \, \|\mathcal{D}_{\theta_v}(\mathcal{G}_{\theta_r}(z_{v_0})) - \hat{\boldsymbol{v}}^n\|_2^2 + \mathrm{Reg}(\theta_r, \theta_v), \quad s.t. \text{ Eq. (1)}, \qquad (3)$$

where λ and η are positive weighting parameters to balance the image matching term and the geodesic loss, and $\mathrm{Reg}(\cdot)$ is a network regularization.

Latent Geodesic Diffusion Model. With a pre-trained geodesic learning network, we are now ready to introduce a latent geometric diffusion module to simulate the distribution of latent geodesic flows for image transformations, $\{z_{v_t}^n\}$. To simplify the notation, we define $\boldsymbol{z}^n := \{z_{v_t}^n\}_{t=0}^{1}$ for the n-th subject, where all time-dependent components $\{z_{v_t}^n\}$ are concatenated to form a unified representation. In contrast to previous approaches that perform the diffusion process in the image space [8,16,21], our latent geometric diffusion operates in the deformation space. Specifically, our diffusion process directly samples geodesics of topology-preserving diffeomorphic transformations, which are then applied to deform the input template to generate the final output.

Geometric Conditioning in the Latent Geodesic Space. Following the principles of video diffusion models [16], we perform forward diffusion process by progressively adding noise, $\epsilon \sim \mathcal{N}(0, \mathbf{I})$, to the initial latent representations $z_0^n \sim p(z_0^n), z_0^n \leftarrow z^n$ with a number of T steps. The forward diffusion process results in a distribution $q(z_\tau^n | z_0^n)$, where $\tau \in [1, \cdots, T]$ [16].

In the reverse diffusion process, we first introduce a novel geometric conditioning in the latent geodesic space. This enables our model to adapt more effectively to specific contexts; hence improving its predictive performance based on observed template images and provided text instructions. More specifically, our geometric conditioning is achieved by concatenating three components: image embedding - downsampled template image gradient, $\{\nabla s^n\}$; the text embedding, encoded using a CLIP text encoder [33], $\{\Lambda^n\}$; and the geodesic deformation embedding, represented by the latent velocity fields, $\{z_\tau^n\}$. The reverse diffusion process then involves sampling the denoised latent geodesic flow of velocity fields $\hat{z}_0^n$, starting from the noisy state z_τ^n at the τ-th time-step and progressively removing noise to reconstruct the original latent representation. This process is mathematically expressed as

$$p_\theta(z_{\tau-1}^n | z_\tau^n, \nabla s^n, \Lambda^n, \tau) = \mathcal{N}(z_{\tau-1}^n; \mu_\theta(z_\tau^n, \nabla s^n, \Lambda^n, \tau), \Sigma_\theta(z_\tau^n, \nabla s^n, \Lambda^n, \tau)), \quad (4)$$

where μ_θ [16] and Σ_θ represent the mean and variance of the Gaussian distribution at each time step of the reverse diffusion process. The parameters μ_θ [16] and Σ_θ are learned functions, and the reverse process is practically implemented using a denoising network, $\mathbf{Z}_\theta$. In our experiments, we employ a 3D UNet architecture [35] as the backbone for $\mathbf{Z}_\theta$, which predicts the noise component to be removed from z_τ^n to reconstruct $z_{\tau-1}^n$.

The network loss of our proposed latent geodesic diffusion module is formulated as

$$\mathcal{L}_\theta = \frac{1}{N} \sum_{n=1}^{N} \|\epsilon_\tau^n - \mathbf{Z}_\theta(z_\tau^n, \nabla s^n, \Lambda^n, \tau)\|_2^2 + \text{reg}(\theta), \tau \in \mathrm{U}[1, T], \quad (5)$$

where $\text{reg}(\cdot)$ is a regularization of the network parameter θ, and U denotes a uniform distribution.

To balance the trade-off between sample quality and diversity of velocity fields in the latent space after training, we optimize the combination of conditional and unconditional diffusion models. This is achieved by leveraging image-conditioned guidance with scale δ_I, text-conditioned guidance with scale δ_T, and a null-condition $\varnothing$. Following similar principles from the classifier-free guidance approach [15], we first jointly train a conditional and an unconditional IGG model. The resulting score estimates from these models are then combined to predict the noise $\hat{Z}$ as

$$\begin{aligned}\hat{\mathbf{Z}}_\theta(z_\tau^n, \nabla s^n, \Lambda^n, \tau) &= \mathbf{Z}_\theta(z_\tau^n, \varnothing, \varnothing, \tau) + \delta_\mathrm{I} \cdot [\mathbf{Z}_\theta(z_\tau^n, \nabla s^n, \varnothing, \tau) - \mathbf{Z}_\theta(z_\tau^n, \varnothing, \varnothing, \tau)] \\ &\quad + \delta_\mathrm{T} \cdot [\mathbf{Z}_\theta(z_\tau^n, \nabla s^n, \Lambda^n, \tau) - \mathbf{Z}_\theta(z_\tau^n, \nabla s^n, \varnothing, \tau)] \\ &= (1 - \delta_\mathrm{I}) \cdot \mathbf{Z}_\theta(z_\tau^n, \varnothing, \varnothing, \tau) + (\delta_\mathrm{I} - \delta_\mathrm{T}) \cdot \mathbf{Z}_\theta(z_\tau^n, \nabla s^n, \varnothing, \tau) \\ &\quad + \delta_\mathrm{T} \cdot \mathbf{Z}_\theta(z_\tau^n, \nabla s^n, \Lambda^n, \tau). \end{aligned} \quad (6)$$

3.2 Network Optimization

We first optimize the loss of representation learning of geodesic transformations (Eq. (3)), followed by the latent geodesic diffusion model (Eq. (5)). The training process of IGG is summarized in Algorithm 1. During testing, the procedure for sampling the geodesic of latent velocity fields, conditioned on a given template image and text instructions, is outlined in Algorithm 2.

Algorithm 1. IGG Training
1: **Input:** data $\{s^n, f^n, \Lambda^n\}_{n=1}^{N_{train}}$
2: Pretrain $\mathcal{E}_{\theta_v}, \mathcal{E}_{\theta_r}, \mathcal{D}_{\theta_v}$ via Eq. 3.
3: **repeat**
4: $\quad z_0^n \leftarrow \mathcal{G}_{\theta_r}(\mathcal{E}_{\theta_v}(s^n, f^n))$
5: $\quad$ Sample $\tau \sim \mathrm{U}(1, \mathrm{T}), \epsilon \sim \mathcal{N}(0, \mathbf{I})$
6: $\quad z_\tau^n \leftarrow q(z_\tau^n | z_0^n)$
7: $\quad$ Take gradient descent step on Eq. 5
8: **until** converge

Algorithm 2. IGG Sampling
1: **Input:** data $\{s^n, \Lambda^n\}_{n=1}^{N_{test}}$
2: Initialize $z_T^n \sim \mathcal{N}(0, \mathbf{I})$
3: **for** $\tau = \mathrm{T}, ..., 1$ **do**
4: $\quad \hat{z}_{\tau-1}^n \leftarrow p_\theta(z_{\tau-1}^n | z_\tau^n, \nabla s^n, \Lambda^n, \tau)$
5: **end for**
6: $\{\hat{v}^n\} = \mathcal{D}_{\theta_v}(\hat{z}_0^n)$
7: $\{\phi^n\} \leftarrow \{\hat{v}^n\}$ via Eq. 1
8: **return** $\{\phi^n\}$

---

4 Experimental Evaluation

We evaluate the proposed IGG framework using a diverse set of real-world image datasets that capture deformable shape changes over time. We first assess the quality of the learned latent representations of geodesics by comparing them to numerical solutions of the EPDiff equation obtained via Euler integration.

Next, we evaluate the quality of the generated images deformed by the sampled geodesics guided by given text instructions. These results are compared with state-of-the-art generative models that synthesize image sequences/videos with publicly available training code or fine-tuning options, including the video diffusion model (VDM) [16], CogVideoX [46], and DynamiCrafter [44]. Note that all baselines are trained using sequences of deformed images generated along the geodesic transformations derived from the numerical solutions of EPDiff equation. This approach ensures that the results are independent of the learned geodesics predicted by the first module of the IGG framework, providing a consistent and unbiased evaluation.

4.1 Dataset

Komatsuna Plant. We include 300 frames of RGB-D label-maps representing five Komatsuna plants from the publicly available data repository [38]. The label maps cover different leaves that emerge from the bud and grow in size over time, where the plant growth was monitored between 228–236 h. All data frames were resampled to 128^2 and pre-aligned with affine transformations.

Longitudinal Brain MRI. We include a total of 2618 T1-weighted longitudinal (time-series) brain MRIs sourced from the Open Access Series of Imaging Studies (OASIS-3) dataset [23]. This experiment aims to validate our method using longitudinal data that includes scans at varying time intervals for individuals spanning both healthy subjects and Alzheimer's diseases (AD), aged 60-90. Given the scenario that many existing image generation/editing methods with text instructions focus on 2D natural images [8, 27], we specifically utilize 2D scans derived from this 3D brain data for comparison with state-of-the-art baselines. All MRIs undergo pre-processing, including resizing to 128^2, with isotropic voxels of $1\,\text{mm}^2$, skull-stripping, intensity normalization, bias field correction, and pre-alignment using affine transformations.

Text Instructions. Our text condition consists of a description of the template image, followed by details of the specific edits or progressions applied to it. For brain MRIs, this includes biological variables such as age, sex, and gender. In contrast, for plants, it primarily focuses on the growth timeline. All information is sourced directly from the original dataset available in the data repository.

4.2 Experimental Design and Implementation Details

Evaluate Learned Latent Representations of Geodesics. We evaluate the model's ability to predict geodesic dynamics by comparing the learned geodesics with numerical solutions to the EPDiff equation using Euler integration. The assessment focuses on errors in velocity fields, transformations, and deformed images between the two approaches. Quantitative metrics, such as the mean absolute error of velocity fields, are computed at each time integration step.

Evaluate Generated Samples. We compare the quality of samples generated by IGG with three baseline models: VDM [16], CogVideoX [46], and DynamiCrafter [44]). In particular, we first assess individual synthesized images at each time step using standard image synthesis metrics, including the Fréchet Inception Distance (FID) [14] and Kernel Inception Distance (KID) [6] to measure the distributional similarity between generated and real images. Additionally, we use the Structural Similarity Index (SSIM) [41] to evaluate structural, contrast, and luminance similarities. We then evaluate the continuity and smoothness of generated video sequences by computing the Fréchet Video Distance (FVD) [39] and providing qualitative visual comparisons. Finally, we assess the perceptual quality of the generated samples using the Inception Score (IS) [37].

Evaluate the Preservation of Object Geometry and Topology. To demonstrate the efficiency of our IGG model in preserving fine-grained geometric structure and topology, we first generate image sequences by deforming a template image using the sampled geodesic flow of deformations produced by IGG and compare these results with all baseline models. Secondly, we present the corresponding determinant of the Jacobian (DetJac) maps for the deformations. The DetJac values reveal important patterns of volume change: a value of 1 indicates no volume change, DetJac < 1 reflects volume shrinkage, and DetJac > 1 implies volume expansion. A DetJac value below zero indicates artifacts or singularities in the transformation field, highlighting a failure to maintain the topological integrity. Note that only our IGG model enables such a metric to quantify the topological changes in the generated samples.

textbfEvaluate Reliability of Model Predictions. We evaluate the reliability and confidence of IGG in learning the geodesic deformations of geometric shapes over time by computing pixel-wise mean and standard deviation for individual sampled image sequences along the geodesic path. We visualize confidence intervals to highlight regions with 95% certainty in growth patterns. These intervals are defined by the mean of 1000 generated samples with pixel-wise bounds that are two standard deviations above and below the mean.

Parameter Setting. We split all datasets into 80%, 10%, and 10% for training, validation, and testing. All experiments are conducted on NVIDIA A100 GPUs. For training the autoencoder for geodesic learning, we alternatively update the encoder-decoder and the latent geodesic neural operator, and finally jointly train both components for 2000 epochs per stage. We set the weight decay as $1e^{-4}$, batch size as 64, and a learning rate as $5e^{-4}$. For training the latent geodesic diffusion model, we set the learning rate as $1e^{-4}$, the number of epochs as 6000, batch size as 36, and diffusion steps as 500. All models are trained using the Adam optimizer [22].

4.3 Experimental Results

The left panel of Fig. 2 visualizes the predicted transformations, velocity fields, and deformed images from IGG's autoencoder, compared with numerical solutions to the EPDiff equation. These methods show a high degree of similarity. The right panel of Fig. 2 presents quantitative results of the differences between the real numerical solution and IGG's latent representation learning module at each time step. These results collectively highlight IGG's ability to effectively learn geodesic mapping functions comparable to real numerical solutions.

Figure 3 compares the generated images from our model IGG (along with the associated DetJac maps of the deformations) with the baseline generative models. The visualized images and DetJac values of IGG suggest that our model effectively preserves the topological structure of objects within the generated

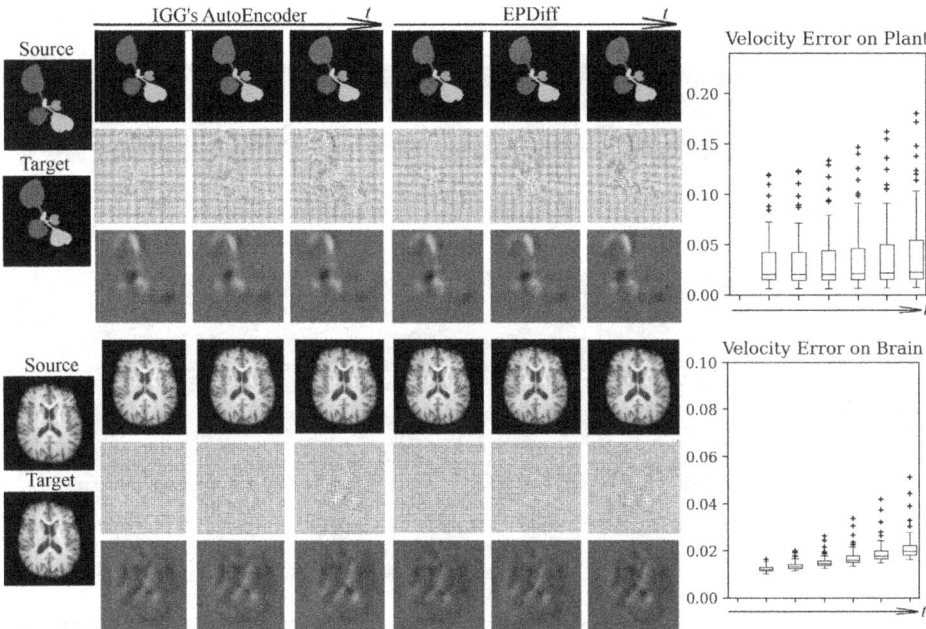

Fig. 2. Comparison of predicted geodesics by IGG vs. real numerical solutions from the EPDiff equation. Left: Visualization of predicted deformed images, deformations, and velocities along time. Right: Mean absolute error of predicted velocities over time compared to numerical integration of EPDiff.

images with well-captured progression of geometric shape changes over time. In contrast, samples generated by the baselines fail to maintain the geometric integrity of various structures. For instance, in plant growth images, the leaves appear to merge or overlap unnaturally, disrupting their biological topology. Similarly, the baselines inaccurately predict parts of the hippocampus, resulting in regions that do not correspond to the original brain structure.

Table 1 reports the evaluation metrics of sample quality and diversity for the generated images from our model IGG and the baselines. While all models achieve similar IS, IGG shows significantly lower FVD, FID, and KID scores, and higher SSIM scores. This indicates that while all models generate diverse images (leading to a good IS), the baselines lack realistic features or proper distribution alignment with real images. The observation of IGG achieving approximately 10 times better scores demonstrates its ability to generate more realistic images.

Figure 4 visualizes exemplary confidence maps of plant growth and brain progression images generated from IGG. It suggests that our model effectively captures the growth pattern of plants over time, primarily focusing on the boundaries of the leaves. The confidence maps of brain images highlight expanding patterns in ventricles, demonstrating the model's ability to capture dynamic changes over time.

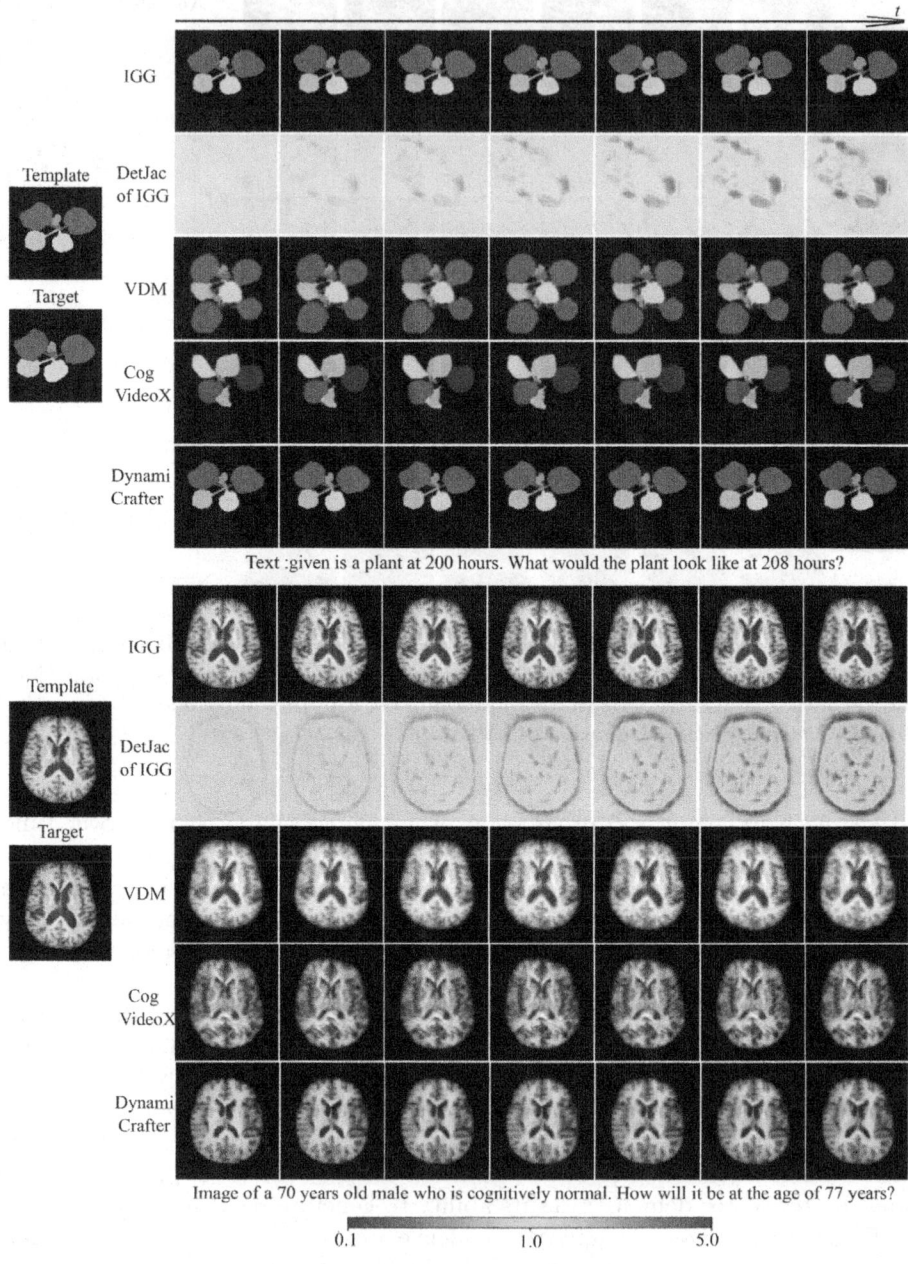

Fig. 3. A comparison of images generated by IGG (with corresponding DetJac) against all baseline models across different time frames. Given an input template image and text instructions, all models generate samples of target images. The ground truth "target" along with input template images are provided on the left side of the panel for reference.

Table 1. A comparison of performance metrics across all methods for various datasets.

Dataset	Model	FVD ↓	FID ↓	KID ↓	SSIM ↑	IS ↑
Plant	IGG	**48.29**	**9.23**	**0.007 ± 0.02**	**0.98 ± 0.02**	**1.06 ± 1.2%**
	VDM [16]	357.56	71.79	0.23 ± 0.03	0.23 ± 0.10	1.05 ± 1.0%
	CogVideoX [46]	471.49	153.93	0.63 ± 0.07	0.69 ± 0.19	1.05 ± 0.6%
	DynamiCrafter [44]	466.70	80.78	0.41 ± 0.04	0.89 ± 0.04	1.03 ± 0.3%
Brain	IGG	**26.39**	**6.23**	**0.006 ± 0.008**	**0.97 ± 0.03**	1.02 ± 0.08%
	VDM [16]	247.84	69.23	0.46 ± 0.03	0.89 ± 0.02	1.02 ± 0.07%
	CogVideoX [46]	302.20	114.50	0.46 ± 0.04	0.69 ± 0.27	**1.06 ± 2.8%**
	DynamiCrafter [44]	288.62	142.40	1.05 ± 0.03	0.88 ± 0.03	1.03 ± 0.20%

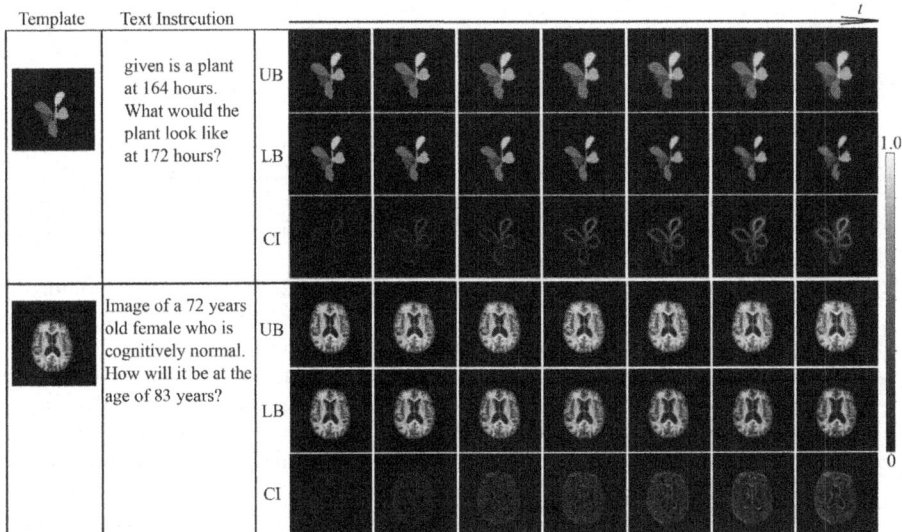

Fig. 4. Left to right: input template images with text instructions, followed by confidence maps illustrating the lower bounds (LB), upper bounds (UB), and confidence intervals (CI), which represent regions representing 95% of ideal growth patterns, based on 1000 samples generated by our IGG model across different time frames.

5 Conclusion and Discussion

This paper introduces IGG, a novel approach to image generation that deforms a given template along learned random geodesics in deformation spaces, guided by text instructions. In contrast to current generative models that focus on manipulating image intensity and texture, IGG explicitly learns dynamic geodesics of image transformations during the diffusion process. This enables natural, interpretable transformations and provides quantitative metrics to assess topological changes in generated samples. Our key contributions include a geodesic-informed network to learn latent representations of time-dependent diffeomorphic transformations and a latent geometric diffusion model that captures sequential deformation features conditioned on text inputs. Experimental results demonstrate the effectiveness of IGG with significantly improved sample fidelity and diversity compared to the state-of-the-art.

Our IGG model advances image generation by integrating geometric principles, offering tools for topology assessment, and establishing a framework for generating anatomically consistent image samples. Future research will further (i) leverage IGG's outputs to benefit downstream tasks, such as image classification, segmentation, and object recognition in the medical domain; and (ii) thoroughly validate and expand its utility across diverse applications.

Acknowledgments. This work was supported by NSF CAREER Grant 2239977.

References

1. Amit, T., Shaharbany, T., Nachmani, E., Wolf, L.: SegDiff: image segmentation with diffusion probabilistic models. arXiv:2112.00390 (2021)
2. Arnold, V.: Sur la géométrie différentielle des groupes de lie de dimension infinie et ses applications à l'hydrodynamique des fluides parfaits. Annales de l'institut Fourier **16**, 319–361 (1966)
3. Avants, B.B., Epstein, C.L., Grossman, M., Gee, J.C.: Symmetric diffeomorphic image registration with cross-correlation: evaluating automated labeling of elderly and neurodegenerative brain. Med. Image Anal. **12**(1), 26–41 (2008)
4. Azizi, S., Kornblith, S., Saharia, C., Norouzi, M., Fleet, D.J.: Synthetic data from diffusion models improves imagenet classification. arXiv:2304.08466 (2023)
5. Beg, M.F., Miller, M.I., Trouvé, A., Younes, L.: Computing large deformation metric mappings via geodesic flows of diffeomorphisms. Int. J. Comput. Vision **61**(2), 139–157 (2005)
6. Bińkowski, M., Sutherland, D.J., Arbel, M., Gretton, A.: Demystifying mmd GANs. arXiv:1801.01401 (2018)
7. Brock, A., Donahue, J., Simonyan, K.: Large scale GAN training for high fidelity natural image synthesis (2019)
8. Brooks, T., Holynski, A., Efros, A.A.: InstructPix2Pix: learning to follow image editing instructions. In: Proceedings of the IEEE/CVF Conference on Computer Vision and Pattern Recognition, pp. 18392–18402 (2023)
9. Chung, H., Ye, J.C.: Score-based diffusion models for accelerated MRI. Med. Image Anal. **80**, 102479 (2022)
10. Fernandez, V., et al.: Can segmentation models be trained with fully synthetically generated data? In: International Workshop on Simulation and Synthesis in Medical Imaging, pp. 79–90. Springer (2022)
11. Gupta, S., Samaras, D., Chen, C.: TopoDiffusionNet: a topology-aware diffusion model. arXiv:2410.16646 (2024)
12. Hamamci, I.E., et al.: Diffusion-based hierarchical multi-label object detection to analyze panoramic dental X-rays. In: International Conference on Medical Image Computing and Computer-Assisted Intervention, pp. 389–399. Springer (2023)
13. Hertz, A., Mokady, R., Tenenbaum, J., Aberman, K., Pritch, Y., Cohen-Or, D.: Prompt-to-prompt image editing with cross attention control (2022)

14. Heusel, M., Ramsauer, H., Unterthiner, T., Nessler, B., Hochreiter, S.: GANs trained by a two time-scale update rule converge to a local nash equilibrium. Adv. Neural Inf. Process. Syst. **30** (2017)
15. Ho, J., Salimans, T.: Classifier-free diffusion guidance. arXiv:2207.12598 (2022)
16. Ho, J., Salimans, T., Gritsenko, A., Chan, W., Norouzi, M., Fleet, D.J.: Video diffusion models. Adv. Neural. Inf. Process. Syst. **35**, 8633–8646 (2022)
17. Hu, J., et al.: Topology-aware latent diffusion for 3d shape generation. arXiv:2401.17603 (2024)
18. Ijishakin, A., Abdulaal, A., Hadjivasiliou, A., Martin, S., Cole, J.: Interpretable Alzheimer's disease classification via a contrastive diffusion autoencoder. arXiv:2306.03022 (2023)
19. Joshi, S., Davis, B., Jomier, M., Gerig, G.: Unbiased diffeomorphic atlas construction for computational anatomy. Neuroimage **23**, S151–S160 (2004)
20. Kawar, B., et al.: Imagic: text-based real image editing with diffusion models. In: Proceedings of the IEEE/CVF Conference on Computer Vision and Pattern Recognition, pp. 6007–6017 (2023)
21. Kim, B., Han, I., Ye, J.C.: DiffuseMorph: unsupervised deformable image registration using diffusion model. In: Computer Vision–ECCV 2022: 17th European Conference, Tel Aviv, Israel, 23–27 October 2022, Proceedings, Part XXXI, pp. 347–364. Springer (2022)
22. Kingma, D.P., Ba, J.: Adam: a method for stochastic optimization. arXiv:1412.6980 (2014)
23. LaMontagne, P.J., et al.: OASIS-3: longitudinal neuroimaging, clinical, and cognitive dataset for normal aging and Alzheimer disease. medrxiv pp. 2019–12 (2019)
24. Li, D., Li, J., Hoi, S.: Blip-diffusion: pre-trained subject representation for controllable text-to-image generation and editing. Adv. Neural Inf. Process. Syst. **36** (2024)
25. Mao, Y., Jiang, L., Chen, X., Li, C.: DisC-Diff: disentangled conditional diffusion model for multi-contrast MRI super-resolution. In: International Conference on Medical Image Computing and Computer-Assisted Intervention, pp. 387–397. Springer (2023)
26. Mazé, F., Ahmed, F.: Diffusion models beat GANs on topology optimization. In: Proceedings of the AAAI Conference on Artificial Intelligence, vol. 37, pp. 9108–9116 (2023)
27. Meng, C., et al.: SDEdit: guided image synthesis and editing with stochastic differential equations. arXiv:2108.01073 (2021)
28. Miller, M.I., Trouvé, A., Younes, L.: On the metrics and Euler-Lagrange equations of computational anatomy. Annu. Rev. Biomed. Eng. **4**(1), 375–405 (2002)
29. Mukhopadhyay, S., et al.: Diffusion models beat GANs on image classification (2023)
30. Nguyen, T., Li, Y., Ojha, U., Lee, Y.J.: Visual instruction inversion: image editing via image prompting. Adv. Neural Inf. Process. Syst. **36** (2024)
31. Özbey, M., et al.: Unsupervised medical image translation with adversarial diffusion models. IEEE Trans. Med. Imag. (2023)
32. Patashnik, O., Garibi, D., Azuri, I., Averbuch-Elor, H., Cohen-Or, D.: Localizing object-level shape variations with text-to-image diffusion models. In: Proceedings of the IEEE/CVF International Conference on Computer Vision, pp. 23051–23061 (2023)
33. Radford, A., et al.: Learning transferable visual models from natural language supervision (2021)

34. Ramesh, A., Dhariwal, P., Nichol, A., Chu, C., Chen, M.: Hierarchical text-conditional image generation with clip latents (2022)
35. Ronneberger, O., Fischer, P., Brox, T.: U-Net: convolutional networks for biomedical image segmentation. In: International Conference on Medical Image Computing and Computer-Assisted Intervention, pp. 234–241. Springer (2015)
36. Saharia, C., et al.: Photorealistic text-to-image diffusion models with deep language understanding (2022)
37. Salimans, T., Goodfellow, I., Zaremba, W., Cheung, V., Radford, A., Chen, X.: Improved techniques for training GANs. Adv. Neural Inf. Process. Syst. **29** (2016)
38. Uchiyama, H., et al.: An easy-to-setup 3d phenotyping platform for Komatsuna dataset. In: Proceedings of the IEEE International Conference on Computer Vision Workshops, pp. 2038–2045 (2017)
39. Unterthiner, T., Van Steenkiste, S., Kurach, K., Marinier, R., Michalski, M., Gelly, S.: Towards accurate generative models of video: a new metric & challenges. arXiv:1812.01717 (2018)
40. Vialard, F.X., Risser, L., Rueckert, D., Cotter, C.J.: Diffeomorphic 3d image registration via geodesic shooting using an efficient adjoint calculation. Int. J. Comput. Vision **97**(2), 229–241 (2012)
41. Wang, Z., Bovik, A.C., Sheikh, H.R., Simoncelli, E.P.: Image quality assessment: from error visibility to structural similarity. IEEE Trans. Image Process. **13**(4), 600–612 (2004)
42. Wu, N., Xing, J., Zhang, M.: TLRN: temporal latent residual networks for large deformation image registration. In: International Conference on Medical Image Computing and Computer-Assisted Intervention, pp. 728–738. Springer (2024)
43. Wu, N., Zhang, M.: NeurEPDiff: neural operators to predict geodesics in deformation spaces. In: International Conference on Information Processing in Medical Imaging, pp. 588–600. Springer (2023)
44. Xing, J., et al.: DynamiCrafter: animating open-domain images with video diffusion priors. In: European Conference on Computer Vision, pp. 399–417. Springer (2025)
45. Yang, Y., Fu, H., Aviles-Rivero, A.I., Schönlieb, C.B., Zhu, L.: DiffMIC: dual-guidance diffusion network for medical image classification. In: International Conference on Medical Image Computing and Computer-Assisted Intervention, pp. 95–105. Springer (2023)
46. Yang, Z., et al.: CogVideoX: text-to-video diffusion models with an expert transformer. arXiv:2408.06072 (2024)
47. Younes, L., Arrate, F., Miller, M.I.: Evolutions equations in computational anatomy. Neuroimage **45**(1), S40–S50 (2009)
48. Yu, Z., et al.: SURF-D: generating high-quality surfaces of arbitrary topologies using diffusion models. In: European Conference on Computer Vision, pp. 419–438. Springer (2025)
49. Zhang, M., et al.: Frequency diffeomorphisms for efficient image registration. In: International Conference on Information Processing in Medical Imaging, pp. 559–570. Springer (2017)
50. Zhu, L., et al.: Make-a-volume: leveraging latent diffusion models for cross-modality 3d brain MRI synthesis. In: International Conference on Medical Image Computing and Computer-Assisted Intervention, pp. 592–601. Springer (2023)

Image Enhancement

Cycle-Consistent Zero-Shot Through-Plane Super-Resolution for Anisotropic Head MRI

Samuel W. Remedios[1]( ), Shuwen Wei[1], Aaron Carass[1], Blake E. Dewey[2], and Jerry L. Prince[1]

[1] The Image Analysis and Communications Laboratory, Johns Hopkins University, Baltimore, USA
`samuel.remedios@jhu.edu`
[2] Department of Neurology, Johns Hopkins Hospital, Baltimore, USA

Abstract. Magnetic resonance (MR) images are often acquired as anisotropic volumes in clinical settings. Such volumes have a worse through-plane resolution than in-plane resolution, hampering results in many processing pipelines that expect isotropic resolutions. Super-resolution (SR) is a promising methodology to address this problem, but there is concern whether the estimated high-resolution (HR) image suffers from egregious hallucinations, especially with deep learning methods that produce aesthetically pleasing results. One approach to restrict the impact of hallucinations is to guarantee that the estimated HR image is exactly cycle-consistent with the low-resolution observation. The denoising diffusion null space model (DDNM) achieves this through a range null space decomposition, but the specific design of the forward map is left to the application. In this work, we analyze the forward problem in 2D MR acquisition and construct an appropriate linear map A. We train a denoising diffusion probabilistic model on T_1-weighted (T_1-w) head MR images from multiple datasets and implement DDNM using A for the SR task. We show that the approach yields exact cycle-consistent solutions that are also realistic. We evaluated the approach in a wide variety of T_1-w MR datasets, including withheld subjects from training sites and two sites outside of the training domain. We achieve excellent qualitative and quantitative results according to both distortion and perceptual metrics.

Keywords: super-resolution · MRI · zero-shot · generative model · cycle-consistency

1 Introduction

The acquisition of magnetic resonance (MR) images in 2D is common in clinical applications, striking a balance between signal-to-noise ratio (SNR) and scan time. However, many image processing and analysis pipelines are developed for and tested on 3D-acquired, isotropic, research-quality scans. This hinders the use of such pipelines on clinically acquired data and in pragmatic clinical trials. Specifically, clinically acquired image volumes often have much worse through-plane resolution than in-plane resolution. This is due to two factors. First, 2D multi-slice MR imaging (MRI) commonly acquires images optimized for higher in-plane resolution and lower scan time. The through-plane resolution is then reduced to maintain acceptable SNR for clinical use, making the voxels anisotropic. Second, the spacing between acquired slices may be less than, equal to, or greater than the slice thickness. Due to SNR and scan time concerns, a slice spacing greater than the slice thickness is more common in clinical settings; this is termed "slice gap."

In this work, we consider the relationship between a high-resolution (HR) image—thin slices without slice gap—and a low-resolution (LR) image—thick slices potentially with slice gap. This can be modeled as blurring with a slice selection profile and subsampling at the slice spacing (i.e., a strided convolution). We can write this as the matrix A. Then, the forward model mapping the HR image x to the LR image y is

$$y = Ax. \qquad (1)$$

Since y has fewer slices than x, A is underdetermined and there are many candidate HR images that satisfy Eq. 1.

Super-resolution (SR) is a methodology that aims to produce an estimate of the HR image $\hat{x} \approx x$. Ideally, the $\hat{x}$ estimated in SR is both realistic and cycle-consistent. To be realistic means that, given a probability distribution of real-world HR images $p(x)$, the candidate super-resolved image $\hat{x}$ should be from the distribution $p(x)$. To be cycle-consistent is to satisfy $y = A\hat{x}$; i.e., the super-resolved image maps precisely to the LR image under the forward model. There is a long history of SR applied to MR images. Classically, SR aims to find

$$\hat{x} = \arg\min_{x} ||Ax - y|| + \lambda \mathcal{R}(x), \qquad (2)$$

where the first term optimizes for cycle-consistency and, since there are potentially infinite solutions for a candidate x, the second term regularizes the space of solutions. The form of $\mathcal{R}(\cdot)$ has traditionally been the focus of study, including choices such as sparsity and total variation [4,7,13–15,38]. Alternative denoising approaches have also been developed, including methods such as the plug-and-play denoising prior [22,44,47] and regularization by denoising [35]. A contemporary approach to regularizing the solution space is to leverage deep generative models that characterize the probability distribution of desired images $p(x)$. Works based on compressed sensing using generative models (CSGM) [3,8,9,28] aim to find a result in the range of the generator which maximizes cycle-consistency. While SR is a well-studied field, in Eq. 2 the cycle-consistency term $||Ax - y||$ is not guaranteed to be zero. Thus, harmful hallucinations are a common concern, especially in mission-critical situations such as medical imaging. Guaranteeing cycle-consistency is paramount.

Cycle-consistent SR approaches exist. Deep wavelet SR [17] predicts the missing details lost by the low-pass prototype filter. However, this approach is only cycle-consistent with respect to wavelet prototypes, which may not reflect low-pass filters in real-world image acquisitions. Deep filter bank regression [34] approximates a perfect reconstruction filter bank with A as the low-pass filter and uses a GAN to estimate the missing details. However, the cycle-consistency is only approximate and it is only defined for integer subsampling factors.

The range null space decomposition (RNSD) is $x = A^{\dagger}y + (I - A^{\dagger}A)x$, where $A^{\dagger}$ is the right inverse of A and I is the identity matrix. Several approaches use RNSD to guarantee cycle-consistency. Mardani et al. [26] use RNSD to project a GAN's results onto the cycle-consistent manifold. Their method is framed for compressed sensing MR imaging and A is set to model sparse Fourier sampling. Schwab et al. [40] show that RNSD is a special case of $\mathcal{M}$-regularizations when using the Moore-Penrose pseudoinverse A^+. They define null space networks as deep networks whose outputs are cycle-consistent. The RNSD is also used in hyperspectral image reconstruction [49], wherein the authors analyze and propose the form of A appropriate for their imaging modality.

Wang et al. [50] note that the RNSD does not need a Moore-Penrose pseudoinverse of A; the right-inverse $AA^{\dagger} = I$ is sufficient. With this observation, they set A to model mean downsampling and $A^{\dagger}$ to model mean upsampling. They evaluate their method on natural images using a pre-trained GAN to produce images in the null space of A. The denoising diffusion null space model (DDNM) [51] is an extension of this work with two key improvements. First, they show how to construct A and $A^{\dagger}$ for additional image restoration tasks such as sparse sampling, masking, and natural image colorization. Second, they use a denoising diffusion probabilistic model (DDPM) [20] to generate images in the null space of A and show that the iterative refinement inherent to DDPMs greatly improves the plausibility of the final result.

The contributions of our work are five-fold. **First**, we present an analysis of the forward model in slice-selective acquisitions, including several clinically common slice thicknesses and separations. **Second**, we train a DDPM on 2D sagittal slices from 15000 open-source T_1-w head MR images to approximate $p(x)$. **Third**, based on our analysis of the forward model, we implement DDNM with the Moore-Penrose pseudoinverse, A^+. **Fourth**, we unify the foundations of cycle-consistent approaches. **Fifth**, we conduct several experiments with validation data and withheld site data to show the efficacy of the proposed approach. The result is a generic zero-shot method that super-resolves LR MR images to cycle-consistent solutions.

2 Methods

In this section, we discuss the slice selection profile in 2D multi-slice MRI and analyze the properties in matrix form. Then, we discuss the mathematics behind guaranteed cycle-consistency for SR. Finally, we discuss the implementation of DDNM for 2D MRI, including dataset preparation, the training of a DDPM, and the implementation of the pseudoinverse of the slice selection matrix.

2.1 Preliminaries

We are interested in super-resolving an anisotropic image volume in the through-plane direction. To this end, we will consider signals along the through-plane axis individually in 1D. Throughout this work, we will only consider the discrete, sampled images. Let $x \in \mathbb{R}^n$ be the HR signal and $y \in \mathbb{R}^m$ be the LR signal. Since the LR image volume has fewer slices than the HR image volume, we have $m < n$. Let $A \in \mathbb{R}^{m \times n}$ be the linear map relating x and y as in Eq. 1. We assume that A has rank m; if it does not, then rows can be removed until they are linearly independent and we consider the row-reduced matrix as A.

2.2 Slice Selection Profile

In 2D multi-slice MRI, image volumes are formed by acquiring independent 2D slices and stacking them along the through-plane axis. Although each 2D slice is acquired in k-space, the stacking occurs in the image domain. When each slice is acquired, atomic nuclei must be excited in the desired region. The through-plane magnetic gradient causes the nuclei to resonate at frequencies as a function of position along the through-plane axis. The waveform of the excitation pulse determines the shape excited for each slice. The slice selection is a bandpass filter that limits the region of excited nuclei based on the gradient strength and the shape of the waveform. Because slice selection aims to acquire measurements in a limited field of view, the bandpass filter is ideally a rect; i.e., only the selected region is excited and all others contribute nothing to the signal. In practice, a rect-like, Gaussian, or Shinnar-Le Roux [31,42] slice profile is used. Notably, although the slice selection profile is a bandpass filter, in the context of the through-plane magnetic gradient, local to the selected slice it acts as a low pass filter. It is also applied uniformly to all in-plane voxels. Thus, in this paper, we model the slice selection process as a convolution between an HR signal x and a continuous 1D point-spread function (PSF) h. Like others [16,18,30,32,53], we set h as a Gaussian function with full-width-at-half-max equal to the slice thickness.

We now use h to model the relationship between the HR 3D sampled volume and the corresponding LR volume. Let $X \in \mathbb{R}^{k_1 \times k_2 \times n}$ be the desired HR 3D sampled volume with $k_1 \times k_2$ as the number of in-plane pixels. In this setting, X defines a set of n-length discrete signals. A simulated slice selection of X at location η is written

$$y(\eta) = \sum_{\xi} h(\eta - \xi) x[\xi], \qquad (3)$$

where ξ are discrete indices. The choice of subsequent η determines the amount of slice gap; if the distances between subsequent samples is greater than the FWHM of h, y has slice gap. Similarly if the spacing between subsequent η is equal to or less than the FWHM of h, y has adjacent slices or slice overlap, respectively. Since Eq. 3 resembles convolution, we can write it as a Toeplitz-like matrix A where each entry is $A_{i,j} = h(\eta_i - \xi_j)$. Thus, to model the LR volume $Y \in \mathbb{R}^{k_1 \times k_2 \times m}$, each signal in X is independently degraded by the forward model in Eq. 1 using this A. That is, the set of corresponding m-length signals is $Y = \{Ax | x \in X\}$.

We make two comments on the form of A. First, because h is a non-periodic low pass filter and each row of A is a unique shift of h, rank$(A) = m$. Therefore A is guaranteed to have a right inverse and $AA^\dagger A = A$. Second, null$(A) \neq \{0\}$; i.e., there are missing details that must be recovered.

2.3 Cycle-Consistency

A super-resolved signal $\hat{x}$ is cycle-consistent with respect to Eq. 1 if $A\hat{x} = y$. This can be accomplished by the RNSD, which uses the following equation:

$$\hat{x} = A^\dagger y + (I - A^\dagger A)\bar{x}, \qquad (4)$$

where $\bar{x}$ is an intermediate prediction. Since $(I - A^\dagger A)$ is a projection onto the null space of A, $\hat{x}$ is guaranteed to be cycle-consistent with y.

We consider whether an alternative null space projector exists. Perfect reconstruction of x is possible given an invertible matrix $H = \begin{bmatrix} A \\ R \end{bmatrix}$, where $R \in \mathbb{R}^{(n-m) \times n}$ has rows linearly independent to the rows of A. By construction, rank$(R) = n - m$ and rank$(H) = n$; R therefore has a right inverse $R^\dagger$ and H is invertible. Because the columns of $R^\dagger \in$ null(A) and the columns of $A^\dagger \in$ null(R), $AR^\dagger = RA^\dagger = 0$. Since

$$\begin{bmatrix} A \\ R \end{bmatrix} \begin{bmatrix} A^\dagger & R^\dagger \end{bmatrix} = \begin{bmatrix} AA^\dagger & AR^\dagger \\ RA^\dagger & RR^\dagger \end{bmatrix} = I,$$

we have

$$H^{-1} = \begin{bmatrix} A^\dagger & R^\dagger \end{bmatrix}.$$

Then,

$$I = H^{-1}H = \begin{bmatrix} A^\dagger & R^\dagger \end{bmatrix} \begin{bmatrix} A \\ R \end{bmatrix} = A^\dagger A + R^\dagger R$$

$$(I - A^\dagger A) = R^\dagger R.$$

We conclude that, while many R exist, $R^\dagger R$ is the same projector as $(I - A^\dagger A)$ and any remaining freedom in producing a cycle-consistent estimate must lie in the null space of A.

2.4 Super-Resolution

Any choice $\bar{x}$ in Eq. 4 yields a cycle-consistent $\hat{x}$. In this work, we implement DDNM [51] to produce $\bar{x} \sim p(x)$. DDNM leverages the inherent iterative denoising and refinement of DDPMs to improve the quality of $\bar{x}$ while remaining cycle-consistent. For every denoising step t, the diffusion model first estimates $\bar{x}_t$. Then, DDNM computes the next time step with

$$\hat{x}_{t-1} = c_1^{(t)} \cdot \left(A^\dagger y + (I - A^\dagger A)\bar{x}_t \right) + c_2^{(t)} \cdot \epsilon_\theta^{(t)}(\hat{x}_t),$$

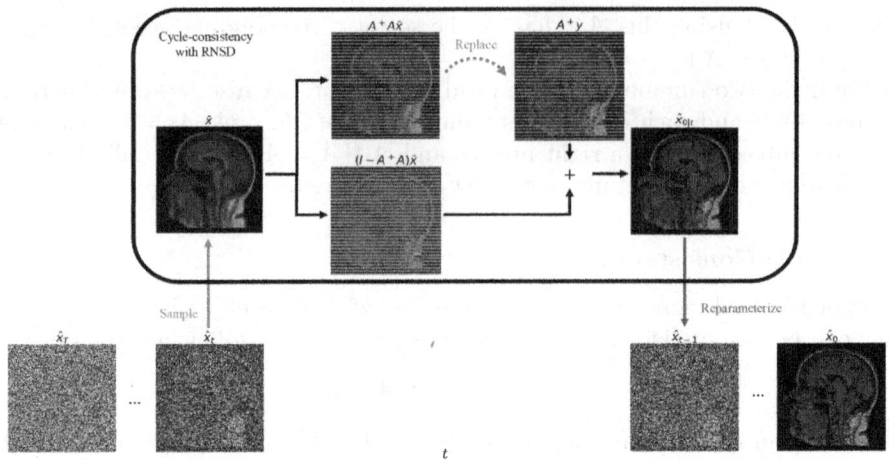

Fig. 1. In DDNM, the RNSD is used within each diffusion timestep to align the sampled $\bar{x}$ with the LR observation y to produce a cycle-consistent $\hat{x}$. At timestep t, the sampled image $\bar{x}$ is split into its complementary components. The component in the range of A is replaced by $A^+ y$ and summed element-wise with the estimated component in the null space of A to produce $\hat{x}_{0|t}$. This is then used as the new mean to reparameterize the sample $\hat{x}_{t-1}$. This process repeats until the final result $\hat{x} = \hat{x}_0$ is produced.

where $c_1^{(t)}$ and $c_2^{(t)}$ are the standard time-varying constants needed for variance preservation in DDPMs and $\epsilon_\theta^{(t)}$ is the denoising function parameterized by θ. This is the conventional implementation of the denoising diffusion implicit model (DDIM) [43], but DDNM alters the predicted x_0 term to be cycle-consistent with y. We illustrate this in Fig. 1. As t approaches 0, the three primary outputs of DDNM approach the true answer. We show this in Fig. 2.

To use DDNM, both $A^\dagger$ and a deep generative model ϵ_θ are required. We use the Moore-Penrose pseudoinverse A^+ of the A described in Sect. 2.2, which satisfies the right-inverse condition needed by RNSD. We train ϵ_θ in Sect. 3.2.

3 Experiments and Results

Our goal is to super-resolve the through-plane axis in 2D-acquired MR volumes. In the following experiments, we consider axial acquisition of human heads, meaning the sagittal and coronal views of the volume are degraded. For the rest of this paper, we will study central sagittal slices in 2D. To this end, the denoising model we used ϵ_θ is a 2D U-Net [36]. The downsampling still occurs in 1D.

3.1 Data and Pre-processing

The data used in this paper were sourced from several open datasets: ABIDE [11], ABIDE II [10], ADNI [21], AIBL [37], COBRE [33], CORR [55], DLBS [6],

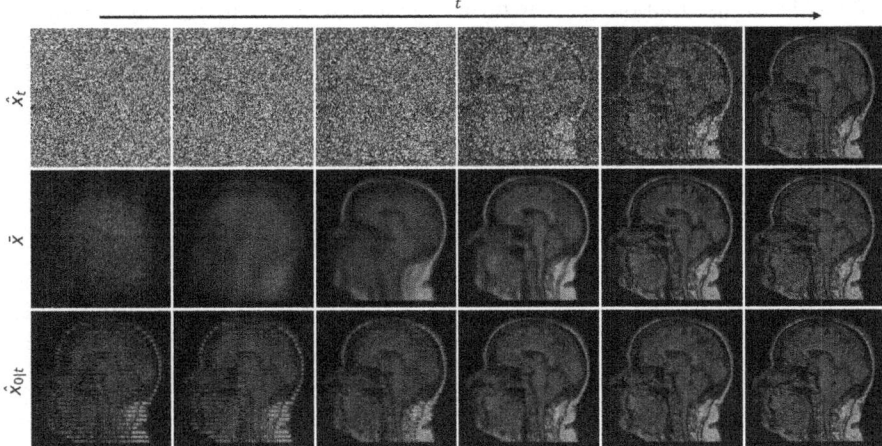

Fig. 2. The three primary results of DDNM as the diffusion time t progresses are shown row-wise. The first row shows the cycle-consistent iterative denoising result $\hat{x}_t$; our final result is when $t = 0$. The second row shows the model's sample $\bar{x}$, which becomes more realistic over time. This also has influence from the observation replacement step starting at time T. The third row shows the cycle-consistent prediction $\hat{x}_{0|t}$, which better aligns with the observation over time.

FCON1000 [2], HCP [46], ICBM [27], MPILEMON [1], NKIROCKLAND [29], OASIS-3 [23], PPMI [12], SALD [52], and SLIM [24]. We selected the T_1-w images from these datasets with resolutions between 0.9 and 1.1 mm in all three cardinal axes. All image volumes were set to a common orientation and were bias-field corrected by N4 [45].

Data were split into training and testing sets. Two datasets, AIBL and ICBM, were entirely withheld from training. For all other sites, approximately 10% of subjects were withheld from training by random selection. For training, the 64 central sagittal slices were extracted from each image volume. For testing, three central slices were extracted from each image volume spaced 13 mm to the right, 5 mm to the left, and 9 mm to the left of the midline. A summary of the data distribution is shown in Table 1.

3.2 Generative Model Training

DDNM only requires a pre-trained DDPM and does not require specialized training. Thus, we trained ϵ_θ using the MONAI framework [5]. The network was a 2D U-Net with 32, 64, 64, and 64 channels per level with eight-headed attention only in the fourth level. One residual block was used per layer. The model was trained using the conventional variance preservation linear noise schedule with $T = 1,000$ steps. The weights θ were optimized with the Adam optimizer and a learning rate of 2.5×10^{-5}. The model was trained for 50 epochs. At inference time, 16 DDIM [43] steps were used. Random non-cherry-picked samples from

Table 1. Data distribution of T_1-w image volumes among training and testing splits. In training, 64 central sagittal slices were extracted. In testing, three central sagittal slices were extracted. AIBL and ICBM were entirely withheld from training.

Dataset	Training		Testing	
	# Subjects	# Slices	# Subjects	# Slices
ABIDE	387	24,768	53	159
ABIDEII	890	56,960	100	300
ADNI	1,687	107,968	177	531
COBRE	1,146	73,344	139	417
CORR	2,230	142,720	247	741
DLBS	279	17,856	36	108
FCON1000	576	36,864	54	162
HCP	1,012	64,768	105	315
MPILEMON	212	13,568	15	45
NKIROCKLAND	2,078	132,992	220	660
OASIS-3	1,336	85,504	132	396
PPMI	1,530	97,920	160	480
SALD	433	27,712	61	183
SLIM	948	60,672	99	297
AIBL	0	0	170	510
ICBM	0	0	727	2,181
Total	14,744	943,616	2,495	7,485

the model using these DDIM steps are shown in Fig. 3. The 2D sagittal slice data used to train ϵ_θ were of approximately 1×1 mm, and slices were center padded and cropped to be of size 256×256 pixels. This resulted in the DDPM modeling the distribution of 1 mm^2 isotropic sagittal slices.

3.3 Simulations of 2D Acquisition

To evaluate the efficacy of both the generative model and the DDNM approach, we simulated 2D axial acquisitions by degrading our test set slices along the superior-inferior axis. Following our reasoning in Sect. 2.2, we chose h as a normalized Gaussian. The FWHM of the Gaussian determines the slice thickness and the separation of the Gaussians when constructing A determines the slice separation. Since slice gap is more common than slice overlap in clinical scenarios, we denote a slice thickness of τ mm and gap of γ mm as $\tau \| \gamma$. We validated the method in three slice thicknesses and gaps common to clinical acquisitions: $3\|1$, $4\|0$, and $5\|1.5$. Since the HR target is 1 mm^2, these correspond to $4\times, 4\times$, and $6.5\times$ 1D scale factors, respectively, though the first two differ in PSF width.

3.4 Experiments

We compared SR results for three methods: traditional bicubic B-spline interpolation, CSGM [3,28], and DDNM. Bicubic B-spline interpolation serves as a baseline, and CSGM is a generative approach that optimizes a variant of Eq. 2 and therefore aims to be cycle-consistent. Other approaches such as CycleGAN [54] and conditional DDPMs [39] do not guarantee cycle-consistency and DDNM has already shown superior performance to similar approaches such as RePaint [25]. Both CSGM and DDNM used the same generative model described in Sect. 3.2, the same number of DDIM inference steps (16), the same matrix A, and the same random seeds for all samples. For CSGM zero-shot SR, we optimized the latent code using the Adam optimizer with a learning rate of 0.01 for 100 gradient update steps (hyperparameters optimized empirically). As such, the runtime of CSGM was 100× slower. For evaluation, we used the testing set described in Tab. 1. We used three quantitative metrics to compare approaches: peak-signal-to-noise-ratio (PSNR), structural similarity (SSIM), and the Frechet inception distance (FID) [19]. We used the scikit-image [48] implementations of PSNR and SSIM and the PyTorch implementation of FID [41].

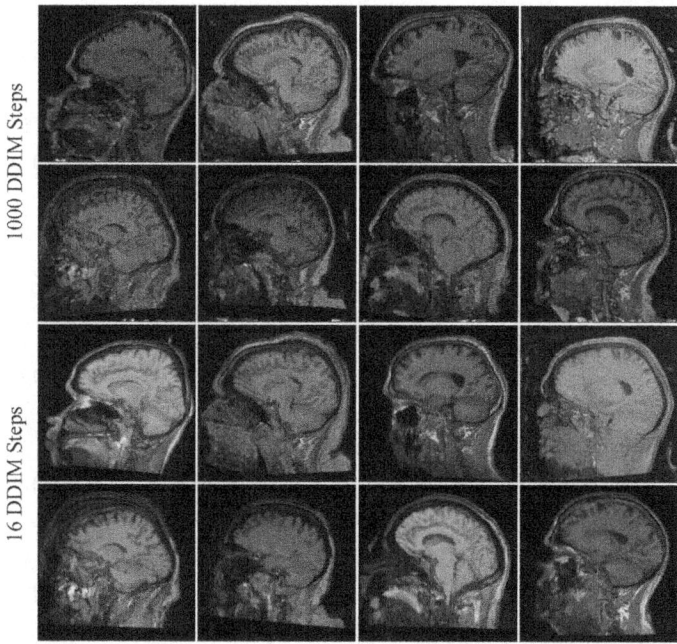

Fig. 3. Uncurated samples from our trained DDPM using 1000 (top two rows) and 16 (bottom two rows) DDIM steps using exactly the same latent noise and random seeds. Using 16-steps yields less noise but is sufficient for realistic super-resolution.

3.5 Results

Figure 4 shows qualitative results across the considered resolutions and methods. DDNM outperforms the other methods qualitatively and also is guaranteed to be cycle-consistent with the observations. Given the same generative model and random seed, CSGM also performs comparably, though the potential of hallucinations is higher without the cycle-consistency guarantee. Additionally, since CSGM requires multiple iterations, it is slower than DDNM. Also, while both DDNM and CSGM produce reasonable results, not all sulci are biologically plausible. This may be better addressed by using a 3D model that accounts for anatomical variation along both in-plane axes.

Table 2. Metrics computed over the test set of 7485 sagittal slices. The top portion of the table reports metrics on the data from the same sites used in training and the bottom portion of the table reports on the two withheld sites. All DDPM-based methods in the table used 16 DDIM inference steps. PSNR and SSIM are reported as mean values ± std. Best results are reported in **bold** and second-best results are underlined.

			Bicubic	CSGM	DDNM
Training Sites	3||1	PSNR↑	28.00 ± 2.11	24.32 ± 2.49	**29.73 ± 2.71**
		SSIM↑	0.88 ± 0.05	0.79 ± 0.05	**0.91 ± 0.04**
		FID↓	83.58	37.40	**9.05**
	4||0	PSNR↑	27.81 ± 2.07	24.19 ± 2.47	**29.89 ± 2.67**
		SSIM↑	0.87 ± 0.05	0.79 ± 0.05	**0.91 ± 0.04**
		FID↓	86.07	38.68	**9.70**
	5||1.5	PSNR↑	25.64 ± 2.00	23.35 ± 2.08	**26.76 ± 2.25**
		SSIM↑	0.80 ± 0.06	0.76 ± 0.05	**0.84 ± 0.06**
		FID↓	172.70	38.60	**18.09**
Withheld Sites	3||1	PSNR↑	27.52 ± 1.15	28.67 ± 1.78	**29.67 ± 1.71**
		SSIM↑	0.86 ± 0.02	0.86 ± 0.03	**0.89 ± 0.02**
		FID↓	87.25	23.02	**10.52**
	4||0	PSNR↑	27.40 ± 1.14	28.46 ± 1.70	**29.78 ± 1.63**
		SSIM↑	0.85 ± 0.02	0.85 ± 0.03	**0.89 ± 0.02**
		FID↓	89.89	26.60	**10.13**
	5||1.5	PSNR↑	25.26 ± 1.12	26.23 ± 1.52	**26.69 ± 1.51**
		SSIM↑	0.77 ± 0.03	0.79 ± 0.04	**0.81 ± 0.03**
		FID↓	191.44	29.12	**22.10**

Table 2 shows the quantitative results, where DDNM outperforms the other methods in terms of both the conventional distortion metrics PSNR and SSIM as well as the perceptual metric FID. These quantitative results also correlate well with the qualitative appearance in Fig. 4. Training and withheld sites are

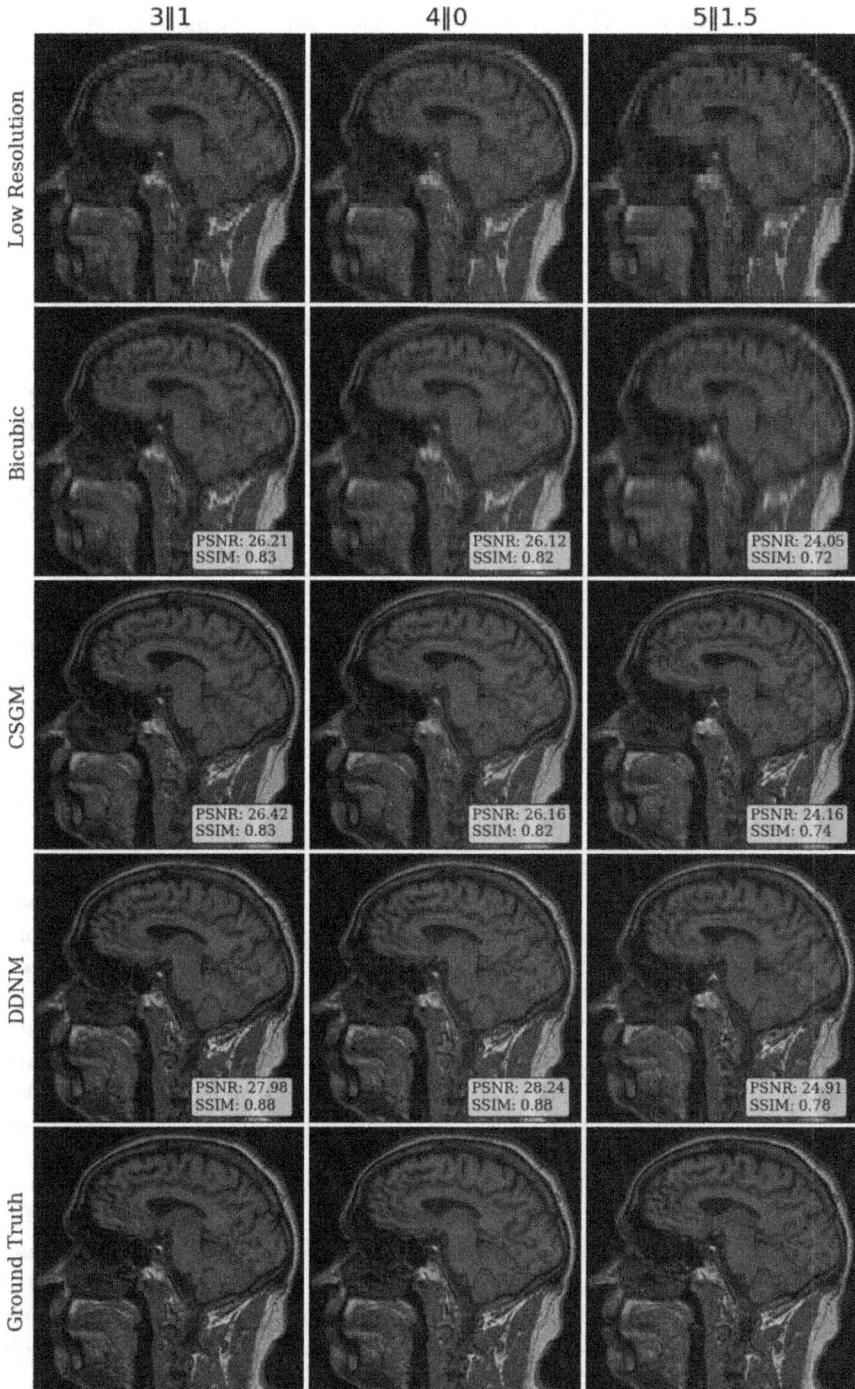

Fig. 4. A representative result at common clinical resolutions. From left to right: the LR image visualized with nearest-neighbor interpolation, bicubic interpolation, CSGM, DDNM, and the HR ground truth image. PSNR and SSIM are calculated for the displayed 2D slice.

separated in the table. While it is expected that withheld metrics are worse, it is surprising that CSGM improves across the board when moving from training sites to withheld sites. For some cases, such as at 5∥1.5, the differences in PSNR between CSGM and DDNM are small; however, DDNM produces cycle-consistent results with improved FID and thus are closer to our ideal case of realistic images that are consistent with the observation.

4 Discussion and Conclusion

In this work, we presented an analysis and empirical evaluation of a cycle-consistent zero-shot super-resolution technique for head MR images. We discussed the model of 2D MR acquisition and showed how to implement this model as a Toeplitz-like matrix for use in the canonical linear observation equation. We also discussed the range null space decomposition and proved that it is equivalent to any other solution of the noiseless observation described by Eq. 1. We then trained a DDPM for 2D T_1-w sagittal slices and implemented DDNM to achieve exact cycle-consistent SR. Our results showed that this method is better in both distortion and perception metrics. Since DDNM is zero-shot, the same pre-trained generative model can be used for a variety of different resolutions without the need for further training or fine-tuning. Since the method is cycle-consistent, the only aspects of the image which are prone to hallucination are the unobserved high-frequencies. These, however, are regularized by the form of the generative model, which we show empirically.

Medical images have long been a focus of SR. Although progress has accelerated alongside the development of data-driven methods, relatively few works in the computer vision community have focused on safety and reliability. DDNM, with its exact cycle-consistency, has shown potential in addressing this. However, there is still more work to do. First, the range null space decomposition relies on knowing A. In our case, this involves knowing both the shape of the slice selection profile h and the separation between slices. While the latter can be derived from the medical image header, the former is often obfuscated by the scanner manufacturer. There is some work in kernel estimation [18], but the estimation of the continuous form of h is still unsolved. Second, this work studied the efficacy of DDNM on 2D sagittal slices. Ideally, the pre-trained DDPM would generate full 3D volumes at the desired resolution. There are two tasks to solve: 1) hardware and data constraints necessary to achieve a sufficient 3D model; and 2) expanding the mathematical model of A to include Fourier acquisition in-plane for images which have lower in-plane resolution than desired. Third, the metrics used in this paper were only with respect to the image domain. Metrics such as PSNR, SSIM, and FID are limited in their ability to describe the goodness of super-resolved images. Downstream task performance, such as classification or segmentation, is needed to evaluate the utility of super-resolved image in real-world scenarios. Fourth, this work showed a model trained on a specific data distribution: T_1-w MR whole-head images without apparent pathology at clinical magnet strengths (1.5T and 3T). Many neuroimaging studies are conducted outside of this distribution: images may have apparent pathology, have non-T_1-w contrasts, are acquired at other field strengths such as 7T or 65mT, etc. Future work should investigate the generalization of MR generative models in such cases. Addressing these tasks will undoubtedly move the field of super-resolution forward. As generative modeling and MRI modeling improve, the utility and practical application of safe super-resolution methods will ideally advance medical image analysis.

Acknowledgements. This material is supported by the National Science Foundation Graduate Research Fellowship under Grant No. DGE-1746891 (Remedios). This work also received support from FG-2008-36966 (Dewey), DoD CDMRP HT9425-24-1-0785 (Dewey) and CDMRP W81XWH2010912 (Prince). Data used in this paper were sourced from open datasets; due to space constraints, please visit the citations of the appropriate data descriptor manuscripts for relevant information.

Disclosure of Interests. The authors have no relevant conflicts to disclose.

References

1. Babayan, A., et al.: A mind-brain-body dataset of MRI, EEG, cognition, emotion, and peripheral physiology in young and old adults. Sci. Data **6**(1), 1–21 (2019)
2. Biswal, B.B., et al.: Toward discovery science of human brain function. Proc. Natl. Acad. Sci. **107**(10), 4734–4739 (2010)
3. Bora, A., Jalal, A., Price, E., Dimakis, A.G.: Compressed sensing using generative models. In: International Conference on Machine Learning, pp. 537–546. PMLR (2017)
4. Candès, E.J., Romberg, J., Tao, T.: Robust uncertainty principles: exact signal reconstruction from highly incomplete frequency information. IEEE Trans. Inf. Theory **52**(2), 489–509 (2006)
5. Cardoso, M.J., et al.: MONAI: an open-source framework for deep learning in healthcare. arXiv preprint arXiv:2211.02701 (2022)
6. Chan, M.Y., Park, D.C., Savalia, N.K., Petersen, S.E., Wig, G.S.: Decreased segregation of brain systems across the healthy adult lifespan. Proc. Natl. Acad. Sci. **111**(46), E4997–E5006 (2014)
7. Danielyan, A., Katkovnik, V., Egiazarian, K.: BM3D frames and variational image deblurring. IEEE Trans. Image Process. **21**(4), 1715–1728 (2011)
8. Daras, G., Dagan, Y., Dimakis, A., Daskalakis, C.: Score-guided intermediate level optimization: fast Langevin mixing for inverse problems. In: Proceedings of the 39th International Conference on Machine Learning. Proceedings of Machine Learning Research, 17–23 July 2022, vol. 162, pp. 4722–4753. PMLR (2022)
9. Daras, G., Dean, J., Jalal, A., Dimakis, A.: Intermediate layer optimization for inverse problems using deep generative models. In: Proceedings of the 38th International Conference on Machine Learning. Proceedings of Machine Learning Research, 18–24 July 2021, vol. 139, pp. 2421–2432. PMLR (2021)
10. Di Martino, A., et al.: Enhancing studies of the connectome in autism using the autism brain imaging data exchange II. Sci. Data **4**(1), 1–15 (2017)
11. Di Martino, A., et al.: The autism brain imaging data exchange: towards a large-scale evaluation of the intrinsic brain architecture in autism. Mol. Psychiatry **19**(6), 659–667 (2014)

12. Dinov, I.D., et al.: Predictive big data analytics: a study of Parkinson's disease using large, complex, heterogeneous, incongruent, multi-source and incomplete observations. PLoS ONE **11**(8), e0157077 (2016)
13. Donoho, D.L.: Compressed sensing. IEEE Trans. Inf. Theory **52**(4), 1289–1306 (2006)
14. Elad, M., Aharon, M.: Image denoising via sparse and redundant representations over learned dictionaries. IEEE Trans. Image Process. **15**(12), 3736–3745 (2006)
15. Figueiredo, M.A., Nowak, R.D.: Wavelet-based image estimation: an empirical Bayes approach using Jeffrey's noninformative prior. IEEE Trans. Image Process. **10**(9), 1322–1331 (2001)
16. Greenspan, H.: Super-resolution in medical imaging. Comput. J. **52**(1), 43–63 (2009)
17. Guo, T., Seyed Mousavi, H., Huu Vu, T., Monga, V.: Deep wavelet prediction for image super-resolution. In: Proceedings of the IEEE Conference on Computer Vision and Pattern Recognition Workshops, pp. 104–113 (2017)
18. Han, S., Remedios, S.W., Schär, M., Carass, A., Prince, J.L.: ESPRESO: an algorithm to estimate the slice profile of a single magnetic resonance image. Magn. Reson. Imaging **98**, 155–163 (2023)
19. Heusel, M., Ramsauer, H., Unterthiner, T., Nessler, B., Hochreiter, S.: GANs trained by a two time-scale update rule converge to a local Nash equilibrium. Adv. Neural Inf. Process. Syst. **30** (2017)
20. Ho, J., Jain, A., Abbeel, P.: Denoising diffusion probabilistic models. Adv. Neural. Inf. Process. Syst. **33**, 6840–6851 (2020)
21. Jack Jr, C.R., et al.: The Alzheimer's disease neuroimaging initiative (ADNI): MRI methods. J. Magn. Reson. Imaging Off. J. Int. Soc. Magn. Reson. Med. **27**(4), 685–691 (2008)
22. Kamilov, U.S., Mansour, H., Wohlberg, B.: A plug-and-play priors approach for solving nonlinear imaging inverse problems. IEEE Signal Process. Lett. **24**(12), 1872–1876 (2017)
23. LaMontagne, P.J., et al.: OASIS-3: longitudinal neuroimaging, clinical, and cognitive dataset for normal aging and Alzheimer disease. medrxiv pp. 2019–12 (2019)
24. Liu, W., et al.: Longitudinal test-retest neuroimaging data from healthy young adults in southwest China. Sci. Data **4**(1), 1–9 (2017)
25. Lugmayr, A., Danelljan, M., Romero, A., Yu, F., Timofte, R., Van Gool, L.: RePaint: inpainting using denoising diffusion probabilistic models. In: Proceedings of the IEEE/CVF conference on Computer Vision and Pattern Recognition, pp. 11461–11471 (2022)
26. Mardani, M., et al.: Deep generative adversarial neural networks for compressive sensing MRI. IEEE Trans. Med. Imaging **38**(1), 167–179 (2018)
27. Mazziotta, J., et al.: A probabilistic atlas and reference system for the human brain: International Consortium for Brain Mapping (ICBM). Philos. Trans. Royal Soc. Lond. Ser. B Biol. Sci. **356**(1412), 1293–1322 (2001)
28. Menon, S., Damian, A., Hu, S., Ravi, N., Rudin, C.: PULSE: self-supervised photo upsampling via latent space exploration of generative models. In: Proceedings of the IEEE/CVF Conference on Computer Vision and Pattern Recognition, pp. 2437–2445 (2020)
29. Nooner, K.B., et al.: The NKI-Rockland sample: a model for accelerating the pace of discovery science in psychiatry. Front. Neurosci. **6**, 152 (2012)
30. Oktay, O., et al.: Multi-input cardiac image super-resolution using convolutional neural networks. In: Ourselin, S., Joskowicz, L., Sabuncu, M.R., Unal, G., Wells,

W. (eds.) MICCAI 2016. LNCS, vol. 9902, pp. 246–254. Springer, Cham (2016). https://doi.org/10.1007/978-3-319-46726-9_29
31. Pauly, J., Le Roux, P., Nishimura, D., Macovski, A.: Parameter relations for the Shinnar-Le Roux selective excitation pulse design algorithm (NMR imaging). IEEE Trans. Med. Imaging **10**(1), 53–65 (1991)
32. Pham, C.H., et al.: Multiscale brain MRI super-resolution using deep 3D convolutional networks. Comput. Med. Imaging Graph. **77**, 101647 (2019)
33. Power, J.D., Barnes, K.A., Snyder, A.Z., Schlaggar, B.L., Petersen, S.E.: Spurious but systematic correlations in functional connectivity MRI networks arise from subject motion. Neuroimage **59**(3), 2142–2154 (2012)
34. Remedios, S.W., et al.: Deep filter bank regression for super-resolution of anisotropic MR brain images. In: International Conference on Medical Image Computing and Computer-Assisted Intervention, pp. 613–622. Springer (2022)
35. Romano, Y., Elad, M., Milanfar, P.: The little engine that could: regularization by denoising (RED). SIAM J. Imag. Sci. **10**(4), 1804–1844 (2017)
36. Ronneberger, O., Fischer, P., Brox, T.: U-Net: convolutional Networks for Biomedical Image Segmentation. In: Navab, N., Hornegger, J., Wells, W.M., Frangi, A.F. (eds.) MICCAI 2015. LNCS, vol. 9351, pp. 234–241. Springer, Cham (2015). https://doi.org/10.1007/978-3-319-24574-4_28
37. Rowe, C.C., et al.: Amyloid imaging results from the Australian Imaging, Biomarkers and Lifestyle (AIBL) study of aging. Neurobiol. Aging **31**(8), 1275–1283 (2010)
38. Rudin, L.I., Osher, S., Fatemi, E.: Nonlinear total variation based noise removal algorithms. Physica D **60**(1–4), 259–268 (1992)
39. Saharia, C., Ho, J., Chan, W., Salimans, T., Fleet, D.J., Norouzi, M.: Image super-resolution via iterative refinement. IEEE Trans. Pattern Anal. Mach. Intell. **45**(4), 4713–4726 (2022)
40. Schwab, J., Antholzer, S., Haltmeier, M.: Deep null space learning for inverse problems: convergence analysis and rates. Inverse Prob. **35**(2), 025008 (2019)
41. Seitzer, M.: pytorch-fid: FID Score for PyTorch (2020). https://github.com/mseitzer/pytorch-fid. Version 0.3.0
42. Shinnar, M., Eleff, S., Subramanian, H., Leigh, J.S.: The synthesis of pulse sequences yielding arbitrary magnetization vectors. Magn. Reson. Med. **12**(1), 74–80 (1989)
43. Song, J., Meng, C., Ermon, S.: Denoising diffusion implicit models. In: International Conference on Learning Representations (2021). https://openreview.net/forum?id=St1giarCHLP
44. Sun, Y., Wohlberg, B., Kamilov, U.S.: An online plug-and-play algorithm for regularized image reconstruction. IEEE Trans. Comput. Imaging **5**(3), 395–408 (2019)
45. Tustison, N.J., et al.: N4ITK: improved N3 bias correction. IEEE Trans. Med. Imaging **29**(6), 1310–1320 (2010)
46. Van Essen, D.C., et al.: The Human Connectome Project: a data acquisition perspective. Neuroimage **62**(4), 2222–2231 (2012)
47. Venkatakrishnan, S.V., Bouman, C.A., Wohlberg, B.: Plug-and-play priors for model based reconstruction. In: 2013 IEEE Global Conference on Signal and Information Processing, pp. 945–948. IEEE (2013)
48. van der Walt, S., et al.: Scikit-image: image processing in Python. PeerJ **2**, e453 (2014). https://doi.org/10.7717/peerj.453
49. Wang, J., Wang, S., Zhang, R., Zheng, Z., Liu, W., Wang, X.: A range-null space decomposition approach for fast and flexible spectral compressive imaging. arXiv preprint arXiv:2305.09746 (2023)

50. Wang, Y., Hu, Y., Yu, J., Zhang, J.: GAN prior based null-space learning for consistent super-resolution. In: Proceedings of the AAAI Conference on Artificial Intelligence, vol. 37, no. 3, pp. 2724–2732 (2023)
51. Wang, Y., Yu, J., Zhang, J.: Zero-shot image restoration using denoising diffusion null-space model. In: The Eleventh International Conference on Learning Representations (2023)
52. Wei, D., et al.: Structural and functional brain scans from the cross-sectional Southwest University adult lifespan dataset. Sci. Data **5**(1), 1–10 (2018)
53. Zhao, C., Dewey, B.E., Pham, D.L., Calabresi, P.A., Reich, D.S., Prince, J.L.: SMORE: a self-supervised anti-aliasing and super-resolution algorithm for MRI using deep learning. IEEE Trans. Med. Imaging **40**(3), 805–817 (2020)
54. Zhu, J.Y., Park, T., Isola, P., Efros, A.A.: Unpaired image-to-image translation using cycle-consistent adversarial networks. In: Proceedings of the IEEE International Conference on Computer Vision, pp. 2223–2232 (2017)
55. Zuo, X.N., et al.: An open science resource for establishing reliability and reproducibility in functional connectomics. Sci. Data **1**(1), 1–13 (2014)

Bayesian Learning with Stochastic Perturbations and Langevin Expectation Maximization for Unsupervised DNN Image Quality Enhancement

Vatsala Sharma[✉] and Suyash P. Awate

Computer Science and Engineering (CSE) Department, Indian Institute of Technology (IIT) Bombay, Mumbai, India
vatsala@cse.iitb.ac.in

Abstract. *Unsupervised* learning of deep-neural-networks (DNNs) for image quality enhancement can overcome the real-world challenge of the lack of high-quality training images. Typical DNNs for weakly/unsupervised image restoration make strong assumptions that are often infeasible or undesirable in clinical scenarios, e.g., they (a) demand multiple acquired degraded instances per scene, (b) simulate degraded instances assuming independent identically-distributed noise per pixel, (c) demand pre-training large diffusion models on large sets of high(er)-quality images, or (d) ignore uncertainty estimation in their outputs. We propose a novel *Bayesian* DNN framework for unsupervised image quality enhancement incorporating (i) *stochastic perturbations* at multiple stages within the DNN architecture, for regularization and data-driven automatic generation of realistic degraded instances, (ii) *variational*/distribution modeling in latent space, (iii) novel Monte-Carlo *expectation maximization* of DNN parameters using *Langevin diffusion* in latent space, and (iv) novel *low-density sampling* for perturbations using normalized Langevin diffusion. Results on publicly available datasets demonstrates the benefits of our DNN framework over existing methods in CT, MRI, and PET.

Keywords: Unsupervised · image quality enhancement · Bayesian · stochastic perturbations · variational · Monte-Carlo EM · Langevin diffusion

1 Introduction

Deep neural network (DNN) learning for image-quality enhancement can play a vital role in many applications in medical imaging and image analysis, e.g., reducing radiation doses in X-ray computed tomography (CT) and positron emission tomography (PET), faster acquisition protocols in magnetic resonance imaging (MRI), improving performance of subsequent processing methods (e.g., segmentation or classification), etc. However, typical DNN methods use supervised learning that needs paired training sets, with low-quality and high-quality image counterparts, which are expensive, time-consuming, or infeasible to acquire in many clinical scenarios. Thus, we focus on *unsupervised* learning with DNNs.

The recent literature presents several methods on weakly-supervised, unsupervised, or zero-shot image quality enhancement. However, such methods make strong *assumptions* about the training images and their degradations, which either (i) become *invalid* in clinical imaging scenarios [15,18,21,29], or (ii) remain limited to one specific medical imaging modality [31]. For example, some methods demand multiple acquired degraded instances per scene [15,24,25] or simulate noisy instances assuming independent identically-distributed (i.i.d.) noise per pixel [18,21]. Zero-shot methods [28], unlike learning-based methods, fail to leverage any information from a training set (high or low quality). A recent class of methods [29] are often (mis)referred to as zero-shot despite relying on pre-training large diffusion models using a large set of high-quality images; also, their training set is often in a non-medical domain with little resemblance to medical images. All aforementioned classes of methods ignore uncertainty estimation in their outputs, when such uncertainty can be vital in high-stake medical scenarios to human experts or subsequent image-processing methods [14,34].

Learned DNN models often face severe challenges in generalizing to out-of-distribution (OOD) data [5], and sometimes even in-distribution (ID) data. DNN model learning (for training-based methods) or model fitting (for zero-shot methods) leverages many regularization schemes to reliably estimate the underlying regressors, e.g., priors on DNN parameters, stochasticity in optimization using batch subsets and algorithms like SGD [9] and Adam [10], dropout [8], etc. In the same vein, we propose stochastic perturbations at multiple DNN layers, within a novel Bayesian learning framework, for improved DNN regularization.

This paper makes many contributions. We propose a novel *Bayesian* DNN-learning formulation for unsupervised image quality enhancement that includes (i) *stochastic perturbations* at multiple stages within the DNN architecture, for regularization as well as data-driven automatic generation of realistic degraded instances, and (ii) *variational*/distribution modeling in latent space that also allows uncertainty estimation for enhanced images. To sample stochastic perturbations, we propose a novel *low-density sampling* method using *normalized Langevin* diffusion. Given the sampled perturbations, to optimize DNN parameters, we propose a novel Monte-Carlo *expectation maximization* (EM) that subsumes (variational) sampling in latent space using Langevin diffusion [20]. Results on publicly available datasets demonstrates that, in CT, MRI, and PET, our DNN framework outperforms existing methods, on ID and OOD images.

2 Related Work

Noise2Noise (N2N) [15] is a weakly-supervised method for image denoising, which trains using multiple independently acquired noisy images of the same scene. N2N showed that even without high-quality data, DNNs can learn to denoise nearly as effectively as fully-supervised methods. However, the need for multiple acquisitions per scene hampers N2N's applicability in medical imaging. Recorrupted-to-recorrupted (R2R) [18] extends N2N to cases when multiple acquisitions are unavailable; given a single noisy image, R2R simulates a pair of degraded images for use during training by recorrupting them, adhocly, by introducing i.i.d. Gaussian noise. UNID [21] further extends R2R to non-blind deconvolution, which can be used for deblurring and super-resolution when given an appropriate degradation kernel. ssDEM [24,26] uses semi-supervised learning for image quality enhancement of low-dose PET-CT, but, like N2N, relies on multiple acquisitions per scene (unlike typical clinical protocols). Unlike N2N and ssDEM that need multiple acquired noisy images per scene, and unlike R2R and UNID that simulate two recorrupted images by introducing (adhocly) i.i.d. Gaussian noise, our framework utilizes a single low-quality image instance per scene and then automatically generates (distributions of) degraded images using (stochastic) data-driven perturbations within a Bayesian learning framework.

Noise2Void (N2V) [13] proposes unsupervised denoising relying on blind-spot masking, to mask out a random set of pixels in the DNN input, and then using these masked pixels at the output end to define the DNN loss. This allows each pixel value to be predicted using only its neighborhood pixel values, allowing for effective denoising for spatially uncorrelated noise. However, the assumption of uncorrelated noise may not hold true in medical applications involving image reconstructions from low-quality/dose or subsampled data. A modified-N2V scheme [27] for PET images uses group-level pre-training followed by individual fine-tuning (which increases run time) during inference. However, it additionally uses anatomical MRI images from PET-MRI, and in case of unavailability of this additional information, the performance of this method is likely to suffer. Unlike N2N, R2R, UNID, ssDEM, N2V, and modified-N2V, our proposed method models data-driven stochastic perturbations at multiple layers in the DNN architecture to model the effects of spatially uncorrelated as well as correlated degradations in input images; we generate these perturbations using a novel Langevin diffusion scheme within a Bayesian learning framework.

Another class of DNN methods, aiming to overcome the paucity of large training sets, rely on *zero-shot* learning, pioneered by deep image prior (DIP) [28]. DIP is an unsupervised method for image restoration devoid of any training stage, but relies on DNN-model fitting during inference itself. Similar methods have been proposed for super-resolution/deconvolution tasks for medical images [1,22,23]. While zero-shot learning eliminates a training stage, thereby also eliminating the need for any training data, it comes with several drawbacks compared to training-based methods that use fully/weakly/un-supervised learning, e.g., (i) risk of model overfitting, (ii) challenging hyperparameter tuning stemming from the absence of a validation dataset, (iii) very long inference times resulting from iterative algorithms (at test time) involving multiple passes through the DNN, etc. Some recent methods relying on large DNN models such as *diffusion models* [2,7,16,29] and transformers [12] (mis)refer to themselves as "zero-shot" methods but, unlike earlier zero-shot methods, e.g., DIP, these typically rely on priors learned from pre-trained diffusion/transformer models that have been trained using supervised learning on large high-quality image sets. Pre-training such large diffusion/transformer models can itself be challenging in the medical domain where data availability is a challenge. Moreover, directly using a prior trained on non-medical images can result in their comparatively poor performance for medical images (as our results empirically demonstrate).

3 Methodology

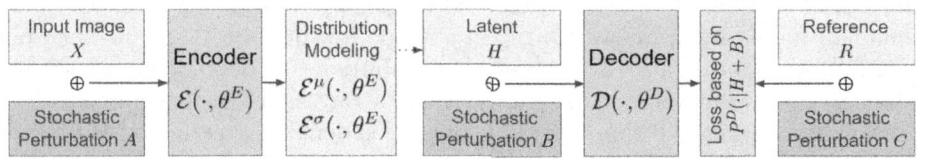

Fig. 1. Bayesian Learning with Stochastic Perturbations and Langevin EM. We model *distributions* on the stochastic perturbations (A, B, C) and the latent representation H, within a novel joint generative model $P(X, R, H, A, B, C, \theta)$. We propose: sampling (A, B, C) from *low-density* regions using normalized Langevin diffusion, and maximum-a-posteriori estimation of θ using Monte-Carlo EM with Langevin sampling of H. Variational modeling on H leads to uncertainty estimates for decoder outputs.

We propose a DNN framework for unsupervised image quality enhancement (Fig. 1), with Bayesian statistical modeling incorporating (i) variational or distribution modeling in latent space, (ii) stochastic perturbations for regularization at multiple stages within the DNN architecture, (iii) novel low-density sampling for perturbations using normalized Langevin diffusion, and (iv) novel EM optimization of DNN parameters using Langevin diffusion in latent space.

3.1 Bayesian Variational Modeling with Stochastic Perturbations

Let X be the *degraded* image (observed) that is input to the DNN for image quality enhancement, with an associated probability density function (PDF) $P(X)$. Associated with X, let Y be the corresponding *high-quality* image (unknown/unobserved), and let R be the corresponding *reference* image (observed) used to compute the loss with respect to the DNN output. In typical supervised learning, $R = Y$ (observed) and the DNN directly learns a mapping from X to Y. However, in weakly/self-supervised scenarios, R can be a lower-quality image or some variant of X itself; some such methods model R as another degraded image that is acquired (e.g., in N2N [15]) or simulated (e.g., in R2R [18]) where X and R are assumed to be independently degraded versions of Y.

For *variational* modeling, let H represent the *hidden/latent* random variable/representation associated with X. Image quality enhancement is an ill-posed inverse problem, and the distribution of H also implicitly represents the distribution/uncertainty associated with the estimated enhanced image.

For *regularized Bayesian learning*, we propose to inject data-driven *stochastic perturbations* [6,19], during DNN learning, at three different stages within the DNN architecture. Let A, B, and C model independent zero-mean additive perturbations associated with X, H, and R, respectively, with *prior* PDFs $P(A;\alpha)$, $P(B;\beta)$, and $P(C;\alpha)$ parameterized by α and β. The role of the random perturbations (A, B, C) is similar, in spirit, to the use of dropout [8] or stochasticity in optimization (in SGD [9] or Adam [10]) which also infuse random fluctuations in DNN layers and parameter estimates. Indeed, better regularization can improve generalization of the DNN regressor for ID as well as OOD images.

We propose an encoder-decoder DNN, with (i) an encoder $\mathcal{E}(\cdot;\theta^E)$, parameterized by θ^E, mapping the perturbed input image $X + A$ to a Gaussian PDF $P(H|X + A)$ on the latent H, and (ii) a decoder $\mathcal{D}(\cdot;\theta^D)$, parameterized by θ^D, aiming to map the perturbed latent variable $H + B$ to the high-quality image Y. The Gaussian latent PDF $P^E(H|X+A)$ is represented by an encoder-mapped mean $\mathcal{E}^\mu(\cdot;\theta^E)$, and an encoder-mapped $\mathcal{E}^\sigma(\cdot;\theta^E)$ that is $\sqrt{2}$ times the standard deviation. We propose a decoder to model a PDF $P^D(Y|H + B)$ on high-quality images Y; this paper models $P^D(Y|H + B)$ as a Gaussian with mean $\mathcal{D}(H + B;\theta^D)$ and standard deviation 1. DNN training evaluates the decoder output PDF $P^D(\cdot|H + B)$ at the perturbed reference image $R + C$, i.e., $P^D(R + C|H + B)$. Let $\theta := \theta^E \cup \theta^D$ model all DNN parameters, with the standard isotropic-Gaussian prior $P(\theta)$ leading to the squared-norm regularization penalty. We design the joint PDF, and the underlying generative model, as

$$P(X, H, R, A, B, C, \theta; \alpha, \beta) := P(X)P(\theta)P(A;\alpha)P^E(H|X + A;\theta^E)$$
$$P(B;\beta)P(C;\alpha)P^D(R + C|H + B;\theta^D). \quad (1)$$

3.2 Bayesian Learning Strategy

Let $\mathcal{S}^{\text{XR}} := \{(X_i, R_i)\}_{i=1}^I$ be the training sample set. Associated with (X_i, R_i), let H_i be the latent/hidden variable, and let (A_i, B_i, C_i) be the random perturbations variables. We propose maximum-a-posteriori (MAP) estimation for θ as $\operatorname{argmax}_\theta P(\theta|\mathcal{S}^{\text{XR}}) = \operatorname{argmax}_\theta P(\theta, \mathcal{S}^{\text{XR}})$ that integrates out, from the joint PDFs $P(X_i, H_i, R_i, A_i, B_i, C_i, \theta; \alpha, \beta)$, (i) latent variables H_i: by Monte-Carlo sampling each H_i, which would tend to produce more samples from the high-density regions of the joint PDF, and (ii) perturbations (A_i, B_i, C_i): by Monte-Carlo sampling from the *low-density* regions of the joint PDF. We propose an iterative learning strategy that alternately performs the following: (i) given sample sets of the triplets (A_i, B_i, C_i), we propose EM to update θ (Sect. 3.3) while sampling each H_i using Langevin diffusion (Sect. 3.4), and (ii) given θ and the sample sets for H_i, we propose to sample each triplet (A_i, B_i, C_i) using Langevin diffusion with normalized gradient vectors (Sect. 3.5).

3.3 Optimizing DNN Parameters with Monte-Carlo EM

Given a sample set $\mathcal{S}^{\text{ABC}} := \{\{(A_{ij}, B_{ij}, C_{ij})\}_{j=1}^J\}_{i=1}^I$, we propose to update θ to maximize the posterior $P(\theta|\mathcal{S}^{\text{XR}}, \mathcal{S}^{\text{ABC}})$ that has integrated out latent variables

H_i from the joint PDF as $\int_{H_1,\cdots,H_I} P(\theta, H_1, \cdots, H_I | \mathcal{S}^{\text{XR}}, \mathcal{S}^{\text{ABC}}) dH_1 \cdots dH_I$. We propose EM optimization to solve this problem elegantly. At iteration t of the EM algorithm, with parameter estimates θ^t, EM designs a tight lower bound of the log-posterior PDF $\log P(\theta | \mathcal{S}^{\text{XR}}, \mathcal{S}^{\text{ABC}})$ as the expectation of the *complete-data* log-posterior over the posterior PDF of the hidden variable, i.e.,

$$Q(\theta; \theta^t) := E_{P(H_1,\cdots,H_I | \mathcal{S}^{\text{XR}}, \mathcal{S}^{\text{ABC}}, \theta^t)}[\log P(\theta, H_1, \cdots, H_I | \mathcal{S}^{\text{XR}}, \mathcal{S}^{\text{ABC}})], \quad (2)$$

where the complete-data posterior $P(\theta, H_1, \cdots, H_I | \mathcal{S}^{\text{XR}}, \mathcal{S}^{\text{ABC}})$ is proportional to the product of the complete-data likelihood $P(\mathcal{S}^{\text{XR}}, H_1, \cdots, H_I | \theta, \mathcal{S}^{\text{ABC}})$ and the prior $P(\theta)$. We propose Monte Carlo expectation using a sample set $\mathcal{S}^H := \{\{H_{ik}\}_{k=1}^K\}_{i=1}^I$ where each H_{ik} is drawn independently from the posterior PDF $P(H_i | \mathcal{S}^{\text{XR}}, \mathcal{S}^{\text{ABC}}, \theta^t)$ (Sect. 3.4). Then, the Monte-Carlo estimate of the $Q(\theta; \theta^t)$ function, including only those terms dependent on the variables (H, θ), is

$$\widehat{Q}(\theta, \mathcal{S}^{\text{H}}; \theta^t) := \log P(\theta^E) + \sum_{i=1}^I \sum_{j=1}^J \sum_{k=1}^K -\left(\frac{H_{ik} - \mathcal{E}^{\mu}(X_i + A_{ij}; \theta^E)}{\mathcal{E}^{\sigma}(X_i + A_{ij}; \theta^E)}\right)^2$$

$$+ \log P(\theta^D) + \sum_{i=1}^I \sum_{j=1}^J \sum_{k=1}^K -\|R_i + C_{ij} - \mathcal{D}(H_{ik} + B_{ij}; \theta^D)\|_{\text{F}}^2, \quad (3)$$

where $\|\cdot\|_{\text{F}}$ denotes the Frobenius norm of the underlying matrix/tensor.

Given $\mathcal{S}^{\text{H}}$, $\mathcal{S}^{\text{ABC}}$, and $\mathcal{S}^{\text{XR}}$, EM aims to update θ^t to increase $\widehat{Q}(\theta; \theta^t)$ so that $\widehat{Q}(\theta^{t+1}; \theta^t) \geq \widehat{Q}(\theta^t; \theta^t)$. We use iterative gradient based updates using Adam, which introduces stochasticity to avoid local maxima and also implicitly applies the squared-norm prior/regularization stemming from $\log P(\theta^E)$ and $\log P(\theta^D)$.

3.4 Sampling in Latent Space Using Langevin Diffusion

We obtain the sample set $\mathcal{S}^H := \{\{H_{ik}\}_{k=1}^K\}_{i=1}^I$ using Langevin diffusion that relies on the gradients (with respect to each H_i) of the *log*-posterior PDF for each H_i, i.e., $P(H_i | \mathcal{S}^{\text{XR}}, \mathcal{S}^{\text{ABC}}, \theta^t) = P(H_i, \theta^t | \mathcal{S}^{\text{XR}}, \mathcal{S}^{\text{ABC}}) / P(\theta^t | \mathcal{S}^{\text{XR}}, \mathcal{S}^{\text{ABC}})$ where the denominator is independent of H_i. The desired gradient equals the gradient of the complete-data log-posterior PDF $P(H_i, \theta^t | \mathcal{S}^{\text{XR}}, \mathcal{S}^{\text{ABC}})$, which equals the gradient of $\widehat{Q}(\theta, \mathcal{S}^{\text{H}}; \theta^t)$ at $\theta = \theta^t$. Langevin diffusion sampling for each H_i first initializes each H_{ik} randomly, and then iteratively updates it (T times) as

$$H_{ik} \leftarrow H_{ik} + \sqrt{2\epsilon_H} Z_H + \epsilon_H \nabla_{H_{ik}} \widehat{Q}(\theta^t, \mathcal{S}^{\text{H}}; \theta^t), \quad (4)$$

where $\epsilon_H > 0$ is the step-size, and Z_H is a random vector with values drawn independently from $\mathcal{N}(0, 1)$. This paper chooses $K := 1$ for computational efficiency and to infuse sampling-related stochasticity (unlike $K \to \infty$ that would remove Monte-Carlo-sampling related stochasticity), in the same vein as Adam/SGD and batches introduce stochasticity that is beneficial for learning/optimization.

3.5 Low-Density Sampling of Perturbations Using Projected-and-Normalized Langevin Diffusion

Given a sample set $\mathcal{S}^{\mathrm{H}} := \{\{H_{ik}\}_{k=1}^{K}\}_{i=1}^{I}$ and an estimate for θ, we propose to sample the perturbations (A, B, C) to create the sample set $\mathcal{S}^{\mathrm{ABC}}$ from the low-density regions of their posterior PDF $P(A, B, C|X, H, R, \theta)$ in order to regularize the learning of the regressor underlying the DNN mapping and lead to better generalization. Langevin diffusion relies on the gradients (with respect to each A_i, B_i, C_i) of the *log*-posterior PDF $P(A_i, B_i, C_i|\mathcal{S}^{\mathrm{XR}}, \mathcal{S}^{\mathrm{H}}, \theta) = P(A_i, B_i, C_i, \mathcal{S}^{\mathrm{XR}}, \mathcal{S}^{\mathrm{H}}, \theta)/P(\mathcal{S}^{\mathrm{XR}}, \mathcal{S}^{\mathrm{H}}, \theta)$ where the denominator is independent of (A_i, B_i, C_i). Thus, the desired gradients equals the gradients of the log-joint PDF $\log P(A_i, B_i, C_i, \mathcal{S}^{\mathrm{XR}}, \mathcal{S}^{\mathrm{H}}, \theta)$ that equals $\log P(\mathcal{S}^{\mathrm{XR}}, \mathcal{S}^{\mathrm{H}}, \theta|A_i, B_i, C_i) + \log P(A_i, B_i, C_i)$; the gradient of the first term equals the gradient of $\widehat{Q}(\theta, \mathcal{S}^{\mathrm{H}}; \theta^t)$. We propose to use a variant of Langevin diffusion sampling where we rescale the gradient of the log-PDF by its norm; this helps stabilize the gradient computation in low-PDF regions where the log-PDF gradient is equivalent to the ratio of the PDF gradient to the PDF value; if the PDF value tends to zero, then, even small non-zero PDF-gradient vectors can lead to instability in the ratios. So, our *normalized* Langevin diffusion sampling for each A_i, B_i, C_i first initializes each A_{ij}, B_{ij}, C_{ij} randomly, and then iteratively updates them (T times) as

$$A'_{ij} \leftarrow -\nabla_{A_{ij}} \left(\frac{H_{ik} - \mathcal{E}^{\mu}(X_i + A_{ij}; \theta^E)}{\mathcal{E}^{\sigma}(X_i + A_{ij}; \theta^E)} \right)^2 + \nabla_{A_{ij}} \log P(A_{ij}; \alpha), \tag{5}$$

$$B'_{ij} \leftarrow -\nabla_{B_{ij}} \|R_i + C_{ij} - \mathcal{D}(H_{ik} + B_{ij}; \theta^D)\|_{\mathrm{F}}^2 + \nabla_{B_{ij}} \log P(B_{ij}; \alpha), \tag{6}$$

$$C'_{ij} \leftarrow -\nabla_{C_{ij}} \|R_i + C_{ij} - \mathcal{D}(H_{ik} + C_{ij}; \theta^D)\|_{\mathrm{F}}^2 + \nabla_{C_{ij}} \log P(C_{ij}; \alpha), \tag{7}$$

$$A'_{ij} \leftarrow A'_{ij}/\|A'_{ij}\|_{\mathrm{F}}, \qquad A_{ij} \leftarrow A_{ij} - \sqrt{2\epsilon_A} Z_A - \epsilon_A A'_{ij}, \tag{8}$$

$$B'_{ij} \leftarrow B'_{ij}/\|B'_{ij}\|_{\mathrm{F}}, \qquad B_{ij} \leftarrow B_{ij} - \sqrt{2\epsilon_B} Z_B - \epsilon_B B'_{ij}, \tag{9}$$

$$C'_{ij} \leftarrow C'_{ij}/\|C'_{ij}\|_{\mathrm{F}}, \qquad C_{ij} \leftarrow C_{ij} - \sqrt{2\epsilon_B} Z_C - \epsilon_C C'_{ij}, \tag{10}$$

where $\epsilon_A > 0$, $\epsilon_B > 0$, and $\epsilon_C > 0$ are the step sizes, and Z_A, Z_B, Z_C are random vectors with values drawn independently from $\mathcal{N}(0,1)$. This paper chooses $J := 1$ for computational efficiency and for retaining Monte-Carlo-sampling related stochasticity ($J \to \infty$ would remove such stochasticity), in the same spirit as Adam, the notion of batch subsets, and the sampling of the latent vectors H_i introduce stochasticity that is beneficial for the learning process.

This paper chooses the priors $P(A, \alpha)$, $P(B, \beta)$, and $P(C, \alpha)$ as per-component uniform distributions over the ranges $[-\alpha, +\alpha]$ and $[-\beta, +\beta]$. In that case, (i) the gradients of their log-prior terms become zero, (ii) the support of (A_{ij}, B_{ij}, C_{ij}) becomes restricted over their specific ranges, and (iii) we apply projections on (A_{ij}, B_{ij}, C_{ij}) following their Langevin updates as $\cdot_{ij} \leftarrow \min(\alpha, \max(-\alpha, \cdot_{ij}))$. This leads to *projected-and-normalized-gradient* Langevin diffusion sampling.

3.6 Architecture, Unsupervised Training, and Inference

Our DNN framework can incorporate generic encoder-decoder architectures, with or without skip connections between the encoder and the decoder. If skip connections are absent (e.g., in a VAE-like architecture), then the output of the bottleneck layer is the latent representation H. If skip connections are present (e.g., in a UNet-like architecture), the combined outputs of all the encoder blocks forms our latent representation H. This paper uses an encoder-decoder architecture with skip-connections, as used by N2N [15], and sets the number of Langevin-diffusion iterations $T := 5$. The parameters (α, β) of the stochastic perturbations are free parameters that we tune using the validation set.

In our unsupervised learning scenario, in the absence of an independently acquired reference image R, we set $R \leftarrow X$ such that the perturbed DNN input image $X + A$ and the perturbed reference image $R + C$ effective act as the two independently degraded versions of the underlying (unknown) high-quality image Y. This strategy improves over the N2N strategy that cannot perform without multiple independent medical image acquisitions of the same scene, and also improves over the R2R denoising strategy that adhocly (not data-driven) introduces simulated Gaussian perturbations to the input image X. Moreover, the effect of our latent-space perturbation B is akin to spatially correlated perturbations in the input image X, which aim to mimic spatial correlations in real-world degradations stemming from low-dose in CT/PET or subsampling in MRI. Training uses the Adam optimizer with learning rate 1^{-5}.

The inference scheme, for a test input X', (i) sets all perturbations (A, B, C) to zero, (ii) gets the quality-enhanced image as $\widehat{Y} := \mathcal{D}(\mathcal{E}^\mu(X'; \theta^E); \theta^D)$, and (iii) gets the uncertainty $\widehat{U}$ associated with $\widehat{Y}$ by sampling $\{H_s \sim P(H|X')\}_{s=1}^{100}$ from the encoder-modeled Gaussian in latent space, and computing the per-voxel standard deviation of the set of decoder mappings $\{\mathcal{D}(H_s; \theta^D)\}_{s=1}^{100}$. The inference time required to obtain the quality enhanced image is at par with any state-of-the-art method as no sampling for perturbations is required. However, the inference time can increase if the uncertainty estimates are also needed.

4 Results

Data. We use three publicly available datasets for performance evaluation: (i) a *lung CT* dataset with 550 standard-dose images from the RIDER [3,17] dataset, where we generate degraded reduced-dose images [4,32] by adding Gaussian-Poisson noise to scaled down sinograms and finally applying a standard CT-reconstruction algorithm; (ii) a *brain PET* dataset with 10000 images from the Bern-Shanghai dataset [30], which contains degraded reduced-dose images; (iii) a *knee MRI* dataset with 5600 high-quality images from the fastMRI [11] dataset, where we generate realistic degraded images by Cartesian subsampling and Gaussian noise addition in k-space. For all datasets, we use two disjoint sets of degraded images to represent ID and OOD data. For CT data, (i) ground-truth images correspond to a beam intensity of 10^6, (ii) the equivalent beam

Table 1. Results: ID, OOD. Values are means across test sets. pLEM− is the ablated version. Best-performing methods in bold; second-best-performing methods underlined.

Datasets →	CT (ID)			MRI (ID)			PET (ID)		
Methods ↓	SSIM↑	PSNR↑	LPIPS↓	SSIM↑	PSNR↑	LPIPS↓	SSIM↑	PSNR↑	LPIPS↓
N2N*	0.792	32.346	0.136	0.850	23.493	0.523	0.764	34.724	0.043
N2V	0.745	34.754	0.125	0.832	<u>23.653</u>	<u>0.352</u>	0.739	34.326	0.042
DIP	0.768	33.984	0.138	0.714	20.932	0.683	0.743	35.052	<u>0.026</u>
DDNM+	0.807	32.825	0.119	0.844	23.508	0.457	0.789	**36.375**	0.027
R2R	0.801	33.771	<u>0.105</u>	0.852	23.518	0.523	0.781	35.169	0.037
pLEM:Ours	**0.882**	**35.759**	**0.104**	**0.868**	**24.499**	**0.339**	**0.827**	<u>36.286</u>	**0.021**
pLEM−	<u>0.831</u>	<u>34.780</u>	0.113	<u>0.853</u>	23.503	0.489	<u>0.790</u>	35.412	0.032
Datasets →	CT (OOD)			MRI (OOD)			PET (OOD)		
Methods ↓	SSIM↑	PSNR↑	LPIPS↓	SSIM↑	PSNR↑	LPIPS↓	SSIM↑	PSNR↑	LPIPS↓
N2N*	0.661	31.355	0.180	<u>0.834</u>	22.360	0.554	0.722	34.562	0.047
N2V	0.502	29.651	0.200	0.832	<u>22.386</u>	<u>0.404</u>	0.698	33.774	0.047
DIP	0.679	31.336	0.172	0.701	20.215	0.708	0.691	33.993	<u>0.034</u>
DDNM+	0.673	29.365	**0.145**	0.822	21.864	0.484	0.719	<u>35.050</u>	0.039
R2R	<u>0.723</u>	<u>33.258</u>	0.128	0.833	22.376	0.554	0.734	34.913	0.042
pLEM:Ours	**0.728**	**33.725**	<u>0.155</u>	**0.844**	**23.057**	0.376	**0.777**	**35.612**	**0.027**
pLEM−	0.689	33.060	0.162	0.832	22.236	0.523	<u>0.744</u>	35.045	0.037

intensity for images termed as ID is 5×10^3 (200× dose reduction), and (iii) the equivalent beam intensity for images termed as OOD is 10^3 (1000× dose reduction). For PET data, the standard dose images are the ground-truth, ID images have dose level $1/50$, and OOD images have dose levels $1/100$ and $1/20$. For MRI, ground-truth images use full sampling in k-space without any noise addition; ID and OOD images use, respectively, 40% and 35% random-lines Cartesian sampling. In all the three applications, the degraded images naturally exhibit spatially correlated noise stemming from their standard image-reconstruction methods applied to transform-domain data. We train separate models for each modality. We use the ID set for training, validation (hyperparameter tuning), and ID testing (70%, 10%, and 20% split). We use OOD data solely for testing.

Methods. We refer to our Langevin EM framework coupled with stochastic perturbations as **pLEM**. We create an *ablated* version of our framework devoid of any perturbations (i.e., A, B, C all fixed to zero) that we call as *pLEM−*. We compare pLEM with 5 baselines: (i) N2N* [15], (ii) N2V [13], (iii) DIP [28], (iv) DDNM+ [29] pre-trained on ImageNet, (v) R2R [18]. N2N* is a modified unsupervised version of N2N [15] that trains using a single degraded image instance per training sample instead of multiple degraded instances like N2N. Performance evaluation uses (i) structural similarity index (SSIM), (ii) peak-signal-to-noise ratio (PSNR), (iii) learned perceptual image patch similarity (LPIPS) [33].

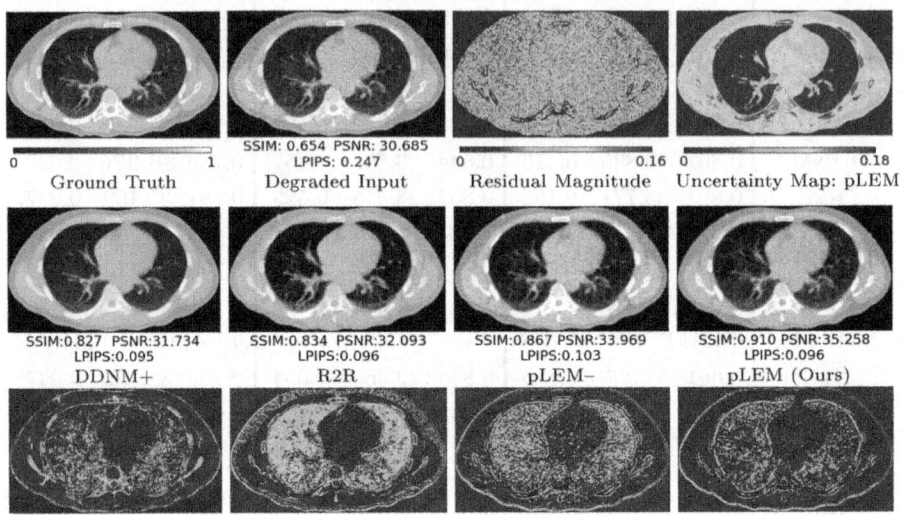

Fig. 2. Results: Qualitative, CT. The ground truth, degraded test image, residual magnitude maps, DNN output images, and per-voxel uncertainty map from pLEM.

Results: Quantitative and Qualitative. Quantitative results (Table 1) clearly indicate that pLEM outperforms all baseline methods, and consistently so across different modalities and across ID and OOD data. Some baselines perform well for specific imaging modalities, but they fail to generalize across all modalities possibly because of the variations in the spatial correlations underlying the degradation models, e.g., N2N* and N2V, designed for spatially-uncorrelated noise, perform much worse for PET compared to MRI. DIP, which is zero-shot, performs quite inconsistently across modalities and datasets, probably because it is devoid of any learning stage that can leverage training data. In contrast, DDNM+, which is "zero-shot" but leveraging pre-training using a large number of high-quality images, usually performs better than DIP. Nevertheless, DDNM+ suffers from the mismatch of the image characteristics between DDNM's pre-training dataset (ImageNet) and CT/MRI/PET images. The perturbation-based learning within R2R clearly offers benefits over its unperturbed version, i.e., N2N*. However, R2R recorrupts the input image adhocly by introducing i.i.d. Gaussian noise that is devoid of spatial correlations present in the degradations in the CT/MRI/PET images. The three different data-driven stochastic perturbations (A, B, C) introduced during pLEM learning include a combination of spatially-uncorrelated perturbations (A, C) to training images and implicit spatially-correlated perturbations by B (termed B^* in Fig. 5; discussed later). By controlling hyperparameters α and β, we can tune the spatial correlations

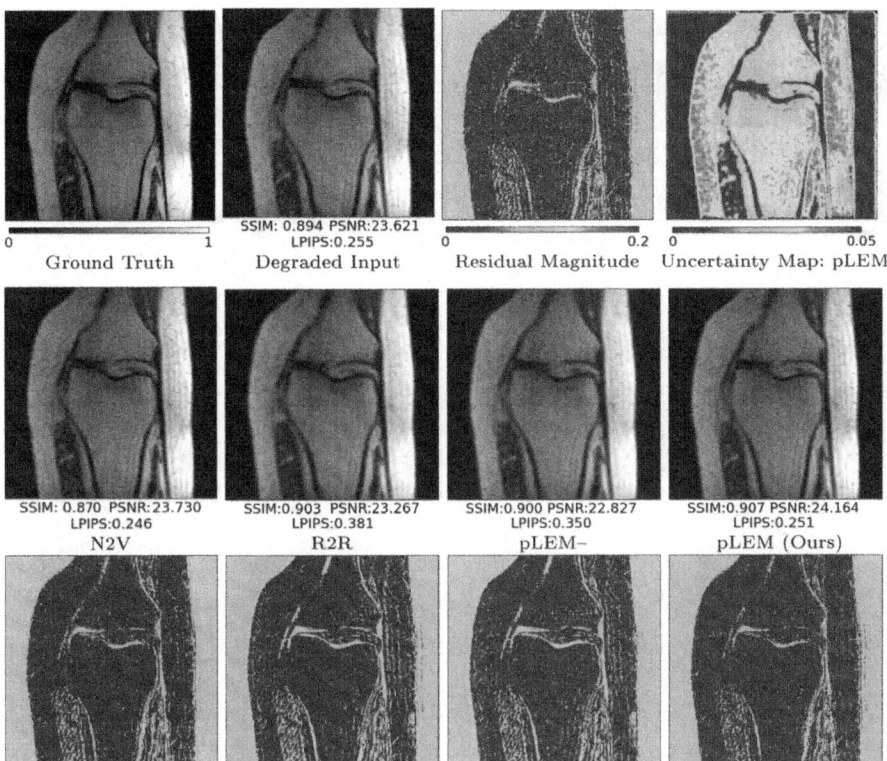

Fig. 3. Results: Qualitative, MRI. The ground truth, degraded test image, residual magnitude maps, DNN output images, and per-voxel uncertainty map from pLEM.

to match those in the dataset of interest. This improves pLEM's performance over R2R, and also allows pLEM to adapt well to varying degradation models in CT/MRI/PET images, leading to consistent performance across all modalities. Our ablated version pLEM–, using EM but without modeling any perturbations (A, B, C fixed to zero), is able to perform well for ID data, owing to the variational learning combined with Langevin diffusion, but, for OOD data, the performance of pLEM– suffers more than that of pLEM. This clearly indicates that stochastic perturbation modeling leads to more robust learning.

Figures 2, 3, and 4 show the outputs produced by the four best-performing methods, along with their residual magnitude (absolute difference between the output and the ground-truth) maps, and the uncertainty for the pLEM outputs.

Visualizing Perturbations in Spatial Domain. For any specific input image X, we visualize the perturbations (A, C) directly as images, and visualize effect of B (defined in latent space) in the spatial domain as follows. For an input image X with latent encoding $H := \mathcal{E}(X; \theta^E)$, we propose to visualize the effect of the latent-space perturbation B by optimizing an equivalent input-domain perturbation B^* as $\arg\min_{B'} \|\mathcal{E}(X + B'; \theta^E) - (H + B)\|_2^2$ using gradient descent.

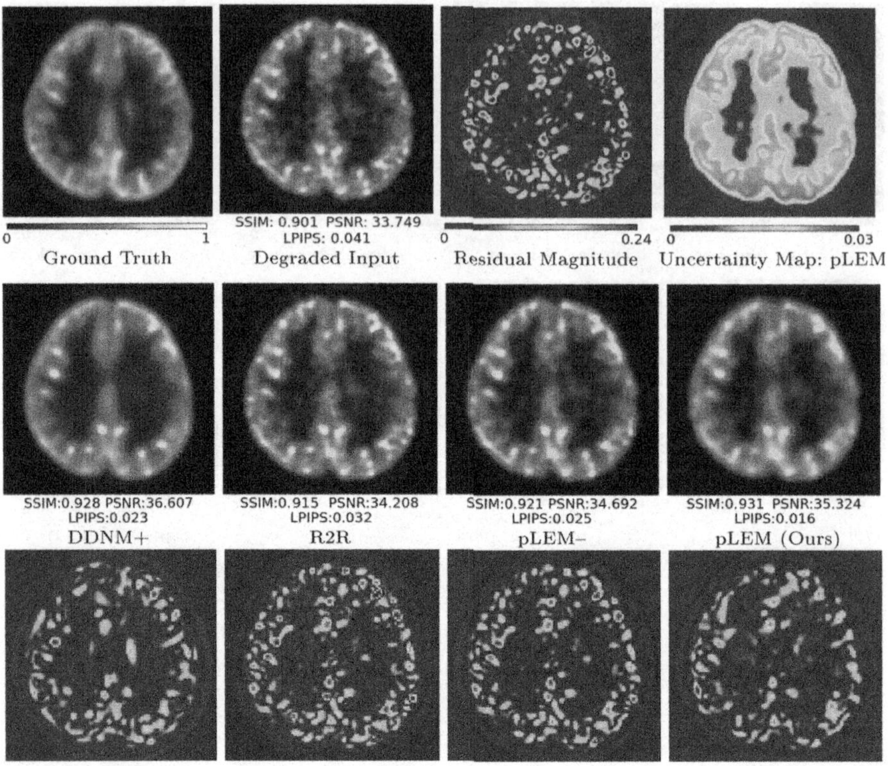

Fig. 4. Results: Qualitative, PET. The ground truth, degraded test image, residual magnitude maps, DNN output images, and per-voxel uncertainty map from pLEM.

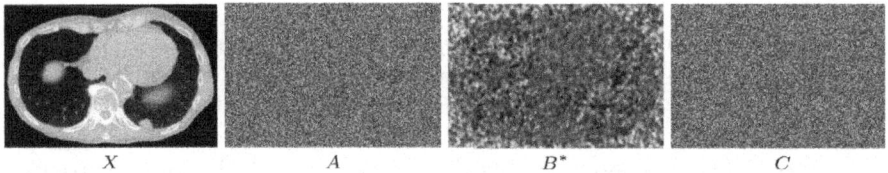

Fig. 5. Visualizing Perturbations in Spatial Domain. For an X, the effect B^* of perturbations B in latent space has much stronger spatial correlations than (A, C).

The convolutional structure within the encoder leads to high spatial correlations in B^* (Fig. 5) compared to A and C. pLEM can adapt perturbations to specific applications/degradations by controlling α, β; a relative increase in β leads to larger magnitudes in the spatially-correlated B^* compared to A, C. Indeed, pLEM's results in this paper use different values of α, β for CT, MRI, PET (all voxel values scaled to $[0, 1]$): (i) for CT: $\alpha = 0.05$ and $\beta = 0.05$, (ii) for MRI $\alpha = 0.05$ and $\beta = 0.1$, and (iii) for PET: $\alpha = 0.01$ and $\beta = 0.5$.

5 Conclusion

We propose a brand-new formulation for DNN learning founded in Bayesian modeling and inference that incorporates (i) low-density sampling of perturbations, for regularization, and (ii) high-density (the usual) sampling of the latent variable, for variational modeling and uncertainty estimation. Furthermore, we leverage perturbation sampling for data-driven automatic generation of realistic degraded instances, thereby improving over adhoc degradation models (e.g., i.i.d. Gaussian perturbations) in existing methods like R2R, and eliminating the demand for multiple degraded images per scene like N2N. The overarching Bayesian framework subsumes distribution modeling in latent space, enabling uncertainty estimation, and DNN-parameter estimation using a novel EM algorithm with Langevin sampling in latent space. The proposed learning framework is theoretically extendable to other image quality enhancement or image-to-image translation tasks. Results on publicly available datasets demonstrates that, in CT, MRI, and PET, our DNN framework outperforms existing methods, on ID and OOD images, and is likely to positively impact any downstream tasks such as segmentation, classification, treatment planning or decision making.

Acknowledgements. The authors express their gratitude to Microsoft Research India for the support provided through Microsoft Research Lab's 2024 PhD Award.

Disclosure of Interests. The authors have no competing interests to declare that are relevant to the content of this article.

References

1. Chen, Z., Yang, L., Lai, J., Xie, X.: CuNeRF: cube-based neural radiance field for zero-shot medical image arbitrary-scale super resolution. In: IEEE/CVF Int. Conf. on Comp. Vision, pp. 21185–21195 (2023)
2. Chung, H., Sim, B., Ryu, D., Ye, J.: Improving diffusion models for inverse problems using manifold constraints. Adv. Neural. Inf. Process. Syst. **35**, 25683–25696 (2022)
3. Clark, K., et al.: The Cancer Imaging Archive (TCIA): maintaining and operating a public information repository. J. Digital Imaging **26**(6), 1045–1057 (2013)
4. Ding, Q., Long, Y., Zhang, X., Fessler, J.: Modeling mixed Poisson-Gaussian noise in statistical image reconstruction for X-ray CT. In: Imag. Form. X-ray Comp. Tomo, pp. 399–402 (2018)
5. Farquhar, S., Gal, Y.: What 'Out-of-distribution' is and is not. In: Adv. Neural Inform. Process. Syst. ML Safety Workshop, pp. 1–7 (2022)
6. Hazan, T., Papandreou, G., Tarlow, D.: Perturbations, Optimization, and Statistics. MIT Press (2016)
7. He, C., et al.: Diffusion models in low-level vision: a survey. arXiv:2406.11138 (2024)
8. Hinton, G.: Improving neural networks by preventing co-adaptation of feature detectors. arXiv preprint arXiv:1207.0580 (2012)
9. Kiefer, J., Wolfowitz, J.: Stochastic estimation of the maximum of a regression function. Ann. Math. Stat. **23**(3), 462–466 (1952)

10. Kingma, D.P., Ba, J.: Adam: a method for stochastic optimization. Int. Conf. Learn. Rep. (2015)
11. Knoll, F., et al.: fastMRI: a publicly available raw k-space and DICOM dataset of knee images for accelerated MR image reconstruction using machine learning. Radiol. Artif. Intell. **2**(1) (2020)
12. Korkmaz, Y., Dar, S., Yurt, M., Özbey, M., Çukur, T.: Unsupervised MRI reconstruction via zero-shot learned adversarial transformers. IEEE Trans. on Med. Img. **41**(7), 1747–1763 (2022)
13. Krull, A., Buchholz, T., Jug, F.: Noise2Void - learning denoising from single noisy images. In: IEEE/CVF Conf. Comp. Vis. Pattern Recog., pp. 2124–2132 (2019)
14. Lambert, B., Forbes, F., Doyle, S., Dehaene, H., Dojat, M.: Trustworthy clinical AI solutions: a unified review of uncertainty quantification in deep learning models for medical image analysis. Artif. Intell. Med. **150**, 102830 (2024)
15. Lehtinen, J., et al.: Noise2Noise: learning image restoration without clean data. In: Int. Conf. Mach. Learn., vol. 80, pp. 2965–2974 (2018)
16. Li, X., et al.: Diffusion models for image restoration and enhancement – a comprehensive survey. arXiv:2308.09388 (2023)
17. Muzi, P., Wanner, M., Kinahan, P.: Data from RIDER Lung PET-CT (2015)
18. Pang, T., Zheng, H., Quan, Y., Ji, H.: Recorrupted-to-recorrupted: unsupervised deep learning for image denoising. In: Proceedings of the IEEE/CVF Conference on Computer Vision and Pattern Recognition (CVPR), pp. 2043–2052 (2021)
19. Papandreou, G., Yuille, A.L.: Perturb-and-map random fields: using discrete optimization to learn and sample from energy models. In: 2011 International Conference on Computer Vision, pp. 193–200 (2011)
20. Parisi, G.: Correlation functions and computer simulations. Nucl. Phys. B **180**(3), 378–384 (1981)
21. Quan, Y., Chen, Z., Zheng, H., Ji, H.: Learning deep non-blind image deconvolution without ground truths. In: Computer Vision – ECCV 2022, pp. 642–659 (2022)
22. Remedios, S., Wei, S., Dewey, B., Carass, A., Pham, D., Prince, J.: Pushing the limits of zero-shot self-supervised super-resolution of anisotropic MR images. In: Medical Imaging 2024: Image Processing, vol. 12926, p. 1292606. SPIE (2024)
23. Sample, C., et al.: Neural blind deconvolution for deblurring and supersampling PSMA PET. Phys. Med. Biol. **69**(8), 085025 (2024)
24. Sharma, V., Khurana, A., Yenamandra, S., Awate, S.: Semi-supervised deep expectation-maximization for low-dose PET-CT. In: IEEE Int. Symposium on Biomedical Imaging, pp. 1–5 (2022)
25. Sharma, V., Awate, S.P.: Adversarial EM for partially-supervised image-quality enhancement: application to low-dose PET imaging. In: 2024 IEEE International Conference on Image Processing (ICIP), pp. 4007–4013 (2024)
26. Sharma, V., Awate, S.P.: Adversarial EM for variational deep learning: application to semi-supervised image quality enhancement in low-dose PET and low-dose CT. Med. Image Anal. **97**, 103291 (2024)
27. Song, T., Yang, F., Dutta, J.: Noise2Void: unsupervised denoising of PET images. Phy. Med. Biol. **66**(21), 214002 (2021)
28. Ulyanov, D., Vedaldi, A., Lempitsky, V.: Deep image prior. In: IEEE/CVF Conf. Comp. Vis. Pattern Recog., pp. 9446–9454 (2018)
29. Wang, Y., Yu, J., Zhang, J.: Zero-shot image restoration using denoising diffusion null-space model. In: Int. Conf. on Learning Representations (2023)
30. Xue, S., et al.: A cross-scanner and cross-tracer deep learning method for the recovery of standard-dose imaging quality from low-dose PET. Eur. J. Nucl. Med. Mol. Imaging **49**, 1843–1856 (2022)

31. Yuan, N., Zhou, J., Qi, J.: Half2Half: deep neural network based CT image denoising without independent reference data. Phy. Med. Biol. **65**(21), 215020 (2020)
32. Zeng, D., et al.: A simple low-dose X-ray CT simulation from high-dose scan. IEEE Trans. Nucl. Sci. **62**(5), 2226–2233 (2015)
33. Zhang, R., Isola, P., Efros, A., Shechtman, E., Wang, O.: The unreasonable effectiveness of deep features as a perceptual metric. In: IEEE/CVF Conf. Comp. Vis. Pattern Recog., pp. 586–595 (2018)
34. Zou, K., Chen, Z., Yuan, X., Shen, X., Wang, M., Fu, H.: A review of uncertainty estimation and its application in medical imaging. Meta-Radiology **1**(1), 100003 (2023)

Segmentation

MC-NuSeg: Multi-Contour Aware Nuclei Instance Segmentation with Segment Anything Model

Hyun Namgung[1], Siwoo Nam[2], Soopil Kim[2], and Sang Hyun Park[1,2(✉)]

[1] Department of Interdisciplinary Studies of Artificial Intelligence, Daegu Gyeongbuk Institute of Science and Technology (DGIST), Daegu, Korea
{hyun.namgung,shpark13135}@dgist.ac.kr
[2] Department of Robotics and Mechatronics Engineering, Daegu Gyeongbuk Institute of Science and Technology (DGIST), Daegu, Korea

Abstract. Accurate nuclei instance segmentation is critical in digital pathology image analysis, facilitating disease diagnosis and advancing medical research. While various methods have been proposed, recent approaches leverage foundation models like the Segment Anything Model (SAM) for their robust representational power. However, existing models face challenges in handling the unique characteristics of histopathology images, particularly dense nuclei clusters, and complex morphological and staining variations. To address these issues, we propose a novel method, Multi-Contour Aware Nuclei Instance Segmentation (MC-NuSeg) framework, which incorporates the hierarchical boundary structure of nuclei for precise segmentation. MC-NuSeg predicts multiple segmentation maps corresponding to different contour layers, allowing for accurate separation of densely clustered nuclei and those with high morphological variance. Furthermore, we introduce an auxiliary instance counting loss that directly supervises the number of nuclei, significantly enhancing segmentation accuracy by reducing false positives and missed cases. Extensive evaluations on four public pathology datasets demonstrate that MC-NuSeg achieves state-of-the-art performance, effectively addressing the challenges of nuclei instance segmentation.

Keywords: Nuclei Instance Segmentation · Multi-Contour Prediction · Nuclei Counting · Segment Anything Model

1 Introduction

Nuclei instance segmentation plays a fundamental role in disease analysis, treatment planning, and research advancement through digital pathology [1,19]. Digital pathology advancement has driven an urgent need for automated nuclei instance segmentation. However, formidable challenges such as containing tens of thousands of nuclei with high morphological variations, low generalization arise from tissue samples spanning diverse sources from the brain to colon tissue, inconsistent staining, and low quality of image gives the barrier of automated nuclei instance segmentation development.

Existing approaches to these challenges primarily rely on insufficient shape representations through multi-branched decoders. These approaches generate various surrogate representation maps such as horizontal-vertical distance map [9], centripetal distance map [10], inter-nuclear distance map [23], ordinal distance map [8], and kernel map [32]. Despite improved accuracies, they often fail to effectively handle both high morphological variations and generalization across diverse data. These limitations persist even in Segment Anything Model (SAM) based models [17,20], despite their remarkable potential in medical image analysis, with outstanding generalization capabilities, stable representation extraction, and adaptability, as evidenced in several studies [7]. Multiple studies [6,7,14,31] have highlighted performance degradation, prompting the exploration of various approaches, such as two-step detection [27], prompt generation [3], and fine-tuning [4]. Despite these efforts, existing solutions have yet to fully address the complex challenges posed by the intricate, irregular-shaped, and densely clustered nuclear structures characteristic of histopathology images.

In this paper, we introduce a *Multi-Contour Aware Nuclei Instance Segmentation* (MC-NuSeg) framework that models the geometric structure of cells. Built upon SAM, our framework leverages multi-contour layers, which represent distinct boundaries defined by their distance from the cell center. Through this hierarchical boundary structure learning, each layer inherently encodes priors about the shape and size of nuclei instances, enabling robust segmentation that preserves the morphological characteristics of nuclei. In addition, we propose an auxiliary instance counting to effectively regularize the model training by predicting the number of cells within the image. To further enhance the accuracy of SAM while maintaining its generalization capabilities, we incorporate a convolution-LoRA [13] hybrid adapter and an attention module with H&E stain deconvolution feature. Our main contributions are summarized as follows:

1. We introduce a novel SAM-based nuclei instance segmentation framework that leverages multi-contour layer masks to capture the hierarchical boundary structure of nuclei. Unlike traditional approaches, our method achieves precise separation of densely clustered nuclei by learning progressive hierarchical boundary representations from outer contours to central regions.
2. We propose an instance counting module for accurate nuclei counting while effectively regularizing model training, thereby enhancing overall instance segmentation performance.
3. Extensive experimental validation confirms that our MC-NuSeg surpasses state-of-the-art nuclei instance segmentation models and effectively separates the clustered and irregularly shaped nuclei on four diverse public pathology datasets.

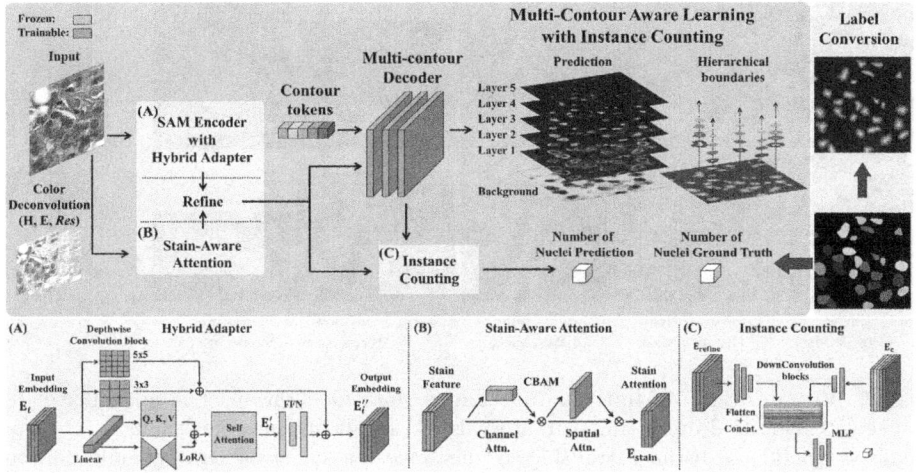

Fig. 1. The overview of our proposed segmentation framework.

2 Methods

2.1 Overview

Our MC-NuSeg framework extends SAM to learn the hierarchical boundary structures of nuclei. To achieve this, we introduce several key components as shown in Fig. 1: (A) a SAM encoder with a hybrid adapter, (B) a stain-aware attention module, and (C) an auxiliary instance counting module, and a multi-contour mask decoder. During training, the framework processes input images through two pipelines. The main pipeline utilizes the SAM encoder with the hybrid adapter, and an auxiliary pipeline incorporating stain deconvolution with Convolutional Block Attention Module (CBAM) [29]. The output of two pipelines then undergoes refinement to generate image embeddings, which are then processed by the multi-contour decoder with contour tokens to produce both multi-contour predictions and multi-contour embeddings. For the auxiliary instance counting, an instance counting module operates on these embeddings to predict nuclei counts, providing extra supervision. During inference, the framework applies the identical processing pipeline to generate both multi-contour predictions and nuclei counts for a given input image. The innermost region of the prediction is used as a marker for the watershed algorithm for the final instance segmentation.

2.2 Multi-Contour Layer Generation

Before training the model, we pre-process the ground truth to generate a multi-contour layer as illustrated in Fig. 2. Given the instance mask (a), we compute the normalized distance of pixels within each nucleus (b), define k levels of contours, and generate discrete labels (c). To be specific, we compute the Euclidean

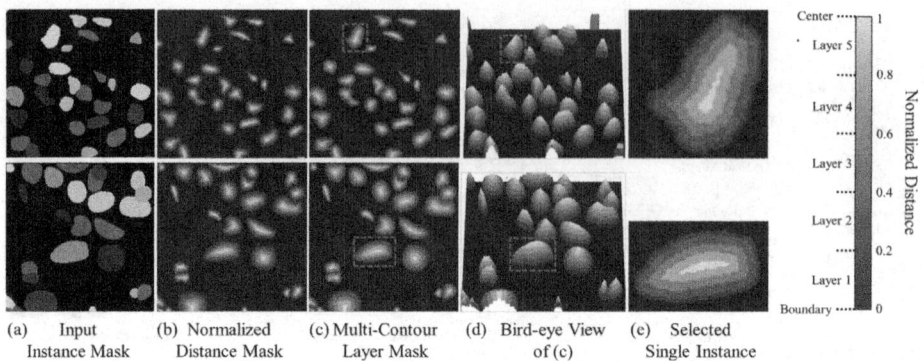

(a) Input Instance Mask (b) Normalized Distance Mask (c) Multi-Contour Layer Mask (d) Bird-eye View of (c) (e) Selected Single Instance

Fig. 2. Visualization of Multi-Contour Layer Mask Generation. The normalized distance was separated into 5 uniform intervals. (a) Input instance mask sample, (b) Normalized Euclidean distance transformed instance mask, (c) Generated multi-contour layer mask, (d) Bird-eye view of multi-contour layer mask, (e) Selected single instance mask from red dotted box in (c) (Color figure online)

distance of foreground pixels within each nucleus from the closest background pixels and denote the transformed mask as $\mathcal{M}_{\text{edt}}$. This distance map is then normalized to the range of [0, 1], denoting $\mathcal{M}_{\text{norm}}$. We partition the normalized distance range into k uniform intervals, denoted as $\mathcal{N}_K = \{n_1, n_2, \cdots, n_K\}$, where each interval n_i is defined as:

$$n_i = \begin{cases} [\frac{i-1}{K}, \frac{i}{K}) & i = 1, 2, \ldots, K-1 \\ [\frac{i-1}{K}, \frac{i}{K}] & i = K \end{cases} \quad (1)$$

Given these intervals, we assign the ith interval as a label i, and generate final multi-contour layer mask $\mathcal{M}_{\text{contour}}$ as follows:

$$\mathcal{M}_{\text{contour}}(x, y) = i, \quad \text{if} \quad \mathcal{M}_{\text{norm}}(x, y) \in n_i, \quad (2)$$

where (x, y) denotes the pixel coordinates. In our hierarchical contour structure, the outermost boundary layer serves as the initial separator between nuclei and surrounding tissue. As we progress inward, each subsequent layer inherits spatial information from its predecessor, enabling progressive refinement of nuclear boundaries. This unique structure allows the model to make precise decisions about nuclei separation.

2.3 Multi-Contour Learning

To learn the hierarchical structure of nuclei, MC-NuSeg predicts $K+1$ distinct maps through the multi-contour mask decoder as shown in Fig. 3. Each map is generated using dedicated learnable query tokens within the decoder architecture. The prediction process requires solving a fundamental imbalance challenge.

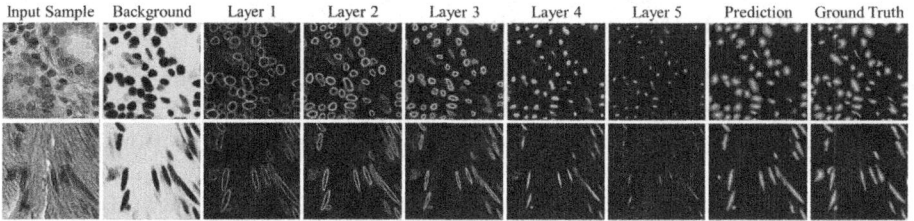

Fig. 3. Example of multi-contour layer prediction from input sample image. From left to right, input sample, background prediction, 5-level contour layer predictions, final prediction map, ground truth of 5-level contour layer mask.

While background pixels heavily outnumber contour layer pixels, the outer contour layers naturally contain more pixels than the following inner layers. To address this imbalanced pixel-wise distribution, we employ a weighted focal loss $\mathcal{L}_{\text{focal}}$ to optimize contour layer label prediction.

Following softmax probability prediction $p_i \in \mathbb{R}^{b \times (K+1) \times h \times w}$ from the mask decoder, where layer number i, batch size b, and h, w for height and width respectively, the weighted focal loss $\mathcal{L}_{\text{focal}}$ computed as:

$$\mathcal{L}_{\text{focal}} = -\frac{\sum_i^{K+1} w_i \left(\alpha \log(p_i)(1-p_i)^\gamma + (1-\alpha) \log(1-p_i)(p_i^\gamma) \right)}{K+1}. \qquad (3)$$

for a given layer i, the corresponding weight element w_i is defined as:

$$w_i = \begin{cases} w_0 = 0.5 \\ w_l = 1.0 + 0.2 \cdot l, \end{cases} \qquad (4)$$

where l represents the layer number. With a layer-specific weight approach for each layer, we prevent the model from being dominated by an outnumbered class while maintaining sensitivity to inner contours which are hard to learn.

Although focal loss effectively handles class imbalance in pixel-wise predictions, capturing progressive layer structure requires more elaboration with the $\mathcal{M}_{\text{contour}}$. Moving from the boundary towards the center mass of each nucleus, the contour layers exhibit a gradual progression where each layer i shares stronger relationships with its neighbors $i \pm 1$ compared to further distanced layers $i \pm 2$. To preserve these spatial relationships and ensure smooth transitions between adjacent layers, we incorporate regression-based loss $\mathcal{L}_{\text{reg}}$. With spanning softmax probability prediction $p_{\text{span}} \in \mathbb{R}^{b \times (K+1) \times (h \times w)}$, hierarchical contour level $\mathcal{H}_{\text{contour}} \in \mathbb{R}^{b \times (h \times w)}$ is modeled and $\mathcal{L}_{\text{reg}}$ is computed as follows:

$$\mathcal{L}_{\text{reg}} = \frac{\sum_{xy}^{MN} \left(\mathcal{H}_{\text{contour}}(x,y) - \mathcal{M}_{\text{contour}}(x,y) \right)}{MN}, \qquad \mathcal{H}_{\text{contour}} = A_{K+1} p_{\text{span}} \quad (5)$$

where M and N are the height and width of output prediction respectively. $A_{K+1} = [0, 1, \cdots, K] \in \mathbb{R}^{b \times (K+1)}$ is probability amplification and background suppression vector multiplied to p_{span} to modeling $\mathcal{H}_{\text{contour}}$.

2.4 Auxiliary Instance Counting

Complex histopathology images present significant challenges for precise nuclei separation, particularly in regions where nuclei boundaries are ambiguous or irregular in shape. These challenges often lead to either false detections or missed nuclei, resulting in inaccurate nuclei counting. To address this, we incorporate an auxiliary counting loss to regularize the model training. As shown in Fig. 1(C), refined embedding $\mathbf{E}_{\text{refine}}$ and contour decoder embedding is fed into the instance counting module. In detail, two embeddings are flattened and concatenated to output a single count prediction $\hat{n}_c$. We define the count loss $\mathcal{L}_{\text{count}}$ between $\hat{n}_c$ and real number n_c by utilizing HuberLoss [15] as below:

$$\mathcal{L}_{\text{count}} = \begin{cases} \frac{1}{2}(n_c - \hat{n}_c)^2, & \text{if } |n_c - \hat{n}_c| < 1 \\ |n_c - \hat{n}_c| - \frac{1}{2}, & \text{otherwise} \end{cases} \quad (6)$$

The final optimization loss $\mathcal{L}_{\text{seg}}$ for training MC-NuSeg is defined as:

$$\mathcal{L}_{\text{seg}} = \mathcal{L}_{\text{focal}} + \mathcal{L}_{\text{reg}} + \mathcal{L}_{\text{count}}. \quad (7)$$

2.5 Feature Refinement

Convolution-LoRA Hybrid Adapter. Drawing inspiration from recent architectural studies [22,24,30], and insufficient pre-trained knowledge of SAM [6,7,31], we introduce a hybrid adapter to fine-tune SAM, by combining depth-wise convolution and low-rank matrix adaptation within the image encoder transformer block. As illustrated in Fig. 1(A). For a given input image embedding $\mathbf{E}_i$ in i^{th} transformer block, we implement two parallel depth-wise convolution (DWC$_\alpha$) block. Each DWC block consisted Gaussian Error Linear Unit (GELU) [11], Layer Normalization (LN) [2], and Depthwise Convolution [5] with kernel size α, sequentially. Also, we incorporate a Low-Rank Adaptation (LoRA) [13] layer into the self-attention mechanism, requiring only a small fraction of parameter updates. The entire process of generating the updated output embedding $\mathbf{E}_i''$ within i^{th} transformer block can be represented as:

$$\mathbf{E}_i' = \text{MHSA}_{\text{LoRA}}(\text{LN}(\mathbf{E}_i)) + \mathbf{E}_i \quad (8)$$
$$\mathbf{E}_i'' = \mathbf{E}_i' + \text{FF}(\text{LN}(\mathbf{E}_i')) + \text{DWC}_3(\mathbf{E}_i) + \text{DWC}_5(\mathbf{E}_i) \quad (9)$$

where FF represents the Feed-Forward layer.

Stain-Aware Attention. To further enhance the model's ability to extract meaningful features from stained images, we incorporate a stain-aware attention mechanism into our framework. We apply color deconvolution to the input H&E stained images using a predefined HED matrix [26], separating the image into its constituent stain components (H, E, *Res*). These decomposed features are then processed through a Convolutional Block Attention Module (CBAM) [29], which generates both channel-wise and spatial attention features to produce a stain-aware feature embedding $\mathbf{E}_{\text{stain}}$ as illustrated in Fig. 1(B). This explicit modeling of stain characteristics helps the model better distinguish nuclear structures even in challenging staining conditions.

Refinement. Finally, we refine the output embedding $\mathbf{E}_o$ of the hybrid image encoder for enhanced delineation of nuclei representations and for the subsequent segmentation and counting process. First, intermediate feature embedding $\mathbf{E}_m$ is incorporated from the hybrid image encoder with CBAM attention. Second, stain-aware feature embedding $\mathbf{E}_{\text{stain}}$ is concatenated in channel dimension, and refined through a second refinement process with a convolution block (Conv). This block comprises 1×1 convolution, Layer Normalization (LN), GELU activation, and 1×1 convolution sequentially. The first and second refinement steps can be described as follows:

$$\mathbf{E}_{\text{attn}} = \text{CBAM}([\mathbf{E}_o + \mathbf{E}_m]) \qquad \mathbf{E}_{\text{refine}} = \text{Conv}([\mathbf{E}_{\text{attn}} \oplus \mathbf{E}_{\text{stain}}]) \qquad (10)$$

where $+$ and $\oplus$ represent element-wise addition and concatenation operation.

3 Experiments

Datasets. We evaluate our method on four public pathology nuclei segmentation datasets: MoNuSeg, CoNSeP, CPM17, and TNBC. **MoNuSeg** [18] contains total 30 H&E stained pathology images of 1000×1000 pixel dimension with total 21,623 annotated nuclei. We use 19 images for training and 11 images for testing. **CoNSeP** [9] contains 41 H&E stained images of 1000×1000 size, and total 24,319 annotated nuclei are included. We divide 41 images into 27 images for training and 14 images for testing. **CPM17** [28] contains 64 H&E stained pathology images of 500×500 or 600×600 size. We use 32 images for training and 32 images for testing. **TNBC** [21] contains 50 H&E stained pathology images of 512×512 size. We use 32 images for training and 18 images for testing. All four datasets are color-normalized and we extract patches of 250×250 size for MoNuSeg, CoNSeP, and CPM17 datasets, and 256×256 for the TNBC dataset with no overlap. All comparison methods and ablation study were conducted using identical dataset preparation procedures.

Implementation Details. All experiments were performed on a single NVIDIA A6000 GPU with PyTorch framework. For a fair comparison, the SAM-H backbone was utilized for all comparison methods using SAM. Our MC-NuSeg framework is trained for a maximum 200 epochs with an Adam optimizer [16]. The learning rate is decayed with an initial learning rate of 1×10^{-3} and a momentum of 0.9 and 0.99. The number of levels of the multi-contour layer was set to 5. We set each of α and γ from Eq. 3 as 0.75 and 2.0 respectively. Since our framework does not require prompt interactions, no additional prompts were used during training and inference. For the comparison methods, we reproduce the results based on corresponding authors implementations.

Table 1. Quantitative comparison with SOTAs on four different datasets. The best results are in bold. The rows below the middle line in the table represent models that utilize the SAM.

Method	Dataset							
	MoNuSeg		CoNSeP		CPM17		TNBC	
	Dice ↑	AJI ↑	Dice	AJI	Dice	AJI	Dice	AJI
HoVer-Net (Graham, 2019)	82.47	61.84	**81.20**	51.94	86.33	68.91	78.89	60.28
FullNet (Qu, 2019)	81.29	60.78	76.28	48.81	84.52	66.55	81.13	61.96
CD-Net (He, 2021)	81.71	61.35	77.51	49.59	85.54	68.02	81.90	64.42
NC-Net (Rashid, 2023)	80.89	50.49	77.06	45.03	84.86	54.16	75.42	49.40
PointNu-Net (Yao, 2023)	80.53	59.03	70.56	47.22	85.16	64.52	80.51	60.98
SONNET (Tan, 2023)	81.26	60.22	80.85	48.39	86.26	67.87	75.71	58.60
PromptNucSeg (Shui, 2023)	79.74	61.41	64.62	41.59	84.13	68.64	76.87	62.45
CellViT$_{SAM-H}$ (Horst, 2024)	80.49	58.59	79.03	50.00	85.74	63.16	75.35	52.07
Ours(MC-NuSeg)	**83.48**	**64.49**	81.03	**53.29**	**86.59**	**69.11**	**82.81**	**65.76**

Evaluation Metric. Following the previous works [8–10,12,23,25,27,32], we employ Dice Coefficient (Dice) and Aggregated Jaccard Index (AJI) for the segmentation task. Mean Absolute Error (MAE) metric is used for the evaluation of nuclei counting tasks.

3.1 Results

Comparison with State-of-The-Art Methods. Table 1 shows the quantitative results of Ours(MC-NuSeg) against state-of-the-art methods. MC-NuSeg improved upon previous best Dice scores, gaining 1.01 on MoNuSeg, 0.26 on CPM17, and 0.91 on TNBC. MC-NuSeg improved AJI scores, surpassing second-best models by 2.65 on MoNuSeg, 1.35 on CoNSeP, 0.2 on CPM17, and 1.34 in TNBC. As noted, ours consistently outperforms all the comparisons across four datasets, while only marginally falling behind HoVer-Net by 0.17 Dice score on the CoNSeP dataset. A qualitative comparison also supports these results. Figure 4 shows examples of instance segmentation results. In detail, on MoNuSeg, only MC-NuSeg correctly segmented clustered and irregular shaped nuclei while maintaining their shapes. We can find this performance superiority in the other three datasets. Other comparison methods, HoVer-Net and Sonnet failed to detect small and irregularly shaped nuclei on the TNBC dataset. CDNet also missed small irregularly shaped nuclei particularly when with the blurry nuclei in the CPM17 dataset. PromptNucseg also struggled to capture nuclei, missing several irregularly shaped and blurry nuclei in the CoNSeP dataset.

Domain Generalization. To validate the generalization capabilities and resilience to distribution shifts across various tissue types, we conducted cross-domain experiments on four datasets. Table 2 shows the generalization

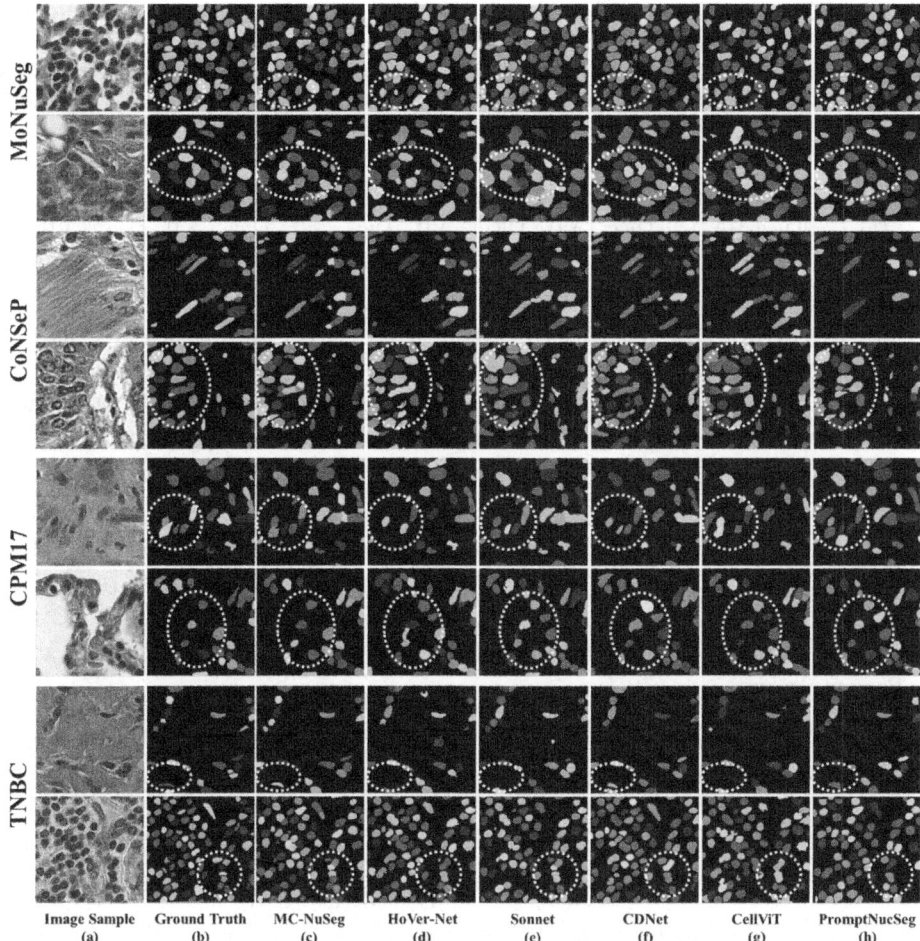

Fig. 4. Visualization of instance segmentation results with comparison methods on four public datasets. (a) Input sample image, (b) Ground Truth, (c) Our MC-NuSeg, (d) HoVer-Net [9], (e) Sonnet [8], (f) CDNet [10], (g) CellViT [12], (h) PromptNucSeg [27]. Only our method shows proper separation and shape-preserved segmentation of nuclei on all four datasets.

performance of our method and state-of-the-art comparisons. The proposed model, evaluated on the MoNuSeg dataset encompassing various organs, demonstrated state-of-the-art generalization performance with AJI improvements of 0.03, 2.56, and 2.48 on $\mathcal{D}_2$, $\mathcal{D}_3$, and $\mathcal{D}_4$ datasets, respectively. Notably, when trained on the $\mathcal{D}_2$ dataset, which includes diverse nuclei classes, sizes, and shapes, the model achieved top performance across most cases when applied to other datasets. This result highlights the effectiveness of the proposed feature refinement structure, showing that even when trained on single-organ data, the model operates robustly on multi-organ datasets. Furthermore, it maintains robust

Table 2. Cross-domain comparison with SOTAs on four different domain MoNuSeg, CoNSeP, CPM17, and TNBC as $\mathcal{D}_1$, $\mathcal{D}_2$, $\mathcal{D}_3$, and $\mathcal{D}_4$, respectively. $\mathcal{D}_S$ and $\mathcal{D}_T$ indicate the training source domain and testing target domain, respectively. The best results are in bold.

Domain		Ours		HoVer-Net		Sonnet		NC-Net		CellViT		Pr.NucSeg[a]	
$\mathcal{D}_S$	$\mathcal{D}_T$	Dice ↑	AJI ↑	Dice	AJI	Dice	AJI	Dice	AJI	Dice	AJI	Dice	AJI
$\mathcal{D}_1$	$\mathcal{D}_2$	69.72	**44.53**	70.84	44.07	**72.86**	42.78	67.76	35.10	70.41	42.09	62.97	38.55
	$\mathcal{D}_3$	82.11	65.59	82.97	66.35	**84.95**	**68.43**	81.38	60.89	79.18	60.69	81.42	65.04
	$\mathcal{D}_4$	**77.57**	**61.81**	70.55	54.68	69.05	51.77	70.71	49.58	72.31	55.36	75.55	60.67
$\mathcal{D}_2$	$\mathcal{D}_1$	**82.44**	**60.44**	80.86	59.58	78.89	53.27	78.43	52.19	77.90	52.45	78.75	60.41
	$\mathcal{D}_3$	**82.41**	61.64	81.17	60.25	80.15	61.42	79.83	60.68	80.06	59.32	79.34	**62.83**
	$\mathcal{D}_4$	**79.05**	**59.96**	70.81	52.21	61.96	43.57	76.59	57.71	75.06	57.50	75.74	59.91
$\mathcal{D}_3$	$\mathcal{D}_1$	**81.48**	**60.21**	79.06	56.20	79.13	54.55	76.16	35.63	72.02	42.42	78.82	57.65
	$\mathcal{D}_2$	69.77	**41.68**	**72.56**	41.62	70.21	39.21	66.07	27.01	72.15	34.78	65.05	36.85
	$\mathcal{D}_4$	**73.74**	55.63	69.37	51.45	65.53	47.89	59.49	30.13	67.76	46.70	73.52	**57.31**
$\mathcal{D}_4$	$\mathcal{D}_1$	**81.43**	**61.94**	81.04	58.10	80.47	58.36	77.46	40.21	72.89	37.37	78.70	59.46
	$\mathcal{D}_2$	67.57	**42.61**	66.95	40.01	68.11	41.40	70.68	32.92	**71.29**	34.21	62.11	38.31
	$\mathcal{D}_3$	81.52	65.07	81.50	63.00	**83.02**	**66.48**	82.34	57.70	81.96	55.77	80.37	64.85

[a] PromptNucSeg

Table 3. Instance Counting performance on four different datasets with and without auxiliary instance counting ($\mathcal{C}$).

Method	Dataset			
	MoNuSeg	CoNSeP	CPM17	TNBC
	MAE ↓	MAE	MAE	MAE
with $\mathcal{C}$	6.177	9.943	4.438	5.619
without $\mathcal{C}$	6.386	10.594	8.890	7.371

performance when trained on challenging datasets with irregular shapes. While Sonnet showed stronger performance in specific cases on $\mathcal{D}_3$ dataset, this can be attributed to dataset characteristics favoring concentric layer of nuclei center predictions for uniform and near-circular cells. These comprehensive results validate the flexibility and generalization capability of our model in managing diverse organs and complex cellular morphologies.

Auxiliary Instance Counting. The results in Table 3 validate the effectiveness of our auxiliary instance counting strategy in reducing incorrect predictions. Particularly in the CPM17 dataset, which features consistent cell morphologies, the model tended to misclassify tissue regions with cell-like structures. However, the instance counting module acted as a regularizer, leading to a notable error reduction of 4.452. Additionally, the module achieved error reductions of

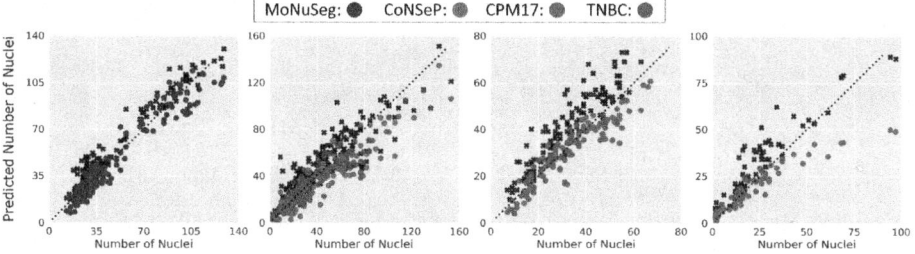

Fig. 5. Regression plot of instance counting prediction results on four datasets. The x-axis is the ground truth number of nuclei and the y-axis is the predicted number of nuclei. The black "x" marker indicates the prediction without $\mathcal{C}$. The black dotted line represents a straight line with a slope of 1 and a y-intercept is 0.

Table 4. Ablation study of key components of our MC-NuSeg on four different datasets. Where base, $\mathcal{H}$, $\mathcal{S}$, and $\mathcal{C}$ indicate baseline method, hybrid adapter module, stain aware attention, and auxiliary instance counting each respectively. The best results are in bold.

	Dataset							
	MoNuSeg		CoNSeP		CPM17		TNBC	
Method	Dice ↑	AJI ↑	Dice	AJI	Dice	AJI	Dice	AJI
base	81.50	60.95	79.65	50.19	83.76	64.49	80.24	60.11
base+$\mathcal{H}$	81.33	61.51	80.05	50.94	84.82	65.72	81.46	61.45
base+$\mathcal{H}$+$\mathcal{S}$	81.64	62.60	80.97	**53.71**	86.30	67.75	81.34	61.69
base+$\mathcal{H}$+$\mathcal{C}$	83.29	62.68	80.94	50.62	84.95	67.38	82.00	62.74
Ours	**83.48**	**64.49**	**81.03**	53.29	**86.59**	**69.11**	**82.81**	**65.76**

1.752 and 0.651 on the MoNuSeg and TNBC datasets, respectively, demonstrating stable performance improvements across diverse datasets. Figure 5 visualizes the prediction outcomes with and without the counting module, clearly highlighting the absence of the module results in over-prediction due to numerous false positives and false negatives.

Ablation Study. We summarized the results of ablation experiments in Table 4 to evaluate three key components: hybrid fine-tuning ($\mathcal{H}$), stain aware attention ($\mathcal{S}$), and auxiliary instance counting ($\mathcal{C}$). For baseline comparison, we adopt SAM-H, which fuses intermediate feature and outputfeature of encoder without prompts. Our ablation study revealed dataset-specific performance patterns. The introduction of $\mathcal{H}$ and $\mathcal{S}$ enable more robust feature extraction, resulting in improvement with 3.52 AJI score in CoNSeP dataset where most of nuclei exhibit irregular shapes. While $\mathcal{C}$ yielded better improvements with 1.17 and 4.07 AJI scores on MoNuSeg and TNBC datasets compared to $\mathcal{S}$. In contrast, $\mathcal{S}$ demonstrated superior performance on CoNSeP and CPM17 datasets, which we attribute to its effectiveness in handling irregular shapes and densely

Table 5. Effect of different number of contour layer on four different datasets MoNuSeg, CoNSeP, CPM17, and TNBC.

Number of Contour layer k	Dataset							
	MoNuSeg		CoNSeP		CPM17		TNBC	
	Dice ↑	AJI ↑	Dice	AJI	Dice	AJI	Dice	AJI
$k = 3$	83.60	64.38	81.54	56.66	85.84	68.41	82.69	67.55
$k = 5$	83.48	64.49	81.03	53.29	86.59	69.11	82.81	65.76
$k = 7$	82.65	61.50	65.20	28.08	85.37	65.04	76.35	54.10

clustered nuclei through explicit extraction of shape and boundary characteristics. The integration of all components (ours) yielded the best overall performance across all four datasets, validating our comprehensive approach to nuclei instance segmentation.

Hyperparameter Analysis. Further ablation study, we perform a key hyperparameter analysis of our MC-NuSeg. Selecting a proper number of k, which is the number of contour layers, plays a critical role in the separation of clustered nuclei. Table 5 revealed that increasing the number of contour layers beyond five adversely affects segmentation performance. Specifically, with seven layers, the reduced interval between layers creates challenges for the model in interpreting the complex multi-layered nuclear structures. This increased complexity leads to error propagation, where inaccuracies in one layer cascade through subsequent layers, particularly in nuclei with irregular morphologies. The empirical evaluation demonstrated that using five contour layers achieves an optimal balance between performance and stability. Additional layers beyond this number introduced unnecessary complexity without performance benefit.

4 Conclusion

In this paper, we have proposed MC-NuSeg (Multi-Contour Aware Nuclei Instance Segmentation). To effectively guide the SAM in understanding the hierarchical boundary of morphological high-variance nuclei, we utilized multi-contour layer representation. With preserving structural boundary awareness towards the central mass of nuclei, we further leveraged the instance counting task as a guidance module. Finally, by combining a hybrid adapter and stain-aware attention module, we enhanced the robustness of our method. Our comprehensive experimental results demonstrated that MC-NuSeg achieves state-of-the-art performance (Table 1) in nuclei instance segmentation, and maintains consistent performance across diverse tissue types and staining conditions (Table 2). Our framework successfully addresses the challenges of accurate nuclei separation with proper shape preservation while maintaining robust performance across diverse unseen domains, demonstrating the adequacy and robustness of MC-NuSeg.

Acknowledgements. This work was supported by Institute of Information & communications Technology Planning & Evaluation (IITP) grant funded by the Korea government(MSIT) (No. RS-2021-II212068, Artificial Intelligence Innovation Hub), (No. RS-2024-00439264, Development of High-Performance Machine Unlearning Technologies for Privacy Protection).

References

1. Alsubaie, N., Sirinukunwattana, K., Raza, S.E.A., Snead, D., Rajpoot, N.: A bottom-up approach for tumour differentiation in whole slide images of lung adenocarcinoma. In: Medical Imaging 2018: Digital Pathology, vol. 10581, pp. 104–113. SPIE (2018)
2. Ba, J.L.: Layer normalization. arXiv preprint arXiv:1607.06450 (2016)
3. Chen, Z., Xu, Q., Liu, X., Yuan, Y.: UN-SAM: universal prompt-free segmentation for generalized nuclei images. arXiv preprint arXiv:2402.16663 (2024)
4. Cheng, J., et al.: SAM-Med2D. arXiv preprint arXiv:2308.16184 (2023)
5. Chollet, F.: Xception: deep learning with depthwise separable convolutions. In: Proceedings of the IEEE Conference on Computer Vision and Pattern Recognition, pp. 1251–1258 (2017)
6. Cui, C., et al.: All-in-SAM: from weak annotation to pixel-wise nuclei segmentation with prompt-based finetuning. J. Phys. Conf. Ser. **2722**, 012012 (2024)
7. Deng, R., et al.: Segment anything model (SAM) for digital pathology: assess zero-shot segmentation on whole slide imaging. arXiv preprint arXiv:2304.04155 (2023)
8. Doan, T.N., Song, B., Vuong, T.T., Kim, K., Kwak, J.T.: SONNET: a self-guided ordinal regression neural network for segmentation and classification of nuclei in large-scale multi-tissue histology images. IEEE J. Biomed. Health Inform. **26**(7), 3218–3228 (2022)
9. Graham, S., et al.: Hover-Net: simultaneous segmentation and classification of nuclei in multi-tissue histology images. Med. Image Anal. **58**, 101563 (2019)
10. He, H., et al.: CDNet: centripetal direction network for nuclear instance segmentation. In: Proceedings of the IEEE/CVF International Conference on Computer Vision, pp. 4026–4035 (2021)
11. Hendrycks, D., Gimpel, K.: Gaussian error linear units (gelus). arXiv preprint arXiv:1606.08415 (2016)
12. Hörst, F., et al.: CellViT: vision transformers for precise cell segmentation and classification. Med. Image Anal. **94**, 103143 (2024)
13. Hu, E.J., et al.: LoRa: low-rank adaptation of large language models. arXiv preprint arXiv:2106.09685 (2021)
14. Huang, Y., et al.: Segment anything model for medical images? Med. Image Anal. **92**, 103061 (2024)
15. Huber, P.J.: Robust estimation of a location parameter. In: Breakthroughs in Statistics: Methodology and Distribution, pp. 492–518. Springer (1992)
16. Kingma, D.P., Ba, J.: Adam: a method for stochastic optimization. In: Bengio, Y., LeCun, Y. (eds.) 3rd International Conference on Learning Representations, ICLR 2015, San Diego, CA, USA, 7–9 May 2015, Conference Track Proceedings (2015). http://arxiv.org/abs/1412.6980
17. Kirillov, A., et al.: Segment anything. arXiv:2304.02643 (2023)

18. Kumar, N., Verma, R., Sharma, S., Bhargava, S., Vahadane, A., Sethi, A.: A dataset and a technique for generalized nuclear segmentation for computational pathology. IEEE Trans. Med. Imaging **36**(7), 1550–1560 (2017)
19. Madabhushi, A., Lee, G.: Image analysis and machine learning in digital pathology: challenges and opportunities. Med. Image Anal. **33**, 170–175 (2016)
20. Nam, S., et al.: InstaSAM: instance-aware segment any nuclei model with point annotations. In: International Conference on Medical Image Computing and Computer-Assisted Intervention, pp. 232–242. Springer (2024)
21. Naylor, P., Laé, M., Reyal, F., Walter, T.: Segmentation of nuclei in histopathology images by deep regression of the distance map. IEEE Trans. Med. Imaging **38**(2), 448–459 (2018)
22. Park, N., Kim, S.: How do vision transformers work? arXiv preprint arXiv:2202.06709 (2022)
23. Qu, H., Yan, Z., Riedlinger, G.M., De, S., Metaxas, D.N.: Improving nuclei/gland instance segmentation in histopathology images by full resolution neural network and spatial constrained loss. In: Shen, D., et al. (eds.) MICCAI 2019. LNCS, vol. 11764, pp. 378–386. Springer, Cham (2019). https://doi.org/10.1007/978-3-030-32239-7_42
24. Raghu, M., Unterthiner, T., Kornblith, S., Zhang, C., Dosovitskiy, A.: Do vision transformers see like convolutional neural networks? Adv. Neural. Inf. Process. Syst. **34**, 12116–12128 (2021)
25. Rashid, S.N., Fraz, M.M.: Nuclei probability and centroid map network for nuclei instance segmentation in histology images. Neural Comput. Appl. **35**(21), 15447–15460 (2023)
26. Ruifrok, A.C., Johnston, D.A., et al.: Quantification of histochemical staining by color deconvolution. Anal. Quant. Cytol. Histol. **23**(4), 291–299 (2001)
27. Shui, Z., Zhang, Y., Yao, K., Zhu, C., Sun, Y., Yang, L.: Unleashing the power of prompt-driven nucleus instance segmentation. arXiv preprint arXiv:2311.15939 (2023)
28. Vu, Q.D., et al.: Methods for segmentation and classification of digital microscopy tissue images. Front. Bioeng. Biotechnol. **7**, 433738 (2019)
29. Woo, S., Park, J., Lee, J.Y., Kweon, I.S.: CBAM: convolutional block attention module. In: Proceedings of the European Conference on Computer Vision (ECCV), pp. 3–19 (2018)
30. Wu, B., et al.: Visual transformers: where do transformers really belong in vision models? In: Proceedings of the IEEE/CVF International Conference on Computer Vision, pp. 599–609 (2021)
31. Xu, K., Goetz, L., Rajpoot, N.: On generalisability of segment anything model for nuclear instance segmentation in histology images. arXiv preprint arXiv:2401.14248 (2024)
32. Yao, K., Huang, K., Sun, J., Hussain, A.: PointNu-Net: keypoint-assisted convolutional neural network for simultaneous multi-tissue histology nuclei segmentation and classification. IEEE Trans. Emerg. Top. Comput. Intell. **8**(1), 802–813 (2023)

Pitfalls of Topology-Aware Image Segmentation

Alexander H. Berger[1,2,5](✉)[iD], Laurin Lux[1,2,3,5][iD], Alexander Weers[1,2], Martin J. Menten[1,3,4][iD], Daniel Rueckert[1,2,3,4][iD], and Johannes C. Paetzold[5][iD]

[1] Chair for AI in Healthcare and Medicine, Technical University of Munich, Munich, Germany
[2] School of Computation, Information and Technology, TUM, Munich, Germany
{a.berger,laurin.lux}@tum.de
[3] Munich Center for Machine Learning (MCML), Munich, Germany
[4] BioMedIA, Department of Computing, Imperial College London, London, UK
[5] Weill Cornell Medicine, Cornell University, New York City, USA
jpaetzold@med.cornell.edu

Abstract. Topological correctness, i.e., the preservation of structural integrity and specific characteristics of shape, is a fundamental requirement for medical imaging tasks, such as neuron or vessel segmentation. Despite the recent surge in topology-aware methods addressing this challenge, their real-world applicability is hindered by flawed benchmarking practices. In this paper, we identify critical pitfalls in model evaluation that include inadequate connectivity choices, overlooked topological artifacts in ground truth annotations, and inappropriate use of evaluation metrics. Through detailed empirical analysis, we uncover these issues' profound impact on the evaluation and ranking of segmentation methods. Drawing from our findings, we propose a set of actionable recommendations to establish fair and robust evaluation standards for topology-aware medical image segmentation methods. (Code is available at https://github.com/AlexanderHBerger/topo-pitfalls).

Keywords: Topology-Aware Segmentation · Model evaluation

1 Introduction

Quantitative imaging biomarkers are of increasing importance in modern medicine. The development and evaluation of these biomarkers are directly linked to the emerging capabilities of artificial intelligence (AI) models [38]. Rapid progress has been made in medical image segmentation with architectures such as the Unet [40] or the Vision Transformer [13]. Despite these advances, achieving perfect pixel-wise accuracy often remains impossible in image segmentation tasks. Consequently, many studies investigate the quality of these segmentations beyond purely pixel-based performance metrics [7].

An important property in image segmentation is topological correctness, which considers structural integrity aside from volumetric accuracy. Its significance has been identified separately in several areas of medical imaging, such

---

A. H. Berger and L. Lux—Equal Contribution.

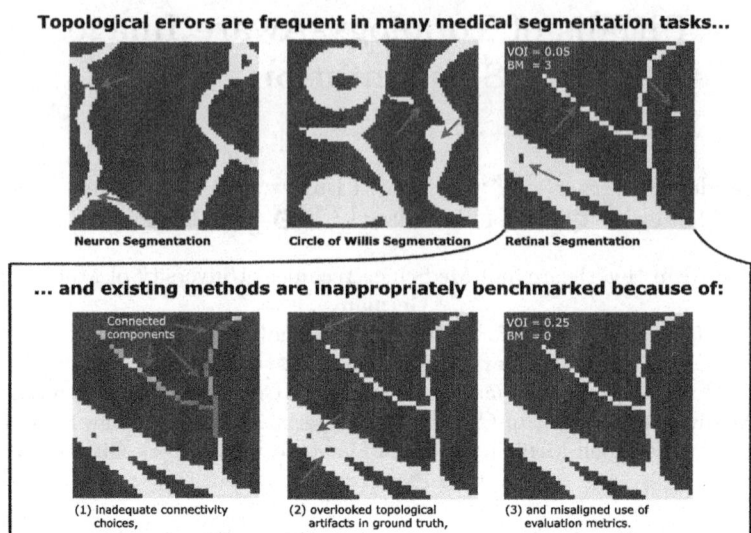

Fig. 1. Topological errors are present across distinct medical image segmentation tasks, e.g., in neuron, Circle of Willis, and retinal segmentation (top). We identify three critical pitfalls (bottom) in the evaluation of topology-aware segmentation methods. These include inadequate connectivity choices that misrepresent a dataset's semantics (left, e.g., representing a single vessel as multiple components), overlooked topological artifacts that skew evaluation results (center), and misaligned use of evaluation metrics that lack expressive power (right, e.g., VOI entangles volumetric and topological information).

as radiology, angiology, and neurology. For example, in lesion detection, metrics such as *lesion true positive rate* and *lesion false positive rate* are reported on an instance level, which are closely related to the topological features of a segmentation. [9,47,49]. In ophthalmology, the *connectivity* metric was proposed by Gegundez et al. [16] to measure the essential connectivity information in a predicted vessel segmentation. In neuron segmentation, community efforts led to the discovery that two topology-related metrics, *adapted rand error (ARE)* and *variation of information (VOI)*, are closely tied to expert assessment of the quality of predicted neuron segmentations [2]. Moreover, *split/merge errors* are used as essential quality estimates for neuron segmentation [24].

The importance of topological correctness in all these medical fields has led to numerous methods that focus on preserving the topology of a target structure in image segmentation. These methods aim to find general solutions for topologically accurate image segmentation and have achieved impressive results in diverse medical tasks [5,19–21,27,29,36,43]. However, we uncover that many works overlook the particularities of topological evaluation and the corresponding downstream tasks when comparing different methods. In particular, we demonstrate that the benchmarking of topology-aware segmentation methods is negatively affected by (1) *connectivity choices that are inadequate for the underlying data*, (2) *overlooked topological artifacts in the ground truth labels that can bias results*, and (3) *inappropriate use of evaluation metrics*. These practices result in the reporting of inaccurate absolute and relative performance of different methods and hinder a fair and robust assessment of their suitability to a specific task or domain. This work provides a detailed overview of these issues and analyses the impact of the most common pitfalls in an extensive empirical study (Fig. 1). Based on our findings, we propose several solutions to these issues.

2 Related Work

Topology-Aware Image Segmentation Methods. Numerous studies have tried to enhance segmentations' topological integrity in highly task-specific settings. Our study focuses on works that propose general-purpose solutions for improving topological correctness in multiple domains. Many of these works build on persistent homology (PH) and its differentiability to define loss functions for neural network optimization [5,8,11,21,43]. These methods are based on the computation of persistence diagrams using filtered cubical complexes from either T- or V-construction. In the persistence diagram, the birth and death cells can be identified and used to calculate a loss. Notably, for all these methods, the choice of cubical complex construction strongly impacts the persistence diagrams and, thus, the loss [6]. Other methods use discrete Morse theory (DMT) to achieve topology-aware image segmentation [4,20]. Apart from PH and DMT methods, other methods, including delineation [36], post-processing [27,28], homotopy warping [20], skeletonization [41], graph-based cortical segmentation [17], and component graphs [29] were proposed.

Topological Evaluation Metrics. Most studies on topologically accurate segmentation use a combination of different metrics to showcase their method's effectiveness. These metrics usually consist of pixel-wise metrics and topological metrics. A pixel-wise metric (e.g., cross-entropy or Dice coefficient) measures pixel-wise agreement between the prediction and label, disregarding topological characteristics. Although a perfect pixel-wise agreement implies identical topology, these metrics are not interpretable for values $\neq 1$ from a topological perspective.

Several topological metrics have been specifically proposed for neuron segmentation. They mainly revolve around detecting **split/merge errors**, which are concepts related to topological errors in dimension 0. Two widespread metrics are Rand Index (RI)-based metrics (e.g., ARE or adjusted rand index (ARI)) [1,22,45], and the VOI metric [32–34], both of which were originally proposed to measure cluster similarity. They have been adapted to image segmentation by partitioning a likelihood map to an instance map, where all pixels belonging to the same connected component are viewed as one cluster. Then, RI-related metrics measure pair-wise pixel agreement in the ground truth and prediction [1,45]. VOI measures how much information about a pixel's instance in the prediction can be gained by the pixel's instance in the ground truth and vice versa [37].

Lastly, purely topological metrics have been proposed for measuring topological accuracy. These metrics include the Betti number error [21] (e.g., B0 and B1 for topological features in dimension 0 and 1, respectively), the Betti matching error (e.g., BM0 and BM1) [43], and the DIU metric [29]. These metrics transfer the images to topological spaces and measure the number of topological errors. They give an interpretable quantification of the topological accuracy and provide various degrees of rigor: while the Betti number error captures global agreement between the number of topological features in each dimension [21], the Betti matching error takes their spatial correspondence into account [43]. The DIU metric additionally measures the correspondence between the topological features in union and the intersection of an image pair [29].

Commonly Used Benchmarking Datasets. The importance of different topological characteristics is deeply tied to the semantics of a specific domain and the associated downstream tasks. Some datasets frequently appear in method papers on topology-aware segmentation. In medical imaging, neuron and vessel segmentation datasets are commonly used as benchmarks, whereas in computer vision, aerial imaging datasets are often used. The most prominent datasets are:
DRIVE. The DRIVE dataset [42], used in [3,19–21,41,46], consists of color fundus images of the human retina. The blood vessels are represented as the foreground (FG) components. The background (BG) components can be interpreted as inter-vessel areas. However, it is important to note that the loops around the background components are often a result of the 2D projection of the vasculature and commonly do not encode physiologically connected vessels.
CREMI. The CREMI dataset [15], used in [19–21,27,29,41,43], contains brain images visualized using transmission electron microscopy. Boundary maps are commonly used as data representations to evaluate topology-aware methods. Here, foreground components resemble neuron boundaries, while the background components resemble two structures: foremost neurons, and a smaller fraction of the background components resemble synaptic clefts. Some works also use the inverse map, where the foreground resembles neurons and synaptic clefts.
Roads. The Roads dataset [35], used in [19,21,29,36,41,43], contains satellite images where the labels represent road networks. It has become a popular benchmark dataset in medical image segmentation due to its complex topological properties. In the Roads dataset, preserving connectivity—captured by dimension 1 features—is crucial for ensuring access between different areas of the map.

Topological information is essential for many other datasets that are not commonly used for benchmarking. Examples of lesion detection include the MSSEG2 [12] and ISBI2015 [9] challenge datasets. In neuron segmentation, the SNEMI3D [26] and ISBI12 [2] are other popular datasets. In ophthalmology, the FIVES [25], STARE [18], and ROSE [30] are common datasets to benchmark segmentation performance on color fundus and optical coherence tomography angiography images. The TopCow [48] and VesSAP [44] datasets are frequently used human and murine brain vessel datasets. In this work, we focus on the three most commonly used datasets (CREMI, DRIVE, Roads) as examples to investigate the impact of the identified pitfalls: *(1) Wrong Connectivity Choice, (2) Ground Truth Artifacts,* and *(3) Wrong Evaluation Metrics.*

3 Common Pitfalls

3.1 *Connectivity Choices* Distort the Performance Ranking Between Different Methods

Voxel connectivity strongly affects topological representations and, hence, directly impacts model training and evaluation. Most prior works do not report which connectivity they use or appear to use an unfavorable choice where topological features lose their semantic meaning (see Fig. 2).

To translate a discrete 2D or 3D binary image $\mathcal{I}$ into a topological space, defining the connectivity between voxels (i.e., whether diagonal adjacent voxels belong to the same component) is necessary. Voxel connectivity in a D-dimensional image can be defined by *direct* connectivity, i.e., a voxel has $2 \times D$ neighbors (e.g., 4 for 2D and 6 for 3D images) or *all* connectivity, i.e., a voxel has $3^D - 1$ neighbors (e.g., 8 for 2D and 26 for 3D images). In 2D, direct connectivity connects pixels that share an edge, and all connectivity additionally connects diagonally adjacent pixels, with their boundary only sharing a vertex. For 3D data, direct connectivity connects pixels that share a surface cell, and all-connectivity also connects pixels whose boundaries only intersect in a line or a vertex on the boundary. To maintain a meaningful notion of the *interior* and *exterior* of structures, it is necessary to use opposite connectivity choices for foreground and background (Jordan-Curve Theorem), e.g., all connectivity for the foreground and direct connectivity for the background. In this work, we use the letter **A** to denote the setting where all connectivity is applied to the foreground and direct connectivity is applied to the background. We denote the inverted connectivity choice with the letter **D**.

In cubical complexes, which are often used to describe the topology of digital images [11,21,29,43], the complex's construction choice implicitly encodes a specific connectivity. V-construction $V(\mathcal{I})$ is closely related to direct connectivity for the foreground and all connectivity for the background. The T-construction $T(\mathcal{I})$ is closely related to the inverted connectivity choice. Bleile et al. [6] describe the relationship between the two constructions.

Making the correct connectivity choice is essential to accurately capture the domain-specific semantics of the underlying data. Investigating the semantics of

Table 1. Dependence of the number of connected components on the foreground connectivity choice. The background connectivity is set to the opposite connectivity choice to satisfy the Jordan closed curve theorem. **A** denotes that all connectivity is selected for the foreground and direct for the background. **D** is used to denote the inverted connectivity setting.

	DRIVE-2D		CREMI-2D		Roads-2D		MSSEG2-3D	
	FG	BG	FG	BG	FG	BG	FG	BG
A	132	2362	705	49592	641	8839	153	29
D	18850	1113	712	44815	644	7535	155	29
Ratio	0.8%	47.1%	99.0%	90.4%	99.5%	85.2%	98.7	100

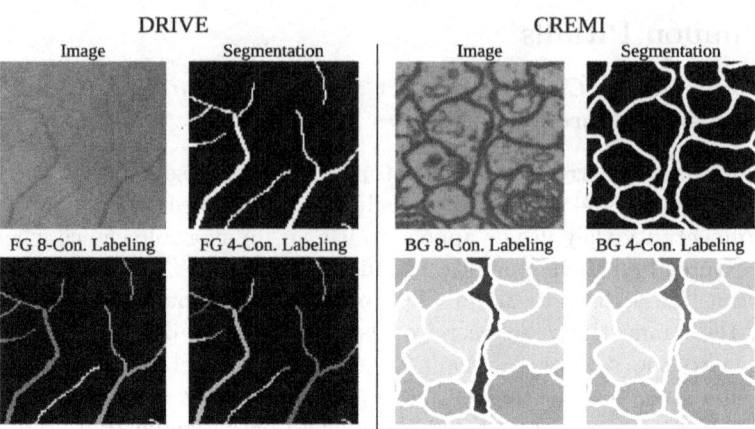

Fig. 2. Example of the importance of making the correct connectivity choices for the DRIVE and CREMI datasets. In the DRIVE dataset, small vessels are disconnected with 4-connectivity for the FG. In the CREMI dataset, synaptic clefts can become disconnected with 4-connectivity for the BG.

the most commonly used datasets, we find that often one connectivity choice is favorable over the other. For example, in the frequently used DRIVE dataset, the smallest vessels are represented through pixels that are only linked through all connectivity (see Fig. 2 (a)). There, choosing direct connectivity instead of all connectivity results in a more than 100-fold increase in the number of foreground connected components (see Table 1) because single vessel segments are split into multiple connected components. Among the most important previous works, only one paper reports their connectivity choice, which is the semantically unfavorable **D**-connectivity for the DRIVE dataset; most other works do not explicitly state their connectivity choices [3,5,19–21,27,28,43,46]. This is problematic for the commonly used datasets, as their topological representation strongly depends on the connectivity choices, as displayed in Table 1. However, not all datasets exhibit this problem, e.g., MSSEG2 [12].

Experimental Investigation of the Problem: To demonstrate the effect of the connectivity choice on topology-aware segmentation methods, we perform an experiment in which we train and evaluate different methods with the two distinct connectivity choices **A** and **D**. The results are displayed in Table 2. We observe that the variation originating from connectivity choices dominates the inter-method variation. The ranking of the methods changes drastically depending on the connectivity choice. We conclude that making the semantically meaningful connectivity choice is essential for a robust evaluation. High performance under the unfavorable **A** connectivity does not indicate that high performance can be expected under the more meaningful **D** connectivity. Interestingly, we mostly observe negative correlations for the scores under the distinct connectivity choices. A potential explanation is that methods that are independent of the connectivity choice during training (Dice, Mosin, clDice) perform very well com-

Table 2. Results of the connectivity experiment on the CREMI dataset. The first block of results stems from evaluation and training with direct connectivity for the foreground and all connectivity for the background (D). The next block contains the results for the inverted setting (A). Numbers in brackets () behind the results contain the method's ranking. The bottom block contains an analysis of the comparability of a method's rank under the two connectivity choices.

	Loss	DICE ↑	B0 ↓	B1 ↓	BM0 ↓	BM1 ↓	VOI ↓	ARE ↓
	Dice	.9466 (1)	1.52 (6)	2.09 (2)	2.38 (6)	4.14 (1)	.4212 (1)	.1129 (1)
	ClDice	.9404 (6)	1.45 (5)	2.44 (4)	2.32 (5)	4.31 (4)	.4580 (5)	.1271 (5)
D ✓	HuTopo	.9455 (2)	1.34 (4)	2.11 (3)	2.08 (4)	4.24 (2)	.4270 (2)	.1129 (2)
	BettiM	.9438 (4)	1.21 (1)	2.06 (1)	1.95 (2)	4.30 (3)	.4336 (4)	.1140 (3)
	Mosin	.9435 (5)	1.30 (3)	2.74 (6)	1.90 (1)	4.59 (6)	.4881 (6)	.1370 (6)
	TopoG	.9444 (3)	1.26 (2)	2.46 (5)	2.01 (3)	4.33 (5)	.4324 (3)	.1183 (4)
	Dice	.9435 (3)	0.46 (2)	3.91 (4)	0.51 (1)	7.39 (3)	.4397 (3)	.1203 (3)
	ClDice	.9441 (2)	0.53 (4)	3.57 (1)	0.58 (3)	7.20 (1)	.4336 (2)	.1169 (2)
A ✗	HuTopo	.9418 (4)	0.46 (1)	4.36 (6)	0.54 (2)	7.66 (5)	.4481 (4)	.1247 (6)
	BettiM	.9411 (6)	0.57 (6)	3.96 (5)	0.63 (6)	7.81 (6)	.4547 (6)	.1219 (4)
	Mosin	.9450 (1)	0.53 (3)	3.66 (2)	0.61 (4)	7.23 (2)	.4282 (1)	.1141 (1)
	TopoG	.9417 (5)	0.56 (5)	3.86 (3)	0.62 (5)	7.50 (4)	.4489 (5)	.1242 (5)
Comparison of Method Ranking for D and A Connectivity Choices
Spearman's ρ		-0.37	-0.85	-0.66	-0.77	-0.37	-0.43	-0.70
Kendall's τ		-0.07	-0.79	-0.47	-0.60	-0.20	-0.20	-0.55
Pearsons's r		-0.34	-0.70	-0.69	-0.76	-0.39	-0.77	-0.87

pared to methods affected by the connectivity choice (HuTopo, BettiM, TopoG) even in the semantically unfavorable setting. However, when the favorable connectivity choice is made, the latter methods capture the semantically expected topology and, therefore, achieve higher performance.

Strategies to Identify and Resolve Connectivity Problems: We encourage future works to make individual and semantics-oriented connectivity choices for every individual dataset. For the three common benchmarking datasets, we determine **A** connectivity for DRIVE (to maintain vessel connectivity), **D** connectivity for CREMI (to avoid separation of synaptic clefts), and **D** connectivity for Roads (to minimize connectivity artifacts in the background, see Sect. 3.2) as the sensible connectivity choices. The choices should be made transparent to the reader when reporting metrics; e.g., for Betti numbers β_{0_A} versus β_{0_D}; alternatively β_{0_T} versus β_{0_V} for an explicit notation of the cubical complex construction. This notation can be extended to all metrics where connectivity choices are influential, e.g., ARE_A or ARE_D. Moreover, we propose a simple method to see how susceptible a dataset is to connectivity choices. We consider the two different partitions

$$P_D = \{F_1, F_2, ..., F_k, B_1, B_2, ..., B_l\}$$
$$P_A = \{F'_1, F'_2, ..., F'_n, B'_1, B'_2, ..., B'_m\}$$

Table 3. Summary of the susceptibility to connectivity choices, introduced by, e.g., using the T- or V-construction, for different datasets. High scores indicate that a dataset, combined with a specific metric, is highly susceptible to connectivity choice.

Dataset	$\beta_{0_{D vs. A}}^{err}$	$\beta_{1_{D vs. A}}^{err}$	$VOI_{D vs. A}$	$ARE_{D vs. A}$
DRIVE	467.95	32.23	0.1171	0.8608
CREMI	0.056	38.216	0.0025	0.0004
Roads	0.0242	10.5161	0.0030	0.0009

of a label G that result from connected component labeling with the **D** and **A** connectivity. The two partitions are the basis for calculating metrics (e.g., β, ARE, VOI). We define the connectivity susceptibility for a metric, e.g., β_0^{err}, as

$$\beta_{0_{D vs. A}}^{err} = |\beta_0(P_D) - \beta_0(P_A)| \qquad (1)$$

where $\beta_0(P)$ measures the number of foreground components in P. Note that because β_0^{err} is a *true* metric, the results are independent of the order of P_D and P_A. Other susceptibility metrics are defined accordingly, and large values indicate high susceptibility. Table 3 shows the results of this analysis for different datasets and metrics. The results of the connectivity experiment (see Table 2) show that high susceptibility values are indicators for a large discrepancy of scores under **D** and **A** connectivity.

3.2 Topological Artifacts in the Label Skew Evaluation Results

In the context of topology-aware image segmentation, we propose to denote *topological artifacts* as topological features existing in the ground truth labels but *conveying no or a wrong semantic meaning*. We observe three prominent causes of such artifacts in the investigated datasets: (1) connectivity issues, (2) label noise, and (3) insufficient resolution (see Fig. 3).

Connectivity Artifacts: In many datasets, neither **A** nor **D** connectivity resolve all connectivity issues. In these cases, we define connectivity artifacts as topological artifacts that occur due to the choices made in connectivity. Connectivity artifacts can be found in the DRIVE dataset (see Fig. 3). While the foreground 8-connectivity is absolutely crucial to capture the semantics of the small vessels for the DRIVE dataset, the then required background 4-connectivity divides a single inter-vessel area into numerous separate components. In the DRIVE dataset, these artifacts result in almost a doubling of background components for 4-connectivity instead of 8-connectivity (see Table 5).

Label Noise Artifacts: Label noise artifacts occur due to annotation errors during the generation of the ground truth segmentation mask. While some types of label noise do not interfere with the topological representation, others can lead to topological artifacts. Single-pixel label noise is a common phenomenon in

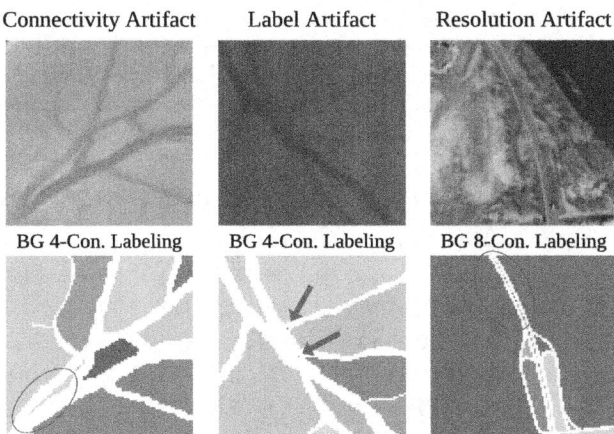

Fig. 3. Examples of topological artifacts. The two left columns show connectivity artifacts and label noise in the DRIVE dataset. The right column shows resolution artifacts in the Roads dataset. Arrows and circles indicate topological artifacts.

segmentation labels. These errors only have a negligible impact on pixel-wise metrics. However, the impact on the topological representation can be dramatic.

Resolution Artifacts: We describe resolution artifacts as artifacts caused by the representation of the underlying semantics with insufficient resolution. Connectivity artifacts are, in fact, a special case of the insufficient resolution issue. An example of resolution artifacts can be found in the Roads dataset, where the insufficient resolution of the aerial images causes the separating strip of two lanes to appear as dozens of separated background components, as displayed in Fig. 3. These artifacts often have negligible effects on pixel-wise accuracy but can drastically affect the evaluation with common topological metrics and interfere with the optimization mechanisms of topology-aware methods.

Experimental Investigation of the Problem: We compare the performance of different topology-aware methods before and after removing small background components using the DRIVE dataset. Visual inspection revealed that these components are predominantly caused by topological label noise and connectivity artifacts. The results of this experiment indicate a strong effect on the Betti and Betti matching metrics (see Table 4). Particularly for Betti number 1 error ($\sim -29\%$) and Betti matching 1 error ($\sim -43\%$), we see extreme changes caused by the artifacts (see Fig. 3 for examples). In comparison, the artifacts only induce minute changes for the Dice score and VOI and ARE, which have traits of both topological and pixel-wise metrics.

Strategies to Mitigate the Problem: A visual inspection of the dataset is paramount to identify topological artifacts and identify ways to remove them. While a visual inspection of the DRIVE dataset reveals that small connected components are mostly artifacts, in the CREMI dataset, such components often

Table 4. Results of the artifacts experiment on DRIVE. The first results block stems from training and evaluating models with adapted labels where small components (≤ 5 pixel) are removed. The next block contains the results for the original labels. Numbers in brackets () contain the method's ranking. The bottom block contains an analysis of the comparability of a method's rank before and after removing small components.

	Loss	DICE ↑	B0$_A$ ↓	B1$_A$ ↓	BM0$_A$ ↓	BM1$_A$ ↓	VOI$_A$ ↓	ARE$_A$ ↓
Corrected Label ✓	Dice	.8808 (4)	9.05 (5)	6.35 (4)	9.59 (6)	11.88 (2)	.5187 (6)	.2550 (6)
	ClDice	.8875 (1)	7.29 (4)	6.61 (6)	7.81 (4)	11.75 (1)	.4869 (1)	.2277 (1)
	HuTopo	.8816 (3)	2.65 (1)	4.50 (2)	3.12 (1)	13.72 (6)	.5022 (2)	.2473 (4)
	BettiM	.8786 (6)	2.70 (2)	4.34 (1)	3.17 (2)	12.25 (5)	.5129 (5)	.2518 (5)
	Mosin	.8823 (2)	9.39 (6)	6.47 (5)	9.39 (5)	12.21 (4)	.5069 (3)	.2380 (2)
	TopoG	.8804 (5)	4.68 (3)	6.11 (3)	5.08 (3)	11.91 (3)	.5078 (4)	.2413 (3)
Original Label ✗	Dice	.8873 (1)	7.92 (4)	11.02 (4)	8.44 (4)	16.38 (2)	.4910 (2)	.2336 (2)
	ClDice	.8827 (4)	9.76 (6)	11.47 (6)	10.31 (6)	16.67 (4)	.5026 (4)	.2375 (4)
	HuTopo	.8827 (5)	4.33 (2)	8.30 (2)	4.87 (2)	18.78 (6)	.4984 (3)	.2375 (3)
	BettiM	.8831 (2)	2.44 (1)	7.56 (1)	2.91 (1)	18.16 (5)	.4894 (1)	.2277 (1)
	Mosin	.8828 (3)	8.35 (5)	11.39 (5)	8.88 (5)	16.57 (3)	.5056 (5)	.2384 (5)
	TopoG	.8746 (6)	5.11 (3)	10.38 (3)	5.65 (3)	16.36 (1)	.5355 (6)	.2658 (6)
Comparison of Performance Label ✗ and Fixed Label ✓								
Avg. Difference		.0003	-0.36	-4.29	-0.48	-4.87	.0022	.0034
Avg. Rel. Change		0.38%	-5.86%	-42.91%	-6.99%	-28.33%	0.53%	1.72%

reflect the beginning of neurons or synaptic clefts, and removing them would destroy essential topological information. For the DRIVE and Roads dataset, a simple image processing that removes those isolated pixels is an effective remedy and allows for a focus on the semantically meaningful topologic structures.

Table 5. Effect of removal of components up to a specific size on the topological representation of the commonly used CREMI, roads, and DRIVE dataset. We use the semantically favorable connectivities for each dataset; Drive: **A**, CREMI and Roads: **D**

Components	DRIVE$_A$		Roads$_D$		CREMI$_D$	
	FG	BG	FG	BG	FG	BG
No Removal	132	2362	644	7535	712	44815
1 Pix. Removal	125	1839	641	7124	712	42826
2 Pix. Removal	120	1706	635	6905	712	42134
5 Pix. Removal	117	1629	613	6664	712	40795
Min/Max Ratio	88.6%	69.0%	95.0%	88.4%	100%	91.0%

3.3 Evaluation Metrics, as Commonly Reported, Lack Expressive Power

The evaluation of topology-aware segmentation typically employs distributional (e.g., VOI or ARE/ARI) and/or topological metrics, such as the Betti number or Betti matching error in combination with pixel-wise/overlap-based metrics (e.g., Dice or cross-entropy). We find that current reporting practices often fail to provide an expressive and interpretable characterization of topological correctness.

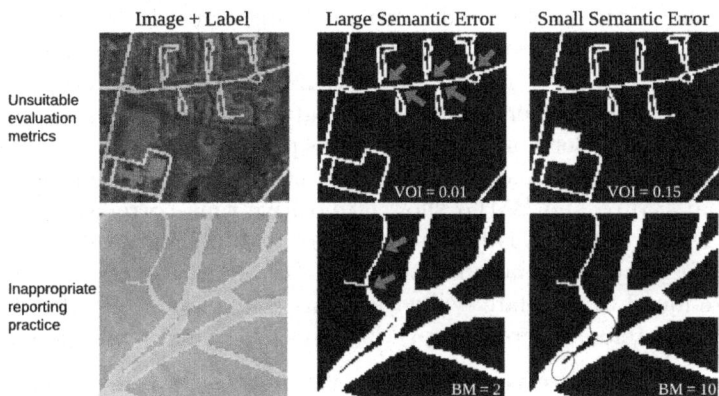

Fig. 4. Examples for the impact of *inappropriate reporting practice* on DRIVE (top) and *unsuitable evaluation metrics* on the Roads dataset (bottom). Left: Image with an overlay of the segmentation label. Middle: Unfavorable predictions, with detached vessels $BM = 2$ (top) and disconnected residential blocks $VOI = 0.01$ (bottom). Right: Favorable predictions with missing background components without semantic meaning $BM = 10$ (top) and additional segmentation of parking areas $VOI = 0.15$ (bottom).

Experimental Investigation of the Problem: Distributional metrics such as VOI and ARI/ARE irreversibly entangle topological and volumetric errors [14,37]. Therefore, reporting only these metrics in addition to pixel-wise performance metrics does not allow for an expressive evaluation of topological accuracy. Figure 4 shows an example where volumetric deviations largely dominate the score of these metrics while small yet critical topological errors are marginalized. Empirically, we validate this shortcoming and find that the distributional metrics do not always correlate with topological accuracy. For example, we find no correlation between $BM0_A$ and ARE_A ($\rho = -0.03$) or VOI_A ($\rho = 0.31$) in the DRIVE dataset. Here, $BM0_A$ is the most important topological metric because it captures disconnected or incorrect vessel segments. However, in other cases, we find a positive correlation. For CREMI, $BM1_D$ is an important topological metric since it captures false splits and merges of neurons. Here, $BM1_D$ correlates well with ARE_D (Spearman's $\rho = 0.94$) and VOI_D ($\rho = 0.83$).

Betti number errors are a better alternative to distributional metrics as they disentangle any volumetric effects and provide topologically interpretable values. However, Stucki et al. [43] show that merely comparing the Betti numbers can be misleading as it disregards the spatial correspondence of invariants and propose the Betti matching error. We investigate this statement empirically and mostly find a good correspondence of performance rankings between Betti number and Betti matching errors, e.g. in CREMI (B0$_D$ with BM0$_D$ ($\rho = 0.83$), and B1$_D$ with BM1$_D$ ($\rho = 0.94$)). However, we also observe a negative correlation between B1$_A$ and BM1$_A$ ($\rho = -0.77$) for the DRIVE dataset. Here, the Betti number 1 error of some methods is reduced by features without spatial correspondence.

While the Betti number and Betti matching error disentangle volumetric effects and provide interpretable values, they are often aggregated across dimensions (e.g., $\beta^{err} = \beta_0^{err} + \beta_1^{err}$) [3,5,19–21,29,43]. Figure 4 illustrates how the combined Betti matching error can be misleading. The segmentation to the right has a 5-times higher Betti number error but is preferable for downstream network analysis tasks because it maintains vessel connectivity. In our empirical analyses, we find that BM0$_D$ and BM1$_D$ show no positive rank correlation ($\rho = -0.77$, see Table 2). Here, BM1$_D$ is the most important topological metric as it shows almost perfect rank correlation with ARI$_D$ ($\rho = 0.94$) and VOI$_D$ ($\rho = 0.83$), which were found to correlate well with neuron segmentation quality by domain experts [2]. Therefore, aggregating BM0$_D$ and BM1$_D$ reduces expressive power.

Strategies to Mitigate the Problem: We propose three important reporting practices for evaluating topology-aware image segmentation models. (1) We recommend always reporting at least one disentangled pair of purely topological and volumetric metrics. Due to the occasional issues with spatial correspondence, we propose to use Betti matching errors instead of Betti number errors (e.g., BM0, BM1, and Dice). (2) We recommend reporting topological errors always without aggregation across dimensions (e.g., BM0, BM1, instead of BM). (3) Finally, we recommend a problem-aware selection of additional metrics [31] (e.g., ARI and VOI for Neuron Segmentation and clDice for Vessel Segmentation).

4 Discussion

The presented work sheds light on common pitfalls during the evaluation of topology-aware image segmentation methods. We identify (1) connectivity definition, (2) label artifacts, and (3) misaligned use of evaluation metrics as major problems that heavily impact previous studies. In dedicated experiments, we show that these pitfalls are a major limitation for meaningful benchmarking in topology-aware image segmentation. Specifically, our results indicate that topological performance rankings are vulnerable to connectivity choices, showing on average negative correlations of the rankings (avg. Spearman's $\rho = -0.63$) of **A** and **D** connectivity. We find that label artifacts can comprise up to 43% of measured topological errors in some metrics. Finally, we find that flawed evaluation practices, such as an aggregation of Betti numbers, drastically impair the expressivity of the evaluation.

Based on our analysis, we conclude with the following recommendations for future works: (1) *Connectivity choices must be made on a dataset and not on a method basis. The choices have to be transparent for the reader.* We introduce a new method to quantify the connectivity susceptibility of datasets, providing a measure of the importance of connectivity choices. (2) *Topological artifacts must be considered in topology-aware image segmentation.* We provide a definition for topological artifacts that were previously overlooked due to their negligible influence on pixel-wise evaluation. (3) *Evaluation metrics should disentangle volumetric and topological information and topological errors of different dimensions. Other metrics should be added in a problem-oriented manner.* Ultimately, this paper should ignite a discussion on the current state and best practices of topology-aware image segmentation.

Limitations. While the presented issues are valid for 2D as well as 3D images, an empirical evaluation of 3D datasets is still warranted because the commonly used slicing of 3D volumes in 2D images changes a segmentation's topological requirements as well as the possible connectivity definitions (e.g., 6–18 connectivity for 3D). Furthermore, our work deliberately does not discuss general pitfalls in model evaluation that are not specific to topology-aware image segmentation. These general pitfalls include insufficient variation analysis [10], unfair baseline comparisons [23], or general unsuitability of metrics [39].

Acknowledgements. AB is supported by the Stiftung der Deutschen Wirtschaft. MM is funded by the German Research Foundation under project 532139938.

Disclosure of Interests. The authors have no competing interests to declare that are relevant to the content of this article.

References

1. Arbelaez, P., Maire, M., Fowlkes, C., Malik, J.: Contour detection and hierarchical image segmentation. IEEE Trans. Pattern Anal. Mach. Intell. **33**(5), 898–916 (2010)
2. Arganda-Carreras, I., et al.: Crowdsourcing the creation of image segmentation algorithms for connectomics. Front. Neuroanat. **9**, 152591 (2015)
3. Attari, M., Nguyen, N.P., Palaniappan, K., Bunyak, F.: Multi-loss topology-aware deep learning network for segmentation of vessels in microscopy images. In: 2023 IEEE Applied Imagery Pattern Recognition Workshop (AIPR), pp. 1–7. IEEE (2023)
4. Banerjee, S., et al.: Semantic segmentation of microscopic neuroanatomical data by combining topological priors with encoder-decoder deep networks. Nat. Mach. Intell. **2**(10), 585–594 (2020)
5. Berger, A.H., et al.: Topologically faithful multi-class segmentation in medical images. In: International Conference on Medical Image Computing and Computer-Assisted Intervention, pp. 721–731. Springer (2024)
6. Bleile, B., Garin, A., Heiss, T., Maggs, K., Robins, V.: The persistent homology of dual digital image constructions. In: Research in Computational Topology 2, pp. 1–26. Springer (2022)

7. Bohlender, S., Oksuz, I., Mukhopadhyay, A.: A survey on shape-constraint deep learning for medical image segmentation. IEEE Rev. Biomed. Eng. **16**, 225–240 (2021)
8. Byrne, N., Clough, J.R., Valverde, I., Montana, G., King, A.P.: A persistent homology-based topological loss for CNN-based multiclass segmentation of CMR. IEEE Trans. Med. Imaging **42**(1), 3–14 (2022)
9. Carass, A., et al.: Longitudinal multiple sclerosis lesion segmentation: resource and challenge. Neuroimage **148**, 77–102 (2017)
10. Christodoulou, E., et al.: Confidence intervals uncovered: are we ready for real-world medical imaging AI? In: International Conference on Medical Image Computing and Computer-Assisted Intervention, pp. 124–132. Springer (2024)
11. Clough, J.R., Byrne, N., Oksuz, I., Zimmer, V.A., Schnabel, J.A., King, A.P.: A topological loss function for deep-learning based image segmentation using persistent homology. IEEE Trans. Pattern Anal. Mach. Intell. **44**(12), 8766–8778 (2020)
12. Commowick, O., Cervenansky, F., Cotton, F., Dojat, M.: MSSEG-2 challenge proceedings: multiple sclerosis new lesions segmentation challenge using a data management and processing infrastructure. In: MICCAI 2021-24th International Conference on Medical Image Computing and Computer Assisted Intervention, p. 126 (2021)
13. Dosovitskiy, A.: An image is worth 16x16 words: transformers for image recognition at scale. arXiv preprint arXiv:2010.11929 (2020)
14. Funke, J., Klein, J., Moreno-Noguer, F., Cardona, A., Cook, M.: TED: a tolerant edit distance for segmentation evaluation. Methods **115**, 119–127 (2017)
15. Funke, J., et al.: Large scale image segmentation with structured loss based deep learning for connectome reconstruction. IEEE Trans. Pattern Anal. Mach. Intell. **41**(7), 1669–1680 (2018)
16. Gegúndez-Arias, M.E., Aquino, A., Bravo, J.M., Marín, D.: A function for quality evaluation of retinal vessel segmentations. IEEE Trans. Med. Imaging **31**(2), 231–239 (2011)
17. Han, X., Xu, C., Braga-Neto, U., Prince, J.L.: Topology correction in brain cortex segmentation using a multiscale, graph-based algorithm. IEEE Trans. Med. Imaging **21**(2), 109–121 (2002)
18. Hoover, A., Kouznetsova, V., Goldbaum, M.: Locating blood vessels in retinal images by piecewise threshold probing of a matched filter response. IEEE Trans. Med. Imaging **19**(3), 203–210 (2000)
19. Hu, X., Wang, Y., Li, F., Samaras, D., Chen, C.: Topology-aware segmentation using discrete Morse theory. In: International Conference on Learning Representations (ICLR) (2021)
20. Hu, X.: Structure-aware image segmentation with homotopy warping. Adv. Neural. Inf. Process. Syst. **35**, 24046–24059 (2022)
21. Hu, X., Li, F., Samaras, D., Chen, C.: Topology-preserving deep image segmentation. Adv. Neural Inf. Process. Syst. **32** (2019)
22. Hubert, L., Arabie, P.: Comparing partitions. J. Classif. **2**, 193–218 (1985)
23. Isensee, F., et al.: nnU-Net revisited: a call for rigorous validation in 3d medical image segmentation. In: International Conference on Medical Image Computing and Computer-Assisted Intervention, pp. 488–498. Springer (2024)
24. Januszewski, M., et al.: High-precision automated reconstruction of neurons with flood-filling networks. Nat. Methods **15**(8), 605–610 (2018)
25. Jin, K., et al.: FIVES: a fundus image dataset for artificial intelligence based vessel segmentation. Sci. Data **9**(1), 475 (2022)

26. Kasthuri, N., et al.: Saturated reconstruction of a volume of neocortex. Cell **162**(3), 648–661 (2015)
27. Li, L., et al.: Robust segmentation via topology violation detection and feature synthesis. In: International Conference on Medical Image Computing and Computer-Assisted Intervention, pp. 67–77. Springer (2023)
28. Li, L., et al.: Universal topology refinement for medical image segmentation with polynomial feature synthesis. In: International Conference on Medical Image Computing and Computer-Assisted Intervention, pp. 670–680. Springer (2024)
29. Lux, L., et al.: Topograph: an efficient graph-based framework for strictly topology preserving image segmentation. arXiv preprint arXiv:2411.03228 (2024)
30. Ma, Y., et al.: ROSE: a retinal oct-angiography vessel segmentation dataset and new model. IEEE Trans. Med. Imaging **40**(3), 928–939 (2020)
31. Maier-Hein, L., et al.: Metrics reloaded: recommendations for image analysis validation. Nat. Methods **21**(2), 195–212 (2024)
32. Meilă, M.: Comparing clusterings by the variation of information. In: Learning Theory and Kernel Machines: 16th Annual Conference on Learning Theory and 7th Kernel Workshop, COLT/Kernel 2003, Washington, DC, USA, 24–27 August 2003. Proceedings, pp. 173–187. Springer (2003)
33. Meilă, M.: Comparing clusterings-an information based distance. J. Multivar. Anal. **98**(5), 873–895 (2007)
34. Meilă, M.: Comparing clusterings: an axiomatic view. In: Proceedings of the 22nd International Conference on Machine Learning, pp. 577–584 (2005)
35. Mnih, V.: Machine learning for aerial image labeling. Ph.D. thesis, CAN (2013). aAINR96184
36. Mosinska, A., Marquez-Neila, P., Koziński, M., Fua, P.: Beyond the pixel-wise loss for topology-aware delineation. In: Proceedings of the IEEE Conference on Computer Vision and Pattern Recognition, pp. 3136–3145 (2018)
37. Nunez-Iglesias, J., Kennedy, R., Parag, T., Shi, J., Chklovskii, D.B.: Machine learning of hierarchical clustering to segment 2d and 3d images. PLoS ONE **8**(8), e71715 (2013)
38. Panayides, A.S., et al.: Ai in medical imaging informatics: current challenges and future directions. IEEE J. Biomed. Health Inform. **24**(7), 1837–1857 (2020)
39. Reinke, A., et al.: Understanding metric-related pitfalls in image analysis validation. Nat. Methods **21**(2), 182–194 (2024)
40. Ronneberger, O., Fischer, P., Brox, T.: U-Net: convolutional networks for biomedical image segmentation. In: Navab, N., Hornegger, J., Wells, W.M., Frangi, A.F. (eds.) MICCAI 2015, Part III. LNCS, vol. 9351, pp. 234–241. Springer, Cham (2015). https://doi.org/10.1007/978-3-319-24574-4_28
41. Shit, S., et al.: clDice-a novel topology-preserving loss function for tubular structure segmentation. In: Proceedings of the IEEE/CVF Conference on Computer Vision and Pattern Recognition, pp. 16560–16569 (2021)
42. Staal, J., Abramoff, M., Niemeijer, M., Viergever, M., van Ginneken, B.: Ridge based vessel segmentation in color images of the retina. IEEE Trans. Med. Imaging **23**(4), 501–509 (2004)
43. Stucki, N., Paetzold, J.C., Shit, S., Menze, B., Bauer, U.: Topologically faithful image segmentation via induced matching of persistence barcodes. In: International Conference on Machine Learning, pp. 32698–32727. PMLR (2023)
44. Todorov, M.I., Piraud, M., et al.: Machine learning analysis of whole mouse brain vasculature. Nat. Methods **17**(4), 442–449 (2020)

45. Unnikrishnan, R., Pantofaru, C., Hebert, M.: Toward objective evaluation of image segmentation algorithms. IEEE Trans. Pattern Anal. Mach. Intell. **29**(6), 929–944 (2007)
46. Wu, Q., Chen, Y., Liu, W., Yue, X., Zhuang, X.: Deep closing: enhancing topological connectivity in medical tubular segmentation. IEEE Trans. Med. Imag. (2024)
47. Yan, K., Wang, X., Lu, L., Summers, R.M.: DeepLesion: automated mining of large-scale lesion annotations and universal lesion detection with deep learning. J. Med. Imaging **5**(3), 036501 (2018)
48. Yang, K., et al.: Benchmarking the CoW with the TopCoW challenge: topology-aware anatomical segmentation of the circle of Willis for CTA and MRA. arXiv preprint arXiv:2312.17670 (2023)
49. Zhang, H., et al.: Geometric loss for deep multiple sclerosis lesion segmentation. In: 2021 IEEE 18th International Symposium on Biomedical Imaging (ISBI), pp. 24–28. IEEE (2021)

GeoT: Geometry-Guided Instance-Dependent Transition Matrix for Semi-supervised Tooth Point Cloud Segmentation

Weihao Yu[1], Xiaoqing Guo[2,3], Chenxin Li[1], Yifan Liu[1], and Yixuan Yuan[1(✉)]

[1] Department of Electronic Engineering, The Chinese University of Hong Kong, Hong Kong, China
yxyuan@ee.cuhk.edu.hk
[2] Department of Computer Science, Hong Kong Baptist University, Hong Kong, China
[3] Department of Engineering Science, University of Oxford, Oxford, UK

Abstract. Achieving meticulous segmentation of tooth point clouds from intra-oral scans stands as an indispensable prerequisite for various orthodontic applications. Given the labor-intensive nature of dental annotation, a significant amount of data remains unlabeled, driving increasing interest in semi-supervised approaches. One primary challenge of existing semi-supervised medical segmentation methods lies in noisy pseudo labels generated for unlabeled data. To address this challenge, we propose GeoT, the first framework that employs instance-dependent transition matrix (IDTM) to explicitly model noise in pseudo labels for semi-supervised dental segmentation. Specifically, to handle the extensive solution space of IDTM arising from tens of thousands of dental points, we introduce tooth geometric priors through two key components: point-level geometric regularization (PLGR) to enhance consistency between point adjacency relationships in 3D and IDTM spaces, and class-level geometric smoothing (CLGS) to leverage the fixed spatial distribution of tooth categories for optimal IDTM estimation. Extensive experiments performed on the public Teeth3DS dataset and private dataset demonstrate that our method can make full utilization of unlabeled data to facilitate segmentation, achieving performance comparable to fully supervised methods with only 20% of the labeled data. The source code is available at https://github.com/CUHK-AIM-Group/GeoT.

Keywords: Semi-supervised learning · Point cloud segmentation · Geometric prior

1 Introduction

Computer-aided design (CAD) systems have facilitated the widespread adoption of intra-oral scans (IOS) in essential tasks such as dental restoration, treatment planning, and orthodontic diagnosis [1,8]. Accurate tooth point cloud segmentation is crucial for these applications [10,15]. However, manual annotation of IOS point clouds is prohibitively labor-intensive, driving research toward automated solutions. Fully-supervised methods [7,9,12,21,23] have made significant progress through strategies like multi-scale local-global feature integration [12] and confidence-aware cascading [7], but they rely heavily on large-scale labeled data, which remains scarce due to the expertise required for dental annotations.

Recently, semi-supervised methods [4,14,20] have garnered increased attention as they exploit unlabeled data by leveraging pseudo labels. The main challenge lies in noise within these pseudo labels. Many existing approaches employ confidence thresholds to filter unreliable data [16,24]. In order to fully utilize unlabeled data, the noise transition matrix (NTM) offers an alternative by modeling the label noise distribution and integrating all available pseudo labels [11,19]. Specifically, NTM estimates a matrix where each entry represents the probability that a sample from the true class is mislabeled as another class. By multiplying NTM with the learning objective, the network can be trained on noisy labels while accounting for their uncertainty. It preserves all available training signals rather than enforcing a hard threshold that may eliminate informative samples. Some works [5,6,18] have further extended this concept to instance-dependent transition matrix (IDTM)[1], which capture how noise patterns vary across different samples, providing a more flexible and accurate model of label noise in real-world scenarios.

Motivated by this, in this paper, we make the first attempt to utilize IDTM to handle pseudo label noise for tooth point cloud segmentation. However, it is not trivial to extend IDTM to such a segmentation task. Firstly, estimating IDTM requires appropriate assumptions to constrain the degrees of freedom of its solution space [6,18]. For instance, in image classification tasks, Xia *et al.* [18] assumed that the IDTM can be decomposed into a series of partially correlated basic transition matrices, and then reconstructed the IDTM through a convex combination of these matrices. But this approach is not applicable to dense tasks such as dental segmentation, where the instance is a point and cannot be further decomposed. Secondly, each point possesses its own IDTM, while an IOS contains tens of thousands of points. In this situation, the network needs to estimate such a large number of IDTMs simultaneously without access to ground truth labels, thus presenting an exceptionally formidable optimization problem. Compared with other tasks, teeth possess unique anatomical structures including fixed spatial arrangements of dental categories and local geometric consistency between neighboring points, which can better assist in accurate segmentation.

To solve these challenges, we propose GeoT, a geometry-guided framework that leverages IDTM for semi-supervised tooth segmentation. Specifically, to align IDTM estimation with dental geometric characteristics, we exploit the relatively fixed geometric distribution of tooth categories and design class-level geometric smoothing (CLGS). It encodes the anatomical relationships between

---

[1] For convenience, unless otherwise specified, NTM in the following discussions will refer to the instance-dependent transition matrix.

tooth categories defined in the FDI notation [17] (e.g., a molar adjoins premolars and other molars), and constructs a Gaussian prior over transition probabilities, ensuring IDTMs respect these topological constraints. To address the optimization complexity arising from excessive number of IDTMs, we propose point-level geometric regularization (PLGR) to smooth the solution space of IDTM. It minimizes the norm of transition matrix differences between neighboring points of the same category. In this way, IDTM varies gradually within a small neighborhood, ensuring the entire space maintains Lipschitz continuity. This smoothness property significantly simplifies the optimization landscape by reducing the number of local minima and creating more gradual transitions between solutions.

Our main contributions can be summarized as follows: (1) We present GeoT, a practical semi-supervised tooth segmentation framework that learns from noisy pseudo labels. Instead of discarding low-confidence pseudo labels, we use IDTM to model label noise distribution and utilize all noisy labels, representing the first utilization of IDTM for handling pseudo label noise in semi-supervised point cloud segmentation. (2) We leverage both point-level and class-level geometric priors through point-level geometric regularization and class-level geometric smoothing to guide the optimization process of IDTM, significantly enhancing the estimation of IDTM. (3) With extensive experiments on the public Teeth3DS dataset and private dataset, our method achieves state-of-the-art results, attaining performance comparable to fully supervised methods trained on the entire labeled dataset with only 20% of the labeled data.

2 Method

An overview of our proposed method is illustrated in Fig. 1. For labeled data, the network's segmentation results are directly supervised by the ground truth labels. For unlabeled data, we apply both weak and strong data augmentations. The segmentation results from weakly augmented data serve as pseudo-labels. For strongly augmented data, we estimate the IDTM of the prediction results to rectify the supervision signal from the noisy pseudo-labels. To enhance IDTM estimation, we develop point-level geometric regularization and class-level geometric smoothing. PLGR imposes constraints on IDTM by utilizing the geometric priors of dental points in 3D space, while CLGS samples the transfer relationships among categories based on the spatial distribution of tooth classes and incorporates this information into the predicted IDTM. More details will be presented in the following sections.

2.1 Preliminary

Formally, the training dataset $\mathcal{D}$ comprises L labeled data and M unlabeled data $(L \ll M)$, which can be expressed as $\mathcal{D} = \mathcal{D}^l \cup \mathcal{D}^u$. Let C and N denote the number of classes and points of each point cloud data, respectively, then $\mathcal{D}^l = \{(X_i^l \in \mathbb{R}^{N \times 3}, Y_i^l \in \mathbb{R}^{N \times C})\}_{i=1}^{L}$ and $\mathcal{D}^u = \{(X_i^u \in \mathbb{R}^{N \times 3})\}_{i=L+1}^{L+M}$. Given a labeled data pair (X^l, Y^l)[2] and an unlabeled data X^u, following [16], we first

---
[2] For the sake of convenience, we omit the subscripts in the subsequent discussion.

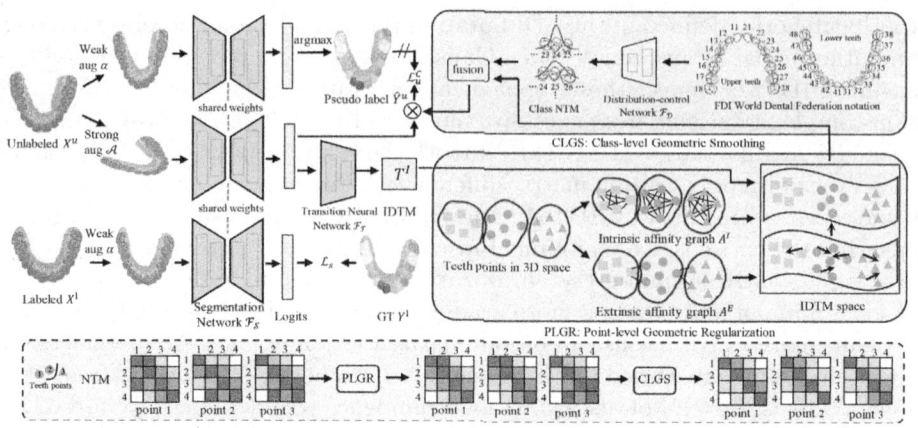

Fig. 1. Overview of the proposed GeoT framework. The top portion presents the flowchart of the method, which encompasses the proposed point-level geometric regularization and class-level geometric smoothing. The lower portion illustrates the impact of these two modules on the noise transition matrix. Point 1 and 2 are the same class, point 3 is different. The 4 × 4 grid represents NTM with 4 classes, where a darker color indicates a higher transition probability. Categories 1 and 2, 2 and 3, as well as 3 and 4 are spatially positioned close to each other. PLGR aligns NTMs of the same class and separates different ones. CLGS corrects NTMs violating distribution law.

perform weak and strong augmentations on them, denoted as $\alpha(\cdot)$ and $\mathcal{A}(\cdot)$, respectively. Then the prediction outputs can be computed by the segmentation network $\mathcal{F}_\mathcal{S}(\cdot)$, i.e., $P^l = \mathcal{F}_\mathcal{S}(\alpha(X^l))$, $P^u = \mathcal{F}_\mathcal{S}(\mathcal{A}(X^u))$, and $Q^u = \mathcal{F}_\mathcal{S}(\alpha(X^u))$. Through $argmax$ operation on Q^u, pseudo label $\hat{Y}^u$ is determined. Finally, the objective function is the combination of supervised loss $\mathcal{L}_s$ and unsupervised loss $\mathcal{L}_u$ as

$$\mathcal{L} = \mathcal{L}_s + \mathcal{L}_u = \sum_{k=1}^{N} \mathcal{H}(p_k^l, y_k^l) + \sum_{k=1}^{N} \mathcal{H}(p_k^u, \hat{y}_k^u), \qquad (1)$$

where $\mathcal{H}(\cdot)$ is the commonly used cross-entropy loss or focal loss [13], $p_k^{l/u}$ is the k-th element of $P^{l/u}$, and y_k^l ($\hat{y}_k^u$) is the k-th element of Y^l ($\hat{Y}^u$).

2.2 Noise Transition Matrix for Semi-supervised Segmentation

Since $\hat{Y}^u$ is noisy, direct optimizing Eq. (1) may lead to model degradation as the segmentation network $\mathcal{F}_\mathcal{S}(\cdot)$ becomes susceptible to label noise. To this end, we model the noise in pseudo labels through instance-dependent transition matrix (IDTM) $T_k^I \in [0,1]^{C \times C}$, which is leveraged to correct the objective function $\mathcal{L}_u$ for robust training. Note that $k \in [1, N]$ indicates the noise transition matrix of k-th point x_k^u. Given that T_k^I relies on x_k^u, we introduce a transition neural network $\mathcal{F}_\mathcal{T}(\cdot)$ (e.g., one fully-connected layer) to estimate the transition matrix via $T_k^I = \mathcal{F}_\mathcal{T}(p_k^u)$, where p_k^u denotes the predicted probability distribution over classes for point xk^u. T_k^I specifies the point-level probability of clean label m being transformed into noisy label n by $T_k^I[m,n] = p(\bar{Y} = n|Y = m, X = x_k^u)$. Thus, the probability of x_k^u being classified as $\bar{Y} = n$ is calculated via $p(\bar{Y} = n) = \sum_{m=1}^{C} p(Y = m) \cdot T_k^I[m,n]$, where $p(Y)$ is the clean class distribution. Then the modeled label noise distribution is utilized to rectify the supervision signal (i.e., $\mathcal{L}_u$) obtained from noisy pseudo labels. Consequently, the objective function is refined to

$$\mathcal{L} = \mathcal{L}_s + \mathcal{L}_u^C = \sum_{k=1}^{N} \mathcal{H}(p_k^l, y_k^l) + \sum_{k=1}^{N} \mathcal{H}(p_k^u T_k^I, \hat{y}_k^u). \qquad (2)$$

During training, the predicted clean distributions p_k^u are transformed through the estimated IDTM T_k^I into noisy distributions that can be directly compared with the noisy pseudo-labels $\hat{y}_k^u$. This process allows the output of $\mathcal{F}_S(\cdot)$ to gradually converge toward clean label distributions, even when only noisy supervision is available. Once properly trained, $\mathcal{F}_S(\cdot)$ directly outputs clean class distributions without requiring any further correction by transition matrices at inference time.

2.3 Point-Level Geometric Regularization

Estimating IDTM of a large number of points without appropriate constraints is an ill-posed problem. We propose point-level geometric regularization to addresses this challenge by introducing a geometric prior that constrains the solution space based on a fundamental observation: adjacent points belonging to the same tooth category should exhibit similar noise transition patterns due to their similar local contexts. We encode this prior knowledge by constructing two types of affinity graphs in 3D space: (1) an intrinsic graph connecting each point to its k_1-nearest neighbors within the same category, and (2) an extrinsic graph connecting each point to its k_2-nearest neighbors from different categories:

$$A_{ij}^I = \begin{cases} e^{-\frac{\|x_i^u - x_j^u\|^2}{\sigma^2}}, & if \ x_j^u \in \mathcal{N}(x_i^u, k_1) \ and \ \hat{y}_i^u = \hat{y}_j^u, \\ 0, & else, \end{cases} \qquad (3)$$

$$A_{ij}^E = \begin{cases} e^{-\frac{\|x_i^u - x_j^u\|^2}{\sigma^2}}, & if \ x_j^u \in \mathcal{N}(x_i^u, k_2) \ and \ \hat{y}_i^u \neq \hat{y}_j^u, \\ 0, & else, \end{cases} \qquad (4)$$

where $\mathcal{N}(x_i^u, k_{1/2})$ denotes the $k_{1/2}$-nearest neighbors, $\hat{y}_i^u = \hat{y}_j^u$ indicates that x_i^u and x_j^u belong to the same tooth, otherwise not. $A_{ij}^{I/E}$ is the intrinsic/extrinsic affinity graph matrix weighted by Gaussian kernel distance. These structural relationships in 3D space are then used to guide the relationships between the corresponding transition matrices as

$$\mathcal{M}_I = \sum_{i,j=1}^{N} A_{ij}^I \|T_i^I - T_j^I\|^2, \ \mathcal{M}_E = \sum_{i,j=1}^{N} A_{ij}^E \|T_i^I - T_j^I\|^2, \qquad (5)$$

$$\mathcal{L}_m = \mathcal{M}_I - \mathcal{M}_E. \tag{6}$$

Here, $\mathcal{M}_{I/E}$ models the relationship between noise transition matrices based on intrinsic/extrinsic affinity graphs in 3D space. Obviously, minimizing $\mathcal{L}_m$ encourages the learned IDTM to exhibit proximity if their corresponding points are close in the same class, otherwise be distant from each other. PLGR smooths the IDTM space in a manner that aligns with the physical structure of teeth, reducing the complexity of the optimization problem.

As we can see in the bottom of Fig. 1, after PLGR, NTMs of the adjacent points in 3D space of one category (i.e., point 1 and 2) are closer. This is evident in the upper left portions of the NTMs for points 1 and 2, which exhibit increased similarity after PLGR compared to their prior states. Concurrently, PLGR effectively guides the NTMs of points belonging to different classes (i.e., points 2 and 3) to become more distant. As shown in the lower right sections of the NTMs for points 2 and 3, the implementation of PLGR enhances the separation between the NTMs of point 2 and point 3, while simultaneously drawing point 2's NTM closer to that of point 1. This mechanism effectively ensures that the structure within the IDTM space remains consistent with that in the geometric space, thereby mitigating the complexity of the optimization process.

2.4 Class-Level Geometric Smoothing

Due to the high degree of freedom in the IDTM solution space, it is insufficient to impose soft constraints solely through PLGR. Following FDI World Dental Federation notation in the upper right corner of Fig. 1, we can observe that each tooth category is associated with its own specific neighbors. This implies that points within each category exhibit a significantly high probability of transition to some specific categories associated with that category, rather than to other categories. Consequently, we further employ the fixed geometric distribution among tooth categories to guide the optimization of IDTM. Specifically, we leverage Gaussian distributions to model the probability distribution law of transition from the current class to other classes, and utilize a distribution-control network $\mathcal{F}_\mathcal{D}(\cdot)$ to generate the standard deviation σ of the Gaussian distribution for each class, i.e., $\sigma_m = \mathcal{F}_\mathcal{D}(c_m)$. Here, c_m is the m-th class, and $1 \leq m \leq C$. Then we can compute the probability distribution of transition for the m-th category:

$$P_m(c_n) = \frac{1}{\sqrt{2\pi}\sigma_m} e^{-\frac{(c_n - c_m)^2}{2\sigma_m^2}}. \tag{7}$$

After calculating each class's distribution, we can sample from them to construct a noise transition matrix T^C that incorporates the geometric prior distribution of the categories via $T^C[m,n] = P_m(c_n)$. Through a fusion operation on T_k^I and T^C, we inject the prior at class level into the transition matrix at point level:

$$T_k^F = (1-\lambda)T_k^I + \lambda T^C, \tag{8}$$

where λ is the weighting factor. The fusion helps to alleviate noise in the point-level estimation by incorporating robust class-level geometric priors.

As illustrated in the bottom of Fig. 1, NTMs that violate the class distribution law (i.e., point 3) are effectively rectified after CLGS. More precisely, we can examine the fourth row of point 3's NTM prior to CLGS, which indicates the transition probabilities of category 4. Notably, it shows a higher likelihood of transitioning to category 2 instead of category 3, despite the spatial proximity of categories 3 and 4, thereby contradicting the expected distribution pattern of tooth classes. In this scenario, CLGS modifies the NTM, and it becomes evident that, post-application of CLGS, point 3's NTM aligns appropriately with the distribution patterns among the categories.

Given that T_k^F incorporates different levels of geometric priors from teeth, it represents an improved estimation of IDTM, and thus can be utilized to replace T_k^I in $\mathcal{L}_u^C$ in Eq. (2). Finally, the overall loss function can be formulated as

$$\mathcal{L} = \mathcal{L}_s + \alpha \mathcal{L}_u^C + \beta \mathcal{L}_m = \sum_{k=1}^{N} \mathcal{H}(p_k^l, y_k^l) + \alpha \sum_{k=1}^{N} \mathcal{H}(p_k^u T_k^F, \hat{y}_k^u) + \beta(\mathcal{M}_I - \mathcal{M}_E), \quad (9)$$

where α and β are weights of $\mathcal{L}_u^C$ and $\mathcal{L}_m$, respectively.

3 Experiments and Results

3.1 Dataset and Implementation Details

We evaluated the proposed method on the public Teeth3DS dataset [3] and our private dataset. Teeth3DS dataset contains 1800 labeled IOS data with 34 classes collected from 900 patients. Following official split, the dataset was divided into 1200 and 600 scans as training and test sets. We randomly selected 0.5%(6), 1%(12), 5%(60), 10%(120), and 20%(240) of training set as labeled data. Our private dataset comprises 1,200 instances of IOS data collected from 600 patients, which were used as unlabeled data to further validate the effectiveness of the proposed method.

We set α, β, and λ to 1.0, 0.1, and 0.9, respectively. σ of Gaussian kernel distance in PLGR was 1.0. PointTransformer [25] was adopted as the segmentation network and the transition neural network consisted of one fully-connected layer. The distribution-control network only contained learnable parameters σ_m. Focal loss was used as $\mathcal{H}(\cdot)$ for $\mathcal{L}_s$ and $\mathcal{L}_u^C$. We implemented our method with PyTorch on Nvidia 4090. The model was trained using Adamw optimizer with a learning rate of 1e-3 for 220 epochs and then 1e-4 for another 30 epochs. Following [15], we randomly sampled 16,000 points per scan as the input and used a knn-based voting mechanism to upsample the output to original size for better efficiency. Three evaluation metrics were used, including mIoU, Dice Similarity Coefficient (DSC), and point-wise classification accuracy(Acc).

3.2 Evaluations

Table 1 compares the quantitative results of our proposed GeoT with different semi-supervised methods. For a fair comparison, we reimplemented these

Table 1. Comparison of our GeoT with different semi-supervised methods on the Teeth3DS Dataset (%).

Ratio	Method	Maxillary			Mandible			All		
		mIoU	DSC	Acc	mIoU	DSC	Acc	mIoU	DSC	Acc
100%	Upper Bound	85.27	90.63	93.41	83.34	88.84	90.89	84.31	89.74	92.15
0.5%	SupOnly	53.65	66.40	77.89	41.77	54.77	66.07	47.71	60.59	71.98
	UA-MT [22]	57.06	69.21	79.94	46.64	59.61	69.63	51.85	64.41	74.78
	FixMatch [16]	59.41	71.24	79.98	45.24	58.07	68.82	52.33	64.45	74.40
	SLC-Net [14]	58.11	69.97	79.53	48.49	61.36	69.46	53.30	65.66	74.49
	DMD [20]	59.38	71.18	80.08	48.81	61.62	70.04	54.09	66.40	75.06
	DC-Net [4]	58.06	70.04	80.26	46.94	59.92	69.11	52.50	64.98	74.69
	BCP [2]	58.66	70.45	79.53	48.60	62.19	69.83	53.63	66.32	74.68
	Ours (GeoT)	**61.47**	**72.81**	**81.02**	**52.15**	**64.72**	**72.28**	**56.81**	**68.76**	**76.65**
1%	SupOnly	56.30	68.20	79.27	55.08	67.35	74.31	55.69	67.78	76.79
	UA-MT [22]	61.80	72.64	81.72	58.62	70.16	76.04	60.21	71.40	78.88
	FixMatch [16]	63.86	74.51	82.77	60.60	71.47	78.55	62.23	72.99	80.66
	SLC-Net [14]	62.66	73.25	82.62	61.19	72.19	78.07	61.93	72.72	80.34
	DMD [20]	61.88	72.66	81.70	59.34	70.62	76.88	60.61	71.64	79.29
	DC-Net [4]	61.20	71.63	81.39	59.14	70.25	76.33	60.17	70.94	78.86
	BCP [2]	64.02	74.27	83.13	62.06	72.74	78.53	63.04	73.51	80.83
	Ours (GeoT)	**66.90**	**76.31**	**84.47**	**64.07**	**74.10**	**79.98**	**65.48**	**75.21**	**82.23**
5%	SupOnly	72.76	81.18	87.95	69.76	78.82	83.94	71.26	80.00	85.95
	UA-MT [22]	77.79	84.97	90.15	72.77	80.61	85.27	75.28	82.79	87.71
	FixMatch [16]	75.43	83.01	89.06	72.57	80.83	85.21	74.00	81.92	87.13
	SLC-Net [14]	77.02	84.53	89.78	72.16	80.41	85.04	74.59	82.47	87.41
	DMD [20]	77.81	85.07	90.23	73.44	81.42	85.71	75.62	83.25	87.97
	DC-Net [4]	76.90	84.21	89.73	71.67	79.58	84.91	74.29	81.89	87.32
	BCP [2]	78.15	85.17	90.38	73.37	81.06	85.82	75.76	83.11	88.09
	Ours (GeoT)	**80.55**	**86.96**	**91.36**	**77.43**	**84.38**	**87.69**	**78.99**	**85.67**	**89.52**
10%	SupOnly	78.11	85.36	90.34	75.58	83.38	86.76	76.85	84.37	88.55
	UA-MT [22]	79.38	85.96	90.87	77.41	84.43	87.81	78.40	85.20	89.34
	FixMatch [16]	80.17	86.77	91.25	78.21	85.31	88.27	79.19	86.04	89.76
	SLC-Net [14]	79.92	86.73	91.04	77.35	84.65	87.80	78.64	85.69	89.42
	DMD [20]	81.10	87.48	91.52	78.85	85.64	88.29	79.98	86.56	89.91
	DC-Net [4]	80.53	86.91	91.38	78.20	85.03	88.26	79.36	85.97	89.82
	BCP [2]	80.04	86.79	91.11	77.63	84.87	87.96	79.83	86.33	89.93
	Ours (GeoT)	**82.37**	**88.18**	**92.04**	**80.98**	**86.97**	**89.51**	**81.68**	**87.57**	**90.77**
20%	SupOnly	79.57	85.22	88.11	77.47	83.35	85.45	78.53	84.28	86.78
	UA-MT [22]	81.22	86.67	89.54	79.10	84.74	86.78	80.16	85.70	88.16
	FixMatch [16]	81.98	87.42	90.25	80.11	85.65	87.65	81.04	86.53	88.95
	SLC-Net [14]	80.91	86.35	89.21	79.60	85.09	86.91	80.25	85.72	88.06
	DMD [20]	83.09	88.52	91.44	81.90	87.37	89.18	82.49	87.95	90.31
	DC-Net [4]	82.94	88.39	91.24	81.20	86.77	88.73	82.07	87.58	89.98
	BCP [2]	82.86	88.30	91.18	81.61	87.08	88.86	82.24	87.69	90.02
	Ours (GeoT)	**84.59**	**90.19**	**93.11**	**83.08**	**88.83**	**90.69**	**83.83**	**89.51**	**91.90**

Table 2. Comparison of our GeoT with different tooth-specific methods on the Teeth3DS Dataset (%).

Ratio	Method		Maxillary			Mandible			All		
			mIoU	DSC	Acc	mIoU	DSC	Acc	mIoU	DSC	Acc
0.5%	MeshSegNet [12]	+ FixMatch [16]	60.10	70.61	79.96	49.30	62.35	70.54	54.70	66.48	75.25
		+ BCP [2]	60.16	71.14	80.01	49.66	62.68	70.79	54.91	66.91	75.40
	TSegNet [7]	+ FixMatch [16]	60.56	71.01	80.32	50.02	63.47	71.06	55.29	67.24	75.69
		+ BCP [2]	60.08	71.21	80.63	48.98	62.01	70.23	54.53	66.61	75.43
	Ours (GeoT)		**61.47**	**72.81**	**81.02**	**52.15**	**64.72**	**72.28**	**56.81**	**68.76**	**76.65**
1%	MeshSegNet [12]	+ FixMatch [16]	65.88	74.58	83.22	61.46	73.48	78.98	63.67	74.03	81.10
		+ BCP [2]	64.91	74.87	83.65	62.75	73.25	78.87	63.83	74.06	81.26
	TSegNet [7]	+ FixMatch [16]	66.01	75.77	83.22	62.33	72.93	79.36	64.17	74.35	81.29
		+ BCP [2]	64.64	74.49	83.50	62.12	73.01	78.40	63.38	73.75	80.95
	Ours (GeoT)		**66.90**	**76.31**	**84.47**	**64.07**	**74.10**	**79.98**	**65.48**	**75.21**	**82.23**
5%	MeshSegNet [12]	+ FixMatch [16]	77.89	85.11	90.15	75.21	82.97	86.33	76.55	84.04	88.24
		+ BCP [2]	79.19	86.05	89.54	74.35	82.19	87.04	76.77	84.12	88.29
	TSegNet [7]	+ FixMatch [16]	78.88	85.66	90.62	76.02	83.44	86.78	77.45	84.55	88.70
		+ BCP [2]	78.76	85.54	90.29	73.76	81.88	86.03	76.26	83.71	88.16
	Ours (GeoT)		**80.55**	**86.96**	**91.36**	**77.43**	**84.38**	**87.69**	**78.99**	**85.67**	**89.52**
10%	MeshSegNet [12]	+ FixMatch [16]	81.03	87.09	91.40	79.17	85.75	88.42	80.10	86.42	89.91
		+ BCP [2]	81.82	88.03	91.70	79.46	86.11	88.62	80.64	87.07	90.16
	TSegNet [7]	+ FixMatch [16]	81.83	87.86	91.81	79.93	86.56	89.09	80.88	87.21	90.45
		+ BCP [2]	82.03	87.75	91.65	77.79	84.93	88.13	79.91	86.34	89.89
	Ours (GeoT)		**82.37**	**88.18**	**92.04**	**80.98**	**86.97**	**89.51**	**81.68**	**87.57**	**90.77**
20%	MeshSegNet [12]	+ FixMatch [16]	83.88	89.11	92.23	82.06	87.45	89.55	82.97	88.28	90.89
		+ BCP [2]	83.61	89.05	92.18	82.63	88.03	90.08	83.12	88.54	91.13
	TSegNet [7]	+ FixMatch [16]	84.31	89.98	92.85	82.41	88.32	90.27	83.36	89.15	91.56
		+ BCP [2]	82.99	88.46	91.49	81.87	87.28	89.35	82.43	87.87	90.42
	Ours (GeoT)		**84.59**	**90.19**	**93.11**	**83.08**	**88.83**	**90.69**	**83.83**	**89.51**	**91.90**

methods using the same backbone. Among these methods, UA-MT [22] and FixMatch [16] failed to achieve satisfactory results, especially in scenarios with extremely limited labeled data (e.g., 0.5%). This is because they filtered pseudo labels, resulting in insufficient utilization of unlabeled data. BCP [2] and DMD [20] enriched the supervision signals from their respective perspectives, but they are still suffering from noisy pseudo labels. Hence, the robustness of both methods are weakened. As we can see, DMD only obtained 60.61 mIoU with 1% labeled data and BCP performed suboptimally on 0.5% split. Compared to all the above methods, our approach consistently achieves significant improvements across all metrics in all settings, demonstrating the effectiveness of the proposed method.

We also compared GeoT with several segmentation methods specifically designed for tooth. Since there are no other semi-supervised tooth-specific methods, we chose fully-supervised tooth segmentation networks [7,12] as backbones

Table 3. Comparison of our GeoT with different semi-supervised methods on our private dataset and Teeth3DS Dataset (%).

Unlabeled	Method	Maxillary			Mandible			All		
		mIoU	DSC	Acc	mIoU	DSC	Acc	mIoU	DSC	Acc
0	Lower Bound	85.27	90.63	93.41	83.34	88.84	90.89	84.31	89.74	92.15
600	UA-MT [22]	86.69	91.28	93.70	83.23	89.04	91.12	84.96	90.16	92.41
	FixMatch [16]	87.36	91.91	93.87	83.40	89.07	91.37	85.38	90.49	92.62
	SLC-Net [14]	87.29	91.99	93.77	83.71	89.25	91.57	85.50	90.62	92.67
	DMD [20]	87.80	92.57	94.20	84.48	89.89	91.92	86.14	91.23	93.06
	DC-Net [4]	87.06	91.90	93.87	84.36	89.74	91.83	85.71	90.82	92.85
	BCP [2]	87.46	92.13	94.06	84.52	90.03	91.96	85.99	91.08	93.01
	Ours (GeoT)	**88.35**	**93.12**	**94.57**	**85.11**	**90.26**	**92.25**	**86.73**	**91.69**	**93.41**
1200	UA-MT [22]	87.07	91.97	94.24	83.63	89.33	91.00	85.35	90.65	92.62
	FixMatch [16]	88.87	93.26	94.23	85.01	90.06	91.83	86.94	91.66	93.03
	SLC-Net [14]	88.41	92.57	94.11	84.85	89.81	91.67	86.63	91.19	92.89
	DMD [20]	89.52	93.54	95.13	85.94	91.04	92.29	87.73	92.29	93.71
	DC-Net [4]	88.56	93.12	94.52	85.70	90.60	92.22	87.13	91.86	93.37
	BCP [2]	88.98	93.21	94.62	86.12	91.13	92.46	87.55	92.17	93.54
	Ours (GeoT)	**89.72**	**93.61**	**95.35**	**86.58**	**91.37**	**92.61**	**88.15**	**92.49**	**93.98**

and combined them with semi-supervised algorithms [2,16]. The results are presented in Table 2. Note that GeoT still used PointTransformer [25] as its backbone, rather than tooth-specific methods. Nevertheless, GeoT achieved SOTA performance, outperforming other methods in all cases, particularly when labeled data is only 0.5% and 1%, leading by at least 2.7% and 2.0% in the mIoU metric, respectively. This demonstrates the strong superiority of our approach.

To further explore the potential of GeoT, we utilized the complete training set of the Teeth3DS dataset as labeled data, while employing the data from our private dataset as unlabeled data for experimentation. The results are presented in Table 3. It is clear that, compared to other semi-supervised methods, GeoT maximizes improvements in results due to its effective utilization of unlabeled data. Specifically, When employing 600 instances of unlabeled data, GeoT achieves an increase of 2.9% in mIoU, while the use of 1,200 instances leads to an enhancement of 4.6%. These findings suggest that GeoT possesses cross-dataset generalizability, enabling it to leverage unlabeled data to significantly improve segmentation outcomes.

Qualitative results with 10% labeled data are displayed in Fig. 2 to demonstrate the superior performance of our model. As observed, most methods tend to confuse in adjacent categories during segmentation, especially in the boundary regions of teeth. In contrast, our approach utilizes NTM to handle pseudo-label noise, resulting in clearer supervision signals and achieving better results.

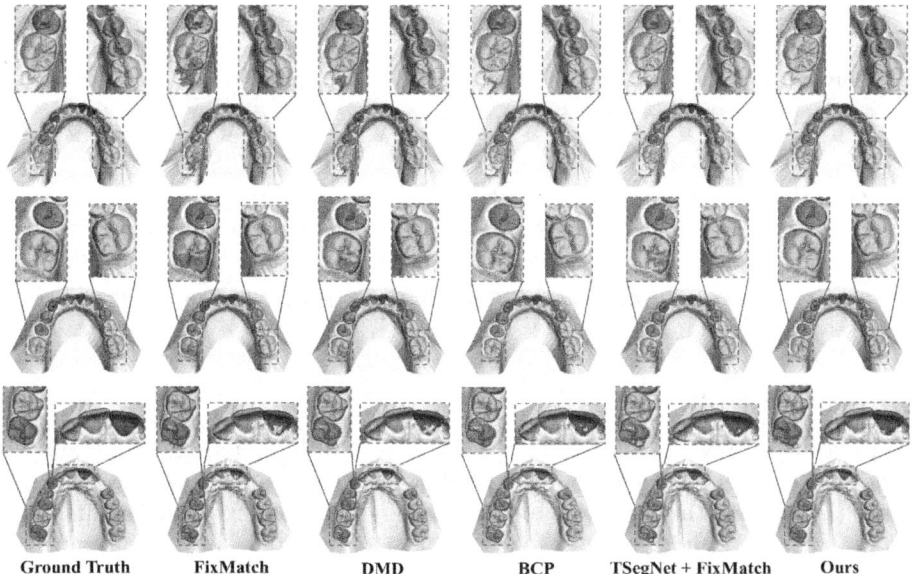

Fig. 2. Visualization of the segmentation results of different methods. The first four rows are the lower jaws, and the last two rows are the upper jaws.

3.3 Ablation Study

We further conducted ablation studies on different ratios (from 0.5% to 5%) to verify the effectiveness of each component of GeoT and Table 4 reports the results. FixMatch is used as the baseline. By modeling pseudo-label noise using IDTM instead of simply discarding low-confidence samples, the model has shown performance improvements across all four settings. PLGR smooths the solution space of IDTM, reducing the optimization difficulty and further enhancing performance. CLGS leverages prior knowledge of the tooth class distribution to reduce the complexity of IDTM, resulting in significant improvements as well. As the amount of labeled data increases, we observe that the influence of PLGR gradually surpasses that of CLGS. This is attributed to the network gradually capturing the distribution law of the classes under the guidance of labels. Combining all the three components, GeoT demonstrates the most powerful ability for semi-supervised tooth point cloud segmentation. The experimental results indicate that these proposed modules do contribute to the satisfactory performance of our method.

Table 4. Ablation study of key components (%).

Components			0.5%			1%			5%		
IDTM	PLGR	CLGS	mIoU	DSC	Acc	mIoU	DSC	Acc	mIoU	DSC	Acc
✗	✗	✗	52.33	64.45	74.40	62.23	72.99	80.66	74.00	81.92	87.13
✓	✗	✗	54.13	66.21	75.03	63.12	73.52	80.80	76.04	83.68	88.17
✓	✓	✗	55.45	67.48	75.80	63.94	74.18	81.32	77.75	84.81	88.84
✓	✗	✓	55.91	67.85	76.32	64.67	74.85	81.69	77.60	84.65	88.81
✓	✓	✓	**56.81**	**68.76**	**76.65**	**65.48**	**75.21**	**82.23**	**78.99**	**85.67**	**89.52**

3.4 Discussion

Hyperparameters. We investigated the impact of hyperparameters under different partitions (from 0.5% to 5%), as shown in Fig. 3 and Fig. 4. Figure 3 illustrates the effect of the weight factor λ on the segmentation results. As λ increases (from 0.1 to 0.9), more geometric priors are incorporated into the optimization process of IDTM, significantly reducing its estimation difficulty. However, excessive priors ($\lambda = 0.99$) can cause the optimization of IDTM to get stuck in local minima, which affects segmentation accuracy. In Fig. 4, the optimal value of β is 0.1. This is because $\mathcal{L}_m$ is designed to maintain the structure of the IDTM space consistent with the 3D space, serving as a spatial regularization term, but it cannot replace the dominant role of the segmentation loss.

Training Cost of IDTM. We further computed the training cost of IDTM. Using maxillary teeth with 17 classes as an example, we sample 16,000 points. Each point's NTM is represented as a 17×17 matrix, resulting in a total of $16,000 \times 17 \times 17 = 4,624,000$ elements, which is equivalent to a feature size of a 1024×1024 image with 4 channels. Therefore, the computational cost is not prohibitive.

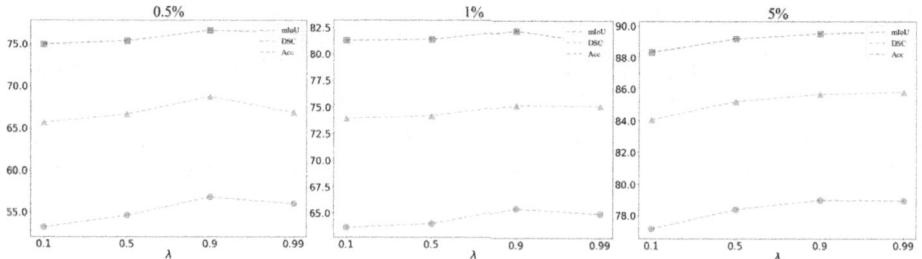

Fig. 3. Experiment resuls with different weighting factors λ for $T^F{}_k$ in different splits.

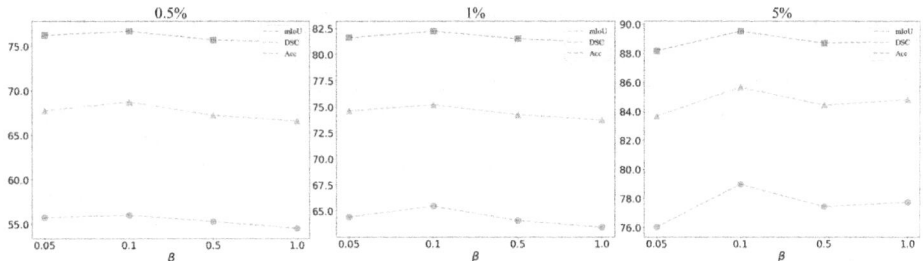

Fig. 4. Experiment resuls with different weights β for $\mathcal{L}_m$ in different splits.

4 Conclusion

This paper presented a novel framework named GeoT for semi-supervised tooth point cloud segmentation by utilizing instance-dependent transition matrix to handle pseudo label noise. To better estimate the noise transition matrix, on one hand, a regularization module incorporating point-level geometric prior is proposed to regularize the solution space of transition matrices. On the other hand, the prior knowledge of tooth class distribution is utilized to design a smoothing module that guides the optimization of the transition matrix. Extensive experiments showed that our proposed method achieved superior performance on the public dataset, leading to an efficient clinical tool for orthodontic diagnosis.

Acknowledgements. This work was supported by CUHK 4055188 and SSFCRS 3136048.

References

1. Almukhtar, A., Ju, X., Khambay, B., McDonald, J., Ayoub, A.: Comparison of the accuracy of voxel based registration and surface based registration for 3D assessment of surgical change following orthognathic surgery. PLoS ONE **9**(4), e93402 (2014)
2. Bai, Y., Chen, D., Li, Q., Shen, W., Wang, Y.: Bidirectional copy-paste for semi-supervised medical image segmentation. In: CVPR, pp. 11514–11524 (2023)
3. Ben-Hamadou, A., et al.: Teeth3ds: a benchmark for teeth segmentation and labeling from intra-oral 3D scans. arXiv preprint arXiv:2210.06094 (2022)
4. Chen, F., Fei, J., Chen, Y., Huang, C.: Decoupled consistency for semi-supervised medical image segmentation. In: MICCAI, pp. 551–561. Springer (2023)
5. Cheng, D., et al.: Instance-dependent label-noise learning with manifold-regularized transition matrix estimation. In: CVPR, pp. 16630–16639 (2022)
6. Cheng, J., Liu, T., Ramamohanarao, K., Tao, D.: Learning with bounded instance and label-dependent label noise. In: ICML, pp. 1789–1799. PMLR (2020)
7. Cui, Z., et al.: Tsegnet: an efficient and accurate tooth segmentation network on 3D dental model. Med. Image Anal. **69**, 101949 (2021)
8. Hajeer, M.Y., Millett, D., Ayoub, A., Siebert, J.: Current products and practices: applications of 3D imaging in orthodontics: Part I. J. Orthod. **31**(1), 62–70 (2004)

9. Hao, J., et al.: Toward clinically applicable 3-dimensional tooth segmentation via deep learning. J. Dent. Res. **101**(3), 304–311 (2022)
10. Im, J., et al.: Accuracy and efficiency of automatic tooth segmentation in digital dental models using deep learning. Sci. Rep. **12**(1), 9429 (2022)
11. Li, X., Liu, T., Han, B., Niu, G., Sugiyama, M.: Provably end-to-end label-noise learning without anchor points. In: ICML, pp. 6403–6413. PMLR (2021)
12. Lian, C., et al.: Deep multi-scale mesh feature learning for automated labeling of raw dental surfaces from 3D intraoral scanners. IEEE Trans. Med. Imaging **39**(7), 2440–2450 (2020)
13. Lin, T.Y., Goyal, P., Girshick, R., He, K., Dollár, P.: Focal loss for dense object detection. In: ICCV, pp. 2980–2988 (2017)
14. Liu, J., Desrosiers, C., Zhou, Y.: Semi-supervised medical image segmentation using cross-model pseudo-supervision with shape awareness and local context constraints. In: MICCAI, pp. 140–150. Springer (2022)
15. Liu, Z., et al.: Hierarchical self-supervised learning for 3D tooth segmentation in intra-oral mesh scans. IEEE Trans. Med. Imaging **42**(2), 467–480 (2022)
16. Sohn, K., et al.: Fixmatch: Simplifying semi-supervised learning with consistency and confidence. NIPS **33**, 596–608 (2020)
17. Standardization, I.O.: Dentistry—designation system for teeth and areas of the oral cavity (1984)
18. Xia, X., et al.: Part-dependent label noise: towards instance-dependent label noise. NIPS **33**, 7597–7610 (2020)
19. Xia, X., et al.: Are anchor points really indispensable in label-noise learning? NIPS **32** (2019)
20. Xie, Y., Yin, Y., Li, Q., Wang, Y.: Deep mutual distillation for semi-supervised medical image segmentation. In: MICCAI, pp. 540–550. Springer (2023)
21. Xu, X., Liu, C., Zheng, Y.: 3D tooth segmentation and labeling using deep convolutional neural networks. IEEE Trans. Vis. **25**(7), 2336–2348 (2018)
22. Yu, L., Wang, S., Li, X., Fu, C.W., Heng, P.A.: Uncertainty-aware self-ensembling model for semi-supervised 3D left atrium segmentation. In: MICCAI, pp. 605–613. Springer (2019)
23. Zanjani, F.G., et al.: Mask-mcnet: instance segmentation in 3D point cloud of intra-oral scans. In: MICCAI, pp. 128–136. Springer (2019)
24. Zhang, B., et al.: Flexmatch: boosting semi-supervised learning with curriculum pseudo labeling. NIPS **34**, 18408–18419 (2021)
25. Zhao, H., Jiang, L., Jia, J., Torr, P.H., Koltun, V.: Point transformer. In: CVPR, pp. 16259–16268 (2021)

RemInD: Remembering Anatomical Variations for Interpretable Domain Adaptive Medical Image Segmentation

Xin Wang[1,2], Yin Guo[2,3], Kaiyu Zhang[2,3], Niranjan Balu[2], Mahmud Mossa-Basha[2], Linda Shapiro[1,4], and Chun Yuan[2,5](✉)

[1] Department of Electrical and Computer Engineering, University of Washington, Seattle, USA
[2] Vascular Imaging Lab, Department of Radiology, University of Washington, Seattle, USA
cyuan@uw.edu
[3] Department of Bioengineering, University of Washington, Seattle, USA
[4] Paul G. Allen School of Computer Science and Engineering, University of Washington, Seattle, USA
[5] Department of Radiology and Imaging Sciences, University of Utah, Salt Lake City, USA

Abstract. This work presents a novel Bayesian framework for unsupervised domain adaptation (UDA) in medical image segmentation. While prior works have explored this clinically significant task using various strategies of domain alignment, they often lack an explicit and explainable mechanism to ensure that target image features capture meaningful structural information. Besides, these methods are prone to the curse of dimensionality, inevitably leading to challenges in interpretability and computational efficiency. To address these limitations, we propose RemInD, a framework inspired by *human adaptation*. RemInD learns a domain-agnostic latent manifold, characterized by several anchors, to memorize anatomical variations. By mapping images onto this manifold as weighted anchor averages, our approach ensures realistic and reliable predictions. This design mirrors how humans develop representative components to understand images and then retrieve component combinations from memory to guide segmentation. Notably, model prediction is determined by two explainable factors: a low-dimensional anchor weight vector, and a spatial deformation. This design facilitates computationally efficient and geometry-adherent adaptation by aligning weight vectors between domains on a probability simplex. Experiments on two public datasets, encompassing cardiac and abdominal imaging, demonstrate the superiority of RemInD, which achieves state-of-the-art performance using a single alignment approach, outperforming existing methods that often rely on multiple complex alignment strategies.

Keywords: Domain Adaptation · Medical Image Segmentation · Interpretability · Variational Inference

1 Introduction

Creating dense annotations for deep medical image segmentation models is labor-intensive. Unsupervised domain adaptation (UDA) addresses this challenge by utilizing a labeled source dataset to improve performance on an unlabeled target domain with differing imaging patterns [8]. Its rationale lies in the shared task-relevant anatomical information between the source and target datasets. Numerous studies have attempted to exploit this invariance through domain alignment. For example, adversarial or semi-supervised approaches [23,24,28,29] align domains implicitly through discriminators or pseudo-labels, which, however, prioritize domain consistency over structural correctness, and thus risk the loss of critical anatomical details. Meanwhile, previous variational or optimal transport works [4,5,27], while effective at aligning global feature distributions, are computationally expensive and may overlook the quality of individual features. To summarize, prior methods face two intrinsic challenges:

1. Alignment in high-dimensional feature spaces requires substantial computational costs and often relies on approximations, leading to imprecise results.
2. The absence of an explicit and explainable mechanism to control image features hinders their ability to capture meaningful anatomical information.

In contrast to these issues, humans learn from labeled examples by forming concepts of *components* that encapsulate physiologically valid shapes [2]. When encountering a new modality, they recall suitable component combinations, and adapt them with moderate warping to account for natural individual-level spatial distortions. This component-driven, memory-based process allows humans to generalize efficiently across domains while maintaining structural consistency. However, a crucial gap remains between prior methods and human cognition.

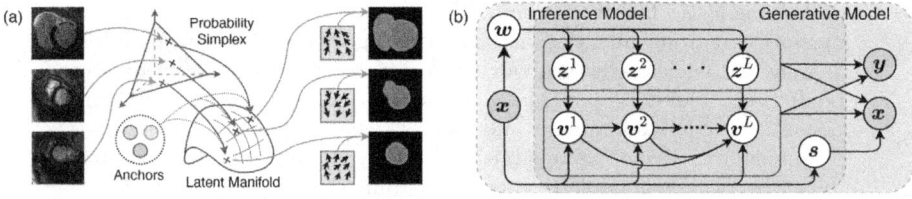

Fig. 1. The proposed framework RemInD. (a) Based on inferred anchor weights (yellow arrows), images are mapped onto a latent manifold (red arrows), and segmentations are warped by spatial transformations to produce final predictions (green arrows). (b) Inference (pink) and generative (green) models (Sect. 2.1), with observations shaded. (Color figure online)

To this end, we make a radical departure from previous works by introducing RemInD, a novel Bayesian framework that *mimics human adaptation*. As shown in Fig. 1(a), RemInD learns a domain-agnostic latent manifold (akin to human memory), characterized by a small set of anchors (akin to components). Each input image corresponds to a vector of anchor weights, which serves as a shape blending mechanism that enables a controlled yet flexible construction of diverse anatomical structures. The low-dimensional weight vectors are further aligned across domains, offering a significantly more efficient and explainable alternative to traditional latent feature alignment used in previous works. Additionally, the segmentation predictions based on anchor weights are composed with spatial transformations to account for natural shape deformations, ensuring adaptability to diverse images. The contributions of our work are as follows:

1. We propose RemInD, a novel Bayesian UDA framework for medical image segmentation, which emulates human cognition by mapping images onto a structured manifold designed to capture the full range of anatomical variations through representative components (anchors).
2. RemInD enables geometrically faithful adaptation within a low-dimensional probability simplex, substantially reducing computational costs while enhancing alignment efficiency and interpretability.
3. We demonstrate the superiority of RemInD over existing state-of-the-art methods, which often rely on multiple complex alignment strategies.

2 Methodology

Suppose the source dataset $(x_i^s, y_i^s)_{i=1}^{N_s}$ and the target dataset $(x_i^t)_{i=1}^{N_t}$ consist of independent samples from the density $p_s(X, Y)$ and the marginal $p_t(X)$ of $p_t(X, Y)$, respectively, where $X : \mathbb{R}^D \supset \Omega \to \mathbb{R}$ represents the image and $Y : \Omega \to \{1, \cdots, K\}$ represents the label, with D the number of pixels, and K the number of classes. We assume that the information in X can be disentangled into a content representation for segmentation and a style representation S for image appearance. Inspired by atlas-based segmentation [10], we further disentangle the content into two independent variables: an atlas representation Z and a spatial transformation $\phi : \Omega \to \Omega$, such that $X \circ \phi$ is registered to Z. Considering that a single atlas is not topologically diverse, we assume the atlas is conditioned on a vector W, i.e., $Z \sim p(Z|W)$. This model allows for a flexible representation of anatomy in observations through controllable factors, i.e., a low-dimensional vector W and a diffeomorphism ϕ.

2.1 Bayesian Inference of Latent Variables

Let (x, y) be an observation of (X, Y). We propose a novel variational Bayesian framework, RemInD, to infer the corresponding latent representations: w, z, s, and the stationary velocity field v parameterizing ϕ [1]. Specifically, we make two independence assumptions: 1) x captures all information about the latent variables, and 2) the style code s is conditionally independent of structure-related latent variables given x. Hence, the joint distribution can be factorized as $p(x, y, w, z, v, s) = p(w)p(s)p(v)p(z|w)p(x|z, v, s)p(y|z, v)$, and similarly, the variational posterior becomes $q(\boldsymbol{w}, \boldsymbol{z}, \boldsymbol{v}, \boldsymbol{s}|\boldsymbol{x}, \boldsymbol{y}) = q(\boldsymbol{s}|\boldsymbol{x})q(\boldsymbol{w}|\boldsymbol{x})q(\boldsymbol{z}|\boldsymbol{w})q(\boldsymbol{v}|\boldsymbol{x},\boldsymbol{z})$. Following the variational Bayes framework [12], the evidence lower bound (ELBO) of the log-likelihood is expressed as

$$\text{ELBO}(\boldsymbol{x}, \boldsymbol{y}) := \mathbb{E}_{q(\boldsymbol{w},\boldsymbol{z},\boldsymbol{v},\boldsymbol{s}|\boldsymbol{x},\boldsymbol{y})} \log \frac{p(\boldsymbol{x},\boldsymbol{y},\boldsymbol{w},\boldsymbol{z},\boldsymbol{v},\boldsymbol{s})}{q(\boldsymbol{w},\boldsymbol{z},\boldsymbol{v},\boldsymbol{s}|\boldsymbol{x},\boldsymbol{y})}$$

$$= \underbrace{\mathbb{E}_{q(\boldsymbol{s}|\boldsymbol{x})q(\boldsymbol{w}|\boldsymbol{x})q(\boldsymbol{z}|\boldsymbol{w})q(\boldsymbol{v}|\boldsymbol{x},\boldsymbol{z})} \log p(\boldsymbol{x}|\boldsymbol{z},\boldsymbol{v},\boldsymbol{s})}_{\mathcal{L}_{\text{recon}}(\boldsymbol{x})} + \underbrace{\mathbb{E}_{q(\boldsymbol{w}|\boldsymbol{x})q(\boldsymbol{z}|\boldsymbol{w})q(\boldsymbol{v}|\boldsymbol{x},\boldsymbol{z})} \log p(\boldsymbol{y}|\boldsymbol{z},\boldsymbol{v})}_{\mathcal{L}_{\text{seg}}(\boldsymbol{x},\boldsymbol{y})}$$

$$- D_{\text{KL}}\left[q(\boldsymbol{s}|\boldsymbol{x}) \parallel p(\boldsymbol{s})\right] - D_{\text{KL}}\left[q(\boldsymbol{w}|\boldsymbol{x}) \parallel p(\boldsymbol{w})\right]$$

$$- \underbrace{\mathbb{E}_{q(\boldsymbol{w}|\boldsymbol{x})} D_{\text{KL}}\left[q(\boldsymbol{z}|\boldsymbol{w}) \parallel p(\boldsymbol{z}|\boldsymbol{w})\right]}_{\mathcal{L}_{\text{atlas}}(\boldsymbol{x})} - \underbrace{\mathbb{E}_{q(\boldsymbol{w}|\boldsymbol{x})q(\boldsymbol{z}|\boldsymbol{w})} D_{\text{KL}}\left[q(\boldsymbol{v}|\boldsymbol{x},\boldsymbol{z}) \parallel p(\boldsymbol{v})\right]}_{\mathcal{L}_{\text{vel}}(\boldsymbol{x})}, \quad (1)$$

where D_{KL} denotes the Kullback-Leibler (KL) divergence. Intuitively, $\mathcal{L}_{\text{recon}}$ and $\mathcal{L}_{\text{seg}}$ correspond to image reconstruction and segmentation, while the other KL terms serve as regularization for the latent variables. For unlabeled images, a similar derivation applies, with the only difference being the omission of $\mathcal{L}_{\text{seg}}$.

To enhance the expressiveness of $\boldsymbol{z}$ and $\boldsymbol{v}$, we decompose them hierarchically [22] as $\boldsymbol{z} = (\boldsymbol{z}^l)_{l=1}^L$ and $\boldsymbol{v} = (\boldsymbol{v}^l)_{l=1}^L$, similar to [25,26], where a larger l correspond to a finer spatial resolution. This construction expresses complex information in $\boldsymbol{z}$ and $\boldsymbol{\phi}$ by simpler components to facilitate learning. The final spatial transformation can then be calculated as $\boldsymbol{\phi} = \boldsymbol{\phi}^1 \circ \cdots \circ \boldsymbol{\phi}^L$, where $\frac{\partial}{\partial t}\boldsymbol{\phi}^l(a,t) = \boldsymbol{v}^l(\boldsymbol{\phi}^l(a,t)), \forall a \in \Omega, t \in [0,1]$. We further assume 1) different levels of $\boldsymbol{z}^l$ are independent given $\boldsymbol{w}$, 2) $\boldsymbol{v}^l$ can be inferred directly from $\boldsymbol{x}$ and $\boldsymbol{z}^l$ at the same level, and 3) the prior $p(\boldsymbol{v}^l|\boldsymbol{v}^{<l}) = p(\boldsymbol{v}^l)$, where $<l$ denotes levels below l. Therefore, the KL terms $\mathcal{L}_{\text{atlas}}$ and $\mathcal{L}_{\text{vel}}$ can be simplified as

$$\mathcal{L}_{\text{atlas}}(\boldsymbol{x}) = \mathbb{E}_{q(\boldsymbol{w}|\boldsymbol{x})} \left\{ \sum_{l=1}^L D_{\text{KL}}\left[q(\boldsymbol{z}^l|\boldsymbol{w}) \parallel p(\boldsymbol{z}^l|\boldsymbol{w})\right] \right\},$$

$$\mathcal{L}_{\text{vel}}(\boldsymbol{x}) = \mathbb{E}_{q(\boldsymbol{w}|\boldsymbol{x})q(\boldsymbol{z}|\boldsymbol{w})} \left[\sum_{l=1}^L \mathbb{E}_{q(\boldsymbol{v}^{<l}|\boldsymbol{x},\boldsymbol{z}^{<l})} \left\{ D_{\text{KL}}\left[q(\boldsymbol{v}^l|\boldsymbol{x},\boldsymbol{z}^l,\boldsymbol{v}^{<l}) \parallel p(\boldsymbol{v}^l)\right] \right\} \right],$$
(2)

with $q(\boldsymbol{v}^{<1}|\boldsymbol{x},\boldsymbol{z}^{<1}) := 1$ for simplicity. Thus, the graphical model corresponding to the ELBO is illustrated in Fig. 1(b). This formulation enables level-wise inference and regularization of $\boldsymbol{z}$ and $\boldsymbol{v}$. Specifically, $\mathcal{L}_{\text{vel}}$ is calculated through the technique introduced in [6] to guarantee a diffeomorphic $\boldsymbol{\phi}$, while $\mathcal{L}_{\text{atlas}}$ and $q(\boldsymbol{z}^l|\boldsymbol{w})$ facilitate retrieving anatomical information akin to human visual recognition, as detailed in the next section.

2.2 Interpretable Anchor-Based Manifold Embedding

In RemInD, the atlas representation $\boldsymbol{z}$ is inferred based on $\boldsymbol{w}$ through the posteriors $q(\boldsymbol{z}^l|\boldsymbol{w})$ for the given image $\boldsymbol{x}$. To make this procedure explainable, we propose learning *anchor* distributions $\{q_m(\boldsymbol{z}^l)\}_{m=1}^M$ for each level l, with M

the length of w. To note, these anchors are not conditioned on x. We further assume $q(z^l|w)$ to be the w-weighted geometric mean [14] of the anchors, i.e., $q(z^l|w) \propto \prod_{m=1}^{M} q_m^{w_m}(z^l)$, where $w \in \Delta := \{w \in \mathbb{R}^M \mid w \succeq 0, \mathbf{1}^\top w = 1\}$, the $(M-1)$-dimensional probability simplex. Therefore, the effects of w can be interpreted as follows: 1) w serves as a shape blending weight, blending distinct shapes represented by the anchors to form an atlas distribution that encodes a new shape to fit x; 2) This process essentially constructs a probabilistic manifold characterized by the anchors, which mimics humans retrieving suitable combinations of learned components [2] to form anatomical shapes for segmentation. Moreover, through anchors not conditioned on x, we impose a strong inductive bias, capturing global anatomical representations and enhancing domain adaptation by providing domain-agnostic information across various images. In sharp contrast, prior works extract representations directly from x, which could be too flexible to ensure reasonable structures for images in the target domain.

We model $q_m(z^l)$ as diagonal Gaussians $\mathcal{N}(\boldsymbol{\mu}_m^l, \boldsymbol{\Sigma}_m^l)$. Therefore, $q(z^l|w) = \mathcal{N}(\boldsymbol{\mu}^l, \boldsymbol{\Sigma}^l)$, with $\boldsymbol{\mu}^l = \boldsymbol{\Sigma}^l \sum_m w_m \left(\boldsymbol{\Sigma}_m^l\right)^{-1} \boldsymbol{\mu}_m^l$, $\left(\boldsymbol{\Sigma}^l\right)^{-1} = \sum_m w_m \left(\boldsymbol{\Sigma}_m^l\right)^{-1}$. Besides, the prior distributions $p(z^l|w)$ are set as standard Gaussians $\mathcal{N}(\mathbf{0}, \boldsymbol{I})$. Consequently, $\mathcal{L}_{\text{atlas}}$ involves calculating KLs between $q(z^l|w)$ and $\mathcal{N}(\mathbf{0}, \boldsymbol{I})$ for each image in a training mini-batch. We further propose replacing $\mathcal{L}_{\text{atlas}}$ with $\mathcal{L}_{\text{anchor}}$, defined as the average of KLs for each anchor, i.e.

$$\mathcal{L}_{\text{anchor}} = \sum_{l=1}^{L} \frac{1}{M} \sum_{m=1}^{M} D_{\text{KL}}\left[q_m(z^l) \,\|\, \mathcal{N}(\mathbf{0}, \boldsymbol{I})\right]. \tag{3}$$

Notably, calculating $\mathcal{L}_{\text{anchor}}$ does not depend on x. The rationale for the replacement is twofold: 1) It can be proven that "$\mathcal{L}_{\text{anchor}} = 0$" $\Leftrightarrow$ "$\forall m, l, q_m(z^l) = \mathcal{N}(\mathbf{0}, \boldsymbol{I})$" $\Rightarrow$ "$\mathcal{L}_{\text{atlas}} = 0$", and thus, minimizing $\mathcal{L}_{\text{anchor}}$ effectively minimizes $\mathcal{L}_{\text{atlas}}$, and 2) $\mathcal{L}_{\text{anchor}}$ is much more computationally efficient, requiring only LM KL calculations, compared to LB calculations for $\mathcal{L}_{\text{atlas}}$, where B is the batch size and typically $B \gg M$. Intuitively, $\mathcal{L}_{\text{anchor}}$ encourages the anchors $q_m(z^l)$ to stay close to a fixed location (standard Gaussian), preventing small variations in w from causing excessive divergence in $q(z^l|w)$.

2.3 Efficient, Geometrically Faithful Domain Alignment

A common approach in UDA involves aligning feature distributions between domains. Previous studies universally operate in high-dimensional latent spaces for this purpose, which is implicit, computationally expensive, and often requires approximations that compromise accuracy. In contrast, RemInD offers a computationally efficient and interpretable alternative: Since anchors are shared across all images, aligning atlas distributions $q(z|w)$ between domains reduces to aligning the shape blending weights w directly. To this end, we propose applying an optimal transport loss to the simplex Δ. Specifically, we assume a deterministic posterior for the weights w, similar to VQ-VAE [19], i.e., $q(w|x) = \delta(w - \widetilde{w}(x))$, with δ the Dirac delta, and $\widetilde{w} \in \Delta$ inferred from x.

Thus, $D_{\mathrm{KL}}[q(\bm{w}|\bm{x}) \parallel p(\bm{w})]$ in Eq. 1 vanishes as a constant by setting $p(\bm{w})$ to the standard Dirichlet $\mathrm{Dir}(\bm{1})$. More importantly, for source and target minibatches, represented by $(\widetilde{\bm{w}}_i^s)_{i=1}^{B_s}$ and $(\widetilde{\bm{w}}_i^t)_{i=1}^{B_t}$, we define the empirical distribution functions $F_s(\widetilde{\bm{w}}) := \frac{1}{B_s}\sum_{i=1}^{B_s} \bm{1}\{\widetilde{\bm{w}}_i^s \preceq \widetilde{\bm{w}}\}$, and similarly F_t, where $\bm{1}\{\cdot\}$ is the indicator function. The alignment loss is then defined as their Sinkhorn divergence [7], i.e.

$$\mathcal{L}_{\mathrm{align}} := \mathrm{OT}_\varepsilon(F_s, F_t) - \frac{1}{2}\mathrm{OT}_\varepsilon(F_s, F_s) - \frac{1}{2}\mathrm{OT}_\varepsilon(F_t, F_t),$$
$$\text{with } \mathrm{OT}_\varepsilon(F, F') := \min_{\zeta \in \prod(F,F')} \int_{\Delta \times \Delta} C \mathrm{d}\zeta + \varepsilon \int_{\Delta \times \Delta} \log\left(\frac{\mathrm{d}\zeta}{\mathrm{d}F\mathrm{d}F'}\right)\mathrm{d}\zeta. \quad (4)$$

Here, ε controls the strength of entropy regularization, ζ is a transport plan, $\prod(F, F')$ denotes all probabilistic couplings over $\Delta \times \Delta$ with marginals F and F', and C is a symmetric non-negative cost function. We impose the Fisher-Rao metric D_{FR} on Δ to reflect its non-Euclidean statistical geometry, i.e.

$$C(\widetilde{\bm{w}}, \widetilde{\bm{w}}') := D_{\mathrm{FR}}(\widetilde{\bm{w}}, \widetilde{\bm{w}}') := 2 \arccos\left(\sum_{i=1}^{M} \sqrt{\widetilde{w}_i \widetilde{w}_i'}\right), \quad \forall \widetilde{\bm{w}}, \widetilde{\bm{w}}' \in \Delta. \quad (5)$$

Therefore, C measures the geodesic distances among $\bm{w}$ on a Riemannian manifold [17], ensuring an intrinsic and geometrically faithful optimal transport. Since Δ is low-dimensional, $\mathcal{L}_{\mathrm{align}}$ can be calculated efficiently with minimal computational overhead. Considering $D_{\mathrm{FR}} \in [0, \pi]$, we set $\varepsilon = \pi/10$.

2.4 Geometry Regularization and Final Loss

We propose regularizing $\widetilde{\bm{w}}$ of labeled source images through an additional loss

$$\mathcal{L}_{\mathrm{geo}} := \sum_{i=1}^{B_s} \sum_{j=i+1}^{B_s} \left[\left(1 - \frac{D_{\mathrm{FR}}(\widetilde{\bm{w}}_i^s, \widetilde{\bm{w}}_j^s)}{\pi}\right) - \mathrm{Sim}\left(\bm{y}_i^s \circ \bm{\phi}_i^s, \bm{y}_j^s \circ \bm{\phi}_j^s\right)\right]^2, \quad (6)$$

where Sim is the Dice similarity coefficient, and the denominator π normalizes D_{FR} to $[0, 1]$. This regularization loss offers several benefits: 1) It establishes a principled association between the distances among $\widetilde{\bm{w}}$ and the structural differences among images, explicitly refining geometry of the latent manifold to better capture anatomical variations, 2) It inherently translates the cross-domain alignment of $\widetilde{\bm{w}}$ to the alignment of segmentation semantics, improving the interpretability of $\mathcal{L}_{\mathrm{align}}$, 3) It accelerates training convergence by preventing $\widetilde{\bm{w}}$ of disparate images from collapsing into a single value, a pervasive issue in other works, e.g., mode a similar regularization due to the unaffordable computational cost of operating on high-dimensional representations.

The final loss is the negative of ELBO, plus $\mathcal{L}_{\mathrm{align}}$ and $\mathcal{L}_{\mathrm{geo}}$. For the remaining terms in the ELBO: We model $p(\bm{x}|\bm{z}, \bm{v}, \bm{s})$ as a Laplacian distribution, factorized for each pixel. Consequently, $-\mathcal{L}_{\mathrm{recon}}$ becomes the scaled L1 loss between the input image and its reconstruction, with pixelwise scales predicted alongside

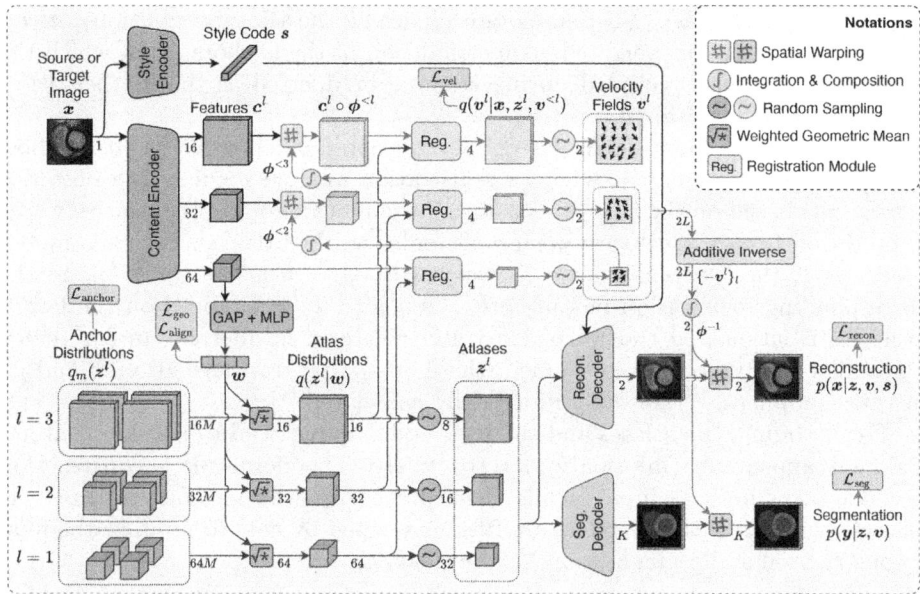

Fig. 2. Network architecture of RemInD, with $L = 3$ and $M = 4$ as an example, including image feature encoding (orange), atlas inference (red), spatial transformation inference (blue), segmentation & reconstruction decoding (green), and loss calculations (purple). The feature maps parameterizing Gaussians have half the channels for the mean and half for the log variance. Values around arrows indicate channel numbers. (Color figure online)

the reconstruction by a decoder. For the segmentation term $-\mathcal{L}_{\text{seg}}$ on the source domain, we utilize a combination of the cross-entropy loss and Dice loss, following prior works [5]. Besides, s is inferred deterministically, similar to w. As a result, the KL for $q(s|x)$ in Eq. 1 vanishes. Therefore, given source and target batches, the final loss, with $\boldsymbol{\lambda}$ controlling term weights, is

$$\mathcal{L} = -\mathcal{L}_{\text{seg}}^{s} - \lambda_1 \mathcal{L}_{\text{recon}}^{s,t} + \lambda_2 \mathcal{L}_{\text{vel}}^{s,t} + \lambda_3 \mathcal{L}_{\text{anchor}} + \lambda_4 \mathcal{L}_{\text{align}} + \lambda_5 \mathcal{L}_{\text{geo}}, \quad (7)$$

where s and/or t indicate if the term is based on source and/or target samples.

2.5 Network Architecture

We design a dedicated variational autoencoder (VAE) to facilitate inference and loss calculation within RemInD, as shown in Fig. 2. Given a source or target image x, the content encoder extracts multilevel features, denoted as $\{c^l\}_{l=1}^{L}$, and the style encoder extracts s as the style code. For atlas inference, a multilayer perceptron (MLP) followed by a Softmax predicts the shape blending weight $w = \tilde{w}$, based on the global average pooling (GAP) of the bottom-level feature c^1. The anchor distributions $\{q_m(z^l)\}_{m=1}^{M}$ are modeled using learnable parameters $\boldsymbol{\mu}_m^l, \log \boldsymbol{\Sigma}_m^l$. Therefore, for each level l, the atlas distribution $q(\boldsymbol{z}^l|\boldsymbol{w})$ is calculated as the $\boldsymbol{w}$-weighted geometric mean of the anchors, from which the atlas $\boldsymbol{z}^l$ is randomly sampled during training or inferred as the mathematical expectation during evaluation.

To infer the velocity fields $\{\boldsymbol{v}^l\}_{l=1}^L$, similar to [15,26], we start at the bottom (coarsest) level $l = 1$, where a registration module predicts the posterior $q(\boldsymbol{v}^1|\boldsymbol{x}, \boldsymbol{z}^1)$ based on the level-1 atlas distribution $q(\boldsymbol{z}^1|\boldsymbol{w})$ and the feature $\boldsymbol{c}^1$. $\boldsymbol{v}^1$ is produced from the velocity posterior similar to the atlas $\boldsymbol{z}^l$. For each higher level $l > 1$, the level-l feature $\boldsymbol{c}^l$ is warped by $\boldsymbol{\phi}^{<l} := \boldsymbol{\phi}^1 \circ \cdots \circ \boldsymbol{\phi}^{l-1}$, and the corresponding registration module predicts $q(\boldsymbol{v}^l|\boldsymbol{x}, \boldsymbol{z}^l, \boldsymbol{v}^{<l})$ based on the level-l atlas distribution and the warped feature, where $\boldsymbol{v}^l$ is inferred similar to $\boldsymbol{v}^1$. Once all velocity fields are obtained, the final spatial transformation $\boldsymbol{\phi}$ and its inverse mapping $\boldsymbol{\phi}^{-1}$ are deterministically calculated [1].

For decoding, the atlases and the style code are required to provide anatomical and appearance information, respectively. Therefore, the reconstruction decoder takes both as input, while the segmentation decoder only utilizes the atlases. Outputs of the decoders are further warped by $\boldsymbol{\phi}^{-1}$ to produce the final reconstruction and segmentation for the image $\boldsymbol{x}$.

3 Experiments and Results

3.1 Datasets

MS-CMR. The MS-CMRSeg challenge [30] provides cardiac MRI images in three sequences: bSSFP, LGE and T2, with labels for the left/right ventricles (LV/RV) and myocardium (Myo). Following [27], we used 35 bSSFP images as the source dataset and 45 LGE images as the target dataset, with 5/40 LGE images allocated for validation/test. All 2D slices were shuffled to be unpaired, resampled to a 0.76-mm spacing, and cropped to 192 × 192.

AMOS. The AMOS challenge [11] provides a multi-center, multi-disease dataset of unpaired abdominal 3D CT and MRI scans. Here, we focused on the segmentation of liver, spleen and right/left kidneys. We randomly selected 25 MRI scans as the source dataset and 35 CT scans as the target dataset, with CT scans randomly split into 25/5/5 for training/validation/test. Axial slices were used for the experiments after being resampled to a 1.5-mm spacing and cropped to ensure a consistent field of view centered on the organs of interest, following [3].

3.2 Experimental Setups

Implementation Details. All images were min-max normalized. For the model architecture, we set $L = 5$ and $M = 6$. Moreover, the content encoder is an attention U-Net [18], and the features $\{\boldsymbol{c}^l\}$ are the outputs of the attention layers. The style encoder is a Conv-LeakyReLU-Pool-Linear sequence, producing a 64-dimensional style code. The reconstruction and segmentation decoders share the same structure as the decoding part of a U-Net [21], while adaptive instance normalizations [9] are used in the reconstruction decoder to modulate feature maps

Table 1. Quantitative results (mean/standard deviation) of the compared methods on the MS-CMR dataset, with the best performance highlighted in bold. #Align denotes the number of loss terms for domain alignment.

Method	#Align	DSC (%) ↑				ASSD (mm) ↓			
		Mean	Myo	LV	RV	Mean	Myo	LV	RV
NoAdapt	0	46.2/15	32.8/19	57.2/18	48.6/15	18/11	15/14	16/15	23/15
ADVENT	3	69.7/17	58.1/17	77.8/17	73.3/20	4.2/12	6.8/29	4.1/6.6	1.8/1.0
CyCMIS	11	79.1/7.9	71.4/7.3	87.2/7.9	78.7/8.5	1.7/1.5	1.5/1.4	**1.3**/0.9	2.3/2.2
VarDA	1	79.8/9.3	73.0/8.3	88.1/4.8	78.5/14.9	2.6/1.3	1.7/0.6	2.6/1.2	3.5/2.2
DARUNet	7	82.0/6.9	75.0/9.6	88.4/5.5	82.7/9.3	1.7/0.9	1.3/0.7	2.2/1.7	1.6/1.1
MAPSeg	3	66.0/11	56.9/9.7	75.5/13	65.6/14	5.2/10	7.0/27	4.0/7.0	4.6/5.0
VAMCEI	3	82.5/5.5	75.8/6.7	88.2/5.3	83.5/8.6	1.4/0.6	1.1/0.3	1.7/1.1	1.5/1.0
RemInD	1	**83.1**/5.3	**77.1**/5.8	**88.6**/4.5	**83.5**/7.5	**1.3**/0.6	**0.9**/0.3	1.6/1.1	**1.4**/0.8

with the style code. Each registration module contains four Conv-BatchNorm-LeakyReLU sequences followed by a final Conv to adjust the channel number. Experiments were conducted with PyTorch [20] on an NVIDIA RTX 4090 GPU.

Compared Methods and Evaluation Metrics. We compared RemInD with state-of-the-art works that utilize various adaptation methods, including VAMCEI [5], MAPSeg [29], DARUNet [28], VarDA [27], CyCMIS [24], and ADVENT [23]. Results from an attention U-Net trained purely on the source domain are also presented as NoAdapt. For evaluation, we reported the mean and per-class Dice Similarity Coefficients (DSCs) and Average Symmetric Surface Distances (ASSDs). For CyCMIS without publicly available code, we reported the results from its publication for the overlapping dataset.

3.3 Results

Quantitative Comparison. The evaluation metrics of the compared methods on the two datasets are presented in Table 1 and Table 2. Unlike the baselines, which typically employ multiple complex alignment strategies, RemInD relies solely on a single alignment term ($\mathcal{L}_{\text{align}}$). Despite this simplicity, RemInD consistently outperforms the baselines in both average DSC and ASSD. Notably, it achieves significant improvements for smaller, more challenging structures. For example, it improves the ASSD of myocardium by 18%, the DSC of right kidney by 7%, and the ASSDs of left and right kidneys by 39% and 42%, respectively, compared to the best-performing baselines. This superior performance can be attributed to RemInD's ability to memorize anatomical structures from labeled images and adapt through shape blending weights, which allows for preserving fine-grained details. In contrast, previous methods directly align image features with spatial dimensions, which may dilute focus on less prominent structures.

Table 2. Quantitative results (mean/standard deviation) of the compared methods on the AMOS dataset, with the best performance highlighted in bold.

Method	DSC (%) ↑					ASSD (mm) ↓				
	Mean	Liver	Left Kidney	Right Kidney	Spleen	Mean	Liver	Left Kidney	Right Kidney	Spleen
NoAdapt	9.4/9.7	29/28	0.2/0.4	4.8/6.8	3.2/6.6	61/19	31/9.1	73/22	61/31	80/27
ADVENT	65.0/6.7	74.9/35	54.2/15	60.0/13	71.0/7.4	5.9/2.4	5.3/6.7	8.0/2.7	5.1/2.2	5.3/3.5
VarDA	81.4/5.3	85.2/7.5	79.8/9.9	78.4/7.6	82.0/8.4	4.0/1.8	5.3/4.6	4.5/0.9	2.7/0.6	**3.4**/3.1
DARUNet	85.8/5.8	**91.6**/4.3	82.4/14	82.1/11	87.1/6.2	4.0/2.4	3.1/2.1	3.0/1.9	5.4/3.6	4.5/4.7
MAPSeg	85.9/4.5	85.1/19	85.3/3.5	82.0/3.9	**91.0**/3.5	8.6/1.8	4.8/7.0	13/9.6	10/6.1	6.1/2.8
VAMCEI	84.8/5.5	90.3/5.0	82.2/13	80.5/9.7	86.2/7.2	3.0/2.0	**2.8**/2.3	2.3/1.4	2.4/1.2	4.3/4.4
RemInD	**87.0**/2.0	85.8/7.2	**87.8**/2.7	**88.1**/2.6	86.3/5.3	**2.8**/1.5	4.7/2.9	**1.4**/0.3	**1.4**/0.2	3.7/2.8

Table 3. Performance (mean/standard deviation) of RemInD without certain components on the MS-CMR dataset. ✓: the loss term was used with optimal strength; ∞: spatial transformations were fixed to identity mappings for all images.

$\mathcal{L}_{vel}$	$\mathcal{L}_{anchor}$	$\mathcal{L}_{align}$	$\mathcal{L}_{geo}$	Mean DSC (%)	Mean ASSD (mm)	Epochs to Converge
∞	✓	✓	✓	60.7/8.5	5.1/1.8	86
✓		✓	✓	80.6/6.5	1.5/0.8	2055
✓	✓		✓	81.1/6.5	1.5/0.9	738
✓	✓	✓		83.0/5.2	1.3/0.7	6619
✓	✓	✓	✓	83.1/5.3	1.3/0.6	3867

Qualitative Comparison. The results for example images by RemInD and the best-performing baselines (VAMCEI for MS-CMR and DARUNet for AMOS) are illustrated in Fig. 3. Visually, RemInD generally achieves better performance, particularly in challenging cases with poor imaging quality or artifacts. Baseline methods often produce fragmented or anatomically implausible structures, such as broken ventricles or myocardium, broken or merged kidneys and spleen, and other irregularities. In contrast, RemInD delivers robust predictions closely aligned with real-world labels. Moreover, baseline methods often rely heavily on intensity information, leading to segmentation errors. For example, as shown in the last column, some ribs and the postcava are misclassified as liver by the baseline (indicated by the two arrows), while RemInD avoids these mistakes. This is notable given that the ribs appear bright and the postcava shares similar intensity with and is connected to the liver. This demonstrates that RemInD effectively learns anatomical knowledge through the anchors and domain alignment via shape blending weights, thereby enhancing prediction quality in the target domain. Additionally, the flexible displacement fields bridge the gap between individual images and the inferred atlases, capturing detailed shape variations and improving the model's capability in an interpretable manner.

Visualization of Shape Blending Weights. The low-dimensional nature of the shape blending weights w allows for more accurate and informative visualization of domain alignment than t-SNE [16] of high-dimensional features com-

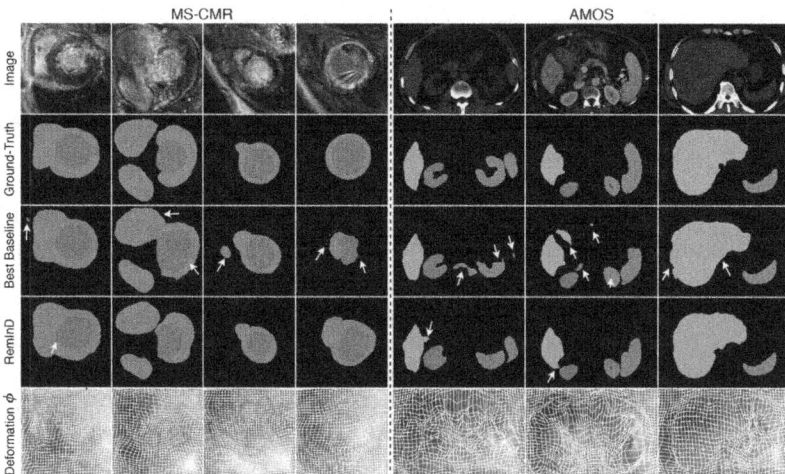

Fig. 3. Qualitative comparison between RemInD and the best baselines. The spatial deformations ϕ from RemInD are also displayed. Yellow arrows indicate regions where one method produces inferior predictions compared to the other. (Color figure online)

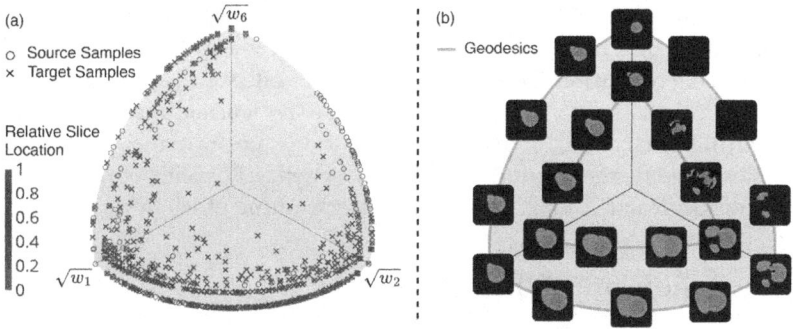

Fig. 4. 3D visualization of the positive orthant of the unit sphere, with the shape blending weights for images in the MS-CMR dataset. (a) Distribution of the transformed weights $w^\dagger$. Each point is based on a 2D image slice, with colors indicating its relative location among the total number of slices for the corresponding patient. (b) Manipulating $w^\dagger$ (thus z) along geodesics induces gradual variations in predicted segmentation (before warped by ϕ^{-1}). Endpoints of the shown six geodesics: (1,0,0), (0,1,0), (0,0,1), (0.99,0.1,0.1), (0.1,0.99,0.1), (0.1,0.1,0.99).

monly used in previous works. In a specific example of RemInD trained on the MS-CMR dataset, three components of the weights, w_3, w_4, w_5, are consistently zero for all source and target images. Given that $M = 6$, this sparsity enables visualizing w in 3D without any information compression, as shown in Fig. 4. To better reflect the geometry imposed by the Fisher-Rao metric, we transform (w_1, w_2, w_6) into $w^\dagger := (\sqrt{w_1}, \sqrt{w_2}, \sqrt{w_6})$, which maps the points onto the positive orthant of the unit sphere, $\{t \in \mathbb{R}^3 | t \succeq \mathbf{0}, t^\top t = 1\}$, where the geodesic distance (arc length) between any two points is equal to half of their Fisher-Rao metric. In Fig. 4(a), alignment between the source and target domains is evident, with most points close to the orthant border and only a few outliers in the target domain. Additionally, the point colors, representing slice locations within the corresponding patient, exhibit a smooth and gradual transition across the orthant. This highlights that the learned manifold effectively captures the spatial continuity of anatomical structures across patients in both domains, even though slice location information was not used for training. Figure 4(b) further indicates that the manifold efficiently encodes a wide variety of segmentations with smooth transitions along geodesics. The displayed examples cover nearly all possible topological patterns observed in the ground-truth labels. These visualizations underscore the advantages of RemInD in developing an explainable and global understanding of anatomical structures, similar to human memory.

Ablation Study. The performance of RemInD with different components removed is summarized in Table 3. The results demonstrate that spatial transformation is crucial for accurate segmentation, as the weighted average of anchors alone lack the flexibility to fit structures in every image. Besides, both $\mathcal{L}_{\text{anchor}}$ and $\mathcal{L}_{\text{align}}$ significantly contribute to segmentation accuracy. Notably, even without the domain alignment term $\mathcal{L}_{\text{align}}$, our model surpasses multiple baseline methods, showcasing the robustness of RemInD in generalizing across domains. This strong generalization likely stems from its ability to retain global segmentation information as prior knowledge through the learned manifold. Moreover, while RemInD without $\mathcal{L}_{\text{geo}}$ achieves comparable performance to the full version, its slower convergence underscores the benefit of explicitly regularizing the geometry of the learned manifold to enhance training efficiency.

4 Conclusion and Discussion

We have introduced a new paradigm for domain-adaptive medical image segmentation that is distinct from all previous works. Our framework, RemInD, constrains image features to a latent probabilistic manifold, effectively capturing accurate anatomical information. This design is interpretable, resembling the human process of retrieving the most appropriate segmentation shape from memory. Furthermore, it enables computationally efficient and geometrically faithful domain alignment, outperforming state-of-the-art methods that usually rely on multiple complicated and ad hoc alignment strategies.

RemInD has the potential to address even more challenging scenarios, such as source-free adaptation [13]. Specifically, during source-domain training, anchor parameters and the shape blending weights w for all source images could be stored and later used to adapt to the target images without requiring access to the source data—a promising direction for future research. A limitation of our method is that sharing anchors across images may reduce the network's flexibility, potentially leading to slightly lower segmentation performance on the source domain. However, as shown in our experiments, this trade-off greatly improved target-domain performance. Besides, this potential limitation could be addressed by increasing the number of anchors (M) and employing more advanced network architectures for the encoders and decoders.

Acknowledgements. This work was funded by National Institute of Health (NIH) grants RO1NS127317 and RO1NS125635.

References

1. Ashburner, J.: A fast diffeomorphic image registration algorithm. Neuroimage **38**(1), 95–113 (2007)
2. Biederman, I.: Recognition-by-components: a theory of human image understanding. Psychol. Rev. **94**, 115–147 (1987)
3. Chen, C., Dou, Q., Chen, H., Qin, J., Heng, P.A.: Unsupervised bidirectional cross-modality adaptation via deeply synergistic image and feature alignment for medical image segmentation. IEEE Trans. Med. Imaging **39**(7), 2494–2505 (2020). https://doi.org/10.1109/TMI.2020.2972701
4. Courty, N., Flamary, R., Tuia, D., Rakotomamonjy, A.: Optimal transport for domain adaptation. IEEE Trans. Pattern Anal. Mach. Intell. **39**(9), 1853–1865 (2017). https://doi.org/10.1109/TPAMI.2016.2615921
5. Cui, H., Li, Y., Wang, Y., Xu, D., Wu, L.M., Xia, Y.: Toward accurate cardiac MRI segmentation with variational autoencoder-based unsupervised domain adaptation. IEEE Trans. Med. Imaging **43**(8), 2924–2936 (2024). https://doi.org/10.1109/TMI.2024.3382624
6. Dalca, A.V., Balakrishnan, G., Guttag, J., Sabuncu, M.R.: Unsupervised learning of probabilistic diffeomorphic registration for images and surfaces. Med. Image Anal. **57**, 226–236 (2019)
7. Feydy, J., Séjourné, T., Vialard, F.X., Amari, S.I., Trouve, A., Peyré, G.: Interpolating between optimal transport and mmd using sinkhorn divergences. In: Chaudhuri, K., Sugiyama, M. (eds.) Proceedings of the Twenty-Second International Conference on Artificial Intelligence and Statistics. Proceedings of Machine Learning Research, vol. 89, pp. 2681–2690. PMLR (2019). https://proceedings.mlr.press/v89/feydy19a.html
8. Guan, H., Liu, M.: Domain adaptation for medical image analysis: a survey. IEEE Trans. Biomed. Eng. **69**(3), 1173–1185 (2022). https://doi.org/10.1109/TBME.2021.3117407
9. Huang, X., Belongie, S.: Arbitrary style transfer in real-time with adaptive instance normalization. In: 2017 IEEE International Conference on Computer Vision (ICCV), pp. 1510–1519 (2017). https://doi.org/10.1109/ICCV.2017.167
10. Iglesias, J.E., Sabuncu, M.R.: Multi-atlas segmentation of biomedical images: a survey. Med. Image Anal. **24**(1), 205–219 (2015)
11. Ji, Y., et al.: Amos: a large-scale abdominal multi-organ benchmark for versatile medical image segmentation. arXiv preprint arXiv:2206.08023 (2022)
12. Kingma, D.P., Welling, M.: Auto-encoding variational bayes. In: 2nd International Conference on Learning Representations, ICLR 2014, Banff, AB, Canada, 14–16 April 2014, Conference Track Proceedings (2014)

13. Liu, Y., Chen, Y., Dai, W., Gou, M., Huang, C.T., Xiong, H.: Source-free domain adaptation with domain generalized pretraining for face anti-spoofing. IEEE Trans. Pattern Anal. Mach. Intell. **46**(8), 5430–5448 (2024). https://doi.org/10.1109/TPAMI.2024.3370721
14. Lorenzen, P., Prastawa, M., Davis, B., Gerig, G., Bullitt, E., Joshi, S.: Multi-modal image set registration and atlas formation. Med. Image Anal. **10**(3), 440–451 (2006)
15. Luo, X., Wang, X., Shapiro, L., Yuan, C., Feng, J., Zhuang, X.: Bayesian unsupervised disentanglement of anatomy and geometry for deep groupwise image registration (2024). https://arxiv.org/abs/2401.02141
16. van der Maaten, L., Hinton, G.: Visualizing data using t-SNE. J. Mach. Learn. Res. **9**(86), 2579–2605 (2008). http://jmlr.org/papers/v9/vandermaaten08a.html
17. Miyamoto, H.K., Meneghetti, F., Pinele, J., Costa, S.: On closed-form expressions for the fisher-rao distance. Inf. Geom. (2024)
18. Oktay, O., et al.: Attention u-net: learning where to look for the pancreas. In: Medical Imaging with Deep Learning (2018). https://openreview.net/forum?id=Skft7cijM
19. van den Oord, A., Vinyals, O., Kavukcuoglu, K.: Neural discrete representation learning. In: Guyon, I., et al. (eds.) Advances in Neural Information Processing Systems, vol. 30. Curran Associates, Inc. (2017). https://proceedings.neurips.cc/paper_files/paper/2017/file/7a98af17e63a0ac09ce2e96d03992fbc-Paper.pdf
20. Paszke, A., et al.: PyTorch: an imperative style, high-performance deep learning library. Curran Associates Inc., Red Hook (2019)
21. Ronneberger, O., Fischer, P., Brox, T.: U-Net: convolutional networks for biomedical image segmentation. In: Navab, N., Hornegger, J., Wells, W.M., Frangi, A.F. (eds.) MICCAI 2015. LNCS, vol. 9351, pp. 234–241. Springer, Cham (2015). https://doi.org/10.1007/978-3-319-24574-4_28
22. Vahdat, A., Kautz, J.: Nvae: a deep hierarchical variational autoencoder. In: Larochelle, H., Ranzato, M., Hadsell, R., Balcan, M., Lin, H. (eds.) Advances in Neural Information Processing Systems, vol. 33, pp. 19667–19679. Curran Associates, Inc. (2020). https://proceedings.neurips.cc/paper_files/paper/2020/file/e3b21256183cf7c2c7a66be163579d37-Paper.pdf
23. Vu, T.H., Jain, H., Bucher, M., Cord, M., Pérez, P.: Advent: adversarial entropy minimization for domain adaptation in semantic segmentation. In: 2019 IEEE/CVF Conference on Computer Vision and Pattern Recognition (CVPR), pp. 2512–2521 (2019). https://doi.org/10.1109/CVPR.2019.00262
24. Wang, R., Zheng, G.: Cycmis: cycle-consistent cross-domain medical image segmentation via diverse image augmentation. Med. Image Anal. **76**, 102328 (2022)
25. Wang, X., et al.: Automated MRI-based segmentation of intracranial arterial calcification by restricting feature complexity. Magn. Reson. Med. **93**(1), 384–396 (2025)
26. Wang, X., Luo, X., Zhuang, X.: Bingo: Bayesian intrinsic groupwise registration via explicit hierarchical disentanglement. In: Frangi, A., de Bruijne, M., Wassermann, D., Navab, N. (eds.) Information Processing in Medical Imaging, pp. 319–331. Springer, Cham (2023)
27. Wu, F., Zhuang, X.: Unsupervised domain adaptation with variational approximation for cardiac segmentation. IEEE Trans. Med. Imaging **40**(12), 3555–3567 (2021). https://doi.org/10.1109/TMI.2021.3090412
28. Yao, K., et al.: A novel 3D unsupervised domain adaptation framework for cross-modality medical image segmentation. IEEE J. Biomed. Health Inform. **26**(10), 4976–4986 (2022). https://doi.org/10.1109/JBHI.2022.3162118

29. Zhang, X., et al.: Mapseg: unified unsupervised domain adaptation for heterogeneous medical image segmentation based on 3D masked autoencoding and pseudo-labeling. In: Proceedings of the IEEE/CVF Conference on Computer Vision and Pattern Recognition (CVPR), pp. 5851–5862 (2024)
30. Zhuang, X., et al.: Cardiac segmentation on late gadolinium enhancement MRI: a benchmark study from multi-sequence cardiac MR segmentation challenge. Med. Image Anal. **81**, 102528 (2022)

Dynamic Allocation Hypernetwork with Adaptive Model Recalibration for Federated Continual Learning

Xiaoming Qi[1], Jingyang Zhang[2], Huazhu Fu[3], Guanyu Yang[2], Shuo Li[4], and Yueming Jin[1(✉)]

[1] Department of Biomedical Engineering and Department of Electrical and Computer Engineering, National University of Singapore, Singapore, Singapore
ymjin@nus.edu.sg
[2] Key Laboratory of New Generation Artificial Intelligence Technology and Its Interdisciplinary Applications (Southeast University), Ministry of Education, Nanjing, China
[3] Institute of High Performance Computing, A*STAR, Singapore, Singapore
[4] Departments Biomedical Engineering, and Computer and Data Science, Case Western Reserve University, Cleveland, USA

Abstract. Federated continual learning (FCL) offers an emerging pattern to facilitate the applicability of federated learning (FL) in real-world scenarios, where tasks evolve dynamically and asynchronously across clients, especially in medical scenario. Existing server-side FCL methods in nature domain construct a continually learnable server model by client aggregation on all-involved tasks. However, they are challenged by: (1) Catastrophic forgetting for previously learned tasks, leading to error accumulation in server model, making it difficult to sustain comprehensive knowledge across all tasks. (2) Biased optimization due to asynchronous tasks handled across different clients, leading to the collision of optimization targets of different clients at the same time steps. In this work, we take the first step to propose a novel server-side FCL pattern in medical domain, Dynamic Allocation Hypernetwork with adaptive model recalibration (**FedDAH**). It is to facilitate collaborative learning under the distinct and dynamic task streams across clients. To alleviate the catastrophic forgetting, we propose a dynamic allocation hypernetwork (DAHyper) where a continually updated hypernetwork is designed to manage the mapping between task identities and their associated model parameters, enabling the dynamic allocation of the model across clients. For the biased optimization, we introduce a novel adaptive model recalibration (AMR) to incorporate the candidate changes of historical models into current server updates, and assign weights to identical tasks across different time steps based on the similarity for continual optimization. Extensive experiments on the AMOS dataset demonstrate the superiority of our FedDAH to other FCL methods on sites with different task streams. The code is available: https://github.com/jinlab-imvr/FedDAH.

Keywords: Federated continual learning · hypernetwork · recalibration

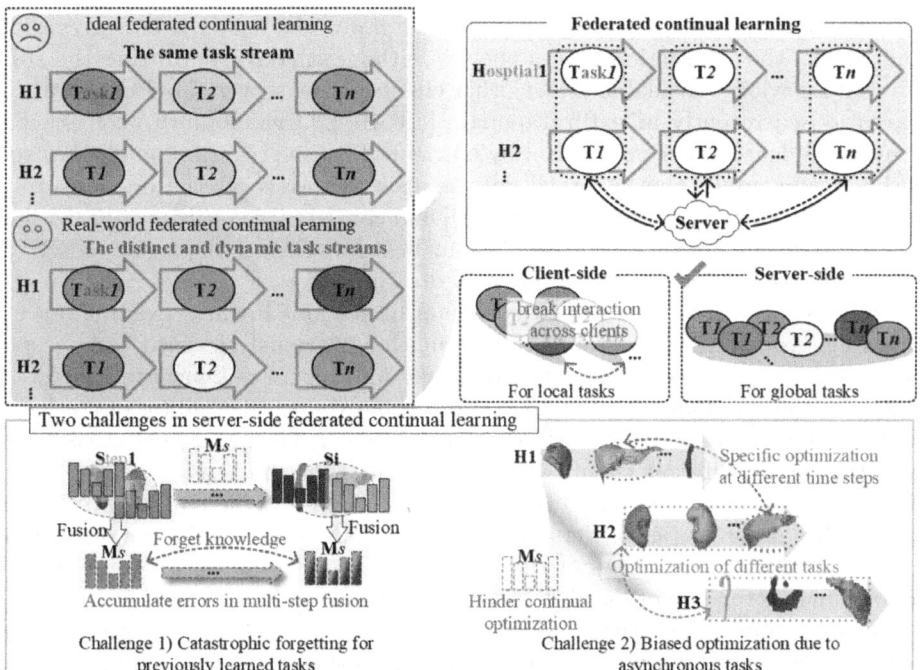

Fig. 1. Task: Since disease evolves and treatment options change, different clients require to continually evolves on different task orders (asynchronous) or add new tasks (dynamic). **Challenge:** The construction of a server-side FCL model is challenged by: 1) Catastrophic forgetting for previously learned tasks. 2) Biased optimization due to asynchronous tasks.

1 Introduction

Federated learning (FL) [6,14,19,22,26] is proposed as a paradigm to learn from decentralized data with privacy protection in different clinical centers (clients) and collaboratively learn a global model in server. However, since disease evolves, the development and deployment of treatment options and medical devices occur at varying rates across different clinical centres, this necessitates that clients continuously learn new tasks dynamically and adapt to varying task orders asynchronously [21]. These realities limit the applicability of FL in real-world clinical scenarios (Fig. 1). Hence, how to make clients adapt to dynamic and asynchronous task learning, while preserving effective collaborative training, is crucial for facilitating the real-world deployment of the FL model.

To this end, we focus on a more practical FL setting where clients handling dynamic tasks with asynchronous evolution, namely federated continual learning (FCL). Some previous studies propose client-side based methods to meet the challenges in FCL [1,20,23], which simply employs the off-the-rack continual learning (CL) methods onto client-side updating in federated learning (Fig. 1). However, the client-side FCL ignores the server-side aggregation and breaks the interaction across clients, without effectively utilizing the substantial knowledge available across other clients. Some server-side FCL methods are proposed recently in natural domain [3,6,16,22], which aim to construct a continually learnable server model by efficient client aggregation on all-involved tasks. For example, the historical data is utilized to recover the previous optimization by knowledge distillation [16,24] and consistency constraints [5,6,12] in the server fusion process. However, to our best knowledge, the server-side FCL method is still underexplored in medical domain.

Meanwhile, we have identified two main limitations in these existing server-side FCL works: 1) Catastrophic forgetting for previously learned tasks, especially historical data is unavailable for server and future unknown task in FCL. The server accumulates error in FCL and can hardly preserve all task knowledge without data in retraining. 2) Biased optimization due to asynchronous tasks handled across different clients. The existing FCL methods assume each client have the same task order in continual learning. However, the real-world medical sites utilize different task orders in FCL. This leads to the collision of optimization targets of different sites at the same time steps, hindering the provision of an optimal server model for all tasks to each client.

To meet above limitations, one main critical factor lies in how to improve the server memory with harmonious optimization. In this work, our core insight and contribution is to effectively equip the hypernetwork [9] onto the server design to achieve this goal. The idea is motivated by the advantage of the hypernetwork, which can learn an task-specific mapping from a task identity to the task model weights, providing a feasible way to replay all task models of clients to reduce server forgetting and thus facilitate the harmonious optimization. However, there exist some challenges to effectively utilize hypernetwork to tackle FCL problems. For asynchronously evolving tasks in each client, the mapping learning by hypernetwork would be confused with the task-hypernetwork correspondence, misguiding the server optimization. In addition, for server updating, hypernetwork should be recalibrated to update faster for new tasks and slower for existing tasks, which further prompts harmonious optimization.

In this paper, we propose a novel server-side FCL pattern, termed dynamic allocation hypernetwork with adaptive prototype recalibration (**FedDAH**), aiming to tackle a more realistic collaborative learning setting where distinct and dynamic task streams present in different clients. Specifically, we first propose a **dynamic allocation hypernetwork (DAHyper)** module. DAHyper presents a continually updated hypernetwork for managing the mapping between task identities and their associated model parameters, enabling dynamic allocation of model parameters across various clients. Through the identity of task I, the hypernetwork is trained to preserve the model parameters of task I. By this setting, the server can establish the mappings between all tasks and model parameters. This enables server updates to leverage all task models learned by clients without accumulating errors. We further design an **adaptive model recalibration (AMR)**. Benefiting from the defined mapping mechanism between

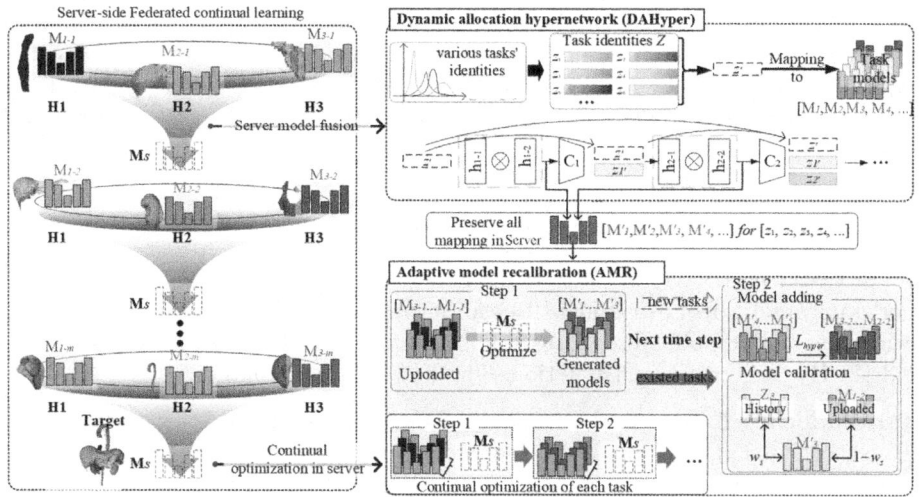

Fig. 2. The framework of FedDAH: (a) Dynamic allocation hypernetwork preserves the mappings (task identity to model weights) by the hypernetwork to avoid knowledge forgetting. (b) Adaptive model recalibration assigns a calibration based on the contrastive similarity for continual optimization on asynchronous tasks.

task and model parameter, the server could obtain the same task parameter in the asynchronous tasks. AMR assigns a calibration to each model optimization based on the contrastive similarity, enabling rapid integration of new knowledge from new task models while retaining previously learned knowledge with less fading. We have conducted extensive experiments on AMOS dataset for abdominal organ segmentation with multi-center, multi-vendor, multi-modality, multi-phase, multi-disease patients. Our FedDAH achieves the substantial improvement compared with the state-of-the-art methods.

Overall, our contributions can be summarized as follows:

1. For the first time, we propose a novel server-side FCL pattern in the medical scenario, FedDAH, to tackle a more practical collaborative training setting where different clinical clients have their distinct and dynamic task streams.
2. A novel server-side model aggregation pattern, DAHyper, is proposed to manage and allocate the model parameters across various clients without error accumulation caused by forgetting in FCL.
3. A novel server-side model optimization strategy, AMR, is proposed to calibrate the continual optimization on asynchronous task streams in FCL.

2 Methodology

We propose a novel server-side FCL framework FedDAH, aiming to tackle the crucial yet challenging scenario that different clients have different task streams

(Fig. 2). FedDAH consists of (1) DAHyper, which is to preserve the mappings of task identities to model weights, to avoid knowledge forgetting and allocate a required model to the client (Sect. 2.1); (2) AMR, which is to assign a calibration to each model optimization for continual optimization on the distinct task streams (Sect. 2.2).

2.1 DAHyper for Knowledge Preservation

In our FedDAH, DAHyper defines a novel hypernetwork to generate the whole model parameter from task identities for each client knowledge preservation and away from catastrophic forgetting in FCL. It contains (1) Task identity definition and (2) Hypernetwork construction. The details are as follows:

Rationale: In CL, a neural network $f(x, \theta)$ with weights θ is given from a set of tasks $\{(X_1, Y_1), ..., (X_T, Y_T)\}$. Instead of retain $f(x, \theta)$ for previous tasks in continual learning, a metamodel $f_h(e, \theta_h)$ (DAHyper) maps a task embedding e to weights θ by weights θ_h. Through training f_h on the acquired input(task embedding e)-output(weights θ) mappings, all the task knowledge can be preserved. Hence, hypernetworks can address catastrophic forgetting in continual learning at the meta level. Different from the generation of one layer parameter in traditional hypernetwork [18], our DAHyper enables to generate the weights for the entire network, and learns the parameters θ_h of a metamodel to output the model parameter θ for a specific task.

Task Identity Definition: Considering the dynamically updated tasks in FCL, DAHyper proposes a unique pattern to distinguish different tasks and associate tasks with the corresponding model weights. In the traditional hypernetwork, a task embedding e is randomly generated during training for a layer's weight. Since e is random value for a layer, this pattern cannot be applied to FCL to distinguish various tasks. In DAHyper, we define the task identity set $Z = \{z_1, z_2, ...\}$ for various tasks in continual learning process. The element of Z is a vector generated from normal distribution ($N(\mu_z, \sigma^2)$) with different μ_z and σ. Considering the different tasks in different clients, DAHyper designs z_i from Z for each task to distinguish the different tasks in server.

Hypernetwork Construction: Since DAHyper generates the entire model weights of Task i, each layer parameter of the model require to be considered. For Task i model each layer, the parameters of a layer are associated with the task identity z_i and the previous layers. Hence, DAHyper defines the hypernetwork according to the task identity and inter-layer consistency.

In Task Identity: We assume the parameters of a layer j in the Task i model are stored in a matrix $K^j \in \mathbb{R}^{N_{in}f_s \times N_{out}f_s}$, where $f_s \times f_s$, N_{in}, and N_{out} are the filter sizes, input size, and output size of the layer. Since the K^j can be viewed as N_{in} slices of a matrix K_{in}^j with $f_s \times N_{out}f_s$, we generate the parameter by two-layer linear network. In the first layer (h_{1-1}), the z_i is projected into the N_{in} vectors a_i, with N_{in} different matrices $W_i \in \mathbb{R}^{d \times N_z}$ and bias $B_i \in \mathbb{R}^d$, where d and N_z are the size of the hidden layer and z_i. The h_{1-2} takes the vector a_i

and projects it into K_{in}^j using weights $W_o \in \mathbb{R}^{f_s \times N_{out} f_s \times d}$ and $B_o \in \mathbb{R}^{f_s \times N_{out} f_s}$. The K^j is a concatenation of every K_{in}^j. The whole process can be expressed as:

$$a_i = W_i z_i + B_i, \quad K_{in}^j = W_o a_i + B_o, \quad K^j = Concat(K_1^j, ..., K_{N_{in}}^j) \qquad (1)$$

In Inter-layer Consistency: The next layer parameters are not only associated with the task identity z_i, but also keep the inter-layer consistency with the previous layer parameters K^j. According to this, DAHyper introduces a mechanism: Firstly, the previous layer parameters K^j is encoder into a vector $z1'$ with the same size of z_i by Encoder C_1. Then the concatenation of z_i and $z1'$ is the input of next layer generation (h_{2-1}&h_{2-2}). The feedforward process of h_{2-1}&h_{2-2} is the same as h_{1-1}&h_{1-2}. Following this operation, the concatenation of z_i and $z1'$ also will be concatenated with the further more outputs ($\{z2', z3', ...\}$). Finally, DAHyper can obtain the parameters θ of model M_i' in server for Task i based on z_i.

2.2 AMR for Continual Optimization

To avoid the optimization bias caused by asynchronous tasks in FCL, AMR treats the first model weights for each task as a basic model (standard) and ensures continual optimization of each basic model by calculating a calibration based on the similarity to the same model weights uploaded at different time steps in FCL.

AMR ensures the continual optimization from two aspects: (1) Continual optimization of different tasks. For the uploaded different tasks, AMR defines the historical calibration to regularize each task in server. (2) Continual optimization of the same task. For the models with the same task yet uploaded at different time steps, AMR defines the similarity among the models and utilizes the similarity as weights to guide optimization.

Continual Optimization on Different Tasks. For the uploaded models of different tasks in server, AMR requires to optimize the DAHyper with the models and balance the optimization on the current tasks and previous tasks in different time steps. Hence, AMR treats each task model as a basic model, and the optimization of each basic model during the following steps should not be degraded by other basic models. AMR takes a two-stage learning ($\mathcal{L}_{hyper}$) on the current task and historical basic models. Firstly, a candidate change $\triangle \theta_h$ is calculated by minimizing the loss on the current task $\mathcal{L}_{task}(\theta_h, z_i, M_i)$, where θ_h, z_i, and M_i are the parameters of DAHyper, task identity, and target model of the current task i. Through the $\mathcal{L}_{task}$, we can guide the DAHyper to obtain the θ for each task. Here, AMR utilizes L2 distance to calculate $\mathcal{L}_{task}$. Secondly, AMR regularizes the historical basic models while attempting to learn the current task by:

$$\mathcal{L}_R = \frac{1}{T-1} \sum_{t=1}^{T-1} ||f_h(z_t, \theta_h^*) - f_h(z_t, \theta_h + \triangle \theta_h)||^2, \qquad (2)$$

where z_t and θ_h^* are the task identity of task t and the set of DAHyper parameters before attempting to learn task T (current task). Since the knowledge of historical basic models is preserved by DAHyper without current task optimizations, the regularization $\mathcal{L}_R$ takes the minimization of difference between updated output and historical knowledge to ensure the DAHyper effective on different basic models (current and previous tasks) at the same time. Hence, the $\mathcal{L}_{hyper} = \mathcal{L}_{task} + \beta \mathcal{L}_R$. The β is a hyperparameter of $\mathcal{L}_R$.

Continual Optimization of the Same Task. Besides the $\mathcal{L}_{hyper}$ controls the optimization on different basic models, the basic models for the same task at different time steps in FCL also require continual optimization. Hence, we further develop a recalibration based on $\mathcal{L}_{hyper}$. The process is shown in Fig. 2: (1) In the step 1, the weight parameters M_1', M_2', M_3' generated by DAHyper are optimized by the uploaded models $M_{1-1}, M_{2-1}, M_{3-1}$ from different clients. (2) Then, the optimized M_1, M_2, M_3 are treated as 3 basic models in server. (3) In the step 2, server receives $M_{1-2}, M_{2-2}, M_{3-2}$ from H1, H2, and H3. M_{1-2} is correspond to the existing basic model M_3. With the new M_3' generated by DAHyper, there are 2 optimization targets (M_3 and M_{1-2}). Considering the convergence of M_{1-2} worse than the historical model M_3, AMR takes the M_{1-2} as regularization to benefit M_3. Hence, AMR calculates the similarity weights of $W_s(M_3', M_3)$ as the basic, and utilize $(1 - W_s)(M_3', M_{1-2})$ as further recalibration. The similarity is measured by JS divergence [8].

According to the weights, the final loss is:

$$\mathcal{L} = W_s[\mathcal{L}_{task}(M_3', M_3) + \beta_1 \mathcal{L}_R 1] + (1 - W_s)[\mathcal{L}_{task}(M_3', M_{1-2}) + \beta_2 \mathcal{L}_R 2]. \quad (3)$$

Through treating the new updated model of existing basic model as the recalibration, AMR ensures the continual optimization of the same task at different time steps.

3 Experiments and Results

3.1 Dataset and Implementation

Dataset and Evaluation Metric: To evaluate the performance of our Fed-DAH, we conduct experiments on the AMOS dataset [10]. AMOS provides 500 CT scans collected from multi-center, multi-vendor, multi-modality, multi-phase, multi-disease patients, each with voxel-level segmentation annotations of 15 abdominal organs. We reconstruct the AMOS dataset to simulate a more realistic clinical FCL. We set 4 clients (C1-C4) with each having 125 CT respectively. The 125 CT are divided into the training and testing sets as 4:1. Each client takes all the 15 organs for testing. Considering the high likelihood that different clinical centers may have some identical tasks at the beginning, we select some organ segmentation as the initialization task existing in all clients (left kidney and right kidney in this work). This can also evaluate the effectiveness of FCL methods on the same task streams. In addition, different clinical centers are likely to tackle the same or varying tasks in differing sequences in

Table 1. The details of the dataset and the settings of each client in FCL.

Clients	Num	Task 1	Task 2-8 (random order)	
			Shared	Unique
C1	125		spleen,	bladder, prostate
C2	125	left kidney,	stomach, pancreas,	aorta, inferior vena cava
C3	125	right kidney	gallbladder,	duodenum, esophagus
C4	125		liver	left, right adrenal gland

Table 2. The mean Dice score of each client evaluates the superior ability of continual learning in FedDAH (the testing of each method is performed on 15 organs).

Method client	FedAvg [17]	FBL [6]	FedWeIT [23]	FedSpace [20]	FedDAH -DAHyper	$-\mathcal{L}_R$	$-W_s$	Full	Local	Centralized
C1	0.019	0.213	0.700	0.763	0.682	0.347	0.432	0.801	0.667	0.831
C2	0.020	0.255	0.723	0.761	0.711	0.338	0.466	0.805	0.631	0.801
C3	0.018	0.236	0.707	0.733	0.679	0.340	0.458	0.812	0.589	0.828
C4	0.016	0.131	0.659	0.744	0.708	0.283	0.414	0.807	0.577	0.820

the upcoming steps. We further divided other organs into shared and unique parts to evaluate FCL methods on the same tasks with different streams, and on distinct tasks with different streams. Each client conducts continual learning by a random order based on the combination of the shared part and the unique part. Details are shown in Table 1. We employ Dice similarity coefficient [4] as the evaluation metric for this segmentation task. We calculate the average Dice of all organs in the continual learning process for a fair comparison.

Implementation: Our FedDAH takes 3D Unet [2] as the basic segmentation network for each client's continual learning. In each client training, the network is based on Pytorch with the learning rate of 1×10^{-3}, Adam [13] optimizer, and a batch size of 1. The communication of FL is conducted after every $E = 5$ in client training until $T = 20$ in total for each task. In each client training, data augmentation (rotation, translation, scale, and mirror) and maximum connected domain are the post-processing. In server, the hypernetwork is optimized by Adam with the learning rate of 1×10^{-3}. All experiments are performed on four NVIDIA A6000 GPUs.

3.2 Experimental Results

Quantitative and Qualitative Analysis. We evaluate our FedDAH from quantitative and qualitative aspects by the comparison with the state-of-the-art FCL methods (FBL [6], FedWeIT [23], and FedSpace [20]) and CL based on FedAvg [17]. FBL utilizes task relations to benefit different clients' optimization

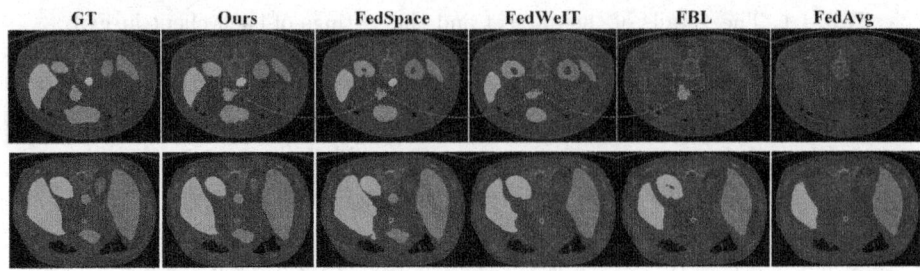

Fig. 3. The visual results indicate the superior performance of FedDAH. Especially in the red box, we show a task optimized by other clients, and our FedDAH provides more complete segmentation. (Color figure online)

to realize the continual learning at each client. FedWeIT designs knowledge distillation at client optimization to benefit different tasks in continual learning. FedSpace benefits different tasks in continual learning by additional task data. The CL based on FedAvg directly utilize continual learning setting at each client and take FedAvg to realize FL setting. Apart from these, we also centralize the training data for model optimization as the upper bound, and just train the local models by the local data with all organ labels. The results are shown in Table 2.

We can see that (1) Our FedDAH achieves the best mean Dice compared with others, peaking at 0.801, 0.805, 0.812, and 0.807 Dice for the four clients. This indicates that our FedDAH can alleviate knowledge forgetting and asynchronous server optimization difficult under this more realistic FCL setting, which different clinical centres have different task streams. (2) FedAvg and FBL show the worse performance than others, showing that directly combining CL with conventional FL method still struggle to tackle the challenges brought by different task streams in FCL. (3) Compared with the centralized training and local training methods, FedDAH achieves better performance than local training model and comparable performance with centralized training model. This indicates that the FL can improve the different client model optimization with different tasks in the real-world by FedDAH. (4) Through the comparison of different clients, the local training could achieve a better performance than FedAvg and FBL. This is caused by the challenges of catastrophic forgetting and asynchronous tasks in FCL. However, the model sharing technology in traditional CL methods hardly overcome these challenges. In FedWeIT and FedSpace, the client localization for each task in client optimization requires additional optimizations in clients and worse than centralized training. This results also indicates that our FedDAH could alleviate the challenges of catastrophic forgetting and asynchronous task streams in FCL.

To further evaluate the performance, we visualized the segmentation of different methods. From the visual results in Fig. 3, it can be found that our FedDAH could achieve the accurate and complete segmentation masks during the continual learning process. Through the comparison of the same regions across different

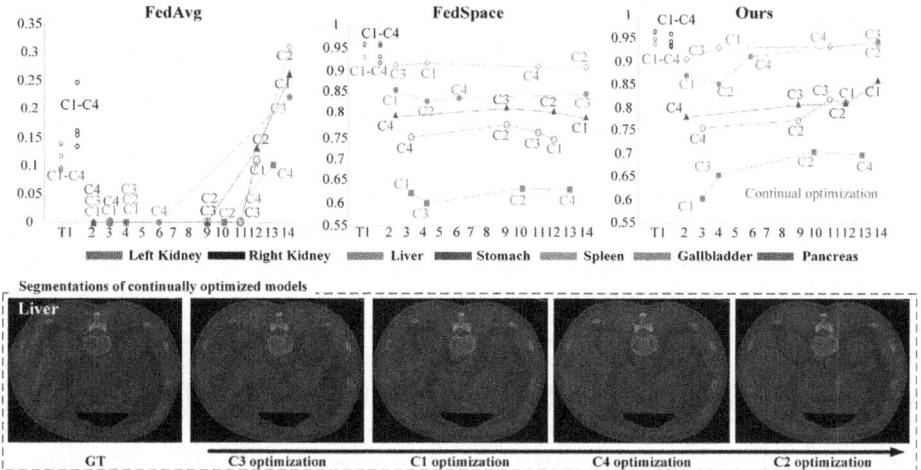

Fig. 4. FedDAH ensures the continual optimization of different task streams in FCL. The horizontal axis is time step and vertical axis is dice score.

methods, we can see that the existing FL and FCL methods tend to omit some regions which are difficult to segment due to catastrophic forgetting.

Ablation Study. To evaluate the contributions of each part in FedDAH, we design different ablation studies based on the following experimental settings. -DAHyper: we remove the DAHyper module to evaluate the effectiveness on overcoming catastrophic forgetting in FCL. -$\mathcal{L}_R$: we remove the historical calibration to evaluate the effectiveness of history in continual optimization of different tasks in FCL. -W_s: we remove the similarity weight to evaluate the effectiveness of similarity calibration on the continual optimization of the same task.

The results are also listed in Table 2. It can be found that: (1) In -DAHyper, the proposed DAHyper is replaced by the average strategy in FedAvg [17] for each task model and the server preserves all models' parameters. The lower Dice achieved by this configuration indicates that the proposed DAHyper can preserve all model knowledge in server model more effectively. (2) In -$\mathcal{L}_R$, we remove the regularization of different tasks continual optimizations. It can be found that the performance suffers severe decrease. This indicates that the $\mathcal{L}_R$ benefits the FedDAH optimization from historical knowledge. (3) In -W_s, we remove the recalibration of continual optimization on the same task and just use the uploaded client model to optimize DAHyper. The lower mean Dice demonstrates that the W_s can provide effective guidance for the model to optimize the same task model at different time steps.

3.3 Detailed Analytical Experiments

Results in Each Continual Step. To validate the ability of continual learning on different task streams in FCL, we visulaize the learning process of FedDAH,

Table 3. The performance of different continual methods on C1 evaluates that FedDAH is able to provide a global FCL model for each sites.

Method	lk	rk	spl	sto	pan	gal	liv	bla	pro	aor	inf	duo	eso	lag	rag	AVG
PLOP	0.962	0.941	0.857	0.850	0.790	0.548	0.915	0.715	0.730	-	-	-	-	-	-	0.487
LISMO	0.956	0.938	0.847	0.829	0.764	0.515	0.899	0.706	0.715	-	-	-	-	-	-	0.478
CLAMTS	0.958	0.964	0.866	0.842	0.808	0.569	0.920	0.724	0.728	-	-	-	-	-	-	0.492
CSTSUA	0.960	0.957	0.877	0.847	0.796	0.550	0.928	0.719	0.754	-	-	-	-	-	-	0.493
FedDAH	0.963	0.944	0.870	0.860	0.811	0.603	0.930	0.732	0.751	0.833	0.857	0.671	0.747	0.743	0.707	0.801

FedSpace, and FedAvg in FCL and the example of continual optimization in our FedDAH learning process (Liver). It can be found that (Fig. 4): (1) Our FedDAH makes each task achieve the best performance at last optimization. FedDAH gradually improves segmentation at different time steps. This benefit from the task identity is preserved in server to maintain the model better than the previous optimized model with the same task. This indicates that FedDAH ensures the continual optimization on a task at different time steps and different clients. (2) In FedAvg, only the last two task can be optimized and achieves poor performance. In FedSpace, the segmentation performance of pancreas become gradually worse in the learning process. This indicates the knowledge forgetting and optimization bias make it difficult to realize FCL in the real-world. (3) Compare the learning process of the different methods, it could be found that organs suffers unstable optimization in the existing FL and FCL methods. This indicates that the asynchronous tasks makes the existing method hardly work in the real-world FCL. (4) From the visual results of liver, it can be found that the segmentation is gradually improved during the optimization on different clients. This evaluates that our FedDAH could maintain the knowledge in continual learning and correct the optimization bias in FCL.

Our FedDAH v.s. CL Methods. Training each local model by using CL methods can also be an option to tackle the challenges of different task streams. To evaluate the superiority of FedDAH over CL methods, we compare FedDAH with several advanced CL methods, including PLOP [7], LISMO [15], CLAMTS [25], and CSTSUA [11]. We train these CL methods on each local data and labels,

Table 4. The details of dataset and the settings of each client using all organs for CL.

Clients	Num	Task 1(initial)	Task 2-14 (random order)	
			Shared	
C1	125		spleen,	bladder, prostate,
C2	125	left kidney,	stomach, pancreas	aorta, inferior vena cava,
C3	125	right kidney	gallbladder,	duodenum, esophagus,
C4	125		liver	left, right adrenal gland

Table 5. The performance of each client evaluates the ability of continual learning in our FedDAH.

Task	1 Kidney		2	3	4	5	6	7
	Left	Right						
C1	0.963	0.944	Spl:0.870	Gal:0.603	Liv:0.930	Duo:0.671	Aor:0.833	Bla:0.732
C2	0.95	0.932	Aor:0.852	Bla:0.724	Spl:0.851	Inf:0.832	Eso:0.706	Duo:0.703
C3	0.966	0.959	Liv:0.906	Aor:0.811	Gal:0.653	Pro:0.688	Eso:0.734	Inf:0.898
C4	0.938	0.936	Sto:0.781	Pan:0.755	Eso:0.693	Lag:0.703	Spl:0.911	Aor:0.883
	8	9	10	11	12	13	14	Avg
C1	Inf:0.857	Eso:0.747	Rag:0.707	Pro:0.751	Pan:0.811	Lag:0.743	Sto:0.86	0.801
C2	Lag:0.694	Pan:0.772	Gal:0.704	Rag:0.703	Sto:0.811	Pro:0.816	Liv:0.939	0.799
C3	Rag:0.685	Sto:0.808	Duo:0.697	Pan:0.819	Bla:0.703	Lag:0.771	Spl:0.943	0.803
C4	Pro:0.700	Inf:0.853	Rag:0.734	Liv:0.933	Duo:0.753	Gal:0.697	Bla:0.756	0.802

and the testing is conducted on 15 organs. We take the Client 1 (C1) as an example and the results are shown in Table 3. It can be found that: (1) considering that one local client may not see all tasks during train (e.g., C1 does not have the labels of aorta organ), the pure CL methods can not segment these unseen organs (marked as '-' in the table). Instead, our FCL based method FedDAH could makes the partially labeled clients obtain the ability of complete segmentation by learning such knowledge from other clients. (2) Through the comparison of Table 3 and Table 2, we find that FedDAH can achieve similar performance on Client 1 as it does on other clients. This indicates that our FedDAH could balance the optimization on different clients and share the information among all clients.

Performance in Task Level. To more comprehensively illustrate the superiority of FedDAH, we conduct another FCL setting that each client has seen all the tasks (e.g., organs) during the training, and we show the segmentation performance in the task level. As shown in Table 4, the four clients still use the left and right kidney segmentation as initialization tasks. Then each client regards the rest 14 organs as task 2 to task 14, but receives the labels in different sequences with random order. We utilize the same test dataset for each time step model for a fair comparison.

The results are listed in Table 5. It indicates the superiority of our FedDAH on real-world FCL with distinct and dynamic task streams. (1) On the shared tasks with the same stream (left and right kidney), all clients can be well optimized with all Dice over 0.9. (2) The same organs at different time steps are continually optimized, such as spleen ('Spl' in the table) is optimized from 0.87 to 0.943. This evaluates that our FedDAH provides the continual learning ability on the asynchronous task streams. (3) At the same step, different task can be well optimized. Taking step 2 for example, the spleen, aorta, liver, and stomach all have been optimized (0.87, 0.852, 0.906, and 0.781). This evaluates that our FedDAH ensures the different tasks' knowledge preservation in server.

4 Conclusion

We propose a novel server-side FCL framework, FedDAH, to enable global knowledge preservation and continually asynchronous task optimization to narrow the gap of deploying FL in real-world application. FedDAH employs a designed hypernetwork to preserve knowledge, incorporates the candidate changes of history, and balances the continual optimization based on similarity. We conduct extensive experiments to validate the effectiveness of our method on the AMOS dataset, outperforming other approaches by a large margin. In the future work, we propose to expand our FedDAH to the scenario of different clients with different organs and tasks in continual learning. This will advance our FedDAH with the ability to eventually train foundation model that is compatible with existing medical foundation models releasing from the data collection.

Acknowledgement. This work was supported by Ministry of Education Tier 1 Start up grant, NUS, Singapore (A-8001267-01-00); Ministry of Education Tier 1 grant, NUS, Singapore (A-8001946-00-00); and the National Natural Science Foundation of China (Grant No. 82441021).

References

1. Casado, F.E., Lema, D., Criado, M.F., Iglesias, R., Regueiro, C.V., Barro, S.: Concept drift detection and adaptation for federated and continual learning. Multimedia Tools Appl. 1–23 (2022)
2. Çiçek, Ö., Abdulkadir, A., Lienkamp, S.S., Brox, T., Ronneberger, O.: 3D U-Net: learning dense volumetric segmentation from sparse annotation. In: Ourselin, S., Joskowicz, L., Sabuncu, M.R., Unal, G., Wells, W. (eds.) MICCAI 2016. LNCS, vol. 9901, pp. 424–432. Springer, Cham (2016). https://doi.org/10.1007/978-3-319-46723-8_49
3. Criado, M.F., Casado, F.E., Iglesias, R., Regueiro, C.V., Barro, S.: Non-IID data and continual learning processes in federated learning: a long road ahead. Inf. Fusion **88**, 263–280 (2022)
4. Dice, L.R.: Measures of the amount of ecologic association between species. Ecology **26**(3), 297–302 (1945)
5. Dong, J., et al.: Federated class-incremental learning. In: Proceedings of the IEEE/CVF Conference on Computer Vision and Pattern Recognition, pp. 10164–10173 (2022)
6. Dong, J., Zhang, D., Cong, Y., Cong, W., Ding, H., Dai, D.: Federated incremental semantic segmentation. In: Proceedings of the IEEE/CVF Conference on Computer Vision and Pattern Recognition, pp. 3934–3943 (2023)
7. Douillard, A., Chen, Y., Dapogny, A., Cord, M.: Plop: learning without forgetting for continual semantic segmentation. In: Proceedings of the IEEE/CVF Conference on Computer Vision and Pattern Recognition, pp. 4040–4050 (2021)

8. Fuglede, B., Topsoe, F.: Jensen-Shannon divergence and hilbert space embedding. In: Proceedings of International Symposium on Information Theory, ISIT 2004, p. 31 (2004)
9. Ha, D., Dai, A.M., Le, Q.V.: Hypernetworks. In: International Conference on Learning Representations (2017)
10. Ji, Y., et al.: Amos: a large-scale abdominal multi-organ benchmark for versatile medical image segmentation. Adv. Neural. Inf. Process. Syst. **35**, 36722–36732 (2022)
11. Ji, Z., et al.: Continual segment: towards a single, unified and non-forgetting continual segmentation model of 143 whole-body organs in CT scans. In: 2023 IEEE/CVF International Conference on Computer Vision (ICCV), pp. 21083–21094 (2023)
12. Jiang, Z., Ren, Y., Lei, M., Zhao, Z.: Fedspeech: federated text-to-speech with continual learning. In: Proceedings of the Thirtieth International Joint Conference on Artificial Intelligence, IJCAI 2021, pp. 3829–3835 (2021)
13. Kingma, D.P., Ba, J.: Adam: a method for stochastic optimization. In: 3rd International Conference on Learning Representations, ICLR 2015, San Diego, CA, USA, 7–9 May 2015, Conference Track Proceedings (2015)
14. Li, T., Sahu, A.K., Talwalkar, A., Smith, V.: Federated learning: challenges, methods, and future directions. IEEE Signal Process. Mag. **37**(3), 50–60 (2020)
15. Liu, P., et al.: Learning incrementally to segment multiple organs in a CT image. In: International Conference on Medical Image Computing and Computer-Assisted Intervention, pp. 714–724 (2022)
16. Ma, Y., Xie, Z., Wang, J., Chen, K., Shou, L.: Continual federated learning based on knowledge distillation. In: Proceedings of the Thirty-First International Joint Conference on Artificial Intelligence, vol. 3 (2022)
17. McMahan, B., Moore, E., Ramage, D., Hampson, S., Arcas, B.A.: Communication-efficient learning of deep networks from decentralized data. Artif. Intell. Stat. 1273–1282 (2017)
18. von Oswald, J., Henning, C., Grewe, B.F., Sacramento, J.: Continual learning with hypernetworks. In: 8th International Conference on Learning Representations (ICLR 2020) (virtual) (2020)
19. Qi, X., Yang, G., He, Y., Liu, W., Islam, A., Li, S.: Contrastive re-localization and history distillation in federated CMR segmentation. In: International Conference on Medical Image Computing and Computer-Assisted Intervention, pp. 256–265 (2022)
20. Shenaj, D., Toldo, M., Rigon, A., Zanuttigh, P.: Asynchronous federated continual learning. In: Proceedings of the IEEE/CVF Conference on Computer Vision and Pattern Recognition, pp. 5054–5062 (2023)
21. Thakur, A., Armstrong, J., Youssef, A., Eyre, D., Clifton, D.A.: Self-aware SGD: reliable incremental adaptation framework for clinical AI models. IEEE J. Biomed. Health Inform. **27**(3), 1624–1634 (2023)
22. Xu, X., Deng, H.H., Gateno, J., Yan, P.: Federated multi-organ segmentation with inconsistent labels. IEEE Trans. Med. Imaging (2023)
23. Yoon, J., Jeong, W., Lee, G., Yang, E., Hwang, S.J.: Federated continual learning with weighted inter-client transfer. In: International Conference on Machine Learning, pp. 12073–12086 (2021)
24. Zhang, J., Chen, C., Zhuang, W., Lyu, L.: Target: federated class-continual learning via exemplar-free distillation. In: Proceedings of the IEEE/CVF International Conference on Computer Vision, pp. 4782–4793 (2023)

25. Zhang, Y., Li, X., Chen, H., Yuille, A.L., Liu, Y., Zhou, Z.: Continual learning for abdominal multi-organ and tumor segmentation. In: International Conference on Medical Image Computing and Computer-Assisted Intervention, pp. 35–45 (2023)
26. Zhang, Y., et al.: Fedsoda: federated cross-assessment and dynamic aggregation for histopathology segmentation. arXiv preprint arXiv:2312.12824 (2023)

Skelite: Compact Neural Networks for Efficient Iterative Skeletonization

Luis D. Reyes Vargas[1(✉)], Martin J. Menten[2,3,4], Johannes C. Paetzold[5], Nassir Navab[1,4], and Mohammad Farid Azampour[1,4]

[1] Computer Aided Medical Procedures, Technical University of Munich, Munich, Germany
ge45dej@mytum.de
[2] Chair for AI in Healthcare and Medicine, Technical University of Munich, Munich, Germany
[3] BioMedIA, Department of Computing, Imperial College London, London, UK
[4] Munich Center for Machine Learning (MCML), Munich, Germany
[5] Weill Cornell Medicine, Cornell University, New York City, NY, USA

Abstract. Skeletonization extracts thin representations from images that compactly encode their geometry and topology. These representations have become an important topological prior for preserving connectivity in curvilinear structures, aiding medical tasks like vessel segmentation. Existing compatible skeletonization algorithms face significant trade-offs: morphology-based approaches are computationally efficient but prone to frequent breakages, while topology-preserving methods require substantial computational resources.

We propose a novel framework for training iterative skeletonization algorithms with a learnable component. The framework leverages synthetic data, task-specific augmentation, and a model distillation strategy to learn compact neural networks that produce thin, connected skeletons with a fully differentiable iterative algorithm.

Our method demonstrates a 100× speedup over topology-constrained algorithms while maintaining high accuracy and generalizing effectively to new domains without fine-tuning. Benchmarking and downstream validation in 2D and 3D tasks demonstrate its computational efficiency and real-world applicability. Code and data available here: https://github.com/luisdavid64/Skelite.

1 Introduction

Skeletonization algorithms reduce the foreground of an image to a skeletal representation that approximates the medial axis, defined as the set of points with more than one closest point on the shape's boundary [5]. These representations condense the topological and geometric properties of an object, making them useful in a variety of medical applications such as blood flow analysis, image registration, and surgical planning [10,15,20]. Skeleton extraction has been widely explored for digital images with the core principles of accuracy and computational efficiency, leading to a vast body of works [3,4,8,23,29,30,40].

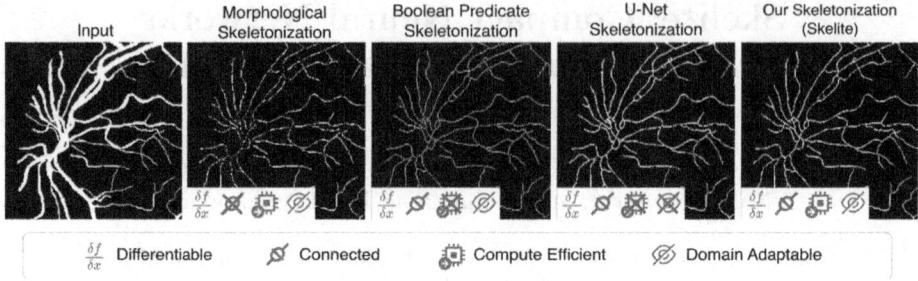

Fig. 1. Existing skeletonization methods trade off speed, accuracy, and adaptability. Morphology-based methods often break skeletons, Boolean predicates are slow but topologically accurate, and U-Nets struggle with domain shifts. Our method is fast, adaptable, and produces thin skeletons with few breakages.

Recently, skeletonization arose as a structural prior in deep learning for semantic segmentation. Studies have shown that loss functions defined using differentiable skeletonization aid in preserving connectivity information in tubular structures [1,22,33]. This can be crucial when analyzing anatomical structures where disruptions can signal health issues, making accurate modeling essential [39]. Skeletonization that effectively captures geometry and topology can enhance the modeling of such structures.

Unfortunately, most established skeletonization methods cannot be trivially adapted to be differentiable, reducing the available algorithms to a select few based on iterative boundary thinning. Inspired by morphological operations, Shit et al. introduced a fast, differentiable skeletonization algorithm based on iterative applications of erosion and dilation operations [33]. However, this approach, which we refer to as the Morphological method, frequently produces breakages in skeletons. In response, a skeletonization algorithm, which we refer to as the Boolean method, was developed that guarantees topological correctness using Boolean predicate kernels for simple point detection at a higher computational cost [25].

The advances in convolutional neural networks (CNNs) have also produced a class of learned approaches that frame skeletonization as a segmentation problem [27,31]. These methods have demonstrated promise in domain-specific challenges [9] supported by advances in semantic segmentation, often leveraging the U-Net architecture [32]. While inherently compatible with backpropagation, these solutions are susceptible to domain shifts and impose resolution constraints on the inputs, complicating their use in downstream tasks where geometric conditions vary.

Neural network solutions follow the trend of increasing parameters for improved performance. However, traditional iterative skeletonization algorithms typically rely on a small set of predefined neighborhood patterns, applied repeatedly, to achieve accurate skeletonization [3,19,40].

Based on this intuition, we propose a novel approach for skeletonization that combines iterative skeletonization and neural networks. This scheme, which we call **Skelite**, requiring only a few learnable convolutional filters coupled with ReLU activations, max-pooling, and matrix additions, which makes it fully compatible with gradient-based optimization.

Our proposed method needs training only on synthetic data, showcasing robust generalization in curvilinear tasks across diverse applications without requiring domain-specific fine-tuning. Unlike other neural network algorithms, our approach allows access to intermediate states of the skeletonization, making the process more transparent and explainable. Our experiments demonstrate that Skelite captures features that improve the connectivity of the resulting skeletons compared to the morphological approach. We evaluate the proposed algorithm's efficacy in the downstream task of segmentation and observe improved performance with minimal computational overhead. This efficiency makes our algorithm more accessible than other methods explicitly addressing topological errors.

To summarize, the contributions of our work include:

- An iterative skeletonization algorithm with a learnable component that is compatible with gradient-based optimization.
- A training framework for the proposed approach, including a synthetic training dataset, task-specific augmentation, and a model distillation strategy.
- Extensive experiments on 2D and 3D, showing computational efficiency and generalization to unseen curvilinear datasets for generating thin skeletons (See Fig. 1).
- Showcasing improvements on four datasets in preserving continuity when integrated into a segmentation pipeline.

2 Method

Given an image I and a target skeleton S, we aim to find a function f_θ, parameterized with a convolutional neural network, that approximates the skeleton in N steps.

$$S^0 = I$$
$$S^{n+1} = f_\theta(S^n) \qquad (1)$$
$$S \approx S^N$$

Following this scheme, Algorithm 1 describes our skeletonization method, where the operator $\odot$ denotes element-wise multiplication. On a high level, each iteration consists of three steps:

1. **Deletion proposal:** We obtain a proposal of points for deletion through morphological boundary extraction. An eroded input is subtracted from the original input to obtain the boundary.

Algorithm 1: Skeletonization Procedure. f_θ is our skeletonization network, I is the input image, and N is the number of iterations.

Input: f_θ, I, N
1. $I \leftarrow \text{binarize}(I)$
2. $S \leftarrow I$
3. **for** $i \leftarrow 1$ **to** N **do**
4. $\quad eroded \leftarrow \text{erode}(\text{erode}(I))$
5. $\quad boundary \leftarrow I - eroded$
6. $\quad delta \leftarrow f_\theta(I, boundary, S)$
7. $\quad delta \leftarrow delta \odot boundary$
8. $\quad S \leftarrow S - delta$
9. $\quad I \leftarrow eroded$

Output: S

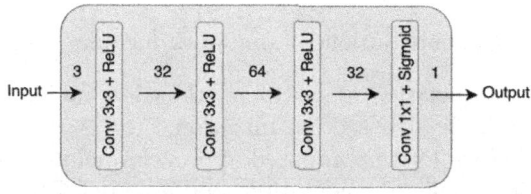

Fig. 2. Our neural network skeletonization module. The arrows indicate the number of output channels after each layer.

2. **Proposal Evaluation**: The network is fed the original input, boundary and current skeleton to evaluate which points can be deleted, producing a delta mask.
3. **Skeleton and Image Update**: The skeleton is updated by subtracting the delta mask, and the input is updated with the eroded input for the next iteration.

Erosion is performed with max-pooling, as described in [33], ensuring that all operations remain differentiable. At each iteration, the CNN receives the image boundary as the deletion target and two contextual images to guide predictions. The first contextual input is the eroded image, which reveals points not currently considered for deletion to encourage the network to remove points in a way that avoids future discontinuities. The second input is the current skeletal representation, helping the network avoid deleting points that would disrupt continuity. Using two erosion operations per pass, instead of one, accelerates the algorithm by processing more points at once. The optimal number of iterations N depends on the maximum radius present in the dataset. While choosing a large N does not affect the quality of the skeleton, it increases computational cost. Conversely, choosing a small N may result in insufficient thinning, leading to an incomplete skeleton.

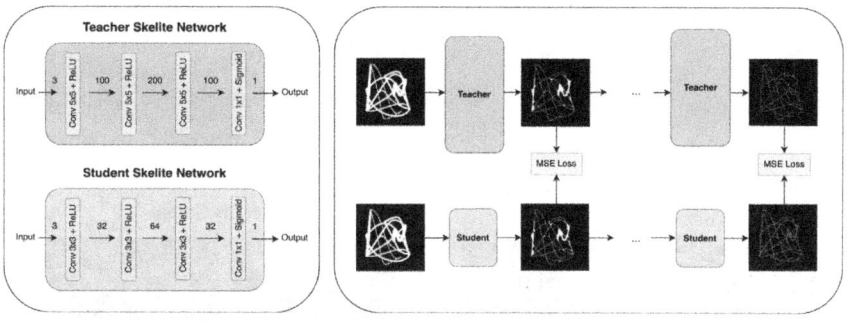

Fig. 3. Example of knowledge distillation strategy. The student network can receive very tight supervision at each skeletonization step from the teacher network.

The method discussed is designed for binary images; however, in many practical applications, the inputs can include matrices of continuous values. To handle these continuous inputs, we adopt the stochastic discretization technique described by Menten et al. [25], denoted as 'binarize' in line 1 of Algorithm 1. This technique enables binarization while preserving gradient flow through the operation with a straight-through estimator, facilitating optimization despite the discrete nature of binarization.

Figure 2 illustrates the architecture of our lightweight convolutional network. It consists of 3×3 convolutional layers, ReLU activations, and a sigmoid activation after the last layer. It does not include downsampling and uses stride-1 convolutions throughout, allowing it to work with inputs of any resolution.

2.1 Model Supervision and Compression

Skeletal data is inherently sparse, with skeletal points constituting less than 5% of the total in our experimental data. This makes classifying skeletal points a highly imbalanced task. Previous works on learnable skeletonization adopt a combination of the focal loss [21] and Dice loss to address the issue [27,31]. However, we find these losses can lead to insufficient thinning in the skeletons.

To address this, we introduce a **neighborhood loss** $L_{neighborhood}$. This loss uses an all ones $m \times m$ (or $m \times m \times m$ in 3D) convolutional kernel K. Given a predicted skeleton S_P and ground truth skeleton S_T, we apply the kernel to obtain the representations $S_{P_K} = S_P * K$ and $S_{T_K} = S_K * K$. We utilize the mean absolute error to obtain $L_{neighborhood} = \text{MAE}(S_{P_K}, S_{T_K})$. We use $m = 5$ in our experiments.

The loss is empirically designed to utilize the sparse neighborhoods of skeletons, encouraging thinner predictions by ensuring similarity between the local neighborhoods of the predicted and ground truth skeletons. To avoid overthinning, we integrate the focal loss L_{focal} and the Dice loss L_{Dice} to account for overlap to obtain the loss on the final outputs of our algorithm.

$$L = L_{focal} + L_{Dice} + L_{neighborhood} \qquad (2)$$

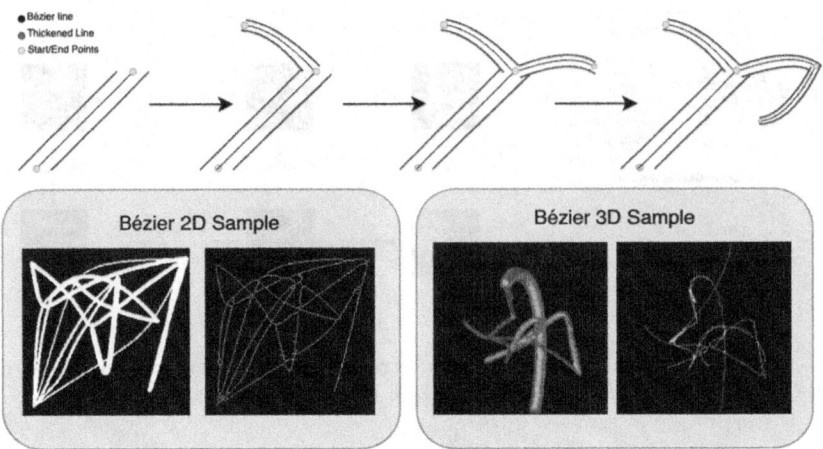

Fig. 4. Top: Illustration of Bézier dataset generation. Bottom: Examples in 2D and 3D.

In our experiments, we observe that large Skelite models are more effective at learning skeletonization from data. We aim to retain this performance in smaller, faster models. Drawing inspiration from knowledge distillation [6,12], we propose a method to transfer the performance of a Skelite model into a more compact version. As shown in Fig. 3, our distillation strategy takes advantage of the iterative nature of the skeletonization process, providing supervision at each step. This fine-grained supervision enables the smaller model to effectively learn skeletonization by leveraging the guidance from each stage of the process of a similarly structured algorithm.

2.2 Synthetic Data: Bézier Dataset

We wish to learn skeletonization from data that mimics the appearance and characteristics of vasculature or other curvilinear structures. To this end, we propose a procedural method to generate synthetic data using cubic Bézier curves, described by the formula:

$$B(t) = (1-t)^3 P_0 + 3(1-t)^2 t P_1 + 3(1-t)t^2 P_2 + t^3 P_3, \ t \in [0,1] \quad (3)$$

where P_0 is the starting point of the curve, P_3 is the end point, P_1 and P_2 are control points and t is the interpolation parameter. The process is detailed in Fig. 4.

The samples are initialized with a set of primary Bézier curves, referred to as trunks, whose control points are randomly sampled within the image boundaries. The thickness of these trunks is uniformly sampled from the integer range $[w_{trunk}, W_{trunk}]$. From the endpoints of these trunks, a branching process begins, in which a number of Bézier curves are sampled to grow from this point in the range $[n_{branches}, N_{branches}]$. The thickness of these branching curves is sampled similarly to the trunks but is constrained such that their thickness cannot exceed that of their parent curve. To simulate hierarchical growth, the process is repeated recursively, with a pre-defined recursion depth serving as the stopping condition for the generation process. We obtain the skeleton labels for the training data by leveraging the Boolean skeletonization algorithm for its topological correctness.

The generation process approximates patterns commonly observed in curvilinear structures, such as branching and smooth curved geometries. We use the same synthetic data generation strategy in 2D and 3D, and generate 5000 samples for 2D and 20000 samples for 3D which we will make publicly available. To further increase the diversity of the data, we introduce a complementary augmentation.

2.3 Thickening Augmentation

Complementary to the diversity of the data generation process, we produce a thickening augmentation to enhance the geometric variety of samples during training. Thickening can be derived as a byproduct of background thinning [4]. However, in our approach we leverage the availability of paired reference images and skeletons for a straightforward process that ensures topology preservation. Specifically, given a reference image I and its corresponding skeleton S, we can obtain intermediate thickened representations S_{thick} through dilation. The process is outlined by the following equation.

$$S_{\text{thick}} = \text{dilate}^k(S) \odot I \qquad (4)$$

where we let $\text{dilate}^k(S)$ represent the result of applying dilation k times to the skeleton. Thus, the thickening is controlled by the number of dilation applications, which is uniformly sampled from the integers between zero and the maximum thickness from the generated samples. Here, the multiplication with the reference image I ensures that the thickened skeleton remains consistent with the original topology and geometry of the object. When $k = 0$, the augmentation returns the skeleton as is, presenting the challenge of skeleton preservation.

3 Experiments and Results

We validate our skeletonization method in three ways. First, we explore skeletonization results in the synthetic data domain and evaluate its transition to real-world data compared to other neural network methods. Next, we assess its topological correctness, geometry, computational efficiency, and overlap with a baseline skeleton obtained using algorithms by Zhang et al. (2D) [40] and Lee et al. (3D) [19], as implemented in the widely-used scikit-image library [36]. Finally, we demonstrate its practical applicability by integrating it with clDice for deep learning-based binary segmentation.

3.1 Datasets

Our experiment data comprises a synthetic dataset and four publicly available datasets containing thin structures.

- Bézier consists of 5000 samples in 2D and 20000 samples in 3D. We set the parameters from Sect. 2.2 to $w_{trunk} = 3$, $W_{trunk} = 10$, $n_{branches} = 1$, $N_{branches} = 3$.
- DRIVE [34] (Digital Retinal Images for Vessel Extraction) consists of 40 2D color fundus images annotated with the visible retinal vessels.
- ROADS [26] (Massachussets Roads) is a dataset for studying the segmentation of roads from aerial images. It consists of 1171 images with annotations.
- ASOCA [11] (Automated Segmentation of Coronary Arteries) consists of 40 3D cardiac CTA scans depicting 20 normal and 20 diseased coronary arteries.
- TopCoW [38] (Topology-Aware Anatomical Segmentation of the Circle of Willis) is a 3D vessel segmentation challenge containing 250 samples of CTA and MRA data depicting the Circle of Willis vessels.

3.2 Evaluation Metrics

We use a diverse set of metrics to benchmark our method and evaluate performance in segmentation, where we consider overlap, connectivity, and topology. For overlap, we use the Dice similarity coefficient, supplemented by the clDice metric to assess connectivity in tubular structure segmentation [33]. We feed the non-differentiable skeletonization results from `scikit-image` to these metrics in our benchmark. Finally, we assess the topological correctness of our method using the Betti number errors β_0 and β_1.

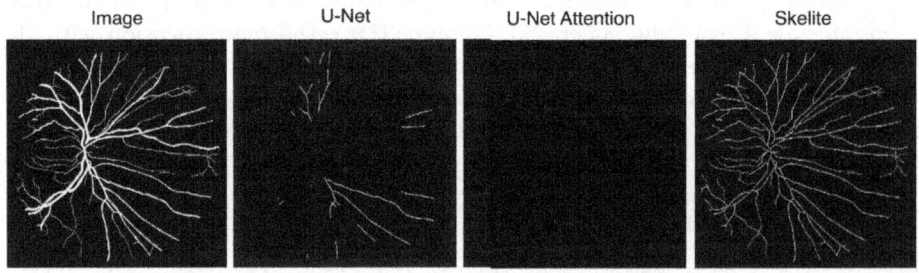

Fig. 5. A Comparison of Bézier trained neural network skeletonization tested on DRIVE. U-Net methods overfit to Bézier-specific features and result in over-thinning. Skelite's controlled iterative thinning addresses this issue, permitting generalization to new domains.

Inspired by previous works on skeleton evaluation, we extend our analysis with proxy metrics to quantify skeleton thickness. [17, 18, 28]. Specifically, we derive two metrics from the Euclidean distance transform, which replaces each foreground pixel in an image with its shortest Euclidean distance to the background. Using this representation, we assess the thinness of the resulting skeletons by taking the 99th percentile and average of the distances. We refer to these metrics as '99th percentile thickness' and 'average thickness' respectively.

3.3 Transitioning from Synthetic to Real World Data

In this experiment, we evaluate the performance of our method on the synthetic dataset and assess its generalizability to unseen real-world data after training on the Bézier dataset. We compare our approach with two other neural network skeletonization architectures, all trained using the same dataset and augmentation strategy, and then tested on DRIVE:

- U-Net: Proposed by Panichev et al. [31], this method leverages a U-Net architecture for skeleton extraction.
- U-Net Attention: Introduced by Nguyen [27], this approach includes the attention mechanism in the U-Net for feature extraction [37].

Table 1. We test the generalization of neural network methods for skeleton extraction by training on synthetic data and testing on real-world data.

Dataset	Method	β_0 error	β_1 error	Avg. Thickness	99th Percentile Thickness	Dice	Run time [ms]
2D Bézier	U-Net	0.07	2.92	1.07	3.77	0.72	43
	U-Net Attention	0.02	4.28	1.03	2.35	0.75	31
	Skelite	1.02	2.35	1.00	1.93	0.76	16
DRIVE	U-Net	24.15	52.15	1.08	1.88	0.26	287
	U-Net Attention	3.40	57.10	0.20	0.20	0.00	558
	Skelite	2.80	3.20	1.00	1.93	0.78	8

Table 1 shows that in the Bézier dataset, our method produces thin skeletons with good topological agreement. The U-Net based approaches also exhibit few Betti errors, but prune less points producing thicker skeletons.

When transitioning to real-world data, U-Net based algorithms can produce over-thinning due to the unseen geometric conditions. This limitation is reflected in their low overlap scores and increased Betti errors. Visual analysis presented in Fig. 5 confirms this. Conversely, Skelite demonstrates robust performance, maintaining results comparable to those on the synthetic data.

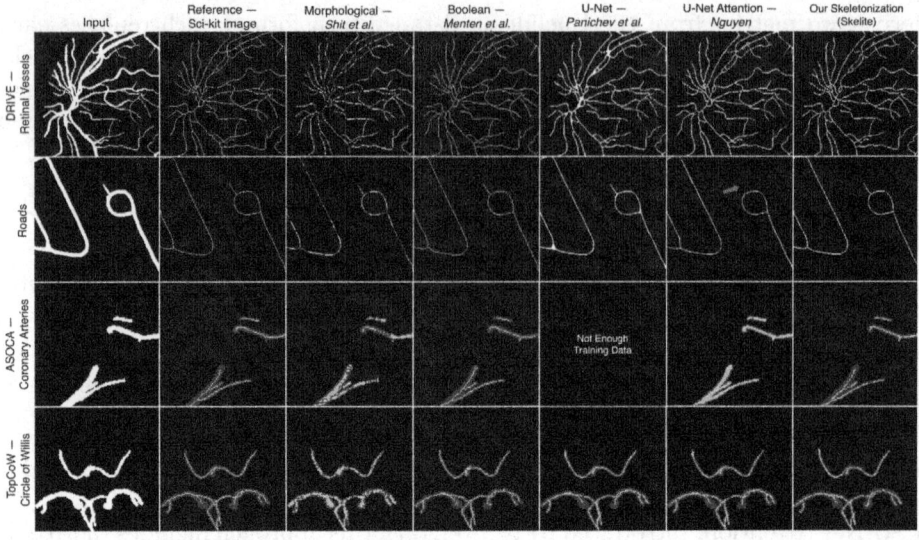

Fig. 6. The result of six different skeletonization algorithms. Despite not being trained on these datasets, our method effectively produces thin skeletons in both 2D and 3D.

3.4 Benchmarking Skeletonization Methods

We compare our skeletonization algorithm to four baselines for differentiable skeletonization: the Morphological method by Shit *et al.*, the Boolean method by Menten *et al.*, and the two U-Net skeletonization algorithms from Panichev *et al.* [31] (denoted as U-Net) and Nguyen [27] (denoted as U-Net Attention) trained on *task-specific data*.

Results in Fig. 6 highlight the robustness of our method. Despite not being exposed to domain samples before testing, our approach generates thin skeletons that closely align with the reference across the diverse datasets. Skelite preserves topology reasonably well with only a few disconnections. This performance is comparable to U-Net based methods, which, unlike ours, have in-domain exposure and significantly more parameters. Even then, the U-Net methods tend to overfit to specific structures in the data so that they miss entire details, as highlighted by the arrow in the ROADS dataset. Furthermore, U-Net methods struggle in low-data domains like ASOCA, producing thick skeletons or failing to converge all together. Morphology-based skeletonization, on the other hand, generates many disconnected points along the medial axis. In 3D, this method exhibits the same disconnections and, in addition, fails to produce thin skeletons.

We support our observations with quantitative data, presented in Table 2. A key highlight is the computational efficiency of our method compared to existing approaches, particularly the Boolean and U-Net methods. While the Boolean method offers topological guarantees, it is computationally expensive. In contrast, our method computes results, on average, two orders of magnitude faster, making it significantly more practical for downstream applications.

Although our method does not explicitly enforce topological guarantees, it achieves a similar alignment to the reference skeleton as the Boolean method, as reflected by the comparable Dice scores. In contrast to the faster Morphological skeletonization, our approach produces significantly thinner and more topologically consistent skeletal representations in 3D. The thickness metrics and Betti errors further emphasize that our method generates skeletons that are both thin and largely connected across various tasks.

3.5 Connectivity-Aware Segmentation

Connectivity-aware and topology-aware segmentation methods have gained significant attention, leveraging approaches such as Persistent Homology [13,35], component graphs [2,24], and skeleton-based techniques [16,33] which relate to our method. We integrate our skeletonization network into clDice for connectivity-aware segmentation, leveraging the stochastic discretization technique from Sect. 2 to handle continuous values in segmentation outputs. For the 2D tasks, we used a vanilla U-Net, and in 3D, we used nnU-Net [14]. We compare our results against a baseline trained solely on Dice loss and models incorporating clDice with skeletal supervision via the Morphological and Boolean methods. Additionally, we include the *Skeleton Recall* loss to incorporate a recent advance that promotes connectivity preservation using non-differentiable skeletonization [16]. The segmentation results are provided in Table 3.

Table 2. Quantitative comparison of topological accuracy, structural properties, and run time of differentiable skeletonization algorithms for deep learning on four datasets.

Dataset	Method	β_0 error	β_1 error	Avg. Thickness	99th Percentile Thickness	Dice	Run time [ms]
DRIVE	Morphological	1014.00	55.10	1.00	1.00	0.75	1.1
	Boolean	0.00	0.00	1.00	1.40	0.74	827.0
	U-Net Panichev	0.55	1.80	1.00	2.98	0.77	51.0
	U-Net Attention	0.50	1.45	1.00	2.18	0.78	400.0
	Ours - Skelite	2.80	3.20	1.00	1.93	0.78	8
ROADS	Morphological	686.76	37.69	1.00	1.22	0.76	1.2
	Boolean	0.00	0.00	1.00	1.00	0.71	2468.0
	U-Net Panichev	0.14	66.10	1.05	4.23	0.72	263.0
	U-Net Attention	5.71	3.71	1.00	1.70	0.88	272.0
	Ours - Skelite	0.73	7.07	1.00	2.16	0.72	15.5
ASOCA	Morphological	19.30	0.20	1.00	1.00	0.29	2.0
	Boolean	0.00	0.00	1.00	1.00	0.40	221.0
	U-Net Panichev	Insufficient Training Data					
	U-Net Attention	0.21	0.05	1.15	1.60	0.21	1269.0
	Ours - Skelite	4.40	0.20	1.00	1.00	0.45	7.0
TOPCOW	Morphological	61.40	0.80	1.00	1.00	0.24	2.0
	Boolean	0.00	0.00	1.00	1.00	0.40	376.0
	U-Net Panichev	6.60	1.20	1.00	1.00	0.36	3253.0
	U-Net Attention	9.00	0.70	1.00	1.00	0.60	2081.0
	Ours - Skelite	19.30	0.90	1.00	1.00	0.43	70.0

Our method performs similarly to the Boolean method in supervising segmentation but with significantly lower computational overhead, closely matching the speed of the Morphological method. Furthermore, our method shows substantial enhancement of connectivity in 3D tasks, showing the lowest β_0 errors, and across 2D and 3D experiments, it demonstrates the highest clDice scores. Our results in terms of spatial accuracy, as shown by the Dice scores, mostly match the ones obtained through other skeletonization algorithms.

3.6 Ablation Studies

In this section, we present ablation studies to evaluate different components of the framework used to train Skelite. Table 4 presents the effect of using the neighborhood loss, which encourages thinner outputs in both seen and unseen domains. Additionally, we evaluate the impact of modifying the input combinations to the CNN module of Skelite. As shown in Table 5, incorporating the skeleton and image reduce disconnections in the outputs and improve topological agreement, as emphasized by the reduced Betti errors.

Table 3. Segmentation performance with Dice loss only (Without) or using skeletal supervision.

Dataset	Method	Dice	clDice	β_0 error	β_1 error	Epoch Time (s)
DRIVE	Without	0.77	0.84	100.00	13.00	0.09
	Morphological	0.82	0.86	45.60	15.00	0.18
	Boolean	0.83	0.86	50.00	13.00	12.00
	Skel Recall	0.63	0.85	48.30	14.80	0.14
	Skelite	0.83	0.86	53.00	14.00	0.17
ROADS	Without (Dice Loss)	0.66	0.83	55.76	21.20	1.57
	Morphological	0.73	0.84	54.4	23.12	3.88
	Boolean	0.74	0.84	37.2	23.20	174.23
	Skel Recall	0.61	0.84	62.80	20.61	4.71
	Skelite	0.75	0.85	46.76	28.96	4.29
ASOCA	Without	0.81	0.79	24.41	1.54	62.45
	Morphological	0.79	0.80	25.29	1.67	66.12
	Boolean	0.78	0.79	27.09	1.63	351.67
	Skeleton Recall	0.72	0.70	33.84	1.96	77.57
	Skelite	0.79	0.80	22.28	1.67	81.50
TOPCOW	Without	0.87	0.93	1.53	1.18	57.70
	Morphological	0.87	0.93	1.49	1.09	62.90
	Boolean	0.86	0.93	1.51	0.91	322.20
	Skeleton Recall	0.83	0.92	1.10	1.56	75.20
	Skelite	0.84	0.95	1.03	1.36	77.80

Table 4. Ablation study on neighborhood loss for skeletonization learning. Adding this term encourages thinner outputs.

Dataset	Loss	β_0 error	β_1 error	Avg. Thickness	99th Percentile Thickness	Dice
2D Bézier	Without	15.36	20.76	1.44	6.13	0.54
	+ Neighborhood Loss	3.57	11.46	1.11	3.35	0.69
DRIVE	Without	4.95	10.15	1.16	3.54	0.66
	+ Neighborhood Loss	7.55	6.85	1.09	3.03	0.71

Table 5. Ablation study on network inputs: boundary, image, and skeleton. Adding the image and skeleton as context improves the connectedness of predictions.

Dataset	Input	β_0 error	β_1 error	Avg. Thickness	99th Percentile Thickness	Dice
2D Bézier	Bound Only	32.34	22.05	1.03	2.28	0.72
	Bound + Image	3.70	9.96	1.02	2.24	0.74
	Image + Bound + Skeleton	2.61	9.48	1.10	3.10	0.70
Drive	Bound Only	96.85	26.2	1.02	2.38	0.75
	Bound + Image	5.85	4.49	1.01	2.27	0.76
	Bound + Image + Skeleton	4.20	4.45	1.07	2.63	0.72

4 Conclusion

We introduce Skelite, a work that bridges the gap between neural networks and iterative skeletonization. We have shown with an extensive set of experiments that Skelite is (a) computationally efficient, (b) learns to produce thin, connected skeletons, and (c) generalizes across domains without retraining.

Skelite abandons the path of explicit topological guarantees of the Boolean method and instead learns to produce skeletal representations from data, obtaining a significant speed-up. The obtained skeletons demonstrate good connectivity and thinness, unlike Morphology-based skeletonization, which presents frequent disconnections and poor thinning in 3D. Finally, whereas U-Net-based methods require retraining for each domain, our approach performs strongly on unseen real-world scenarios despite being trained exclusively on synthetic data. This property is a considerable advantage for domains lacking enough training data.

To the best of our knowledge, this work introduces the first learnable iterative skeletonization method. Future works could research the outcomes of using synthetic datasets emphasizing other properties such as discontinuous and jagged curves. Moreover, the proposed method could benefit from a class of data augmentations that do not introduce aliasing artifacts [7], which can disconnect skeleton labels. Extensions to the network architecture or point proposal system could also be explored.

Acknowledgements. This research was partially supported through the research grant between Brainlab AG and Technical University of Munich. Martin J. Menten is funded by the German Research Foundation under project 532139938.

References

1. Acebes, C., Moustafa, A.H., Camara, O., Galdran, A.: The centerline-cross entropy loss for vessel-like structure segmentation: better topology consistency without sacrificing accuracy. In: International Conference on Medical Image Computing and Computer-Assisted Intervention (2024). https://api.semanticscholar.org/CorpusID:273374503
2. Berger, A.H., Lux, L., Weers, A., Menten, M., Rueckert, D., Paetzold, J.C.: Pitfalls of topology-aware image segmentation. arXiv preprint arXiv:2412.14619 (2024)
3. Bertrand, G.: A boolean characterization of three-dimensional simple points. Pattern Recogn. Lett. **17**(2), 115–124 (1996). https://doi.org/10.1016/0167-8655(95)00100-X. https://www.sciencedirect.com/science/article/pii/016786559500100X
4. Bloomberg, D.S.: Connectivity-preserving morphological image transformations. In: Other Conferences (1991). https://api.semanticscholar.org/CorpusID:122767000
5. Blum, H.: A transformation for extracting new descriptors of shape, pp. 362–380. Models for the Perception of Speech and Visual Form, Cambridge (1967)
6. Bucila, C., Caruana, R., Niculescu-Mizil, A.: Model compression. In: Knowledge Discovery and Data Mining (2006). https://api.semanticscholar.org/CorpusID:11253972
7. Chen, C.C., Peng, C.H.: Topology-preserving downsampling of binary images. arXiv abs/2407.17786 (2024). https://api.semanticscholar.org/CorpusID:271431891
8. Couprie, M., Bertrand, G.: Asymmetric parallel 3D thinning scheme and algorithms based on isthmuses. Pattern Recognit. Lett. **76** (2015). https://doi.org/10.1016/j.patrec.2015.03.014
9. Demir, I., et al.: Skelneton 2019: dataset and challenge on deep learning for geometric shape understanding. In: Proceedings of the IEEE Conference on Computer Vision and Pattern Recognition Workshops (2019)
10. Fridman, Y., Pizer, S.M., Aylward, S.R., Bullitt, E.: Extracting branching tubular object geometry via cores. Med. Image Anal. **8**(3), 169–76 (2004). https://api.semanticscholar.org/CorpusID:469634
11. Gharleghi, R., et al.: Automated segmentation of normal and diseased coronary arteries - the asoca challenge. Comput. Med. Imaging Graph. **97**, 102049 (2022). https://api.semanticscholar.org/CorpusID:246988905
12. Hinton, G.E., Vinyals, O., Dean, J.: Distilling the knowledge in a neural network. arXiv abs/1503.02531 (2015). https://api.semanticscholar.org/CorpusID:7200347
13. Hu, X., Li, F., Samaras, D., Chen, C.: Topology-preserving deep image segmentation. CoRR abs/1906.05404 (2019). http://arxiv.org/abs/1906.05404
14. Isensee, F., Jaeger, P.F., Kohl, S.A.A., Petersen, J., Maier-Hein, K.: nnu-net: a self-configuring method for deep learning-based biomedical image segmentation. Nat. Methods **18**, 203–211 (2020). https://api.semanticscholar.org/CorpusID:227947847
15. Kim, H.C., Min, B.G., Lee, M.M., Seo, J.D., Lee, Y.W., Han, M.C.: Estimation of local cardiac wall deformation and regional wall stress from biplane coronary cineangiograms. IEEE Trans. Biomed. Eng. **BME-32**(7), 503–512 (1985). https://doi.org/10.1109/TBME.1985.325567
16. Kirchhoff, Y., et al.: Skeleton recall loss for connectivity conserving and resource efficient segmentation of thin tubular structures (2024). https://arxiv.org/abs/2404.03010

17. Kruszy, K., van Liere, R., Kaandorp, J.: Quantifying differences in skeletonization algorithms: a case study. In: Proceedings of the 5th IASTED International Conference on Visualization, Imaging, and Image Processing, VIIP 2005 (2005)
18. Lee, S.W., Lam, L., Suen, C.Y.: A systematic evaluation of skeletonization algorithms. Int. J. Pattern Recognit. Artif. Intell. **7**, 1203–1225 (1993). https://api.semanticscholar.org/CorpusID:40522276
19. Lee, T.C., Kashyap, R.L., Chu, C.N.: Building skeleton models via 3-D medial surface/axis thinning algorithms. CVGIP Graph. Model. Image Process. **56**, 462–478 (1994). https://api.semanticscholar.org/CorpusID:35388240
20. Lidayová, K., Frimmel, H., Wang, C., Bengtsson, E., Smedby, Ö.: Skeleton-based fast, fully automated generation of vessel tree structure for clinical evaluation of blood vessel systems (2017). https://api.semanticscholar.org/CorpusID:113501346
21. Lin, T.Y., Goyal, P., Girshick, R., He, K., Dollár, P.: Focal loss for dense object detection (2018). https://arxiv.org/abs/1708.02002
22. Liu, J.J., Hou, Q., Cheng, M.M.: Dynamic feature integration for simultaneous detection of salient object, edge, and skeleton. IEEE Trans. Image Process. **29**, 8652–8667 (2020). https://doi.org/10.1109/TIP.2020.3017352
23. Lobregt, S., Verbeek, P.W., Groen, F.C.A.: Three-dimensional skeletonization: principle and algorithm. IEEE Trans. Pattern Anal. Mach. Intell. **PAMI-2**, 75–77 (1980). https://api.semanticscholar.org/CorpusID:19014532
24. Lux, L., et al.: Topograph: an efficient graph-based framework for strictly topology preserving image segmentation. In: The Thirteenth International Conference on Learning Representations (2025). https://openreview.net/forum?id=Q0zmmNNePz
25. Menten, M.J., et al.: A skeletonization algorithm for gradient-based optimization (2023). https://arxiv.org/abs/2309.02527
26. Mnih, V.: Machine learning for aerial image labeling. Ph.D. thesis, University of Toronto (2013)
27. Nguyen, N.: U-net based skeletonization and bag of tricks, pp. 2105–2109 (2021). https://doi.org/10.1109/ICCVW54120.2021.00238
28. Németh, G., Kovács, G., Fazekas, A., Palagyi, K.: A method for quantitative comparison of 2D skeletons. Acta Polytechnica Hungarica **13**, 123–142 (2016)
29. Palágyi, K., et al.: A sequential 3D thinning algorithm and its medical applications. In: Insana, M.F., Leahy, R.M. (eds.) IPMI 2001. LNCS, vol. 2082, pp. 409–415. Springer, Heidelberg (2001). https://doi.org/10.1007/3-540-45729-1_42
30. Palágyi, K., Kuba, A.: A parallel 3D 12-subiteration thinning algorithm. Graph. Model. Image Process. **61**, 199–221 (1999). https://api.semanticscholar.org/CorpusID:14259397
31. Panichev, O., Voloshyna, A.: U-net based convolutional neural network for skeleton extraction, pp. 1186–1189 (2019). https://doi.org/10.1109/CVPRW.2019.00157
32. Ronneberger, O., Fischer, P., Brox, T.: U-net: convolutional networks for biomedical image segmentation. CoRR abs/1505.04597 (2015). http://arxiv.org/abs/1505.04597
33. Shit, S., et al.: cldice - a topology-preserving loss function for tubular structure segmentation. CoRR abs/2003.07311 (2020). https://arxiv.org/abs/2003.07311
34. Staal, J., Abràmoff, M.D., Niemeijer, M., Viergever, M.A., Van Ginneken, B.: Ridge-based vessel segmentation in color images of the retina. IEEE Trans. Med. Imaging **23**(4), 501–509 (2004)
35. Stucki, N., Paetzold, J.C., Shit, S., Menze, B., Bauer, U.: Topologically faithful image segmentation via induced matching of persistence barcodes (2022). https://arxiv.org/abs/2211.15272

36. Van der Walt, S., et al.: scikit-image: image processing in python. PeerJ **2**, e453 (2014)
37. Woo, S., Park, J., Lee, J.Y., Kweon, I.S.: CBAM: convolutional block attention module (2018). https://arxiv.org/abs/1807.06521
38. Yang, K., et al.: Benchmarking the cow with the topcow challenge: topology-aware anatomical segmentation of the circle of willis for CTA and MRA. arXiv (2023). https://api.semanticscholar.org/CorpusID:267050106
39. Yeganeh, Y., et al.: Scope: structural continuity preservation for medical image segmentation (2023). https://arxiv.org/abs/2304.14572
40. Zhang, T.Y., Suen, C.Y.: A fast parallel algorithm for thinning digital patterns. Commun. ACM **27**, 236–239 (1984). https://api.semanticscholar.org/CorpusID:39713481

VerSe: Integrating Multiple Queries as Prompts for Versatile Cardiac MRI Segmentation

Bangwei Guo[1(✉)], Meng Ye[1], Yunhe Gao[1], Bingyu Xin[1], Leon Axel[2], and Dimitris Metaxas[1]

[1] Rutgers University, New Brunswick, USA
bg654@rutgers.edu, dnm@cs.rutgers.edu
[2] New York University School of Medicine, New York, USA

Abstract. Despite the advances in learning-based image segmentation approach, the accurate segmentation of cardiac structures from magnetic resonance imaging (MRI) remains a critical challenge. While existing automatic segmentation methods have shown promise, they still require extensive manual corrections of the segmentation results by human experts, particularly in complex regions such as the basal and apical parts of the heart. Recent efforts have been made on developing interactive image segmentation methods that enable human-in-the-loop learning. However, they are semi-automatic and inefficient, due to their reliance on click-based prompts, especially for 3D cardiac MRI volumes. To address these limitations, we propose VerSe, a Versatile Segmentation framework to unify automatic and interactive segmentation through mutiple queries. Our key innovation lies in the joint learning of object and click queries as prompts for a shared segmentation backbone. VerSe supports both fully automatic segmentation, through object queries, and interactive mask refinement, by providing click queries when needed. With the proposed integrated prompting scheme, VerSe demonstrates significant improvement in performance and efficiency over existing methods, on both cardiac MRI and out-of-distribution medical imaging datasets.

Keywords: Cardiac MRI · Mutiple Prompts · Versatile Segmentation

1 Introduction

Cardiac magnetic resonance imaging (MRI) can provide comprehensive information for heart disease diagnosis and treatment [30], serving as the gold standard for various clinical applications. For example, cardiac cine MRI [3] enables precise evaluation of cardiac function, while cardiac late gadolinium enhancement (LGE) MRI [32] excels in detecting myocardium infarction (MI) and assessing tissue viability. Despite these advantages, the widespread clinical adoption of cardiac MRI lags behind echo cardiography and computational tomography (CT), in large part due to the challenges in image post-processing, *e.g.*, the accurate segmentation of anatomical structures and lesions.

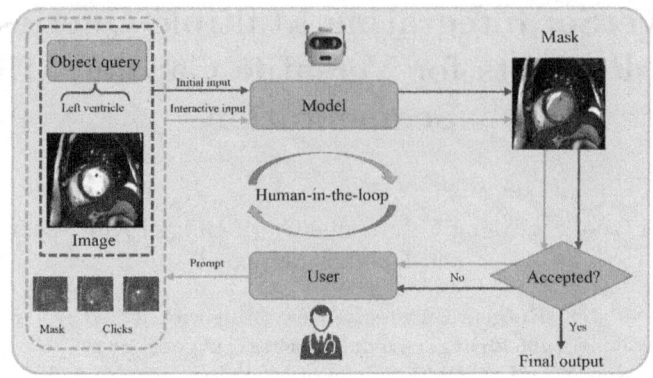

Fig. 1. Illustration of the proposed versatile segmentation framework. Our model accepts an object query, such as left ventricle, to automatically segment the target in the image. If the initial segmentation is unsatisfactory, users can refine the mask by providing corrective clicks until the final output mask can meet clinical accuracy.

Deep learning-based methods have significantly improved automatic cardiac MRI segmentation. U-Net [33] and its variants [15,17,29] remain the most widely used architecture and perform well on both 2D and 3D images, but struggle with intricate object boundaries and complex structures [3]. Vision Transformer (ViT)-based models [7,12,37] effectively capture long-range dependencies but demand extensive datasets for optimal performance. Hybrid models, like UTNet [15] and MedFormer [14] enhance segmentation performance by combining convolutional neural networks (CNNs) and transformers. Despite these advances, current methods still fail to meet clinical precision requirements, particularly in challenging basal and apical regions [4]. These limitations often necessitate time-consuming manual corrections by experts, highlighting the gap between automated methods and clinical precision, especially for large-scale applications.

To address the limitations of fully automatic segmentation approaches, interactive image segmentation methods have been proposed to facilitate efficient segmentation and annotation [27,34,35]. These methods commonly rely on user clicks to guide learning-based models, mimicking human corrections and significantly improving efficiency compared to traditional pixel-by-pixel annotations. Early work [35] first introduced CNN-based interactive segmentation, while RITM [34] enhanced it by integrating mask information from previous iterations. Recently, ViT-based models like SimpleClick [27] and SegNext [26] have achieved state-of-the-art performance by leveraging large-scale ViT architectures and extensive pretraining on natural image datasets. However, these advancements rely heavily on advanced backbones and high-quality data, while their use of point-based prompts remains rudimentary. Notably, no interactive segmentation models have been tailored for cardiac MRIs, despite its pressing clinical demand and heavy reliance on user inputs. This gap highlights the urgent need for more efficient interactive segmentation solutions.

Recent advances in universal image segmentation [8,13] have explored efficient designs that combine learnable queries [11,24,36] with transformer decoders [6]. These works have inspired us to propose a novel image segmentation paradigm that leverages both data priors and human expert intelligence. As illustrated in Fig. 1, we introduce **VerSe**, a **Ver**satile cardiac MRI **Se**gmentation model that integrates automatic and interactive segmentation into a unified framework. Central to our approach is the introduction of multi-query integration, which serves as prompts for a shared segmentation backbone. This design enables our model to operate seamlessly across multiple modes, enhancing its flexibility and adaptability to diverse segmentation tasks.

Our contributions are summarized as follows: (1) We introduce a novel cardiac MRI segmentation paradigm that unifies automatic and interactive segmentation, bridging the gap between learning-based methods and large-scale clinical usage through the fusion of machine and human intelligence. (2) We propose a new image segmentation architecture which can integrate multiply queries, *i.e.*, object query and click query, to prompt a foundational segmentation backbone. (3) We conducted extensive experiments on seven cardiac MRI datasets. The results demonstrate the high accuracy, efficiency, and versatility of our segmentation model. Our method also shows the potential for generalization to other out-of-distribution medical imaging segmentation tasks.

2 Method

2.1 Overview

We propose a novel **Ver**satile image **Se**gmentation model, VerSe, through the integration of object query and click query. An overview of VerSe architecture is shown in Fig. 2. In the following, we give details of the proposed method.

2.2 Multi-query Integration as Prompt

To enable multiple functions within our model, we introduce the **Multi-Query Integration** mechanism, which guides the model to focus on relevant objects. For the automatic segmentation task, instead of using CLIP-based semantic embeddings [25], we employ learnable object queries $\boldsymbol{X}_o$, as they provide stronger task-specific prior knowledge [13]. Each segmentation target is represented by a small group of learnable query vectors. Specifically, for N target objects, we assign N groups of learnable queries $\boldsymbol{X}_o = [\boldsymbol{X}_{o1}, \ldots, \boldsymbol{X}_{oi}, \ldots, \boldsymbol{X}_{oN}]$, where each group $\boldsymbol{X}_{oi} \in \mathbb{R}^{M \times C}$ consists of M query vectors with C channels. These queries are initialized as random parameters and optimized during training.

For the interactive segmentation task based on clicks, inspired by SAM [18], we encode the click locations as sparse positional queries $\boldsymbol{X}_s$. For a positive click set $\boldsymbol{C}_p = \{(x_1, y_1), \ldots, (x_{N_p}, y_{N_p})\}$, we use a Point Encoder to encode each click point to a corresponding positive positional query $\boldsymbol{X}_{sp}$, generating

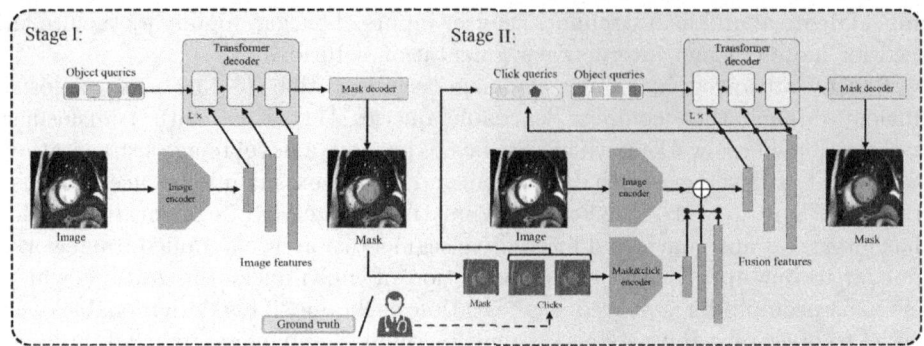

Fig. 2. Overview of the `VerSe` architecture. In stage I, object queries are used to automatically segment a target in the image. In stage II, user provides clicks as prompts to refine the initial segmentation mask. `VerSe` also supports a pure interactive mode, where the initial mask is empty and the object queries aren't activated. The image encoder, transformer decoder and mask decoder are shared across all stages. Implementation details are described in Sect. 2.5.

$N_p \times C$ vectors. The same process applies for the negative click set C_n and the corresponding negative positional queries are denoted as X_{sn}.

To further enhance the effectiveness of click prompts, we propose a novel **Semantic Feature Query** X_f, as illustrated in Fig. 3(a). The image encoder extracts multiple down-scaling features from the input image. For a feature map F_s with shape $(H/s, W/s)$, where s is the down-scaling factor and H, W are the original image height and width, we first map the original click point $P_0 = (x, y)$ to its corresponding coordinates in the down-scaling feature map $P_0' = (x', y')$:

$$x' = \left\lfloor \frac{x}{s} \right\rfloor, \quad y' = \left\lfloor \frac{y}{s} \right\rfloor. \tag{1}$$

We then extract features from a $(2r+1) \times (2r+1)$ window around P_0' and apply average pooling to compute f_{pooled}:

$$f_{\text{pooled}} = \frac{1}{(2r+1)^2} \sum_{i=-r}^{r} \sum_{j=-r}^{r} F_s(x'+i, y'+j). \tag{2}$$

Finally, f_{pooled} is projected into a new feature space via a multilayer perceptron (MLP):

$$X_{f0} = \text{MLP}(f_{\text{pooled}}), \tag{3}$$

where $X_{f0} \in \mathbb{R}^{1 \times C}$ represents the semantic feature query for the click at P_0. Positive and negative click sets generate semantic feature queries X_{fp} and X_{fn}, respectively, at three feature scales $s = 2, 4, 8$.

X_o, X_s, and X_f collaboratively enable the versatile functions of our model. Specifically, click queries X_c are formed by combining sparse positional queries X_s and semantic feature queries X_f. Leveraging multi-query integration, `VerSe` flexibly supports three working modes:

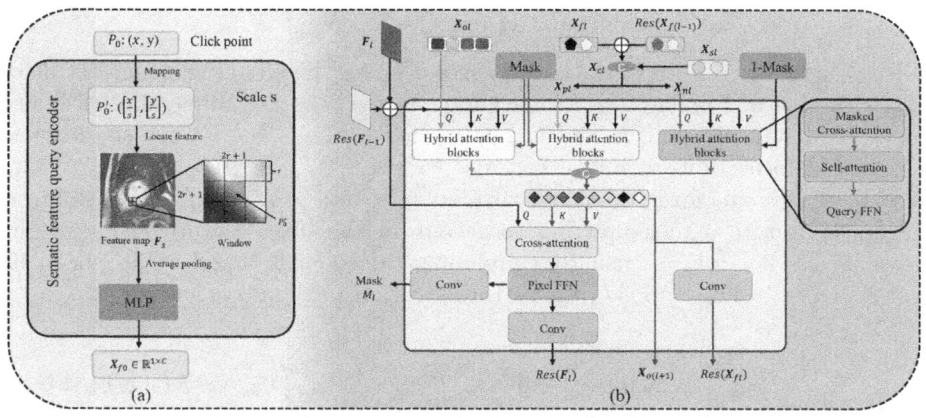

Fig. 3. (a) Process of generating semantic feature query X_{f0} for a specified click point P_0 at scale s. The original click P_0 is mapped to the coordinates P'_0 on the downscaling feature map. A feature patch centered at P'_0 undergoes average pooling, and the resulting feature is transformed via an MLP to produce X_{f0}. (b) Transformer decoder at the l-th layer. Object queries X_{ol}, positive click queries X_{pl} and negative click queries X_{nl} first interact with image features F_l, to capture foreground target and background context. These queries are then concatenated to update the image features. X_{fl}, X_{sl}, and X_{cl} denote semantic feature queries, sparse positional queries, and click queries at the l-th layer, respectively.

- **Mode-1:** Automatic segmentation directed by object queries X_o;
- **Mode-2:** Interactive refinement of the initial segmentation from Mode-1 by combining object queries X_o and click queries X_c;
- **Mode-3**: Purely interactive segmentation guided by user clicks, relying solely on click queries X_c.

2.3 Foreground-Background Masked Attention

After obtaining all queries, we update the image features by attending to the multiple queries. As shown in Fig. 3(b), we use foreground-background masked attention [8,9] to guide multiple queries in focusing on the target-related image features, as well as gathering background information. Unlike [9], we explicitly separate the foreground target from the background using foreground and background click prompts. Specifically, we define positive click queries $X_p = \{X_{sp}, X_{fp}\}$ and negative click queries $X_n = \{X_{sn}, X_{fn}\}$. To ensure X_p and X_n have the same size of $N_1 \times C$, we add dummy points to pad X_p and X_n when click number is smaller than N_1. In our implementation, we set $N_1 = 24$. Our foreground-background masked cross-attention is computed as:

$$X'_{ol} = \mathrm{softmax}(M_{pl} + Q_{ol} K_l^T) V_l + X_{ol}, \qquad (4)$$

$$X'_{pl} = \mathrm{softmax}(M_{pl} + Q_{pl} K_l^T) V_l + X_{pl}, \qquad (5)$$

$$X'_{nl} = \text{softmax}(M_{nl} + Q_{nl}K_l^T)V_l + X_{nl}, \qquad (6)$$

where l is the layer index of the transformer decoder shown in Fig. 2 and Fig. 3(b). Q_{*l} is the learned linear transformation of X_{*l}, where $*$ denotes different query types, i.e., object, positive and negative click queries. K_l, V_l are the learned linear transformations of image features F_l. $M_{pl}, M_{nl} \in \{0, -\infty\}^{N_1 \times H_l W_l}$ are used to control the foreground-background attention masking. At the l-th layer, we first generate a mask prediction M_l from the pixel features F_{l-1} of the previous $(l-1)$-th layer, using a simple mask decoder followed by proper mask resizing. Then, M_{pl} and M_{nl} at feature location (x,y) are calculated as:

$$M_{pl}(x,y) = \begin{cases} 0, & M_l(x,y) \geq 0.5 \\ -\infty, & \text{otherwise} \end{cases}, \quad M_{nl}(x,y) = \begin{cases} 0, & M_l(x,y) < 0.5 \\ -\infty, & \text{otherwise} \end{cases}. \qquad (7)$$

Next, $X'_{ol}, X'_{pl}, X'_{nl}$ are processed separately through self-attention and query feedforward networks (FFN). The resulting queries are concatenated along the query number dimension, then transformed into the **Key** and **Value** spaces using learned linear transformations. A cross-attention layer, with image features as the **Query**, followed by a pixel FFN, updates the image features, which are finally used to generate the segmentation mask.

2.4 Residual Connection with Different Scales

In our multi-scale transformer decoder, the integrated queries forward through different decoder layers, continuously interacting with image features across scales. To enhance the interaction between image features and integrated queries across various scales, we adopt a residual [16] resampling connection similar to the U-Net [33] decoder. For the image feature map F_l at layer l, it's first resampled to align with the resolution of F_{l+1} at next layer $l+1$:

$$F_l^{\text{resampled}} = \text{Resample}(F_l, \text{size}(F_{l+1})). \qquad (8)$$

Here, Resample($\cdot$) can involve upsampling or downsampling depending on the change in scale. Next, we apply a convolutional layer to calculate the residual Res(F_l) before adding it to the feature at the next layer. The updated feature map at scale $l+1$ is computed as:

$$F_{l+1}^{\text{updated}} = F_{l+1} + \text{Res}(F_l) = F_{l+1} + \text{Conv}(F_l^{\text{resampled}}). \qquad (9)$$

A similar operation is applied to semantic feature queries X_f. By iteratively resampling, computing residuals, and updating features at each scale, the model effectively integrates multi-scale context, enhancing the interaction between features and queries across scales.

2.5 Implementation Details

Image Encoder. We utilize UTNet [15] as the image encoder, which is specifically designed for cardiac MRI segmentation. We extract multi-scale features at resolutions of 1/8, 1/4, and 1/2 of the original image, providing a comprehensive feature representation across different scales ($S = 3$).

Mask&Click Encoder. We use simple consecutive convolutional layers to encode the dense mask&click feature map. Starting with original dimensions of $(3, H, W)$, the layers downsample it to 1/8, 1/4, and 1/2 of the original size, matching the feature scales of the image encoder for seamless feature addition.

Click Representation. During training and inference, clicks are represented as disk maps with a fixed radius of 1. Following prior work [27,34], clicks are simulated by comparing current segmentation with ground truth. Unlike RITM [34], we place new clicks at the center of the largest connected component in misclassified regions, better aligning with user interactions in medical imaging. In Mode-1 and Mode-2, the model generates an initial mask, using object queries, and refines it with 2 clicks. In Mode-3, it iteratively uses 3 clicks without object queries to produce 3 segmentation masks.

Transformer Decoder. We use our transformer decoders proposed in Sect. 2 with $L = 2$ (*i.e.*, 6 layers in total). Similar to [8], we use a round robin approach for $L\times$ multi-scale interaction between image features and integrated queries.

Training Settings and Strategy. Our models are trained for 75 epochs with a batch size of 8 on 8 Quadro RTX 8000 GPUs. Images are resized to 256×256 during both training and inference. We use a combination of binary cross-entropy loss and Dice loss [29] to compute the mask loss $\mathcal{L} = \lambda_{ce}\mathcal{L}_{ce} + \lambda_{dice}\mathcal{L}_{dice}$, where $\lambda_{ce} = \lambda_{dice} = 5.0$. To enhance model robustness, we employ the following data augmentation techniques: 1) Random flips along both horizontal and vertical axes; 2) Random 90-degree rotations, 3) Random adjustments to brightness, and 4) Random adjustments to contrast. To support the diverse working modes of Verse during evaluation, we adopt a random training strategy. More specifically, each training batch is randomly assigned to Mode-1&2 or Mode-3.

3 Experiments

Datasets. We conducted experiments on nine publicly available datasets, including seven cardiac MRI datasets and two out-of-distribution medical imaging datasets, as summarized in Table 1. While 3D image volumes are provided in these datasets, we performed all experiments on 2D image slices, excluding slices without any instances. Our combined cardiac training set includes three types of cardiac MRI data: (1) **Balanced Steady-State Free Precession (bSSFP)** sequences, which focus on segmenting the left ventricle (LV), right ventricle (RV), and myocardium wall (Myo); (2) **T2-weighted (T2)** sequences, which highlight myocardial edema; and (3) **LGE** sequences, which highlight myocardial scar. For the M&Ms-2 dataset, we utilized only long-axis (LA) cine MRI

Table 1. Datasets statistics. The upper cardiac MRI datasets are for upstream training and analysis. The bottom two out-of-distribution datasets are for downstream tasks on assessing generalizability of interactive segmentation models.

Dataset	Targets	Modality	3D Volumes		2D Slices	
			Train	Test	Train	Test
ACDC [3]	LV, Myo, RV	bSSFP	200	100	1841	1001
M&Ms [4]	LV, Myo, RV	bSSFP	300	340	2475	2821
M&Ms-2 [4]	LV, Myo, RV	bSSFP	400	320	400	320
MyoPS++ [10,19,31,38]	LV, Myo, RV	bSSFP	90	29	672	194
MyoPS++ [10,19,31,38]	Myocardial Edema	T2	38	12	231	79
MyoPS++ [10,19,31,38]	Myocardial Scar	LGE	56	18	313	98
LASCARQS++ [20-23]	Left Atrial	LGE	98	32	3561	1183
OAIZIB [1]	Knee Bone&Cartilage	DESS	–	–	–	150
BraTS [2]	Brain Tumor	T2	–	–	–	369

images, to enhance data diversity, while other bSSFP datasets were acquired as short-axis images. For out-of-distribution evaluation, we tested our method on OAIZIB [1], a knee MRI dataset acquired with **Double Echo Steady State (DESS)** sequence, and BraTS [2], a brain **T2-weighted** MRI dataset. These datasets allow us to assess the generalizability of our model to different domains.

Baseline Methods. For automatic image segmentation, we compared our method with nnU-Net [17], TransUNet [7], UTNet [15], Swin-Unet [5] and Med-Former [14], all of which are widely recognized and applied in medical image segmentation. For interactive image segmentation, we compared our method with RITM [34], iSegformer [28], SimpleClick [27] and SegNext [26]. Among these, SimpleClick and SegNext are the state-of-the-art (SOTA) methods for interactive segmentation, demonstrating exceptional performance across various benchmarks. We did not experiment with the SAM-base [18] model, as SegNext [26] had already shown superior performance in the click-based task.

Evaluation Metrics. We use the 2D Dice score to evaluate the segmentation accuracy, which is a standard measure in medical image segmentation [3,15]. We use the Number of Clicks (NoC) metric to assess the number of clicks required to reach a specified Dice score. Target Dice scores are set at 80%, 85%, 90%, and 95%, denoted as Noc80, NoC85, NoC90, and NoC95, respectively. Each instance allows a maximum of 20 clicks. In addition, we evaluated segmentation quality using the average Dice score of all instances at a fixed number of clicks, Dice(n).

3.1 Results

Comparison with Automatic Models. Table 2 compares our proposed VerSe framework with several specialized segmentation models designed for specific tasks. Despite lacking any specialized design to enhance automatic segmentation, VerSe achieves competitive performance on large-scale datasets such as

Table 2. Model performance comparison for different working mode settings and different cardiac MRI datasets. Our method is the only one that can work in both automatic and interactive image segmentation modes.

Dataset	Automatic		Interactive				
	Model	Dice ↑	Model	Dice(1) ↑	Dice(20) ↑	NoC85 ↓	NoC90 ↓	NoC95 ↓

Dataset	Model	Dice ↑	Model	Dice(1) ↑	Dice(20) ↑	NoC85 ↓	NoC90 ↓	NoC95 ↓
ACDC (bSSFP)	nnUNet	**89.796**	RITM	83.999	92.911	2.175	4.784	11.055
	TranUNet	87.649	iSegformer	66.595	92.498	4.251	7.371	12.384
	UTNet	89.604	SimpleClick	88.050	92.293	2.402	5.468	11.647
	SwinUNet	87.883	SegNext	87.665	92.265	2.295	5.397	11.794
	Medformer	86.931	VerSe (Mode-3)	89.757	96.698	1.485	2.352	6.480
	VerSe (Mode-1)	89.788	VerSe (Mode-2)	**92.091**	**96.865**	**0.431**	**1.248**	**5.636**
M&Ms (bSSFP)	nnUNet	85.417	RITM	80.583	92.855	2.709	5.772	12.146
	TranUNet	85.212	iSegformer	62.285	91.635	4.947	8.161	13.418
	UTNet	85.998	SimpleClick	85.335	91.289	3.787	6.809	13.216
	SwinUNet	84.363	SegNext	85.197	91.676	3.375	6.502	13.209
	Medformer	84.472	VerSe (Mode-3)	87.460	96.827	1.968	3.074	6.828
	VerSe (Mode-1)	**86.527**	VerSe (Mode-2)	**89.429**	**96.887**	**1.013**	**2.122**	**5.906**
M&Ms-2 (bSSFP)	nnUNet	**90.638**	RITM	85.829	93.606	3.214	6.082	10.474
	TranUNet	89.543	iSegformer	58.925	92.860	6.828	9.066	12.430
	UTNet	90.266	SimpleClick	88.880	93.951	2.004	4.376	9.619
	SwinUNet	89.098	SegNext	87.907	93.599	2.196	5.050	10.366
	Medformer	87.544	VerSe (Mode-3)	88.357	**97.057**	1.800	3.015	7.671
	VerSe (Mode-1)	86.496	VerSe (Mode-2)	**89.830**	97.013	**0.884**	**2.194**	**7.047**
MyoPS++ (bSSFP)	nnUNet	88.316	RITM	84.800	92.694	2.311	5.807	12.635
	TranUNet	86.886	iSegformer	66.061	92.282	5.619	8.951	13.488
	UTNet	87.583	SimpleClick	88.030	92.279	1.868	4.777	12.184
	SwinUNet	86.743	SegNext	86.431	92.422	2.189	5.502	12.551
	Medformer	87.417	VerSe (Mode-3)	88.834	97.079	1.577	2.624	7.047
	VerSe (Mode-1)	**88.657**	VerSe (Mode-2)	**91.071**	**97.141**	**0.595**	**1.526**	**6.032**
MyoPS++ (T2)	nnUNet	71.396	RITM	75.843	91.496	3.494	6.797	18.759
	TransUNet	**71.499**	iSegformer	52.826	90.717	5.076	8.886	19.038
	UTNet	70.251	SimpleClick	**79.231**	90.391	**2.582**	**6.253**	19.481
	SwinUNet	53.753	SegNext	72.492	91.262	4.101	7.873	19.342
	Medformer	63.140	VerSe (Mode-3)	74.096	**94.183**	5.557	7.772	**12.506**
	VerSe (Mode-1)	70.103	VerSe (Mode-2)	74.594	93.131	6.253	8.241	12.683
MyoPS++ (LGE)	nnUNet	44.412	RITM	60.234	86.735	7.694	14.531	19.878
	TransUNet	55.548	iSegformer	38.602	85.691	12.184	18.306	20.000
	UTNet	42.436	SimpleClick	65.232	84.773	8.010	13.541	20.000
	SwinUNet	41.365	SegNext	61.153	86.879	9.194	15.367	19.643
	Medformer	36.821	VerSe (Mode-3)	**66.293**	**95.024**	**6.275**	**8.735**	**13.939**
	VerSe (Mode-1)	**57.879**	VerSe (Mode-2)	65.238	94.752	6.867	9.531	14.235
LAScarQS++ (LGE)	nnUNet	84.827	RITM	82.957	92.704	3.550	6.163	13.866
	TranUNet	83.441	iSegformer	72.017	92.750	4.527	7.546	14.896
	UTNet	**85.275**	SimpleClick	84.056	91.279	3.724	6.433	14.908
	SwinUNet	79.786	SegNext	84.284	94.011	2.558	4.673	10.961
	Medformer	82.378	VerSe (Mode-3)	85.841	**97.452**	2.103	2.940	5.717
	VerSe (Mode-1)	83.159	VerSe (Mode-2)	**86.919**	97.172	**1.414**	**2.270**	**5.077**

ACDC and M&Ms, demonstrating its robustness and adaptability across diverse segmentation scenarios. However, on more challenging datasets like MyoPS++ (T2), where all models struggle to meet clinical requirements, due to the dataset's inherent complexity, the interactive capabilities of VerSe become particularly advantageous. By enabling efficient user-driven refinements through click-based interactions, VerSe provides a practical solution to enhance segmentation accuracy in challenging cases, bridging the gap toward clinical applicability.

Comparison with Interactive Models. Table 2 highlights the performance comparison between VerSe (Mode-3) and previous SOTA interactive models. VerSe (Mode-3) consistently achieves the best Dice scores and lower interaction costs among six out of seven datasets. Notably, on larger datasets, such as ACDC, M&Ms, and LAScarQS++, VerSe achieves the best performance across all metrics, significantly outperforming existing methods. For example, on the M&Ms dataset, which has the largest number of instances, VerSe achieves a Dice(1) score of 89.757%, significantly surpassing SimpleClick (85.335%) and SegNext (85.197%). This demonstrates VerSe's ability to handle complex and large-scale data efficiently. Furthermore, as illustrated in Fig. 4, VerSe not only achieves faster convergence but also demonstrates steady improvements in segmentation accuracy as the number of clicks increases, highlighting the efficiency of its interactive prompting mechanism.

On smaller and more challenging datasets, such as MyoPS++ (LGE) and MyoPS++ (T2), VerSe continues to show significant advantages. On the MyoPS++ (LGE) dataset, VerSe achieves a Dice(20) of 95.024%, far exceeding the ~85% average of competing methods. Similarly, on the MyoPS++ (T2) dataset, while SimpleClick initially leads in the first 10 clicks, VerSe demonstrates more sustained improvements, ultimately achieving the best Dice(20) of 94.183% and a NoC95 of 12.506. These results highlight VerSe's ability to efficiently utilize user interactions to refine segmentation results, even in difficult scenarios.

Overall, VerSe (Mode-3) sets a new benchmark in interactive segmentation, by achieving SOTA performance with minimal interaction costs. Its robust performance across both large-scale datasets and complex segmentation tasks underscores its adaptability and effectiveness in diverse cardiac MRI applications.

VerSe (Mode-2) vs. VerSe (Mode-3). Table 2 presents a detailed comparison between VerSe operating in Mode-2 and Mode-3 across various datasets. VerSe (Mode-2) achieves superior performance in 4 out of 7 datasets compared to VerSe (Mode-3), particularly in the ACDC and M&Ms datasets, where the initial automatic segmentation effectively reduces user interaction while improving accuracy. In contrast, VerSe (Mode-3), which relies solely on interactive segmentation without automatic initialization, excels on datasets where precise

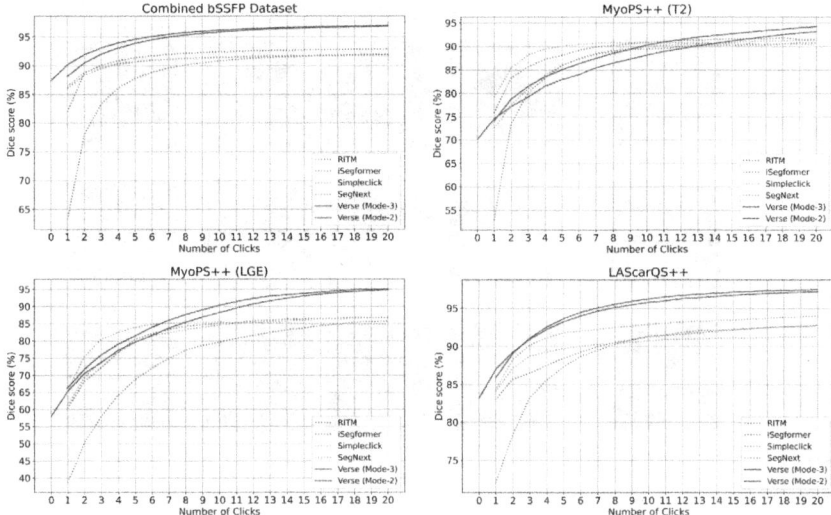

Fig. 4. Convergence analysis for models tested on four types of segmentation targets. The Combined bSSFP Dataset, including ACDC, M&Ms, M&Ms-2, and MyoPS++ (bSSFP), focuses on LV, Myo, and RV structures. VerSe demonstrates consistent accuracy improvements across all tasks as the number of clicks increases.

automatic initialization is particularly challenging, such as MyoPS++ (T2) and MyoPS++ (LGE). These results underscore the robust interactive segmentation performance of VerSe (Mode-3), especially in scenarios lacking reliable automatic initialization. Meanwhile, Mode-2's dependence on object queries highlights the need for larger and more diverse datasets to unlock its full potential. For datasets with limited training samples, Mode-3 serves as a reliable fallback. These results highlight the flexibility and adaptability of VerSe's unified segmentation framework in addressing varying clinical and data-specific needs.

Out-of-Distribution Evaluation. We trained all interactive segmentation models using cardiac MRI datasets, while we evaluated their performance on out-of-distribution datasets. The results are summarized in Table 3. On the BraTS dataset, Verse achieves a remarkable Dice score of 94.492% with just 10 clicks, significantly outperforming other models. In addition, it delivers the highest Dice(20) score of 96.811% and demonstrates the best annotation efficiency. On the OAIZIB dataset, Verse achieves the lowest NoC80 value of 11.787, highlighting its superior efficiency. These results collectively showcase the robust generalization capabilities of Verse across diverse medical imaging domains.

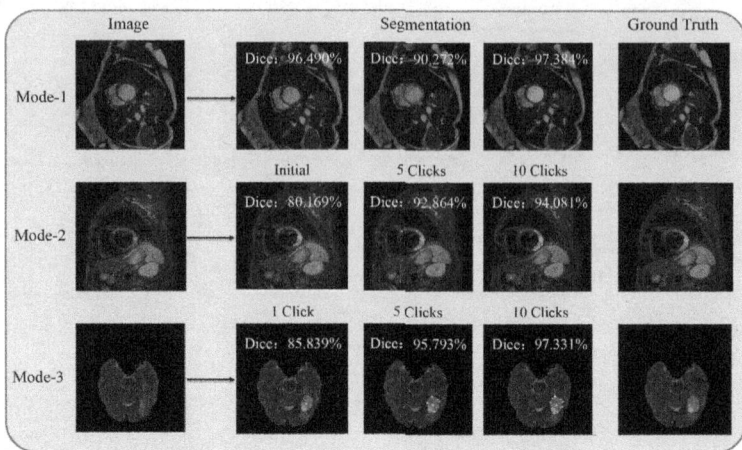

Fig. 5. Segmentation results of VerSe on different medical image segmentation tasks. First row: Automatic segmentation of three structures on cardiac cine MRI. Second row: Interactive refinement of myocardial edema segmentation on cardiac T2-weighted MRI. Third row: Interactive segmentation of a tumor on out-of-distribution brain MRI.

Visualization. In Fig. 5, we demonstrate representative segmentation results of VerSe on cardiac MRI and out-of-distribution brain MRI. As can be seen from the results, our design of VerSe allows it to handle different medical image segmentation tasks with high versatility.

Table 3. Out-of-distribution evaluation on BraTS [2] and OAIZIB [1] datasets. Verse shows strong generalizability in its interactive segmentation.

Model	OAIZIB			BraTS		
	Dice(10) ↑	Dice(20) ↑	NoC80 ↓	Dice(10) ↑	Dice(20) ↑	NoC80 ↓
RITM	70.334	79.071	13.988	88.925	93.898	5.136
iSegformer	65.509	76.784	17.063	86.465	92.281	6.482
SimpleClick	**73.260**	**82.523**	12.976	92.537	93.927	3.436
SegNext	57.211	76.261	17.427	77.448	91.619	9.604
VerSe (Mode-3)	69.516	80.631	**11.787**	**94.492**	**96.811**	**2.777**

Table 4. Ablation study of each model component in VerSe under working Mode-3. 'S': Semantic feature queries. 'F-B': Foreground-background masked attention of click queries. 'R': Residual connection.

Model	S	F-B	R	Dice(1) ↑	Dice(20) ↑	NoC90 ↓	NoC95 ↓
A1	✗	✓	✓	89.563	96.537	2.543	7.015
A2	✓	✗	✓	90.351	96.540	2.112	6.511
A3	✓	✓	✗	88.480	96.172	2.755	7.513
Verse	✓	✓	✓	**91.025**	**96.659**	**2.028**	**6.372**

3.2 Ablation Studies

In this section, we conducted ablation studies to demonstrate the effectiveness of our model design. All of our ablation studies were conducted on the ACDC dataset with VerSe working in Mode-3 only. The results are shown in Table 4. (1) In model A1, we eliminated the semantic feature queries X_f from click queries X_c. (2) In model A2, we eliminated the masked attention branch (see the rightmost hybrid attention blocks in Fig. 3(b)) of negative click queries but combined positive and negative click query masked attention branches as a single one. (3) In model A3, we eliminated the residual connections across different scales. The results show that each of the three model components are necessary for the high performance of our model.

4 Conclusion

In this work, we propose VerSe, a novel framework that unifies automatic and interactive segmentation modes, through a multi-query integration mechanism. By effectively leveraging both object and click queries, VerSe achieves state-of-the-art performance in both segmentation accuracy and interaction efficiency. Our experiments on seven cardiac MRI datasets and two out-of-distribution medical imaging datasets demonstrate the robustness, efficiency, and generalizability of the proposed method. VerSe not only bridges the gap between automatic and interactive segmentation but also sets a new benchmark for versatile image segmentation tasks. Future directions include extending the framework to handle natural data, improving scalability for larger multi-modal datasets, and enhancing interpretability for clinical adoption.

Acknowledgments. This research has been partially funded by research grants to D. Metaxas through NSF: 2310966, 2235405, 2212301, 2003874, 1951890 and NIH 2R01HL127661.

References

1. Ambellan, F., Tack, A., Ehlke, M., Zachow, S.: Automated segmentation of knee bone and cartilage combining statistical shape knowledge and convolutional neural networks: data from the osteoarthritis initiative. Med. Image Anal. **52**, 109–118 (2019)
2. Baid, U., et al.: The RSNA-ASNR-MICCAI brats 2021 benchmark on brain tumor segmentation and radiogenomic classification. arXiv preprint arXiv:2107.02314 (2021)
3. Bernard, O., et al.: Deep learning techniques for automatic MRI cardiac multi-structures segmentation and diagnosis: is the problem solved? IEEE Trans. Med. Imaging **37**(11), 2514–2525 (2018)
4. Campello, V.M., et al.: Multi-centre, multi-vendor and multi-disease cardiac segmentation: the M&MS challenge. IEEE Trans. Med. Imaging **40**(12), 3543–3554 (2021)

5. Cao, H., et al.: Swin-unet: Unet-like pure transformer for medical image segmentation. In: European Conference on Computer Vision, pp. 205–218. Springer (2022)
6. Carion, N., Massa, F., Synnaeve, G., Usunier, N., Kirillov, A., Zagoruyko, S.: End-to-end object detection with transformers. In: European Conference on Computer Vision, pp. 213–229. Springer (2020)
7. Chen, J., et al.: Transunet: transformers make strong encoders for medical image segmentation. arXiv preprint arXiv:2102.04306 (2021)
8. Cheng, B., Misra, I., Schwing, A.G., Kirillov, A., Girdhar, R.: Masked-attention mask transformer for universal image segmentation. In: Proceedings of the IEEE/CVF Conference on Computer Vision and Pattern Recognition, pp. 1290–1299 (2022)
9. Cheng, H.K., Oh, S.W., Price, B., Lee, J.Y., Schwing, A.: Putting the object back into video object segmentation. In: Proceedings of the IEEE/CVF Conference on Computer Vision and Pattern Recognition, pp. 3151–3161 (2024)
10. Ding, W., et al.: Aligning multi-sequence CMR towards fully automated myocardial pathology segmentation. IEEE Trans. Med. Imaging (2023)
11. Ding, Y., Li, L., Wang, W., Yang, Y.: Clustering propagation for universal medical image segmentation. In: Proceedings of the IEEE/CVF Conference on Computer Vision and Pattern Recognition, pp. 3357–3369 (2024)
12. Dosovitskiy, A.: An image is worth 16x16 words: transformers for image recognition at scale. arXiv preprint arXiv:2010.11929 (2020)
13. Gao, Y.: Training like a medical resident: context-prior learning toward universal medical image segmentation. In: Proceedings of the IEEE/CVF Conference on Computer Vision and Pattern Recognition, pp. 11194–11204 (2024)
14. Gao, Y., Zhou, M., Liu, D., Yan, Z., Zhang, S., Metaxas, D.N.: A data-scalable transformer for medical image segmentation: architecture, model efficiency, and benchmark. arXiv preprint arXiv:2203.00131 (2022)
15. Gao, Y., Zhou, M., Metaxas, D.N.: UTNet: a hybrid transformer architecture for medical image segmentation. In: de Bruijne, M., et al. (eds.) MICCAI 2021. LNCS, vol. 12903, pp. 61–71. Springer, Cham (2021). https://doi.org/10.1007/978-3-030-87199-4_6
16. He, K., Zhang, X., Ren, S., Sun, J.: Deep residual learning for image recognition. In: Proceedings of the IEEE Conference on Computer Vision and Pattern Recognition, pp. 770–778 (2016)
17. Isensee, F., Jaeger, P.F., Kohl, S.A., Petersen, J., Maier-Hein, K.H.: nnu-net: a self-configuring method for deep learning-based biomedical image segmentation. Nat. Methods **18**(2), 203–211 (2021)
18. Kirillov, A., et al.: Segment anything. In: Proceedings of the IEEE/CVF International Conference on Computer Vision, pp. 4015–4026 (2023)
19. Li, L., et al.: Myops: a benchmark of myocardial pathology segmentation combining three-sequence cardiac magnetic resonance images. Med. Image Anal. **87**, 102808 (2023)
20. Li, L., et al.: Atrial scar quantification via multi-scale CNN in the graph-cuts framework. Med. Image Anal. **60**, 101595 (2020)
21. Li, L., Zimmer, V.A., Schnabel, J.A., Zhuang, X.: AtrialGeneral: domain generalization for left atrial segmentation of multi-center LGE MRIs. In: de Bruijne, M., et al. (eds.) MICCAI 2021. LNCS, vol. 12906, pp. 557–566. Springer, Cham (2021). https://doi.org/10.1007/978-3-030-87231-1_54
22. Li, L., Zimmer, V.A., Schnabel, J.A., Zhuang, X.: Atrialjsqnet: a new framework for joint segmentation and quantification of left atrium and scars incorporating spatial and shape information. Med. Image Anal. **76**, 102303 (2022)

23. Li, L., Zimmer, V.A., Schnabel, J.A., Zhuang, X.: Medical image analysis on left atrial LGE MRI for atrial fibrillation studies: a review. Med. Image Anal. **77**, 102360 (2022)
24. Lin, J., et al.: Adaptiveclick: click-aware transformer with adaptive focal loss for interactive image segmentation. IEEE Trans. Neural Netw. Learn. Syst. 1–15 (2024). https://doi.org/10.1109/TNNLS.2024.3378295
25. Liu, J., et al.: Clip-driven universal model for organ segmentation and tumor detection. In: Proceedings of the IEEE/CVF International Conference on Computer Vision, pp. 21152–21164 (2023)
26. Liu, Q., Cho, J., Bansal, M., Niethammer, M.: Rethinking interactive image segmentation with low latency high quality and diverse prompts. In: Proceedings of the IEEE/CVF Conference on Computer Vision and Pattern Recognition, pp. 3773–3782 (2024)
27. Liu, Q., Xu, Z., Bertasius, G., Niethammer, M.: Simpleclick: interactive image segmentation with simple vision transformers. In: Proceedings of the IEEE/CVF International Conference on Computer Vision, pp. 22290–22300 (2023)
28. Liu, Q., Xu, Z., Jiao, Y., Niethammer, M.: isegformer: interactive segmentation via transformers with application to 3D knee MR images. In: International Conference on Medical Image Computing and Computer-Assisted Intervention, pp. 464–474. Springer (2022)
29. Milletari, F., Navab, N., Ahmadi, S.A.: V-net: fully convolutional neural networks for volumetric medical image segmentation. In: 2016 Fourth International Conference on 3D Vision (3DV), pp. 565–571. IEEE (2016)
30. Pennell, D.J., Mohiaddin, R.H.: Cardiovascular magnetic resonance: past, present, and future. Circ. Cardiovasc. Imaging **17**(8), e016523 (2024)
31. Qiu, J., et al.: Myops-net: myocardial pathology segmentation with flexible combination of multi-sequence CMR images. Med. Image Anal. **84**, 102694 (2023)
32. Romero R, W.A., et al.: Cmrsegtools: an open-source software enabling reproducible research in segmentation of acute myocardial infarct in CMR images. PLoS ONE **17**(9), e0274491 (2022)
33. Ronneberger, O., Fischer, P., Brox, T.: U-Net: convolutional networks for biomedical image segmentation. In: Navab, N., Hornegger, J., Wells, W.M., Frangi, A.F. (eds.) MICCAI 2015. LNCS, vol. 9351, pp. 234–241. Springer, Cham (2015). https://doi.org/10.1007/978-3-319-24574-4_28
34. Sofiiuk, K., Petrov, I.A., Konushin, A.: Reviving iterative training with mask guidance for interactive segmentation. In: 2022 IEEE International Conference on Image Processing (ICIP), pp. 3141–3145. IEEE (2022)
35. Xu, N., Price, B., Cohen, S., Yang, J., Huang, T.S.: Deep interactive object selection. In: Proceedings of the IEEE Conference on Computer Vision and Pattern Recognition, pp. 373–381 (2016)
36. Yan, K., et al.: Liver tumor screening and diagnosis in CT with pixel-lesion-patient network. In: International Conference on Medical Image Computing and Computer-Assisted Intervention, pp. 72–82. Springer (2023)
37. Zhou, H.Y., et al.: nnformer: volumetric medical image segmentation via a 3D transformer. IEEE Trans. Image Process. (2023)
38. Zhuang, X.: Multivariate mixture model for myocardial segmentation combining multi-source images. IEEE Trans. Pattern Anal. Mach. Intell. **41**(12), 2933–2946 (2019)

Author Index

A
Alshareef, Ahmed II-375
Ashton, Nicholas II-65
Awate, Suyash P. I-265
Axel, Leon I-373
Ayed, Ismail Ben II-278, II-294
Azampour, Mohammad Farid I-357

B
Bai, Xiaoyu II-125
Bai, Yansong I-64
Balu, Niranjan I-327
Bayly, Philip V. II-375
Beizaee, Farzad II-139
Berger, Alexander H. I-297
Bian, Zhangxing II-375
Bocquillon, Constance I-142
Bongratz, Fabian I-187
Bray, Timothy J. P. II-186

C
Cai, Tongan II-390
Cai, Weidong I-79
Carass, Aaron I-249, II-375
Caruyer, Emmanuel I-142
Chang, Heng I-218
Chen, Geng II-125, II-172
Chen, Ke I-154
Chen, Xiang I-108
Chen, Yanbo II-234
Chen, Yanxi II-65
Chen, Yuqian I-79
Cheng, Xinxing I-108
Christodoulidis, Stergios II-278
Chung, Albert C. S. II-111
Collas, Antoine II-79
Corouge, Isabelle I-142
Cournède, Paul-Henry II-278
Cruz, Gastao I-168
Cui, Zhiming II-157, II-361

D
Dan, Tingting II-37, II-51
Desrosiers, Christian II-139
Deutges, Michael I-19
Dewey, Blake E. I-249
Dolz, Jose II-139, II-278, II-294
Dou, Qi II-3
Du, Yuexi II-247
Duan, Jinming I-108
Dvornek, Nicha C. II-247
Dwyer-Hemmings, Louis II-186

E
Elbaroudy, Sama I-187
Elhabian, Shireen II-327
Engemann, Denis A. II-79
Entezari, Alireza I-154

F
Fan, Jianan I-79
Fan, Li II-157
Fan, Yonghui II-65
Farazi, Mohammad II-65
Fillioux, Leo II-278
Fu, Huazhu I-342

G
Gao, Jiahong II-203
Gao, Yaozong II-234
Gao, Yunhe I-373
Ge, Zongyuan II-263
Gong, Shizhan II-3
Guibault, François II-313
Guo, Bangwei I-373
Guo, Xiaoqing I-313
Guo, Yin I-327

H
Hammernik, Kerstin I-168
Han, Linbin II-203

Hanik, Martin I-49
He, Haoyu I-33
He, Mingguang II-263
Hong, Xiaoming I-203
Hossain, Tonmoy II-342
Hosseini, Mahdi S. I-33
Hu, Renjiu I-108
Huang, Haifeng II-51
Huang, Sharon X. II-390
Huang, Wenjian I-3
Huang, Wenqi I-168
Huang, Yawen II-218

I
Iyer, Krithika II-327

J
Jadhav, Kshitij II-263
Jan, Catherine II-263
Jayakumar, Nivetha I-232
Jiang, Caiwen I-94, II-234
Jiang, Haotian II-125, II-172
Jiang, Yiwen II-263
Jin, Yueming I-342
Joshi, Sarang II-327

K
Kikinis, Ron I-79
Kim, Soopil I-283
Kubík, Tibor II-313
Kuestner, Thomas I-168

L
Leicht, Rachel II-390
Li, Chenxin I-313
Li, Gaolei I-108
Li, Hongdong II-361
Li, Shuo I-342
Li, Yingzhen I-127
Li, Yitong I-187
Lian, Chunfeng I-218
Liang, Xiao II-375
Liu, Dongnan I-79
Liu, Hong II-218
Liu, Jingyu II-234
Liu, Min I-108

Liu, Tuo I-218
Liu, Wenbin II-94
Liu, Yifan I-313
Liu, Zhentao II-361
Lodygensky, Gregory II-139
Lombaert, Hervé II-313
Lu, Yuan-Chiao II-375
Luo, Xinzhe I-127
Lux, Laurin I-297

M
Ma, Jianhua I-218
Ma, Qian II-390
Ma, Wenchao II-390
Marr, Carsten I-19
Matinfar, Sasan II-19
Mehta, Deval II-263
Menten, Martin J. I-297, I-357
Metaxas, Dimitris I-373
Minore, Giulio V. II-186
Mossa-Basha, Mahmud I-327

N
Nam, Siwoo I-283
Namgung, Hyun I-283
Navab, Nassir I-19, I-357, II-19
Ni, Haomiao II-390

O
O'Donnell, Lauren J. I-79
Onofrey, John A. II-247

P
Paetzold, Johannes C. I-297, I-357
Paillard, Joseph II-79
Pang, Haowen I-203
Park, Sang Hyun I-283
Peng, Qiong II-218
Pham, Dzung L. II-375
Prieto, Claudia I-168
Prince, Jerry L. I-249
Prince, Jerry L. II-375

Q
Qi, Xiaoming I-342
Qian, Zhen I-3, II-203

Qiao, Yuchuan   I-64
Qiao, Zhi   I-3, II-203
Qin, Chen   I-127

R
Redaelli, Alberto   II-19
Reiman, Eric M.   II-65
Remedios, Samuel W.   I-249
Remedios, Samuel W.   II-375
Reyes Vargas, Luis D.   I-357
Rueckert, Daniel   I-168, I-297
Ruozzi, Veronica   II-19

S
Sadafi, Ario   I-19
Schade, Johannes   I-49
Schnabel, Julia A.   I-168
Schütz, Laura   II-19
Shapiro, Linda   I-327
Sharma, Vatsala   I-265
Shen, Dinggang   I-94, II-157, II-234
Shen, Xin   II-94
Shu, Yiran   II-234
Silva-Rodríguez, Julio   II-278, II-294
Song, Yanli   I-94
Španěl, Michal   II-313
Spieker, Veronika   I-168
Su, Yi   II-65
Sun, Kaicong   II-234

T
Tao, Ze   I-94
Thirion, Bertrand   II-79
Tsang, Colin S. C.   II-111

V
Vakalopoulou, Maria   II-278
Volpi, John   II-390
von Tycowicz, Christoph   I-49
Votta, Emiliano   II-19

W
Wachinger, Christian   I-187
Wang, Fan   I-218
Wang, Haifeng   I-218
Wang, Huayu   II-3

Wang, James Z.   II-390
Wang, Liansheng   II-218
Wang, Mei   II-157
Wang, Qian   II-157
Wang, Rongguang   I-108
Wang, Sheng   II-157
Wang, Xin   I-327
Wang, Xinlu   II-234
Wang, Yalin   II-65
Wang, Yang   I-33
Wang, Yao   I-3
Wang, Yaonan   I-108
Wang, Yi   II-51
Weers, Alexander   I-297
Wei, Dong   II-218
Wei, Shuwen   I-249, II-375
Westin, Carl-Fredrik   I-79
Wiestler, Benedikt   II-19
Wong, Kelvin   II-390
Wong, Stephen T. C.   II-390
Woo, Jonghye   II-375
Wu, Dijia   I-94
Wu, Guorong   II-37, II-51
Wu, Nian   I-232
Wu, Peng   I-94
Wu, Xian   II-218

X
Xin, Bingyu   I-373
Xing, Fangxu   II-375
Xing, Jiarui   I-232
Xiong, Xiaosong   I-94
Xu, Hao   I-79
Xu, Siying   I-168
Xue, Tengfei   I-79
Xue, Yuan   II-390

Y
Yang, Chen   I-19
Yang, Dong   I-3
Yang, Guanyu   I-342
Yang, Liuzhi   I-3
Yang, Yang   II-51
Yang, Zhangsihao   II-65
Ye, Chuyang   I-203
Ye, Meng   I-373

Yu, Mingyang II-234
Yu, Weihao I-313
Yuan, Chun I-327
Yuan, Yixuan I-313

Z
Zha, Ruyi II-361
Zhan, Yiqiang II-234
Zhang, Fan I-79
Zhang, Hang I-108
Zhang, Hui II-186
Zhang, Jingyang I-342
Zhang, Jinwei I-108
Zhang, Kaiyu I-327
Zhang, Miaomiao I-232, II-342
Zhang, Peng I-203
Zhang, Shengjie II-94
Zhang, Weifang II-234
Zhang, Xiao I-94
Zhang, Xiaofan II-3
Zhang, Xinyi I-94
Zhao, Huangxuan II-361
Zhao, Meixin II-234
Zhao, Zihao II-157
Zhen, Xiantong II-203
Zheng, Yefeng II-218
Zheng, Yuxi I-64
Zhong, Shaonan II-234
Zhou, Qian II-157
Zhou, Xiang Sean II-234
Zhou, Yuan II-94
Zhuo, Jiachen II-375

Made in the USA
Monee, IL
03 May 2026